A BOOK OF
THE BEGINNINGS

One of the Mountain-tops of Time
Is left in Africa to climb.

A BOOK OF
THE BEGINNINGS

Containing an attempt to recover and reconstitute the lost origines of
the myths and mysteries, types and symbols, religion and language,
with Egypt for the mouthpiece and Africa as the birthplace.

By
Gerald Massey

Introduction by Charles S. Finch

Volume I

EGYPTIAN ORIGINES IN THE BRITISH ISLES

Lushena Books Publishing

A BOOK OF THE BEGINNINGS

Introduction copyright Charles S. Finch, III

ISBN: 978-1-63923-023-5

A Book of the Beginnings
and *The Natural Genesis*:
The Early Works of Gerald Massey

An Introduction
by Dr. Charles S. Finch

The nations; secretly she brought to birth,
And Egypt still enriches all the earth.

The more one studies ancient history and the correlative discipline of paleontology, the more one is reminded of the iceberg metaphor. That is to say conventional history, as it is taught, is but the visible and most superficial tip of a vast body submerged and hidden in the dense mists of the past. This being the case, it is only a powerless and unconscious people who would insist that history is of no importance to them. No people—in full possession of their lands, their productive capacity, their destiny—ever belittles the centrality of history. In China of old, the first emperors of a new dynasty always ordered the kingdom's annalists to rewrite history to show the ascendant dynasty in the most favorable light, thereby legitimizing its accession. The past is a mirror of the future, and what you hope to be is shaped by what you have been. No nation can face the future with confidence unless it knows what it is capable of. Only history can teach that.

Gerald Massey understood all of these things in the profoundest way possible. Like the molecular biologist applying the most powerful lenses to explore the mysteries of gene structure, Massey sought to unravel the psychocultural strands emanating from the remotest human heritage. This is why he entitled his first two-volume work *A Book of the Beginnings*, containing, as he put it,

> ...an attempt to recover and reconstitute the lost origines of the myth and mysteries, types and symbols, religion and language, with Egypt for the mouthpiece and Africa as the birthplace.[1]

His investigations convinced him that the roots of modern culture went back ages further than admitted and that the keys to discovering these roots lay in myth and symbol, resolvable to their ancient beginnings in Africa. His perception concerning human cultural chronology was much more expansive than that of contemporary historiography, and in current times, we are indeed finding through paleontology and molecular biology that anatomically modern humans and their culture are as old as Massey thought.

These remain contentious issues because there still persists a pronounced tendency to search for a "special creation" for certain human groups. In the last 15 years, for example, Neanderthal Man has been resurrected as a possible Caucasoid ancestor because he seems to have arisen exclusively in Europe. He is therefore held up as an example of "regional continuity," a euphemistic term now employed by the polycentric theorists of human evolution.[2] The polycentrists concede that the prehuman hominid ancestors of modern humanity arose in Africa but claim that one million years ago, these hominids, as the near-human Homo erectus species, migrated to separate parts of the globe and separately became modern humans in as many as five different locations. This explanation then allows the polycentrists to avoid the painful conclusion that all modern human beings—themselves included —are descended from African ancestors.

The polycentrists are notoriously hostile to the paleontological applications of molecular biology because, since the work of Vincent Sarich at Berkeley in 1966, the burgeoning discipline of molecular biology has categorically supported the monocentric model of human origins out of Africa. Most particularly, the work in comparative mitochondrial DNA studies carried forward by Cann and Stoneking at the University of California at Berkeley and Wallace at Emory have cemented the monocentric model. Wallace's recent conclusions in favor of the "out-of-Africa" paradigm are significant because originally, he favored an "out-of-Asia" evolutionary paradigm.[3] Continued work in the field has apparently convinced Wallace to endorse the out-of-Africa thesis.

Wallace's work has even wider implications because he has found that mitochondrial DNA sequences in certain groups of South American Indians cluster more closely to those of Polynesians, Micronesians, and Melanesians than they do other Indian groups. His findings suggest that, in addition to the northern route over the Bering Isthmus, a trans-Pacific migratory route was employed to reach South America as early as 17,000 years ago.[4] Massey took the position that people and culture migrated out of Africa into the rest of the world, including the Americas and that there exists a global cultural unity that is African at its root. Molecular biology is bearing Massey out.

Other evidence concerning the African cultural origins of humanity comes from work done in northeastern Zaire by Alison Brooks and John Yellen, a married George Washington University team of archaeologists. Working at Katanga and Ishango, near the shores of the formerly named Lake Albert, this pair of investigators has discovered a highly-developed "aquatic" culture characterized by finely-crafted harpoon points. These finds have been dated at 70,000 B.P. (Before Present), five times older than similar harpoon points found elsewhere in the world.[5] What is more, similar harpoon cultures at later dates appear more or less sequentially in a northward direction following the flow of the Nile at Khartoum, Upper Egypt, Fayum, and Palestine. These harpoon points seem to indicate the directional flow of culture out of the lacustrine regions of East Africa into the Nile Valley and beyond.

Again, what Massey proclaimed concerning human cultural origins in Africa in *A Book of the Beginnings* has been confirmed by archaeology one hundred years afterward. His insight and prescience highlight the fundamental failure of academic specialization: the human experience is a tapestry properly appreciated only through a panoramic view. Thus, the whole is grasped through a seamless interweaving of disciplines.

The first volume of *A Book of the Beginnings* is devoted to a systematic unveiling of the "Egyptian origines in the British Isles." Massey was not the first nor only antiquarian to discern a southern root for ancient British culture. Godfrey Higgins and MacRitchie devoted whole theses to this premise. [6] But the Masseyan method of comparative linguistics, symbolism, and mythology finds its first evocation in these early volumes. As an example, Massey compares the English **are**, a conjugate of **to be**, to the Egyptian **ar** or **ir**, also meaning **to be** or **being**. Also, the English word **autumn** relates to the Egyptian **Atum** who represents the declining or autumnal sun. Or the English **dad**, i.e., father compares to the Egyptian **tat** for father. Finally the English **nature** may be derived from the Egyptian **ntr** meaning **principle** or **power**. We might also note that the English **desert** comes directly from the Egyptian **teshert** meaning the **red land**, their term for the desert.

What applies to words and names also applies to fairy tales, superstitions, and folk beliefs. For example, Jack-in-the-box and Jack-in-the-beanstalk both represent the child Horus or Atum emerging from the lotus flower and rising on its stalk from the primeval depths. This imagery is that of the morning sun rising from the darkness of Amenta. Both Horus and Atum, as forms of the sun, are **iakhu**, i.e., light itself. This **iakhu**, in Masseyan terms, is the original Jack that in much later times becomes the one who rises on the beanstalk or jumps out of the box. Sleeping Beauty, according to Massey, is the Egyptian moon goddess Hathor who during the dark of the moon is imaged as being "asleep" only to be kissed awake at the beginning of the first quarter by the sun who is Horus.

Similarly the fairy tale of Snow White and the Seven Dwarves derives from the original lore surrounding the Afro-Kamite Great Mother and her Seven Children. In the original, Snow White and the Wicked Queen are unseparated in the Great Mother figure who is both life and death personified. The Seven Children in human form are imaged as seven dwarves because the ancient Egyptians knew that the Twa, i.e., the "Pygmies," were the first human beings. At a later date, the Egyptian Ptah was associated with seven "Pygmies," i.e., the Seven Dwarves of later legend. In ancient Africa the dwarf-people were said to be the first to discover the secrets of mining, minerals, and metals, hence the fairy tale incarnation of the Seven Dwarves as miners. Thus do we see, as Massey showed, that fables and fairy tales in ancient Britain, Europe, and elsewhere evolved as redacted mythology. However, the line of descent could be traced back to the pristine mytho-symbolism born in the cradle lands of East Africa and nurtured to maturity in the Nile Valley.

In Volume II of *A Book of the Beginnings*, we encounter for the first time the essential thrust of the Masseyan oeuvre. In this volume, again through the medium of myth, symbolism, and language he minutely dissects the religion and culture of the ancient Hebrews to reveal them as but an off-shoot of old Kemit and, even more remotely than that, of Africa itself. In contemporary times, it is difficult for us to scan the past with a clear lens because the racial mythology imposed on the world over the last 500 years has obscured everything. Few, whether Jew or non-Jew, are prepared to examine ancient Judaism in terms of its African origins. Insofar as the Hebrew Torah spawned both Christianity and Islam, such an examination would mean acknowledging that the religious faiths of more than a third of the world's population are traceable in a direct line of descent to African sacred thought mediated through its codifier ancient Egypt. Experience shows that such is not an idea that many, of any faith, are prepared to contemplate with equanimity.

Yet the Torah itself, i.e., the Christian Old Testament, is replete with stories detailing the impact of Egypt on the Hebrew religion in its formative centuries. It is after his sojourn in Egypt that Abraham undergoes a series of transformations that, in the fullness of time, lead to a sacred covenant with God. It seems evident, for example, that the establishment of the practice of circumcision among his people was a direct outgrowth of his encounter with Egyptian culture where circumcision had already been customary for centuries. He took the Egyptian woman Hagar as his consort, and she gave birth to his first son Ishmael, the founding ancestor of the Arabian people. It is necessary to point out, however, that Abraham was not himself Hebrew. He is as much the ancestor of the Children of Ishmael (Arabs) as he is the Children of Israel (Hebrews), and in any case, the Hebrew religion does not appear until Moses bestows the Law upon the Children of Israel at Sinai 700 years after Abraham.

The most important historical episode in Hebrew history is unquestionably the 430-year sojourn in Egypt and the Exodus that brought it to a close. In truth, it is difficult to imagine how the Children of Israel could have become a nation and elaborated the Hebrew religion without the prolonged and intimate contact with Egyptian culture. This is in fact the crux of Massey's argument. Moses, unquestionably the single most important figure in Hebrew history, was a trained Temple initiate, as the Torah itself indicates, and therefore the only one qualified to lead the Children of Israel out of Egypt. Concerning the Five Books of Moses, Massey says:

> The Hebrew Books of the Genesis, Exodus, Numbers, Joshua, and Judges are invaluable as a virgin mine of mythology; they are of the utmost importance as an aid in *recovering the primeval types of Egyptian thought*...(emphasis added)[7]

In Massey's mind, the original Children of Israel actually were Egyptian in language, religion, and ethnicity. Doubtless, he first arrived at this conclusion after becoming acquainted with the testimony of Josephus who, in His essay *Against Apion* paraphrases the Egyptian annalists' explanation of the Exodus, indicating that the people who so departed Egypt were themselves Egyptian, though outcasts.[8] Certain classical writers including Tacitus, Diodorus, and Strabo also echoed this assessment. For Massey, the evidence from the monumental inscriptions corroborated and verified these assertions. The very least that can be said is that even if the progenitors of the Children of Israel, i.e., Jacob and his clan, immigrated into Egypt from outside it, in a matter of two or three generations, they would have become fully Egyptianized by race, language, and culture. After 430 years, it would have been an Egyptian multitude that departed Egypt, led by the Temple priest Moses, whom the Egyptians called Osar-siph. Fully trained in priestcraft, Moses the Egyptian would have literally "laid the law down" on the foundation of Egyptian models.

This reliance on Egyptian prototypes continued at least through the reign of Solomon. Solomon allied himself with the Egyptian royal house by marrying one of its princesses. According to James Henry Breasted, when Solomon had the Temple built—the outstanding accomplishment of his reign—his Phoenician architect Hiram constructed it "on the ground plan of an Egyptian temple."[9] Ancient Egypt adumbrates the entire Old Testament.

In *The Natural Genesis*, published in 1883, Massey continued the elaboration of his thought inaugurated in *A Book of the Beginnings*. Indeed, he saw these two-volume sets as constituting a single 4-volume work. However, in *The Natural Genesis* he delved more deeply and searchingly into the source of symbols. Massey, anticipating Carl Jung by a generation, became preoccupied with the role, function, and impact of symbols on human thought and action. He said of symbols:

> Whether for good or for ill the symbol has proved all-powerful. The hold of symbolism is in its way as strong in civilised society as in the savage (sic) world. [10]

Massey understood, as later psychoanalysts would, that the symbol is the "most helpful of servants, most tyrannous of masters."[11] He firmly believed, however, that

> ...so potent is the influence of symbols over the mind that the world's welfare cannot afford to have their indefinable appeal perverted by cunning or ignorance. [12]

Massey was keen to trace the manner in which symbols arose from images taken from the surrounding topography of nature in the inner African cradleland, or "placentaland," (Ta-Kenset) as the Egyptians called it. The spider would, for obvious reasons, be the original type of weaver who wove the strands of creation. The baboon would be the herald or announcer because of its penchant for chattering at dawn just before the rising of the sun. The ape would also be the messenger and the master of the word as seen in the Egyptian figure of Thoth because Massey believed the clicking sounds made by apes were imitated by early African humanity and preserved by the Khoisan people in the click languages of southern Africa. The scarab beetle, because its eggs deposited in a spherical ball of dung buried in the Nile mud transformed into young beetles that flew up out of the mud, became the image of transformation and self evolution.

In *The Natural Genesis*, Massey also examined in depth what he called the "typology of the mythical Serpent." Few animal figures resonated at so many levels of symbolism as the serpent, and there is not a single African culture where serpent-symbolism doesn't figure very prominently. The serpent as Pythoness was an image of the Great Mother as she coiled around and incubated her eggs. Because of the serpentine shape, it was also a masculine, "spermatic" image. The serpent's ability to shed old skin for new symbolized a type of renewal, which is why the serpent became a symbol of healing in Africa and in the Mediterranean world. Because of its sinuous movements, it also symbolized moving water or rivers. Finally, the snake in its venomous aspect was a symbol of death and ultimately of evil. Related to this was the pythonic image of the Swallower or Devourer. In Africa, both serpent and crocodile, being devouring reptiles, were interchangeable symbols. Fusing the two produced the dragon of later mythology.

No earthly animal has or does evoke such a compounded emotional mix of awe, terror, loathing, and veneration as the serpent. The Old Testament enjoined men on the one hand to anathematize the serpent but on the other, to be "as wise as serpents." Most traditional African religions kept and venerated a sacred python and the kingdom of Dahomey took its name from the sacred celestial serpent Da imaged as the rainbow. Serpent power was synonymous with royal power in Egypt, and every pharaoh wore the diadem of the uraeus serpent above his crown.

Finally, time-space itself was configured as the uroboric serpent that forms the circle of eternity by taking its tail in its mouth. The uroborous became the symbol of the totality of time and space. The power of serpent-symbolism has even intruded into modern science. The German chemist Kekule, working on a problem of chemical structure, once dreamed of a serpent with its tail in its mouth. When he awoke, he was able to propose the ring structure of benzene, which was one of the formative discoveries of organic chemistry.

What was done to Hebrew scripture in *A Book of the Beginnings* by way of plumbing its remote origins was also done to Christian origins in *The Natural Genesis*. However, Christianity was more clearly and fully marked out by ancient Kamite astronomical mythology. Indeed, it was Massey's contention that revelation and prophecy consisted solely in knowing the major celestial time cycles that governed the symbolic life and historical consciousness of human beings. In particular, the Precession of the Equinoxes or its twin, the Precession of the Pole Stars, ramifying into a Great Year of 26,000 years, was the heavenly clock by which humanity could measure destiny. What Massey realized was that the ancients had written a *Book* ages anterior to the beginning of conventional history, and that *Book* was inscribed in the heavens. The positions of stars, the shape of constellations, the celestial mechanics—all told the story of the universe in microcosmic and macrocosmic detail.

Christianity was, at bottom, an Essene creation. Christian genesis was astromythical as its generators intended, but along the way, the concept became literalized so as to bear but scant resemblance to its original manifestation. The birth of Christ, the resurrection drama at Easter, and even the adoption of the double-fish symbolism were all originally astromythical events. The Christhood represented the divine spark in human life, a continuation of the process of Osirification whose permutations could be tracked by reference to the Book of the Heavens. Jesus, or Yehusua, surnamed "Pandera," was himself an Essene magus and healer who humanly personified the Osirified ethos, i.e., the Christhood. In Massey's view, the life of Christ, his teachings, and his deeds were meant to serve as a "blueprint" for redemption not as a vehicle for vicarious atonement.

Yehusua's surname Pandera meant "panther" and since the panther-skin was the emblem of the Afro-Kamite priesthood, it can be legitimately wondered if he weren't one of them. Yehusua spent the best part of his 50-odd years in Egypt, and the Essenes, known as "Therapeutae" in Egypt, were undoubtedly connected to the Temple there if only because they seemed to have been Jewish Pythagoreans.

A clue as to why Rome (Roma) eventually became the sacred center of Christianity lies in the very name. The name Roma may have been related to the Egyptian name Remi for the Divine Fish or Remu for the Fish-City. Christianity was the religion of the Double-Fishes which is astronomically configured as the constellation of Pisces, and Roma was the Fish-City of late antiquity.

Gerald Massey's three double-volume works, which include *Ancient Egypt the Light of the World*, constitute his major opus and comprise more than three thousand pages of densely-packed text, virtually every line vibrating with data, much of it obscure. He could have done with an editor, but there were precious few in all England qualified to trim, pare down, and winnow the text. Who, other than the author himself, would know what to edit out?

Not surprisingly it is a challenge to fully grasp the Masseyan material. It often demands numerous attentive re-readings of selected portions of the text. It took the present writer 10 years to wade through the totality of Massey's work, mainly because it was necessary to imbibe it in measured doses. Massey apparently had made the decision, from the first, to put every bit of the massive amount of information he had absorbed into the manuscript. His intent was to assure that the far-reaching conclusions he derived would rest on an unassailable documentary foundation. Because of this foundation, one thing is incontrovertible: few will read the works of Gerald Massey and think about the story of humanity in quite the same way ever again.

Notes:

1. Massey G, *A Book of the Beginnings* (Volume I), London: Williams and Norgate, 1881, Frontispiece.

2. See Finch C, *Echoes of the Old Dark Land: Themes from the African Eden,* Decatur, GA: Khenti Inc., 1991, pp. 12-49 for an extended discussion of the monocentric vs. polycentric debate.

3. See The Search for Adam and Eve, *Newsweek,* January 11, 1988, pp. 46-52 for the then contrasting views of Cann and Wallace on the evolutionary implications of the mitochondrial DNA studies.

4. See Tonor M, Americas' Discovery: How the Indians Arrived, *The Atlanta Journal & Constitution,* February 13, 1993, p. El, for a discussion of Wallace's mitochondrial DNA studies among Amerindians.

5. See Shreeve J, The Dating Game, *Discover,* September 1992, pp. 76-83 for a report on the work of Brooks and Yellen in northeastern Zaire.

6. Higgins G, *Anacalypsis* (Volume I), New Hyde Park: University Books, 1965 and MacRitchie D, *Ancient and Modern Britons,* London: Kegan Paul & Tench Co., 1884.

7. Massey, op. cit., Volume II, p. 363.

8. Whiston W P, translator/editor, *Josephus: The Complete Works,*"Against Apion," Grand Rapids: Kregel Publications, 1981, pp. 610

9. Breasted J H, *Dawn of Conscience,* New York: Charles Scribner's Sons, 1933, p.383.

10. Massey G, *The Natural Genesis,* London: Williams and Norgate, 1883, p. 12.

11. ibid., p. 13.

12. Ibid.

CONTENTS

EGYPT.

EGYPT! how I have dwelt with you in dreams,
So long, so intimately, that it seems
As if you had borne me; though I could not know
It was so many thousand years ago!
And in my gropings darkly underground
The long-lost memory at last is found
Of motherhood—you Mother of us all!
And to my fellow-men I must recall
The memory too; that common motherhood
May help to make the common brotherhood.
Egypt! it lies there in the far-off past,
Opening with depths profound and growths as vast
As the great valley of Yosemité;
The birthplace out of darkness into day;
The shaping matrix of the human mind;
The Cradle and the Nursery of our kind.
This was the land created from the flood,
The land of Atum, made of the red mud,
Where Num sat in his Teba throned on high,
And saw the deluge once a year go by,
Each brimming with the blessing that it brought,
And by that water-way, in Egypt's thought,
The gods descended; but they never hurled
The Deluge that should desolate the world.
There the vast hewers of the early time
Built, as if that way they would surely climb
The heavens, and left their labours without name—
Colossal as their carelessness of fame—
Sole likeness of themselves—that heavenward
For ever look with statuesque regard,

As if some Vision of the Eternal grown
Petrific, was for ever fixed in stone!
They watched the Moon re-orb, the Stars go round,
And drew the Circle; Thought's primordial bound.
The Heavens looked into them with living eyes
To kindle starry thoughts in other skies,
For us reflected in the image-scroll,
That night by night the stars for aye unroll.
The Royal Heads of Language bow them down
To lay in Egypt's lap each borrowed crown.
The glory of Greece was but the After-glow
Of her forgotten greatness Iying low;
Her Hieroglyphics buried dark as night,
Or coal-deposits filled with future light,
Are mines of meaning; by their light we see
Thro' many an overshadowing mystery.
The nursing Nile is living Egypt still,
And as her low-lands with its freshness fill,
And heave with double-breasted bounteousness,
So doth the old Hidden Source of mind yet bless
The nations; secretly she brought to birth,
And Egypt still enriches all the earth.

A

BOOK OF THE BEGINNINGS.

TRAVELLERS who have climbed and stood upon the summit of the Great Pyramid of Ghizeh tell us how all that is most characteristic of Egypt is then and there in sight. To the south is the long Necropolis of the Desert, whose chief monuments are the Pyramids of Abooseer, Dashoor, and Sakkarah. That way lies the granite mountain flood-gate of the waters, which come winding along from the home of the hippopotami to leap down into the Nile-valley at last with a roar and a rush for the Mediterranean Sea. To the north there is desert also, pointed out by the ruined pyramid of Abooroash. To the west are the Libyan Hills and a limitless stretch of yellow sand. Again, there is a grey desert beyond the white line of Cairo, under the Mokattam Hills.

And through these sandy stony desert borders, Egypt runs alongside of its river in a double line of living green, the northward flowing waters and their meadowy margin broadening beneficently into the Delta. Underfoot is the Great Pyramid, still an inscrutable image of might and of mystery, strewn round with reliquary rubbish that every whirl of wind turns over as leaves in a book, revealing strange readings of the past; every chip and shard of the fragments not yet ground down to dusty nothing may possibly have their secret to tell.[1]

The Great Pyramid is built at the northern end of the valley where it relatively overtops the first cataract, nearly 600 miles away to the south, and, as the eye of the whole picture, loftily

[1] *Ancient History from the Monuments*, by Dr. Samuel Birch. *Egypt of the Pharaohs*, Zincke. *Mummies and Moslems*. Warner. P. 92. *Martyrdom of Man*. Reade. Ch. i. *Life and Work at the Great Pyramid*. Smyth, Ch. i. p. 29.

looks down on every part of the whole cultivated land of Egypt. It is built where the land comes to an apex like the shape of the pyramid itself lying flat and pointing south, and the alluvial soil of the Delta spreads out fan-wise to the north. It is near to the centre of the land-surface of the globe. A Hermean fragment shows the earth figured as a woman in a recumbent position with arms uplifted towards heaven, and feet raised in the direction of the Great Bear. The geographical divisions are represented by her body, and Egypt is typified as the heart of all.[1] They set the base of the Great Pyramid very near the heart of all, or about one mile 568 yards south of the thirtieth parallel of latitude.[2]

There, in the stainless air, under the rainless azure, all is so clear that distance cannot be measured, and the remotest past stands up close to you, distinct in its monumental forms and features as it was thousands of years ago ; the colour yet unfaded from its face, for every influence of nature (save man) has conspired to preserve the works of art, and make dead Egypt as it were the embalmed body of an early time eternized.

Once a year the deluge comes down from above, flowing from the lakes lying far away, large as inland seas, and transforms the dry land into a garden, making the sandy waste to blossom and bear the "double-breasted bounteousness" of two harvests a year, with this new tide of life from the heart of Africa. Not only does the wilderness flush with colour, for the waters, which had been running of a dull green hue, are suddenly troubled and turned crimson. The red oxide of iron mixes with the liquid and gives it a gory gleam in the sunlight, making it run like a river of blood.

There is an antithesis to the inundation in another phenomenon almost as unique. This is found in the steady continuance of the north-wind that blows back the waters and spreads their wealth over a larger surface of soil, and enables the boatman to sail up the river right against the descending current. Everything Egyptian is typical, and when we see how the people figured the two truths of mythology as the two factors of being, and how they personified breath and water, we shall more or less perceive the initiatory import of this wonderful arrangement of wind and tide, and its combination of descending and ascending motive power.

The Nile water is highly charged with ammonia and organic matter, which are deposited as manure. It is, for instance, three times as rich in fertilising matter, whether in suspension or in solution, as the Thames at Hampton Court.

The Great Mendes Stele (17) says :—

" The entire wealth of the soil rests on the inundation of the Nile that brings its products." This bounty was spread out for all by

---

[1] Stobœus, *Ecl. Eth.*, p. 992, Ed. Heeren.
[2] Piazzi Smyth.

the breath of the beneficent wind. Num, Lord of the Inundation, is painted on the monument as the Green God, and the limit of the inundation was the measure of Egypt's greenness. The waters that brought the silt clothed the soil with that colour just so far as they were blown.

From the beginning Lower Egypt, the Delta, was a land literally rained down by the inundation as a gift of the gods. For the clouds arise from their several seas and sail off heavily-laden toward Equatorial Africa, and there pour forth their weight of water during a rain of months on mountain slopes that drain into the fresh lakes until these are brimmed to bursting, and their northern outlet of birth is the Nile. The White Nile at first, until the Abyssinian highlands pour into it their rushing rivers of collected rain with force enough to float a mass of silt that is part of a future soil, the presence of which in the waters makes the Blue Nile ; then the river becomes the turbid Red Nile of the inundation, and as it spreads out fanwise towards the Mediterranean Sea, it drops that rich top-dressing. of soil or the very fat of land and unctuous mud-manure, every year renewed and rained down by that phenomenal flood. We shall find the whole of the deluge legends of the world, and all the symbolical deluge language used in astronomical reckoning, are bound up inseparably with this fact of the inundation of Egypt. The universal mythical beginning with the waters, the genesis of creation and of man from the mud, are offspring of this birthplace and parentage. In no other part of earth under heaven can there be found the scenery of the inundation visibly creating the earth, as it is still extant in the land of marvel and mystery. Only in Egypt could such a phenomenon be observed as the periodic overflow of the river Nile, that not only fertilises the fields with its annual flood, but actually deposits the earth, and visibly realises that imagery of the mythical commencement of all creation, the beginning with the waters and the mud, preserved in so many of the myths.

According to Aulus Gellius (Diodorus i. 19), Egypt was named Aeria. The Egyptian Aur (later Aer) is the name of the river Hebrew Iar. Aeria is the land of the river, possibly with the further meaning of the pure, as Ia means to wash, whiten, purify. Another name of Egypt is Tameri. Ta is to drop, heap, deposit, type ; meri is the inundation, Tameri is land thus deposited. The vulgar English to ta is a child's word, and it means to deposit soil ; also Ta-meri reads the gift of the inundation, the gift of the goddess Meri who has a dual form as Meri-Res (South) and Meri-Mehi (North). Egypt is also designated the Land of the Eye.[1] The eye of the cow shedding an emblematic tear was a type of Ta-ing.

It is also called Khemi, the land of the gum-tree, and the acacia

[1] Wilk, 2nd. ser. vol. ii. p. 48.

gum-tree supplied another symbol of shedding substance ; or, of the Kamai-plant, from which the Egyptians obtained a precious oil.

Khemi, Egypt, is personified as a female who wears on her head a sign which Wilkinson thought indicated "cultivated land," but it means the land created from the waters, determined by the sign of marshland or land recovered from the waters. The sign is the determinative of Hat, chaos, or precommencement, and its true value may be found in the Cornish Hatch, a dam.

Egypt is often called Kam, the Black Land, and Kam does signify black ; the name probably applied to the earliest inhabitants whose type is the Kam or Ham of the Hebrew writers. But Kam is likewise to create, and this was the created land ; visibly created like the gum from the tree by droppings. Kam is the root and has the value of the word chemistry, and the land of Kam was the result of Nature's chemistry, aided by the Hatches or Dams.

The Assyrians called Egypt Muzr. Muzau is source, an issue of water, a gathering or collecting. It is the Egyptian mes, the product of a river. Now it is important for the present purpose to wring the meaning out of Egyptian words, drop by drop, every one is portentous and symbolical. For example, mes means mass, cake, chaos, it is the product of the waters gathered, engendered, massed. The sign of this mass was the hieroglyphic cake, the Egyptian ideograph of land.[1] This cake of Mesi was figured and eaten as their bread of the mass, a seed-cake too as the Hieroglyphics reveal. And the cake is extant to-day in the wafer still called by the name of the Mass, as it was in Egypt. Mes, the product of the waters and the cake, is likewise the name for chaos, the chaos of all mythological beginning. Mes, then, the mass or product of the river when caked, is the primeval land, the pure land periodically produced from the waters, the land of Mesr, whether of black mud or red.

We find a word in Æthiopic similar to metzr meaning the earth, land, soil. Mazr, or mizr, is an Arabic name of red mud. There is, however, a mystical reason for this red applied to mud as a synonym of source or beginning.

These derivations of the names from Kam, the created land ; mes, the product of the river ; tameri, the soil and gift of the inundation, show that Lower Egypt was designated from the soil that was shed, dropped, wept, deposited by the inundation of the Nile, and that the natives were in various ways calling it the Alluvial Land.

But the Hebrew name of Egypt, Mitzraim, applies to both lands. For this we have to go farther than Lower Egypt, and mes, the product of a river, the Mud of mythology.

We may rest assured, says Brugsch Bey, that at the basis of the designations Muzur (Assyrian), Mizr (Arabic), Mitzraim (Hebrew), there lies an original form consisting of the three letters *m r s*, all

---

[1] Egyptian room, British Museum, 9,900.

explanations of which have as yet been unsuccessful. His rendering of the meaning as Mazor, the fortified land, the present writer considers the most unsuccessful of all. Mest-ru and Mest-ur are the Egyptian equivalents for the Hebrew Mitzr, plural Mitzraim, and the word enters into the name of the Mestræan Princes of the old Egyptian Chronicle. Mest (Eg.) is the birth-place, literally the lying-in chamber, the lair of the whelp; and ru is the gate, door, mouth of outlet; ur is the great, oldest, chief. Mest-ru is the outlet from the birth-place. In this sense the plural Mitzraim would denote the double land of the outlet from the inland Birth-place.

There is a star "Mizar" in the tail of the Great Bear, the Typhonian type of the Genitrix and the birth-place, whose name is that of Lower Egypt or Khebt. Mest (Eg.) is the tail, end, sexual part, the womb, and ur is the great, chief, primordial. Thus Mitzraim and Khebt are identical in the planisphere as a figure of the birth-place, found in Khebt or Mitzraim below. Mest-ur yields the chief and most ancient place of birth which is not to be limited to Lower Egypt.

It is certain, however, says Feurst, that מצר and מצור meant originally and chiefly the inhabitants. It is here that mest-ur has the superiority over Mest-ru. The ru in mest-ru adds little to the birth-place, whereas the compound mestur expresses both the eldest born and the oldest birth-place. The Hebrew צ represents a hieroglyphic Tes which deposited a phonetic T and S, hence the permutation; mtzr is equated by mstr, and both modify into misr. In the same way the Hebrew Matzebah (מצבה) renders the Egyptian Mastebah. Also, Mizraim is written Mestraim by Eupolemus. Khum, he says, was the father of the Æthiopians, and brother of Mestraim, the father of the Egyptians.[1]

The name Egypt, Greek Aiguptos, is found in Egyptian as Khebt, the Kheb, a name of Lower Egypt. Khebt, Khept or Kheft means the lower, the hinder part, the north, the place of emanation, the region of the Great Bear. The f, p and b are extant in Kuft, a town of Upper Egypt; in Coptos, that is Khept-her or Khept above, whence came the Caphtorim of the Hebrew writings (Gen. x. 14), enumerated among the sons of Mitzraim, and Kheb, Lower Egypt.

The Samaritan Pentateuch (Gen. xxvi. 2) renders Mitzraim by the name of Nephiq נפיק which denotes the birth-land (ka) with the sense of issuing forth; Aramaic נפק to go out. In Egyptian nefika would indicate the inner land of breath, expulsion, going out, or it might be the country of the sailors.

The land of Egypt was reborn annually as the product of the waters was added layer by layer to the soil. Three months inundation and nine months dry made up the year. The nine months

[1] Euseb. Praep. Evang. B. 9.

coincided with the human period of gestation, a fact most fruitful
in suggestion, as everything seems to have designedly been in this
birth-place of ideas. They dated their year from the first quickening heave of the river, coincident with the summer solstice and the
heliacal rising of the Dog-star. The Nile not only taught them to
look up to the heavens and observe and register there the time and
tide of the seasons, but also how to deal with the water by means of
dykes, locks, canals, and reservoirs, until their system of hydraulics
grew a science, their agriculture an art, and they obtained such
mastery over the waters as finally fitted them for issuing forth to
conquer the seas and colonise the world.

The waters of Old Nile are a mirror which yet reflects the earliest
imagery, made vital by the mind of man, as the symbols of his
thought. A plant growing out of the waters is an ideograph of
Sha, a sign and image of primordial cause, and becoming, the
substance born of, the end of one period and commencement of
another, the emblem of rootage in the water, and breathing in
the air, the type of the two truths of all Egypt's teaching, the two
sources of existence ultimately called flesh and spirit, the blood
source and breathing soul.

About the time when the Nile began to rise in Lower Egypt, there
alighted in the land a remarkable bird of the heron kind, which had
two long feathers at the back of its head. This, the *Ardea purpurea*,
was named by the Egyptians the Bennu, from " nu," a periodic type,
and " ben," splendid, supreme—literally, as the crest indicated, tip-
top. It was their Phœnix of the Waters, the harbinger of re-arising
life, and was adopted as an eschatological type of the resurrection
The planet Venus is called on the monuments the " Star of Bennu
Osiris," that is of Osiris *redivivus.*

The beetle appeared on the Nile banks in the month previous to that
of the inundation,[1] the month of re-birth, Mesore, and formed its
ark against the coming flood to save up and reproduce its seed in
due season. This they made their symbol of the Creator by transformation, and type of the only-begotten Son of the Father.[2]

The inundation supplied them with the typical plough. To
plough is to prepare the soil for seed. The inundation was the
first preparer of the soil. The inundation is called Mer, and one
sign of the Mer is a plough. This shows that when they had invented
the primitive hand-plough of the hoe kind they named it after the
water-plough, or preparer of the soil, and the Mer plough is a symbol
of the running water.

The river likewise gave them their first lessons in political economy
and the benefits of barter, by affording the readiest means of exchange.
Its direction runs in that of longitude, or meridian, with all the products ranged on either side like stalls in a street ; so close to the

---

[1] Pettigrew, *History of Mummies*, p. 220.   [2] *Hor-Apollo*, B. i. 10.

water-way was the cultivated soil. It crossed through every degree of Egypt's latitude and became the commercial traveller of the whole land, carrying on their trade for the enrichment of all.

Generally speaking the monuments offer no direct clue to the origin of the people; they bring us face to face with nothing that tells of a beginning or constitutes the bridge over which we can pass to look for it in other lands. Like the goddess Neith, Egypt came from herself, and the fruit she bore was a civilisation, an art, a mythology, a typology, absolutely autochthonous.

We see no sign of Egypt in embryo; of its inception, growth, development, birth, nothing is known. It has no visible line of descent, and so far as modern notions go, no offspring; it is without Genesis or Exodus.

When first seen Egypt is old and grey, at the head of a procession of life that is illimitably vast. It is as if it always had been. There it stands in awful ancientness, like its own pyramid in the dawn, its sphinx among the sands, or its palm amid the desert.

From the first, all is maturity. At an early monumental stage they possess the art of writing, a system of hieroglyphics, and the ideographs have passed into the form of phonetics, which means a space of time unspanable, a stage of advance not taken by the Chinese to this day.

The monuments testify that a most ancient and original civilisation is there; one that cannot be traced back on any line of its rootage to any other land. How ancient, none but an Egyptologist who is also an evolutionist dares to dream. At least twelve horizons will have to be lifted from the modern mind to let in the vistas of Egypt's prehistoric past. For this amazing apparition coming out of darkness on the edge of the desert is the head of an immense procession of life—issuing out of a past from which the track has been obliterated like footprints in the shifting wilderness of sand.

The life was lived, and bit by bit deposited the residual result. The tree rooted in the waste had to grow and lay hold of the earth day by day, year after year, for countless ages. Through the long long night they groped their way, their sole witnesses the watchers on the starry walls, who have kept their register yet to be read in the astral myths, for the heavens are Egypt's records of the past.

The immeasurable journey in the desert had to be made, and was made step by step through that immemorial solitude which lies behind it, as certainly as that the river has had to eat its own way for hundreds of thousands of years through the sandstone, the limestone, and the granite, in order that it might at last deposit the alluvial riches at the outlet, even though the stream of Egypt's long life has left no such visible register on earth as the waters of old Nile have engraved in gulf-like hieroglyphics upon the stony tablets of geological time. The might and majesty of Egypt repose upon a past as reàl as the uplifting rock of the Great Pyramid, and the base, however hidden, must be in proportion to the building.

They are there, and that is nearly all that has been said; how or when they got there is unknown. These things are usually spoken of as if the Egyptians had done as we have, found them there. The further we can look back at Egypt the older it grows. Our acquaintance with it through the Romans and Greeks makes it modern. Also their own growth and shedding of the past kept on modernizing the myths, the religion, the types. It will often happen that a myth of Egyptian creation may be found in some distant part of the world in a form far older than has been retained in Egypt itself.

It is a law of evolution that the less-developed type is the oldest in structure, reckoning from a beginning. This is so with races as with arts. The hieroglyphics are older than letters, they come next to the living gesture signs that preceded speech. But less-developed stages are found out of Egypt than in it because Egypt went on growing and sloughing off signs of age, whilst the Maori, the Lap, the Papuan, the Fijian suffered arrest and consequent decadence. And the earlier myth may be recovered by aid of the arrested ruder form.

Fortunately the whole world is one wide whispering-gallery for Egypt, and her voice may be heard on the other side of it at times more distinctly than in the place where it was uttered first. And looking round with faces turned from Egypt we shall find that language, the myths, symbols, and legendary lore of other lands, become a camera obscura for us to behold in part her unrecorded backward past. The mind of universal man is a mirror in which Egypt may be seen. We shall find the heavens are telling of her glory. Much of her pre-monumental and pre-monarchical past during which she was governed by the gods has to be reflected for us in these mirrors; the rest must be inferred. Vain is all effort to build a boundary wall as her historic starting-point with the *débris* of mythology.

The Stone Age of Egypt is visible in the stone knife continued to be used for the purpose of circumcision, and in the preparation of their mummies. The stone knife was a type persisting from the time of the stone implements. The workers in bronze, iron, and other metals did not go back to choose the flint weapon. The " Nuter " sign of the god is a stone axe or adze of the true Celt type.

The antiquity of Egypt may be said to have ended long before the classical antiquity of the moderns begins, and except in the memorials of myth and language it was pre-monumental. We know that when Egypt first comes in sight it is old and grey. Among the most ancient of the recipes preserved are prescriptions for dyeing the hair. There are several recipes for hair dye or washes found in the Ebers Papyrus, and one of these is ascribed to the lady Skheskh, mother of Teta, the first king on the monuments after Mena.

This is typical. They were old enough more than 6,000 years ago for leprosy to be the subject of profound concern. A manuscript of the time of Rameses II. says :—

"This is the beginning of the collection of receipts for curing leprosy. It was discovered in a very ancient papyrus inclosed in a writing-case under the feet (of a statue) of the god Anubis, in the town of Sakhur, at the time of the reign of his majesty the defunct King Sapti,"[1] who was the fifth Pharaoh of the first dynasty, in the list of Abydus.

Leprosy was indigenous to Egypt and Africa; it has even been conjectured that the white negroes were produced by it, as the Albinos of the Black race.

The most ancient portion known of the ritual, getting on for 7,000 years old, shows that not only was the Egyptian mythology founded on the observation of natural phenomena at that time established, but the mythology had then passed into the final or eschatological phase, and a system of spiritual typology was already evolved from the primordial matter of mythology. The text of chapter cxxx. is said in the annotation to have been found in the reign of King Housap-ti, who, according to M. Déveria, was the Usaphais of Manetho, the fifth king of the first dynasty, and lived over 6,000 years since ; at that time certain parts of the sacred book were discovered as. antiquities of which the tradition had been lost. And this is the chapter of "vivifying the soul for ever."

The Lunar myth with Taht as the Word or logos, is at least as old as the monuments (it is indefinitely older), and we know from the ritual (ch. xlii.) that Sut was the announcer, the Word, and represented the Two Truths before the time of Taht ; Sut was the first form of Hermes the Heaven-born.

From almost the remotest monumental times, dated by Bunsen as the seventh century after Mena, or according to the present reckoning nearly 4,000 years B.C., they had already attained to the point of civilisation at which the sacrifice of human beings as offerings to the gods is superseded by that of animals. There is no doubt of the human victims once sacrificed, as the sacrificial stamp, with the victim bound and knife at the throat, still remains to attest the fact,[2] as well as the Ideographic Kher, the human victim bound for slaughter ; Kher being one with Kill.

There is a stone in the museum at Boulak consecrated to the memory of a noteworthy transaction. We learn from it that in the time of Kufu of the fourth dynasty, and founder of the Great Pyramid, that the sphinx and its recently discovered temple not only existed already, but were then in a state of dilapidation, and

---

[1] Brugsch's *History of Egypt*, vol. i. p. 58, Eng. Tr.
[2] Wilkinson, *Ancient Egypt*, vol. v. p. 352.

it is recorded in the inscription that he restored them. He built the temple of the Great Mother and replaced the divinities in her seat.

" The living Har-Sut[1] King Kufu, giver of life, he founded the house of Isis (Hest) dominating the pyramid near to the house of the sphinx. North-west of the house of Osiris the Lord of Rusta he built his pyramid near to the temple of that goddess, and near the temple of his royal daughter Hantsen. He made to his mother Athor, ruling the monument, the value as recorded on the tablet, he gave divine supplies to her, he built her temple of stone and once more created the divinities in her seat. The temple of the sphinx of Har-Makhu on the south of the house of Isis, ruler of the pyramid, and on the north of (that of) Osiris Lord of Rusta."[2] This was in a climate where, as Mariette observes, there is no reason why monuments should not last for a hundred thousand years, if left alone by man. The dilapidation and overthrowal may, to some extent, have been the work of man.

One of the loveliest things in all poetry is in Wordsworth's lines—

> " And beauty born of murmuring sound
> Shall pass into her face."

Yet this was anticipated by the father of King Pepi of the sixth dynasty in the " Praise of Learning," who says to his son—

> " Love letters as thy mother !
> I make its [learning's] beauty go in thy face."

And in another papyrus, some fifty-five centuries old, the wise sage and thorough artist, Ptah-hept, advises his reader to " beware of producing crude thoughts ; study till thy expression is matured."

Bunsen tried to fix the place of Egypt in history with no clue to the origin and progress of its mythological phenomena ; no grasp of the doctrine of evolution ; no dream of the prehistoric relationship of Egypt to the rest of the world. The Egypt of the lists and dynasties is but the briefest span when compared with the time demanded for the development of its language, its myths, its hieroglyphics. Egypt as an empire with its few thousand years dating from Mena is but as yesterday, it only serves as the vestibule for us to the range of its pre-eval past. The monuments of 5,000 years do not relate a tithe of Egypt's history which is indirectly recorded in other ways. She is so ancient that we are shown nothing whatever of the process of formation in the creation of the earliest mythology. Their dynasties of deities are pre-monumental ; their system is perfected when history begins. A glance at the list of

---

[1] " Har-Sut." I read the *Tie* sign in this passage as " Sut." The name of Suti (Lepsius, Königsbuch) is several times written with the Tie sign.

[2] De Rougé, *Les Monuments qu'on peut attribuer aux six premières Dynasties*, pp. 46, 47. Bunsen, *Egypt's Place*, vol. v. p. 719.

divinities paralleled in the present work will show the significance of this fact, and help to bring the hidden past of Egypt into sight. The traces of evolution and development are not to be expected upon monuments which begin with some of the finest art yet found. Evidence of the primeval nature of the people is not likely to be found contemporary with the perfection of their art. The ascent is out of sight, and the Stone Age of Egypt is buried beneath a hundred feet of Nile sediment, but the opponents of evolution gain nothing by the negative facts. Perfect art, language, and mythology did not alight ready-made in the valley of the Nile; and if the ascent be not traceable here, neither is it elsewhere. There is not a vestige of proof that these were an importation from other lands.

Bunsen assumed that the roots of Egypt were in Asia, although the evidence for that origin has never been given. He speaks of religious records and monuments older than those of Egypt but does not name them.[1] Also of Semitic roots (of language) found in the names of Egyptian gods, whereas no Egyptian roots are found in the names of Semite gods.[2] He affirms that the Egyptian language clearly stands between the Semitic and Indo-Germanic, its forms and roots cannot be explained by either singly, but are evidently a combination of the two.[3] Evidence will be adduced in this work such as will go some way to show that there is originally neither Semitic nor Indo-Germanic independently of the Egyptian or African parentage. Bunsen argued that the religion of Egypt was merely the mummy of the primal religion of Central Asia, its mythology the deposit of the oldest beliefs of mankind which once lived in primeval Asia, and were afterwards found petrified in the valley of the Nile. He has to get rid of the particular country, people, and language, by means of a vast catastrophe in which they sank from sight and left but a few floating fragments.

The foundations for such a theory are entirely mythical, and belong at last to the legends of the Genesis, the primeval home in the Vendidad, and of the deluge, the typology of which subjects will be dealt with in the present work. The Asiatic origin of the Egyptians is necessary to support the delusive theory of the Indo-European migration of races and languages. But not a vestige of evidence has been adduced for the pre-existence of the Egyptian civilisation, art, hieroglyphics, in any other land before these are discovered ready-made in the valley of the Nile.

Brugsch-Bey observes, "Whatever relations of kindred may be found always to exist between these great races of mankind, thus much may be regarded as certain, that the cradle of the Egyptian people must be sought in the interior of the Asiatic quarter of the world. In the earliest ages of humanity, far beyond all historical

---

[1] *Egypt's Place*, vol. iv. p. 71.   [2] *Ibid.*
[3] Vol. i. p. 10, Preface.

remembrance, the Egyptians, for reasons unknown to us, left the soil of their primeval home, took their way towards the setting sun, and finally crossed that bridge of nations, the Isthmus of Suez, to find a new fatherland on the favoured banks of the holy Nile." [1] For these assumptions and assertions there is not a grain of evidence adduced or adducible. The only point of departure for such a wild goose chase as seeking the roots of Egypt in Asia is the fact that on the earliest monuments the ethnological type has changed vastly from what it was tens of thousands of years earlier. Egyptologists as a rule are not evolutionists, and M. Brugsch finds no difficulty in removing a race ready-made bag and baggage in the old cataclysmal fashion. But where, he asks, "is the modern Hir-Seshta to lift the veil which still hides the origin of these men of yore?" Where, indeed! The Egyptian language, he asserts, "shows in no way any trace of a derivation and descent from the African families of speech." [2] He thinks the rude stone monuments of Æthiopia are but a clumsy imitation, an imperfect reproduction of a style of art originally Egyptian. He sees the stamp of antiquity not in primitive but in perfectly polished work. One might as well derive the knife, as a flake of flint, from a model formed in Sheffield steel. Egypt can only be understood by an evolutionist.

Jomard also considered that the monuments of Nubia are more modern than those of Thebes. But, for all who dare trust to the guidance of the laws of evolution, one of which shows us that the ruder the structure the more elementary and antique the art, may see the civilisation of Egypt descending the Nile in the Nubian and Æthiopian Pyramids, which are as certainly prior to those of Egypt as Stonehenge is to Gothic architecture. That is, the type is earlier, no matter when last copied! Such types were not designed by the Egyptians of the monumental period who went back to re-conquer Æthiopia. They mark the height of attainment of the Egyptians before they came down into the Nile valley. Art does not go back in that way, nor is it a question of retrogression but of development. The Æthiopian would still be the older *if the latest built*, because the builders had advanced no farther, and would not be the result of the Egyptians or their imitators going back to the more primitive style. Compared with the Great Pyramid they are pre-monumental so to say, or pre-Egyptian as Egypt is known to us, which means they are yet earlier in type.

If the Egyptians were of Asiatic origin, how comes it that the Camel has no place among the hieroglyphics? No representation of the camel is found on the monuments. It will be seen when we come to consider the origin and nature of the typology and hieroglyphics that this is almost an impossibility for any but an African people. The camel being a late importation into Egypt and other African

---

[1] Brugsch, *History of Egypt*, vol. i. p. 2, Eng. Tr.   [2] *Ibid*, vol. i. p. 3.

countries, will alone explain its absence from the pictures of an African people.

"The origin of the word Nile," says M. Brugsch, "is not to be sought in the old Egyptian language, but as has been lately suggested with great probability, it is derived from the Semitic word Nahar or Nahal, which has the general signification of river." [1] So good an Egyptologist should have remembered that aur or aru (Eg.) is the river and that nai is the definite plural article The, which gives all the Egyptian significance to *the River* as the double stream, the two waters or waterer of the two lands north and south. Naiaru interchanges with Naialu, whence the Nile. But for this combination of nai (plural the) and aru, earlier Karu river, extant in Æthiopic as Nachal, for the Nile, the Semites would have had no such Nahar or Nahal.

M. Brugsch is the Semitizer of Egyptology. He finds a Semitic foundation for doctrines so ancient in Egypt itself as to be almost forgotten there ; so long ago rejected and cast out as to have become the sign and symbol of all that was considered foreign to later Egypt. Nothing is older than the Great Mother and Child ; and as Ta-Urt and Sut, the Great Bear and Dog-Star, these were the gods of the Typhonians within the land long ages before they were re-introduced as the Semite Astarte and Bar-Sutekh. The same deities that were worshipped in Lower Egypt by the Hekshus and Palestino-Asiatic tribes were the divinities of pre-monumental Egypt and of the Shus-en-Har. Sut is called a Semitic divinity and he comes back with his Ass as such. But he returns as an. Exile to the old country, not as a new creation.

The duad of the Mother and Child consecrated in Astarte and Sutekh and execrated as Sut-Typhon was superseded in monumental times by the men who worshipped the Fatherhood as well, or instead of the Son. But the myth of Isis's descent into the under-world in search of Osiris is that of the mother and son who became her consort, and identical with that of Ishtar and Tammuz or any other form of the duad consisting of mother and child. The disk-worship re-introduced by the Amenhepts III. and IV. was looked on as intolerable heresy by the Osirians and Ammonians ; and yet their sun was the same god Har-Makhu of the two horizons, the God of the Sphinx and of the pre-monumental Shus-en-Har, the Har of the double horizon; the Har who was earlier than Ra, and Sabean before the title became solar.

The Goddess Anata appears in Egypt in the time of Rameses II. as if from Syria, a goddess of the bow and spear ; yet this was but a re-introduction of a most ancient divinity who had gone forth from Egypt to become the Anat and Anahita of the Chaldaio-Assyrian religions. She was Neith, the goddess with the bow and

[1] Vol. i. p. 12, Eng. Ed.

arrows whose earliest name on the monuments is Anit. This name is written "Anit-neb-her" in the time of the first dynasty.[1]

The great directress of Amenti, half hippopotamus, half lion, who is pourtrayed in the scenes of the Hades sitting at the fatal corner waiting wide-mouthed for the souls of the dead, is the degraded form in the Egyptian eschatology of the Great Mother who in the earlier mythology had been doubly-first in heaven in her twin starry types of the Great Bear and Sothis the Dog-Star, the types of Sut-Typhon. The dog and hippopotamus in heaven still identify her when in hell as a victim of theology.

In Egypt this oldest form of the Great Mother had been reduced from the status of deity to that of demon, and her idols to that of dolls. Two painted wooden dolls used as children's toys are copied by Wilkinson.[2] One of these reproduces the black negroid Isis with her veil or net thrown back. The other is a womb-shaped figure with the old Typhonian Genitrix the Gestator on it, with the tongue hanging out, but not *enceinte*, not Great, as in the worshipful images. Yet this old hippopotamus goddess was before Nupe, Isis, and Nephthys; hence we find they continued her type, and were imaged in her likeness. Not because she was hideous but primordial, the most ancient genitrix of the gods; and first mother of the sun, as Sut, who became her husband and his own father. She was the good Typhon, and Nephthys takes her place in Plutarch as consort of Typhon, and is found united to Sut.[3] Her knot or noose is the earliest form of the Ankh symbol of life. Her numerous statuettes in the tombs where she accompanies the mummies of the dead, leaning over her emblematic Tie as the image of brooding life, shows the goddess in her ancient rôle as the genitrix of mythology continued in the eschatological phase as the giver of re-birth. Hers is a figure so ancient that it belongs to a typology which preceded both eschatology and mythology, of an order that was set in heaven for use, not for worship; types of time and force, and not of beauty.

Again, Brugsch-Bey says "the Semitic nations used to turn the face to the east, the quarter of the rising sun, and accordingly they called the east the 'front side,' the west the 'hinder side,' the south the 'right,' and the north the 'left.' In opposition to all this the ancient Egyptians regarded the western side as the right, the eastern as the left. Accordingly they turned their face to the south, which they called upwards (her) or foreward, so that the north lay at their back, and hence its appellation of the lower (khar) or hinder (pehu) region. Now having regard to all this, the appellation of khar in the sense of "hinder land" could only have originated with such peoples as had their fixed abodes to the east of the land of khar (Phœnicia) that is, on the banks of the Euphrates. Thus Babel and its famous

---

[1] See Königsbuch, taf. 3, Lepsius.   [2] Vol. ii. p. 426.
[3] Group in the Musée du Louvre, Salle des Dieux.

tower appear unmistakably as the great centre whence the directions of the abodes of nations were estimated in the earliest antiquity."[1] But M. Brugsch does not have regard to "all this;" only to one part of it. He does not touch bottom in the matter, and the inference is fallacious. He begins with the solar nomenclature which is not primary, but the latest of all. Just as it was with the Sabbath, which was Sabean first, the seventh day being that of Sut, before it became Solar as Sunday. True, the Assyrians named the west Akharru, as the Egyptian Akar is the mountain of the west, the entrance to the hinder place, but a far earlier people in the land, the Akkadians, called the east mer-kurra. What then becomes of the Semitic emanation and naming from the centre found in the Biblical Babel? The Egyptian rule of perspective is positively based upon the right hand being the upper, and the left the lower. In the scenes of the Hades the blessed on the right hand are represented as those above, whilst the damned, those on the left hand, are down below.[2] So that in facing the east, their upper land of the south was on the right hand, the lower on the left. In vain we talk of origin until Egypt has been fathomed. The question is whether Phœnicia was named Khal (Kar) the lower and hinder part from the east or the south. The Egyptians employed both reckonings according to the Sabean or Solar starting point. With them Apta was both equatorial and equinoctial. Kheft is the hinder part in the west as well as in the north. This gives a hinder part reckoning from the east and another reckoning from the south. The reckoning from south and north— Sothis and Great Bear—we call Sabean. With the Egyptians south and north were the front and back, the upper and lower heaven, and the lower as the Kar or the Kheft was the hindward part to the south as the front. So in England we go up to the south and down to the north ; people go up to London and down to the shires (or Kars).[3] M. Brugsch assumes that Phœnicia was named Khal the west as the hinder part to the east, and what is here maintained in opposition to his assumption is that Phœnicia is named from the south. The Egyptians called it kefa or kheft, their north and hinder part, the lower of the two heavens and the region of the Great Bear (khepsh) and it was so named as an extension of Lower Egypt, khebt, in still going farther northward. In Fijian Hifo is the name of the underworld or down below; the northern quarter entered at the west, whilst Hev, in English Gipsy, is the Hole or Void. In the annals of Rameses III., Northern Palestine, Taha, is reckoned with the hinder parts of the earth.

Here is another fact, although it does not concern the naming of

---

[1] *Hist. of Egypt*, vol. i. p. 479.   [2] Sarcophagus in the Soane Museum.
[3] Speaking of the lower world, the north, the Osirian says, " I came like the sun through the gate of the lords of Khal" (*Ritual*, ch. cxlvii. Birch). He had come through the celestial Phœnicia, or kheft of the north.

Khal as the lower land from either the south or east, it does apply to the position and age of the primeval namers. The Egyptians had a Sabean orientation still more ancient than either of these two. Time was when the right hand was also the east, Ab, the right hand and the east side. On the other hand the west (sem) is the left side, semhi is the left hand, Assyrian samili, Hebrew samali. Now for the east to be on the right and the west on the left, the namers must face the north, and that this the region of the Great Bear was looked to as the great quarter, the birth-place of all beginning, is demonstrable. Those, then, who looked northward with the east on their right and the west on their left hand were naming with their backs to the south and not their faces; nor were their faces to the east. This mode of orienting was likewise the earliest with the Akkadians, who looked to the north as the front, the favourable quarter, the birth-place. They, like the dwellers in the equatorial regions, had seen the north was the starry turning point, and the quarter whence came the breath of life to the parched people of the southern lands. This accounts for the south being the Akkadian "funereal point" and the hinder part.

When the namers from the south had descended the Nile valley and made out their year, of which the Dog-Star was the announcer, then they looked south as their front, and the west was on the right, the east on the left hand, hence we find Ab for the left hand east.[1] Thus the Egyptians have two orientations, one with the east on the right hand, one with the east on the left, and both of these precede the frontage in the east. Perhaps it was a reckoner by one of the first two methods who asserted that there were in Nineveh "more than 120,000 persons who could not discern between their right hand and their left hand."

The earliest wise men came from the south, not the east, and made their way north. With the Hebrews Kam is a synonym of the south. Their south as Kedar was the place of coming forth, and as negeb it is identical with the Egyptian goddess, Nekheb, who brought forth in the south, and was the opener in that region, as the Chaldean god Negab is the opener in the west.

Khent, the Egyptian name of the south, is the type-word for going back. Khent means to go back, going back, and going up, at the same time that khebt is the hinder part and the place of going down. This may be said to be merely solar. But Khent the southern land, the name for farthest south, which can now be traced as far as Ganda (the U-ganda), means the inner land, the feminine abode, the birth-place, and the lake country. Brugsch-Bey and Bunsen have read from right to left that which was written from left to right.

Egypt, says Wilkinson, was certainly more Asiatic than African. But when? We have to do with the origin, not with the later appearances on the monuments after ages of miscegenation. It is true that a long gradation divides Negroes from the Egyptians,

[1] Pierret, *Vocabulaire*, "Ab, the Left."

but the whole length of that gradation lies behind them in the coloured races of the dark land, and in no other. All the four colours, black, red, yellow, and white, meet upon the monuments, and they all blend in Egyptian types. The red men and the yellow are there as Egyptians, with the background of black out of which the modifications emerge. The problem of origin cannot be worked back again from white, yellow, or red to black, as it can forwards from black, with Egyptians for the twilight dawning out of darkness. The red skin similar to the complexion of the earliest men of the monuments is not an uncommon variety of the Africans. Bowditch asserts that a clear brownish-red is a complexion frequently found among the Ashantis although deep black is the prevailing colour. Dr. Morton, the American craniologist, recanting an earlier opinion says : " I am compelled by a mass of irresistible evidence to modify the opinion expressed in the *Crania Egyptiaca*, namely, that the Egyptians were an Asiatic people. Seven years of additional investigation, together with greatly increased materials, have convinced me that they are neither Asiatics nor Europeans but aboriginal and indigenous inhabitants of the Nile or some contiguous region— peculiar in their physiognomy, isolated in their institutions, and forming one of the primordial centres of the human family."

Professor Huxley has asserted that the aborigines of Egypt were of the same physical type as the original natives of Australia, for, he says, "although the Egyptian has been modified by admixture, he still retains the dark skin, the black silky wavy hair, the long skull, the fleshy lips, and broadish alæ of the nose which we know distinguished his remote ancestors, and which cause both him and them to approach the Australian and the *Dashyu* more nearly than they do any other form of mankind." [1]

A vocabulary of Maori and Egyptian words is given in this work such as will corroborate his statement with the testimony of another witness. This list of words by itself is sufficient to prove the primal identity of the Maori and Egyptian languages. The evidence from mythology is if possible still more conclusive and unique. A reply to Professor Huxley's declaration has been attempted by contrasting two Egyptian heads with two of an Australian type which would show the greatest unlikeness and asking where was the likeness ? [2]

But for aught known to the contrary this unity of Maori and Egyptian may be a thing of twenty thousand years ago. Not that the author has any certain data except that some of the Mangaian myths appear to him to date from the time when the sun was last in the same sign at the spring equinox as it is now, and that the equinox has since travelled once round the circle of precession, which means a period of twenty-six thousand years. And he does not doubt that on their line the ethnologists will ultimately demand

[1] *Journ. of Ethnol. Soc.* 1871.   [2] Owen, *Journ. of Anthrop. Soc.* 1874.

as great a length of time (probably five times as long) to account for the variations of race now apparent.

Professor Owen who replied to Professor Huxley is forced to admit that "The large, patent, indisputable facts of the successive sites of capitals of kings of the ancient race, from the first to the fourteenth dynasties, do not support any hypothesis of immigration; they are adverse to the Asiatic one by the Isthmus. They indicate rather, that Egypt herself, through her exceptionally favourable conditions for an easy and abundant sustenance of her inhabitants, had been the locality of the rise and progress of the earliest civilisation known in the world."[1]

It will be maintained in this book that the oldest mythology, religion, symbols, language, had their birth-place in Africa, that the primitive race of Kam came thence, and the civilisation attained in Egypt emanated from that country and spread over the world.

The most reasonable view on the evolutionary theory—and those who do not accept that have not yet begun to think, for lack of a starting point—is that the black race is the most ancient, and that Africa is the primordial home. It is not necessary to show that the first colonisers of India were negroes, but it is certain that the black Buddha of India was imaged in the negroid type. In the black negro god, whether called Buddha or Sut-Nahsi, we have a datum. They carry in their colour the proof of their origin. The people who first fashioned and worshipped the divine image in the negroid mould of humanity must according to all knowledge of human nature, have been negroes themselves. For the blackness is not merely mystical, the features and the hair of Buddha belong to the black race, and Nahsi is the negro name. The genetrix represented as the Dea Multimammæ, the Diana of Ephesus, is found as a black figure, nor is the hue mystical only, for the features are as negroid as were those of the black[2] Isis in Egypt.

We cannot have the name of Kam or Ham applied ethnologically without identifying the type as that of the black race.

True, the type on the earliest monuments had become liker to the later so-called Caucasian, but even the word Caucasian tells also of an origin in the Kaf or Kaffir. Philology will support ethnology in deriving from Africa, and not from Asia.

The type of the great sphinx, the age of which is unknown, but it must be of enormous antiquity, is African, not Aryan or Caucasian.

The Egyptians themselves never got rid of the thick nose, the full lip, the flat foot, and weak calf of the Nigritian type, and these were not additions to any form of the Caucasian race. The Nigritian elements are primary, and survive all modifications of the old Egyptians made in the lower land. The single Horus-lock, the Rut, worn as a divine sign by the child-Horus in Egypt, is a

---

[1] *Journ. of Anthrop. Soc.* 1874, p. 247.   [2] *Montfaucon*, pl. 47.

distinguishing characteristic of the African people, among whom were the Libyans who shaved the left side of the head, except the single lock that remained drooping down. This was the emblem of Horus the Child, continued as the type of childhood from those children of the human race, the Africans. Yet the Egyptians held the Libyans in contempt because they had not advanced to the status of the circumcised, and they inflicted the rite upon their conquered enemies in death, by excising the Karunata.

The African custom of children going undressed until they attained the age of puberty, was also continued by the Egyptians. Princesses went as naked as commoners; royalty being no exception to the rule. At that age the children assumed the Horus-lock at the left side of the head as the sign of puberty and posterity.

The Egyptians were pre-eminent as anointers. They anointed the living and the dead, the persons of their priests and kings, the statues of their gods; anointing with unguents being an ordinary mode of welcome to guests on visiting the houses of friends. This glorifying by means of grease is essentially an African custom. Among some of the dark tribes fat was the grand distinction of the rich man. According to Peter Kolben, who wrote a century and a half ago, the wealthier the Hottentot the more fat and butter he used in anointing himself and family. A man's social status was measured by the luxury of butter and fat on his body. This glory of grease was only a grosser and more primitive form of Egyptian anointing.[1]

The custom of saluting a superior by going down on the knees and striking the earth with the head is not limited to Africa, but is widely spread in that land. The king of the Brass people never spoke to the king of the Ibos without acknowledging his inferiority by going down on his knees and striking his head against the ground. On the lower Niger, as a mark of supreme homage, the people prostrated themselves and struck their foreheads against the earth. The Coast Negroes are accustomed to fall on their knees before a superior and kiss the earth three times.[2] It is etiquette at Eboe for the chief people to kneel on the ground and kiss it thrice as the king goes by. In the Congo region prostration on the knees to kiss the earth is a mode of paying homage. The custom is Egyptian, and was designated "Senta," for respect, compliment, congratulation, to pay homage; the word signifies literally, by breathing the ground.

Of the Congos, Bastian says, "When they spoke to a superior they might have sat as models to the Egyptian priests when making the representations on the Temple walls, so striking is the likeness between what is there depicted and what actually takes place here."[3] Theirs were the primitive sketches, the Egyptians finished the pictures.

---

[1] Kolben, *State of the Cape of Good Hope*, vol. i. pp. 50, 51 : London, 1731.
[2] Cited by Spencer, *Ceremonial Institutions*, 116, 121-4.
[3] Bastian, *Africanische Reisen*, 143.

The oldest and most peculiar images in the Ideographs point backward toward the equatorial land of the hippopotamus, rhinoceros, giraffe, ostrich, camelopard, ibis, various cranes, the serau, or goat-kind of sheep, the khebsh and oryx, the rock snake and great serpent of the Libyan desert, the cobra, the *octocyon*, a small primitive fox-like dog of South Africa, which has forty-eight teeth, the fox-like type of Anup, the Fenekh, a type of Sut; the caracal lynx and spotted hyæna, the kaf-monkey, or clicking cynocephalus, that typified the word, speech, and language on the monuments, which is now found in Upper Senegal,—as the old home of the aborigines.

The symbolism of Egypt represented in the hieroglyphics has its still earlier phase extant amongst the Bushmen, whose rock-pictures testify to their skill as hieroglyphists, and show that they must have been draughtsmen from time immemorial.

But, beyond this art, just as they have pre-human clicks assigned to the animals, so they have a system of typology of the most primitive nature; one in which the animals, reptiles, birds and insects are themselves the living, talking types, by the aid of which the earliest men of our race would seem to have " thinged" their thoughts in the birthplace of typology. In the " fables " of the Bushmen, the hieroglyphics are the living things that enact the representations. These point to an art that must have been extant long ages on ages before the likenesses of the animals, birds, and insects could be sculptured in stone or pictured in colours on the papyrus and the walls of the tombs and temples of Egypt, or drawn on the rocks by the Bushmen, Hottentots, and Kaffirs.

It was confidently declared by Seeley [1] that the cobra capella, or hooded snake, was unknown to Africa, and that as it appeared amongst the hieroglyphics, these must have been adopted by the Egyptians from some country where the cobra was native. Seeley was wrong, the cobra is African also. The latest testimony is that of Commander Cameron who walked across Africa, and who records the fact of snakes not being numerous and the "greater portion are not venomous, but the cobra capella exists and is much dreaded." [2]

The Egyptians marked the solstices as being in the horizon. The solstices, says Lepsius, " were always considered as in the horizon, and the vernal equinox as up in the sky." There is reason to think this may be the result of astronomical observation made in the equatorial lands. When at the equator the poles of the heaven are both on the horizon, and the North Pole star would furnish there a fixed point of beginning which answers to the starting-point in the north; this would be retained after they had migrated into higher latitudes and the pole of the heaven had risen thirty degrees. The mythology of Egypt as shown in the Ritual, obviously originated in a land of lakes, the lake being and continuing to a late time to be the typical

[1] *Caves of Ellora*, p. 216.   [2] *Across Africa*, vol. ii. p. 289.

great water which dominated after they were aware of the existence of seas. The water, or rather mud of source is a lake of primordial matter placed in the north. Another hint may be derived from the fact that aru is the river in Egyptian, and the anterior form of the word, karu, is the lake or pond.

No geological formation on the whole surface of this earth could have been better adapted for the purpose of taking the nomads as they drifted down from the Æthiopic highlands, into the valley that embraced them, to hold them fast, and keep them there hemmed in by deserts and mountains with no outlet except for sailors, and compressed them until the disintegrating tendencies of the nomadic life had spent their dispersing force and gave them the shaping squeeze of birth that moulded them into civilised men. Divinest foresight could have found no fitter cradle for the youthful race, no more quickening birth-place for the early mind of man, no mouth-piece more adapted, for utterance to the whole world. It was literally a cradle by reason of the narrow limits. Seven hundred miles in length, by seven wide, fruitful and fertile for man and beast. Life was easy there from the first.

" They gather in the fruits of the earth with less labour than other people, for they have not the toil of breaking up the soil with the plough, nor of hoeing, nor of any other work which all other men must labour at to obtain a crop of corn; but when the river has come of its own accord and irrigated their fields, and having irrigated them subsided, then each man sows his own land and turns swine into it, and when the seed has been trodden in by the swine, he waits for harvest time."[1] But the land was a fixed quantity on the surface, however much it increased in depth, and the supply of food was therefore limited by boundaries, as stern as Egypt's double ranges of limestone and sandstone hills. Lane calculated the extent of land cultivated at 5,500 square geographical miles, or rather more than one square degree and a half.[2] And this appears to be a fair estimate in round numbers, of its modern limits.

Such a valley would soon become as crowded with life as the womb of a twin-bearing woman in the ninth month, and as certain to expel the overplus of human energy. It was as if they had gradually descended from the highlands of Africa with a slow glacier-like motion and such a squeeze for it in the valley below as should launch them from the land altogether and make them take to the waters. There was the water-way prepared to teach them how to swim, and float, and sail, offering a ready mode of transit and exit and when overcrowded at home, free carriage into other lands. The hint was taken and acted upon. Diodorus Siculus declares that the Egyptians claimed to have sent out colonies over the whole world in times of the

---

[1] *Herodotus*, ii. 14.
[2] *Encyclopædia Brit.* 8th edition, vol. viii. p. 421.

remotest antiquity.[1] They affirmed that they had not only taught the Babylonians astronomy, but that Belus and his subjects were a colony that went out of Egypt. This is supported by Genesis in the generations of Noah. By substituting Egypt for the mythical Ark we obtain a real starting point from which the human race goes forth, and can even utilise the Hebrew list of names.

Diodorus Siculus was greatly impressed with the assertions of the priests respecting the numerous emigrations including the colonies of Babylon and Greece, and the Jewish exodus, but they named so many in divers parts of the world that he shrank from recording them upon hearsay and word of mouth, which is a pity, as they may have been speaking the truth. He tells us they had sacred books transmitted to them from ancient times, in which the historical accounts were recorded and kept and then handed on to their successors.[2]

In the inscription of Una belonging to the sixth dynasty, we find the earliest known mention of the Nahsi (negroes) who were at that remote period dominated by Egypt, and conscribed for her armies. In this, one of the oldest historical documents, the negroes from Nam, the negroes from Aruam, the negroes from Uaua (Nubia), the negroes from Kau, the negroes from the land of Tatam are enumerated as being in the Egyptian army. Una, the governor of the south, and superintendent of the dock, tells us how the Pharaoh commanded him to sail to some locality far south to fetch a white stone sarcophagus from a place named as the abode of the Rhinoceros. This is recorded as a great feat.

" It came thence, brought in the great boat of the inner palace with its cover, a door, two jambs, and a pedestal (or basin). Never before was the like done by any servant." (Lines 5, 6, 7.) The place named Rumakhu, or Abhat, is an unknown locality. But the performance is considered unparalleled.[3]

The Egyptians literally moved mountains and shaped them in human likeness of titanic majesty. "I dragged as hills great monuments (for statues) of alabaster (for carving) giving them life in the making" says Rameses III., he who built a wall 150 feet in depth, 60 feet below ground, and 90 feet above. They carried blocks of syenite by land and water, weighing 900 tons. It was said by Champollion that the cathedral of Notre Dame might be placed in one of the halls of the Temple at Karnac as a small central ornament; so vast was the scale of their operations. They painted in imperishable colours; cut leather with our knife of the leather-cutters; wove with the same shuttles; used what is with us the latest form of blow-pipe, for the whitesmith. It is the height of absurdity or the profoundest ignorance to suppose they did not build ships and launch navies. The oar-blade or paddle, called the kherp, is the emblem of

---

[1] Book i. 28, 29, 81.   [2] Book i. 44.
[3] *Records of the Past*, vol. ii. p. 3.

all that is first and foremost, excellent and surpassing, the sceptre of majesty, the sign of rule. Thus, to paddle and steer are synonymous with sovereignty. Ship-building yards were extant and are shown to be busy in the time of the pyramid builders. And here is Una, the great sailor, superintendent of the dock, going so far south that the geographical locality is out of sight. But the name shows it was the land of the hippopotamus or rhinoceros.

At this time, then, ships of war were built in the south, of considerable dimensions, over 105 feet long, together with the Uskhs or broad ships in which the armies of Egypt, composed of enrolled Æthiops and negroes, were floated down the river to fight her battles against those on the land.

The crumbling shell of Egypt's past proves it to have been as crowded with life as a fossil formation. But the shell did not merely shed a life that became extinct in the place of its birth. It was a human hive rather, that swarmed periodically ; and the swarms went forth and settled in many parts of the world, leaving the proofs in language, myths, and in one far-off place, the hieroglyphics ; in other lands the religious rites, the superstitions, the symbolical customs and ceremonies are the hieroglyphics still extant amongst races by whom they are no longer read, but which can be read as Egyptian. The parent recognises her offspring when the children have lost all memory of their origin and birthplace.

As Karl Vogt says, " Our civilisation came not from Asia, but from Africa, and Heer has proved that the cultivated plants in the Swiss lake-villages are of African, and to a great extent, of Egyptian origin." According to Logan,[1] the pre-Aryan civilisation of southern India had a partially Egyptian character. The oldest races, he asserts, were of a variable African type who spoke languages allied to the African.

Egypt, and not India is the common cradle of all we have in common, east, west, north, and south, all round the world. The language, beliefs, rites, laws and customs went out to India, but did not return thence by means of the apocryphal Aryan migrations. The Indian affinity with our European folk-lore and fairyology is neither first nor final, 'tis but the affinity of a collateral relationship. Egypt supplied the parent source, the inventive mind, the propagating migratory power. In Egypt alone, we shall find the roots of the vast tree, whose boughs and branches have extended to a world-wide reach.

The greatest difficulty in creation is the beginning, not the finishing, and to the despised black race we have at length to turn for the birth of language, the beginnings of all human creation, and, as the Arabic saying puts it, let us " honour the first although the followers do better."

Among the Æthiops of many thousand years ago there lived and

---

[1] Author of *Languages of the Indian Archipelago.*

laboured the unknown humanisers of our race who formulated the first knowledge of natural facts gathered from the heavens above and the earth beneath, and the waters of the wonderful river which talked to them as with a voice from out the infinite, and who twined the earliest sacred ties of the family-fold to create cohesion and strength and purity of life; men of the dark and despised race, the black blood-royal, that fed the red, yellow, and white races, and got the skin somewhat blanched in Egypt; the men who had dwelt in the Nile valley, and by the fountains of its waters in the highlands above so long in unknown ages past, that the negroid type of form and complexion had modified into the primitive Egyptian; so long, that in this race the conical head of the Gorillidæ had time to grow and bulge into the frontal region and climb into the human crown, until Egypt at length produced and sent forth her long-heads, the melanochroid type found in divers parts of the world. Blackness in the beginning did not depend on, and was not derived merely from, the climatic conditions; these modify, but did not create. Once the black race is extinct it can never be repeated by climature. Its colour was the result of origin from the animal prototype, and not only from nearness to the sun. On the oldest known monuments the Egyptians pourtray themselves as a dusky race, neither negroid nor Caucasian. Livingstone found the likeness of these in the typical negro of Central Africa, or rather he affirms that the typical negro found in Central Africa is to be seen in the ancient Egyptians, not in the native of the west coast.[1]

It is possible that the first intellectual beginnings of the race and of the Egyptians themselves were about the sources of the Nile.

The link between the African and Chinese is yet living in the Hottentot, not to say, on the other side, between man and the ape. Casilis observes [2] :—

"The yellow colour of the Hottentot, his high cheek-bones, half-shut eyes, so wide apart, and set obliquely in his head, his lanky limbs, place him in close connection with the Mongolian race."

And he has the ape's eyes, the negro's woolly hair, and a body that is like the missing link for its anatomy.

The Bushman, as described by Lichtenstein, presents the "true physiognomy of the small blue ape of Kaffraria. What gives the more verity to such a comparison was the vivacity of his eyes and the flexibility of his eyebrows; even his nostrils and the corners of his mouth, nay, his very ears, moved involuntarily."[3]

For this development a consensus of cumulative evidence demands a pre-historic past indefinitely remote, but not to be gauged or guessed at as a period less than tens of thousands of years; and this evidence, consisting of facts instead of the recital of them, is more

---

[1] *Last Travels*, vol. ii. p. 259.   [2] *Basutos*, p. 10.
[3] *Travels in South Africa*, vol. ii. p. 224.

trustworthy than that which has been tampered with by the composers of history. It is difficult, however, to get any time-gauge of Egypt's existence by the ordinary method, nor are those concerned to fight for a little time more or less, who are solely in search after truth which is eternal.

In our researches we shall find that at the remotest vanishing points of the decaying races we continually come upon the passing presence of Egypt, diminishing on the horizon in the far-off distance from the world she once engirdled round with language and laws, rites and customs, mythology and religion. Wheresoever the explorers dig deepest, in Akkad, Karchemish, Palestine, Greece, or Italy they discover Egypt.[1] And the final conclusion seems inevitable, that the universal parent of language, of symbolism, of early forms of law, of art and science, is Egypt, and that this fact is destined to be established along every line of research.

If we find that each road leads back to Egypt, we may safely infer that every road proceeded from Egypt.

In the very morning of the times these men emerged from out the darkness of a pre-historic and pre-eval past from one centre to bear the origins to the ends of the earth. The scattered fragments still remain, whereby they can be traced more or less along each radiating line to prove the common model, the common kinship, and the common centre. As Sir William Drummond observes, if in crossing the desert you find the spring of a watch in one place, an index in another, and pieces of a broken dial-plate in a third, you will scarcely doubt that somebody in the desert had once the whole watch. So is it here. And when the watch is reconstructed it will be found to have been of Egyptian workmanship. Hitherto we have never looked beyond the Phœnicians or Etruscans for a great sea-faring colonising people of the past. But they were Egyptians, not Phœnicians, who were the pioneers of the fore-world whose footprints are indelible wherever they once trod, and of a size to fit no foot that followed after. Only a few ruins of majestic greatness may be left above ground, and these are widely scattered, but they all show the one primeval impression that had no secondary likeness. And there remain the imperishable proofs that live for ever in the myths and the fossils of language, which constitute the geology of pre-historic humanity.

---

[1] "Another remarkable fact became the subject of discussion, and we await with some interest the fuller details which the report will supply. Professor Lieblein, of Christiana, noticed the Egyptian antiquities which had been disinterred in Sardinia, and Signor Fabiani exhibited specimens of others found in a tomb at Rome, under the wall of Servius Tullus. The remains were chiefly Egyptian divinities. It was argued by Fabiani that the site of Rome must have been occupied at a date anterior to the well-known era of ' Urbs Condita.'

"Phœnician remains were also found, supporting the hypothesis that there must have been an Egyptian and Phœnician influence in the pre-historic Italian civilisation."—*Journal of Anthropological Institute*, Feb. 1879.

An opponent of the doctrine of evolution recently wrote of the mythical serpent—

" There is an Aryan, there is a Semitic, there is a Turanian, there is an African serpent, and who but an evolutionist would dare to say that all these conceptions came from one and the same original source, and that they are all held together by one traditional chain ? "[1] No one. But, if the doctrine of development be true, none but an evolutionist will ever get to the origin of anything. And so surely as evolution is true in the development of our earth, so surely is it true for all that has been developed on the earth. The unity of the human race is fast being established, and the present attempt is directed towards establishing the unity of mythology and symbolism, the serpent included. The serpent is but one of a number of types that have the same current value the world over, because, as will be maintained throughout this work, they had one origin in common.

The hare is accounted unclean by Kaffir, Egyptian, Hebrew and Briton alike, because each of these was once in possession of that system of typology in which the hare (Un) was a sign of periodicity, especially in a certain feminine phase called by the name of the hare.

There is an Egyptian, there is a Maori, there is a Hebrew, there is an English, there is an Akkadian mythology, and none but an evolutionist would dream that these have one primary source still extant. Yet this is probable, and the present writer is about to adduce evidence in proof. But then he is among those who think that one of the supreme truths made known to our day and generation is that creative cause is evolutionary everywhere and for ever. Not mindless evolution; evolution without the initial force of purpose, evolution without increase of purpose in the accumulative course; evolution without the fulfilment of purpose as the result of all, is simply inconceivable.

The world is old enough and time has existed long enough for the widest divergencies to have been made from one common centre of mankind, and the proofs of a unity of origin are plentiful enough. What has been wanted is the common centre of the primeval unity. This, it is now suggested, will be found in Africa as the womb of the human race, with Egypt for the outlet into all the world.

Parent of all men give me grace
Our unity from first to trace,
And show the map through all the maze
Of winding, wandering, widening ways :
A shattered looking-glass replace
With wholeness to reflect Thy face,
And help establish for the race
The oneness that shall crown their days.

[1] Max Müller, *Academy*, 1874, p. 548.

The Egyptians identify themselves on the monuments as the *Rut*. A pictorial representation is found on the tomb of Seti I. of four races of people arranged in groups of four men each. These are the Nahsi (negroes); the Hemu, men of a light brown hue, with blue eyes, and hair in a bag; the Tamahu, who are fair as Europeans; and the Rut, who are Egyptians.

These are typical groups, not meant merely for conquered races, as may be gathered from the signification of their names. The Tamahu are light-complexioned people. In Egyptian, tama means people, and created. Hu is white, light, ivory. The Tamahu are the "Created white" people. Na is black, ink. Neh, a blackbird. Su is the person, or birth. The Nahsu one black born, or, in Egyptian phrase, black from the egg (su). Hem is the rudder, to paddle, fish; hemi, to steer. The Hemu thus indicated are the sailors, sea-farers, the people of the isles. The Isles of the Gentiles (Gen. x.) might be rendered in Egyptian by the isles of the Amu or Hemu. The hieroglyphic "hem" is the sign of a water frontier.

The word "rut" has various meanings, all significant when applied to the Egyptians by themselves. Rut is to retain the form, be carved in stone, a footstool, or a pair of feet, cause to do, plant, grow, repeat. One hieroglyphic sign of "rut" is an implement for pegging down and making fast in the earth, to retain animals in one spot. This is symbolic of the Rut as the people who dwelt where they first laid hold and made fast to the earth. These are the Egyptians themselves. They claim to be the root of the race of men or *the* Rut, the Men, knowing of no other point of departure, and being so ancient they have forgotten that their complexion was at one time black. The word passed into Sanskrit as Reta, produced from the seed, and their likeness was the lotus, which contains the seed within itself, and retains the name of the Rut or Reta.

The name of the Ruti went out into other lands as that of the Ruten, upper and lower, and the Ludim; it had become a typical name of race, long-lineage and enduring life in the Assyrian Palatu (pa, is the Egyptian masculine article "the"), the Latins, the various Luds, the Lithuanians, and others who fastened themselves firmly and took root elsewhere. The Egyptians were and are the true Rut, the one primordial people who first took conscious hold of the earth and retained a knowledge of the fact. The land of Egypt is the footstool, rut; the people were the feet (rut) on which the human being first arose erect to attain its full stature. Incidentally however we learn that the Egyptians recognised the black race to be the first of created men. The people of Ra are born of the great one who is in the heavens; the Rut are born of his eye in their persons of superior men. He also created the Aamu and the Tamahu; further he has comforted himself with a multitude who came from him in the shape of negroes, but it is expressly said "*Horus has created them* and he defends their souls." [1] This may be followed mythologically. Har is older than Ra; his type goes back to Sut-Har, and Sut-Nub the negro god who created the Nahsi.

It has been overlooked that the Egyptians do possibly tell us something of their own origin beyond the Ruti of the monuments. We learn from Syncellus (*Chronicon*, li.) and Eusebius (*Chron.* vi.) that among the Egyptians there was a certain tablet called the Old Chronicle, containing thirty dynasties in 113 descents, during the long period of 36,525 years.

" The first series of princes was that of the Auritæ, the second was that of the Mestræans, the third of the Egyptians." [2] The first divine name in the series is one that is earlier than the sun, given through the Greek as Hephæstus, to whom no time is assigned, because this deity was apparent both by night and day. This contains matter of great moment not yet read nor readable until we have seen more of Egypt's mythology. The present point is that the Auritæ princes coincide with the reign of Hephæstus as the beginning.

The Auritæ are of course the Ruti of monumental Egypt, the typical name of the race of men *par excellence*. But the prefixed Au will add something to our knowledge of the pre-monumental Rut. Au in Egyptian means the oldest, the primordial. The word Au is the Egyptian " was," and the Au-rut means the race that was, the first and oldest race of men. Au is a modified form of Af. Both Au and Af signify born of. The name of Africa is derived from this root af or au. The tongue of Egypt tells us that Af-rui-ka is the inner land, born of, literally the birth-place. They knew of no other. Thus the Auritæ were the Af-ritæ, people of the birth-place in Africa. But Af in Egyptian has a still earlier form in Kaf, and the Afritæ become the primordial Kaf-ritæ. The Kaffirs have preserved the primal shape of the word signifying the first, the embryotic, aboriginal root-race of men. The Kaffirs likewise keep the true African colour of the original Ruti or race. There is a sort of pre-Darwinianism in this root Kaf, the name of the Kaffir and the Kafritæ of Egypt. The Kaf is the symbolical monkey, the Cynocephalus. It is apparent from the drawing of the nose of the figures of Taurt [3] that the Kaf enters into her compound character with the hippopotamus and crocodile. Not that the Kaffirs and Kafritæ named themselves from the ape or as the ape-like, or as the beings evolved from the ape. But they were the first, and the name signifies the first; and when the primitive men had advanced and took other names, the prior name was left to the primordial men, the pre-men, so to say, and it remained with the ape and the aborigine. This will help to explain the ape-men and monkey men of various tribes

---

[1] *Records of the Past*, x. 109-10.
[2] Syncellus, *Chronicon*, li., Eusebius, *Chron.* vi.
[3] Wilk. pl. 32, figs. 4, 5, 6, 7, 8, 9, 10.

who retained the earlier status in their names which mixed up the pre-men and apes. So the Assyrian name of the monkey, Udumu, is identical with the Hebrew name Adam for man.

The laws of language prove that the Auritæ, the first princes of this long line of descent given at 36,525 years, were Kaf-ritæ; and the laws of evolution prove the primal race, so far as we can get back, to be the black people. The Kaf is the black, dog-headed, almost human monkey. Ape and Kaf are named as the first preceding man, and there was no other name known for the first than kaf, ap, au or ap. The kaf-ruti name has the same relation to the Ruti of the monuments that the ape, the pre-man, has to man. On another line of modification and development the name Kafruti becomes Karuti, found on the monuments as Karut, natives, autochthones, indigenous inhabitants, also applied to masons, and workers in stone. The Karuti, it is suggested, became the Kaldi and the Keltæ.

Maspero thinks the Egyptians had lost the remembrance of their origin. But for the people who were the recorders and chronologers of mankind that is good evidence of their having no foreign origin to forget; the Ruti were the race. The Auritæ were the oldest race. The Karuti were the natives, inhabitants, indigenes, aborigines, and there was nothing more to be said. They had sloughed the black skin of the Nahsi and repudiated it, as we have denied the ape. Nevertheless, they came out of it as we do from the ape without *remembering* the fact, which has to be re-collected in evidence, "The first series of the princes was that of the Auritæ, the second was that of the Mestræans, the third of the Egyptians." This is a perfect panorama of the descent geographical and ethnological.

By aid of the Auritæ (kafruti) name we ascend to the source whence the race that always was (au) had descended. In Egyptian, mest means the birth-place, the lying-in chamber, to whelp, be born. Mest as the black is the name of kohl or stibium. Ru is the gate or gorge of outlet: Ruan the gorge of a valley.

The Mestræans are the people of the outlet of the birth-place in the stage next to the Auritæ. The Egyptian mest, to be born, whelped, answers to the Hebrew אצמ, to become apparent, to exist, the very first form of to be, the parallel with Au. And the Mestræans are the people of the gorge, the place of outlet, the same that we derived from Mitzraim as the dual land of the outlet from the birth-place, and as mest means stibium, the Mestræans were apparently black. It is in Egyptian alone that mes for birth is the equivalent of mest, and mes-ru the outlet of birth of mestru, whence Mestræan. Now one of the Hindu names of Egypt is Misra-Sthan. But the Sanskrit misra, mixing, does not contain the primary meaning, and the word is supposed to be taken from a lost root which it is here suggested may be the Egyptian mes, generation and birth; the sense of mixing applies to generating. Also, מצר, a *strait*, has a likeness to mest-ru, the road or strait, issuing from the birth-place.

In the old chronicle the Mestræans are placed midway, corresponding to the meaning of mest-ruan the outlet from the place of birth, the human lair, mestruan the gorge from the birth-place, and the mestru (mitzr) we may identify with the country below the first cataract, because at Kartoom and Assouan the inundation was born. Kart signifies the child, um to perceive, and it was there the birth of the child was first perceived. This adds another sense to the words mestruan and mitzraim as the birth-place of the inundation.

Assouan says Mariette [1] "always takes the traveller by surprise. We seem to be quite in a new world—Egypt finishes ; another country begins. The inhabitants of Khartoom especially are remarkable by their grand mien, their black skin, and their finely formed heads, that remind one of the best types of northern races." Here it may be seen that with a skin as black, although more lustrous with light, as that of any negro of the west coast, the African in Nubia at times attained a type of face, and a sculpture of form as noble and refined as any of the white skins of all the vaunted Caucasian and Aryan races. The line of the Mestræans may have begun hereabouts, leaving the Auritæ higher up in the old dark land where the pre-lingual beings clicked away, in equatorial Africa for many a myriad of years ; so long ago that the primeval race can only be known by its radiants and its rootage, and not by any stem extant— so long ago that the Hottentot, the west coast negro, the Nubian and Egyptian are but concentrations of form and colour along the lines diverging from a centre out of sight in the far land of the Auritæ. *Herodotus* [2] says the oracle of Amen pronounced all those who drank the water of the Nile and dwelt north of Elephantine to be Egyptians. They were so in later times when Egypt was both upper and lower, but the name of Khebt (for Egypt) belongs especially to the lower land in the north.

The Egyptians, the people of Khebt come last, they dwelt in the Delta. Three stages are here distinctly marked. The place of the Auritæ above the cataracts. That of the Mestræans in the Ruan, the gorge of the valley, and the Egyptians in Khebt or lower Egypt, called the Delta, and the name of the Delta, Ter-ta, also shows the descent from above, for ter (del) in Egyptian means the extreme limit, the foot, that which ends and completes the whole, and ta is the land.

The Auritæ of this passage, then, are not gods nor are the Mestræans demi-gods, as they have been misrendered. Bunsen assumed that the chronicle was fictitious, and asserted that the number of 36,525 years represents the great year of the world. [3] It does nothing of the kind.

---

[1] *Monuments of Upper Egypt*, pp. 255-6.   [2] B. ii. 18.
[3] *Egypt's Place*, vol. i. p. 215.

The only great year of the world is that of precession—25,868 solar years, and the twenty-five Sothiac cycles (25 × 1461) or 36,525 years do not make any great year whatever. When we read in the Syncellus that "Manetho, the high priest of the Egyptian idols, wrote a *fabulous* work on Sothis under Philadelphus;" that means a book containing what the man uninstructed except in biblical chronology considered mythological, not necessarily a forgery of time-reckonings, for he denied their chronological nature *in toto*.

Jamblichus, in his work on the Mysteries, mentions the number of 36,500 books assigned to Taht, and this agrees very nearly with the number of years given in the old chronicle. This also is denounced by Bunsen as being nothing more than the year of the world in twenty-five Sothiac cycles. The question of interest here is whether these twenty-five Sothiac cycles had been observed and registered ; if so, the records were sure to be entitled the Books of Taht, the earliest forms of which were columns called stelæ; the figure of these is yet extant in our flat roundtopped gravestones. Syncellus describes Manetho as having declared that certain of these stelæ still existing in the *Syriadic* land were his authorities. They were engraved in hierogly-phics and in the sacred tongue, and "after the flood" they were transcribed into Greek in hieroglyphical characters.

These columns, or stelæ, in the Syriadic land are referred to by Josephus.[1] He ascribes them to Seth, and says they were erected to record the peculiar sort of wisdom which concerns the heavenly bodies, that is as chronological tablets, because Adam had predicted that the world was to be at one time destroyed by fire, at another by flood. Plato, in the opening of his *Timæus*, refers to these so-called antediluvian columns, or stelæ, of the Syriadic land. They are likewise mentioned in the "Fragments of Hermes" in Stobæus. The Syriadic land has been accepted unquestioningly as the land of Syria. This in Egyptian is Karua. But there is Karua north and Karua south, and the country of the stelæ is the southern Karua, the Æthiopic country. Now here is a fact unknown to the Greeks, Christians, or Josephus, and one sure not to be recorded by Manetho. We learn from a statement in the *Book of the Dead* that Taht, the recorder of the gods, was otherwise Sut, who had preceded Taht in the character of the scribe. How this was so remains to be unravelled. Enough for the present to say that it was so, and to refer here to the fact that the stelæ assigned to Taht by Manetho are ascribed to Sut by Josephus ; both describing them as standing in the Syriadic land, which we identify with Karua in the south. The Karuans are extant on the monuments as a black people. In both accounts we find the flood. But the monuments of the chronologers of mankind know nothing of the Noachian deluge. The flood of Egypt is the inundation of the Nile. And these antediluvian stelæ in the Karuan land before the flood we

---

[1] *Ant.* B. i. ch. 2.

take to have stood in the Karuan country above the cataracts and the land made by the inundation. The account thus rendered is that a vast pre-monumental period of the Egyptians was represented by the princes of the Auritæ, the ancient and original race under the divine reign of Hephæstus before the time of Ra, and that the stelæ upon which the time was inscribed were set up in the Karuan land beyond the cataracts or the flood, and that as the old dark people came down into the lower parts of the Nile valley to be known as the children of Ham, the black, Mizraim, Kush, and Phut, and the still later Egyptus, they inscribed the same facts on the papyrus in what has been misinterpreted to be the tongue of the Achæan Greeks. Syriadic, then is Karuadic, and we have to look for the Karua south, the land of Sut-Anubis, who was earlier than Taht as the divine registrar. Some of the inscriptions of the time of Tahtmes III. designate the country of Karu or Kalu, in the far south, as the southern boundary.

Brugsch considers this to be the ancient Koloe, which, according to Ptolemy, was situated on the fourth degree, fifteen minutes, of north latitude, in those countries. Ka (Eg.) is an interior region or it may be high land, and rua is the outlet, the place of exit. Karua is also an Egyptian name for a lake, so that the name might mean the lake country.

It is noticeable that the true country of the stelæ and the rock-inscriptions is sharply defined by the first cataract. In the route from Assouan to Philæ the rock-inscriptions abound on all sides. Schayl, a small island in the cataract, is covered with such records, some of which have yielded a clue to historical facts,[1] as it was obviously intended they should do, for the name in Egyptian, read Skha-rui, signifies the island of the writings or inscriptions. One cannot but suppose the rocks in the upper country may still preserve some memorials of the remoter past. The account appears to mean that the stelæ were erected in the time of Sut-Anubis, the first registrar of the gods, and the contents were afterwards transcribed in hieroglyphics and preserved in the Hermean writings assigned to Taht.

The Egyptians, says Hor-Apollo (i. 21), depict three water-pots and neither more nor less, because, according to them, there is a triple cause of the inundation. One of these causes, he tells us, is the rain which prevails in Southern Africa. He further observes, they make the water-pot like a heart having a tongue, because the heart is the ruling member of the body as the Nile is of Egypt, and they call it the producer of existence. The tongue, as he calls it, of the three vases, is the sign of the pouring out the libation; there are two tongues or volumes of water answering to the blue and red Nile, and these issue from the three vases, the triple heart. The vases are three, and we may now fairly infer that this triple symbol with its dual stream, was intended to connect the source of the Nile with the

[1] Mariette, *Monuments of Upper Egypt*, p. 259.

three great lakes in equatorial Africa. It has been rediscovered that the river is fed by two of them, and Tanganyika may easily have been classed with the system. The fact of their existence was known already to the travellers and map-makers of some centuries ago ; the maps of the Arabs in the twelfth and thirteenth centuries; the Portuguese of the sixteenth and seventeenth centuries ; and the Dutch map of the seventeenth century ; show that at least three great lakes were known. It now becomes probable that the triple source and fountain-head of the Nile was known to the Egyptians many thousand years since, and that the fact is stereotyped in the triple ideograph of the inundation.

Professor Oppert assures us from his standpoint, that in all cases we may take for granted that the date of 11,512 B.C., given in his version of the Berosian Chronology, reposes on a real historical tradition, and that the two periods, the Chaldean moon period and the Sothiac period (whether it was Egyptian or not) have the same origin. Professor Oppert, by mathematical calculation, fixes the date of a double phenomenon which struck the sight of men, consisting in an eclipse and in an apparition of Sirius, on Tuesday, 27th April, Julian, or the 28th January, Gregorian. But as at this epoch Sirius was not visible to Northern or Middle Egypt, on account of the equinoctial precession, civilization must start from a more southern point. This, to say the truth, adds Professor Oppert, is a mere hypothesis, and he trusts that further investigation will either confirm or deny this special view of his.

It is now about to be claimed that civilization did start from a more southern point than Mid-Egypt.

For Egyptian, Hebrew, and Greek, there was nothing visible beyond Egypt but the background of blackness in Africa, the land of Ham, the source from which she was fed in secret by tributaries that flowed as stealthily as the hidden fountains of the Nile.

The Hebrew Scriptures, among their other fragments of ancient lore, are very emphatic in deriving the line of Mitzraim from Ham or Kam, the black type coupled with Kush, another form of the black. They give no countenance to the theory of Asiatic origin for the Egyptians. In the Biblical account of the generations of Noah, Mitzraim is the son of Ham, i.e. of Kam, the black race. Thus Ham occupies the place of the Auritæ princes of the old chronicle at the head of the Egyptians, Akkadians, Babylonians, and Assyrians. This account, preserved in the tenth chapter of Genesis, could only have been kept by the black race, the " blameless Æthiopians " and Egyptians as contemporaries of the early time, and recorded in the Hermean books, fragments of which are found in the Hebrew writings.

" Egypt, mother of men, and first born of mortals," the learned Apollonius Rhodius in the Argonautika, calls that country.[1] In the

---

[1] 4, 259.

time of Xerxes, Hippys of Rhegium designated the Egyptians the most ancient of all nations.[1] " For my part," says Herodotus, " I am not of opinion that the Egyptians commenced their existence with the country called the Delta, but that they always were since men have been ; and that as the soil gradually increased many of them remained in their former abodes, and many came lower down. For anciently Thebes was called Egypt."[2] He also tells us that the Egyptians, before the reign of Psammetichus, considered themselves to be the most ancient of mankind.[3]

We see, by the Greek report, the Egyptians knew that Egypt was once all sea or water. Herodotus says the whole of Egypt (except the province of Thebes) was an extended marsh. No part of that which is now situate beyond the Lake Mœris was to be seen, the distance between which lake and the sea is a journey of seven days.[4]

Plutarch also says Egypt was at one time sea. Diodorus Siculus affirms that in primitive times, that which was Egypt when he wrote, was said to have been not a country, but one universal sea.[5]

A persistent Greek tradition asserts that the primitive abode of the Egyptians was in Æthiopia, and mention is made of an ancient city of Meroe, from which issued a priesthood who were the founders of the Egyptian civilisation. Meroe, or in Egyptian Muru, means the maternal outlet, therefore the birthplace, which was typified by the Mount Muru. The modernised form of Muru or Meroe, is Balua.

Ba-rua (Eg.) also yields the place of outlet. And this place may be pursued according to the African naming up to the Rua mountains and the outlets of the lakes. The Hebrew tradition manifestly derives the Egyptians from above and not from lower Egypt. " I will bring again the captivity of Egypt, and will cause them to return (into) the land of Pathros ; into the land of their habitation." (Ezek. xxix. 14.)

Puthrus (פתרוס) is obviously derived from the Egyptian, in which rus is the south, southern. Pet is the name of the crib or birthplace. Puth (Heb.) denotes the opening, the place of birth. Put and Apt (Eg.) interchange, Apt or Aft being the abode and the name of the goddess, the oldest great mother. Pet or Peth is the earlier Beth, and Pathros is the birthplace in the south. " Patoris " is the south country (Upper Egypt.) In the Hebrew writings Æthiopia, Kush, and Zaba are convertible terms for the same country, the Egypt beyond Egypt. Zaba, the first-born son of Kush, is mentioned with Kush and Mitzraim (Is. xliii. 3), and the Zabians are coupled with Kush (xlv. 14). The inhabitants of Meroe were called Zabaim (Ez. xxiii. 42). " The Zabeans from the wilderness."

Zaba, the modified form of Khaba (or Kheft), including the ancient Meroe, has been understood to mean the northern half of Æthiopia.

---

[1] *Schol. to Apoll.*, iv. 262.   [2] B. ii. 15.   [3] B. ii. 2.
[4] B. ii. 4.   [5] *Lib.* 3.

The Æthiopians are called קוּסִין from the reduplicated Pih, קוּסִים (belonging to קוּה), a people of great might or double power (Is. xviii. 27). Kefa (Eg.) means force, puissance, potency. Khefti or Khepti, is a reduplicated Kef equivalent to Kev-Kev, and Kep-Kep is an Egyptian name of Nubia.

In Egyptian Kheb, Kheft, or Khepsh, is the north, as the hindward part, the front being the south. The Khepsh, equivalent of the Hebrew כוּשׁ (Kvsh), is the hieroglyphic hinder part, and the name of the Great Bear, as the constellation of the north.

If the reader will think for a moment of the sight presented by the revolving Great Bear to those who watched it in equatorial lands, he will realise what an arresting image of the north this must have been with its settings and risings in that quarter to a people who did not know the earth was round.

In the title of Psalm vii. mention is made of the Dibrai of Kush, and Dabar (דבר) signifies the hinder part ; it is the exact equivalent for the Egyptian Khepsh. Zaba, as a name of Ethiopia, is the modified Kheba or Kefa. Also in Egyptian Khefti, the North, the Goddess of the North, was worn down to Uati or Uti, and in the time of the Middle Empire a hieroglyphic U passed into an E. Thus we have Eti-opia, Uti-opia, Khefti-opia, and opia from Api is the first, the ancestral land. Cassiopœia, the Lady of the Seat, is the Queen of Æthiopia or Kush, and her name, derived from khus, to found, or earlier khepsh (כוש) the seat, denotes the Queen of the "opia" or "apia" in the north. Thus interpreted, Æthiopia is the first, the ancestral seat, in the north.

This may be followed out. Æthiopia is claimed for the birthplace; khab means to give birth to; khef is the one born of, the place we come out of; af and au mean born of, and uti signifies to emit; emane, go or come out of, and opia or apia is the ancestral land. The diphthong Æ represents an earlier Au and yet earlier Af, hence Æthiopia is Aftiopia or Kheftiopia. Yr Aipht is still the Welsh name of Egypt.

In the celestial north was the mythical birthplace where the Great Mother Taurt, the Goddess of the Great Bear and Seven Stars, was represented as the bringer-forth from the waters in the shape of a hippopotamus, and when Kush, or Æthiopia, was named as the North, the hinder part, literally the hinder thigh, as a sign of the birthplace, *it must have been by a people who were still farther south*, as from no other direction can the naming be applied. Kush, Zaba, or Æthiopia, was once the north land to a people that dwelt above. The name would not be given by a people ascending the Valley of the Nile and going due south. Whereas the name, once given high above Egypt, can be followed down to Coptus, thence to Khebt in Lower Egypt, and then to Kheft, the Egyptian name of Phœnicia. The birthplace was in the north, and the Egyptians identified that with Æthiopia.

A typical title of the Repa, Prince, or heir-apparent to the throne of Ra, was Prince of Æthiopia.

Diodorus (B. I) tells of an annual Æthiopian festival at which the statues of the gods of Egypt, who were represented by Jupiter and Juno, were carried into the Æthiopian land, and then after a certain time of sojourn there they were brought back again to Egypt. Eustathius on Homer,[1] says the Æthiopians used to fetch the images of Zeus and other gods from the Temple at Thebes, and with these they went about at a certain period in Libya and celebrated a splendid feast of twelve gods. This was going back to the ancient birthplace, the backward Khebt ; and the carriage of the later divinities to the old home, to celebrate a festival of the twelve gods, looks like a typical conquest of the earlier domain of the old Mother Khebt, the Hippopotamus Goddess who brought them forth. On this journey they were not following the Celestial North, that lay directly opposite. The Libyans, be it remembered, wore the Horus-lock of the Repa, the Prince of Æthiopia, from whom perhaps they derived their name.

A rock tablet in the neighbourhood of the town of Assouan proclaims to the traveller that to Tahtmes II. (1600 B.C.), came the Asiatics and the Kushite An, the Nomads, "his frontier to the south is at the summit of the world, and his frontier north at the farthest end of Asia," which places the southern limit of the land at the equator. This range southward was no doubt attained in the backward ascent from the lower lands after the Egyptians had become civilised there.

Under the rule of Tahtmes III. the southern boundary of the land was designated Apta, the horn-point or tip-top of the land, the farthest point to the south. This Apta was a type name for the extremity of the land, and means, equal, or equatorial land. Also Ap is first ancestral, aboriginal, and divine. Apta describes the land (ta) or type of all that is initial, primordial, first in being, place, or person. Apt or Aft is the birth-place, the bringer-forth, the genitrix of the gods and the equatorial land ; the horn-point of the world is known and named in Egyptian as Apta. Apt and Khebt are two forms of the same word, as the name of the birthplace and the bringer-forth.

It may seem like considering the matter too curiously, and yet it would not be without warrant from phonetic law if we were to see in Apta the modification of Khepta as the secondary naming from the north, or, with the frontage reversed, to the primal naming of the north as Khebt, when that quarter was the front, and the namers were looking north. Æthiopia includes both Apta and Khepta, and the Great Mother, who was hippopotamus in front and crocodile behind, was both Apt and Khept. Khept also modifies into Sebt. The Ka sign of the crocodile's tail is a later Sa. Sebt (Sothis) passes into Sut, the name of the south. Thus Sebt and Apt are both later forms of Khept. The hinder part as the birthplace was first of all

---

[1] *Il* lib. I, *vide* Smith's *Classical Dictionary*, "Ethiopia."

the region of the Great Bear that brought forth in the north. Then in the solar imagery the Egyptians depicted the heaven by the figure of a woman, whose body arched over the earth and rested on the hands and feet. In this position she brought forth the sun, animal-fashion, at the place of the hinder thigh (Khepsh), and her face was the front, the south. Egyptian shows that the north, as the lower, hinder, the night-side, the lowermost, the dark part, was the *first named*, and the modification of the K into H reflects this fact, corresponding to Khept and Apt; Kak is darkness, and the deity of the dark; Heh and Hui denote the light; Khekh signifies the invisible of being; Heh the visibly manifested. Kar is the lower, Har the upper. Khept and Hut are tail and head. Khar is the first Horus, the child, immature and mortal (Harpocrates). Har is the second, perfected, immortal.

The word Kam is not a primary formation, and in Hebrew the vau of a prior spelling is preserved. The word for Kam or Ham in Eupolemus is written חום (Kvm), as in Hebrew Chush is written כוש (Chvsh). Kvm is extant in Egyptian as Khebm or Kheb-ma. The fuller form of the word is also indicated by Kâm, *i.e.*, Kfm to create, whence the Kamite was the created race. One reduced form occurs in Kâm (Kfm) for the black stone, obsidian, determined by the crocodile's tail.[1] And this carries us beyond the land of the inundation to that of the hippopotamus, whose name is Khebma, the Mother of the Waters, typified by the water-cow. Khab is the name of the black hippopotamus in the Namaqa Hottentot. Khab permutes with Kam in Hebrew for black. And the permutation probably depends on the word Kvm or Kbm dividing in twain to posit two forms.[2]

Kam, then, has to be sought in Kheb-ma, the place of Kheb, or the Mother Kheb.

This takes us back to Ethiopia, the land of Kush or Kvsh, which, in the form of Khepsh, is still the name of the Great Bear, or Hippopotamus. The land of the black Khab, the black Kaf, and the black Kaffir, the Kafruti, the black race of Kvm, Kvsh, and Kvt, the children of Taurt, the Typhonian Genitrix. Etymologically we

---

[1] *Champ. Gram.* p. 90.

[2] So the Gothic *Kvimat* shows the earlier form of *Come*. The Welsh Cwm implies a prior Cfm. So the hieroglyphic "Peh" bifurcates into the P and H of the Setshuana and other languages. This "vm," found in Hebrew, may show the connecting-link between the Egyptian Kef, the hand, a figure of 5, and the supposed original Aryan word quem-quem, for No. 5. Kef, Kep and Kem permute, as is now suggested, on account of a form in Kafm, still found in חום and in Khebma. This abrades in quem and Khamesh for No. 5. Taking Khem as the equivalent of Kef the hand, Khemt (Khemti) for No. 10 (Eg.) is the two hands. Khemsha (Eg.) would read hand the first, or one hand, and Khemt two hands. This is at a depth, however, that philology cannot bottom. It belongs to a primitive system of typology. For example, the first hand was the creative matrix, the Khep in Egyptian. The old Khebma is the paunchy pot-bellied hippopotamus-goddess, and the Hebrew חמש for No. 5, the hand, also means the paunch, and paunch, or panch, is the type-word for hand and No. 5 in other groups of languages.

have the gradation of sounds from the *f* in Kaf, to the *p* in Coptus, the *b* in Khebt. Also we find the later Egyptians calling uncivilised and ignorant men the Khem-Rut. If we read this with a *v* we recover the Khvm-ruti as the equivalent of Kafruti, the Auritæ, the Karti, Kaldi, Keltæ, and the final Ruti. It is noticeable that the Lapps call themselves the *Sabme* people, which is the word Khabma or סחם in a later stage. The Japanese likewise claim to have descended from the Kami or Kafmi.

Egyptian monumental history begins with the name of Mena or Menes. The chronology made out by different Egyptologists is variously recorded thus :—

Bunsen	.	.	.	. 3623	B.C.
Lepsius	.	.	.	. 3892	„
Lauth	.	.	.	. 4157	„
Brugsch	.	.	.	. 4455	„
Unger	.	.	.	. 5613	„
Boeckh	.	.	.	. 5702	„

Brugsch-Bey makes use of the latest data, and also of the investigations of Lieblein into the pedigree of twenty-five Court architects ; he concludes that the year 4455 B.C. is about the nearest approximation that can be made to a correct date for the era of Mena. And, as nearly as can be calculated, the Spring Equinox first occurred in the sign of the Bull, 4560 B.C.

The clue is worth following. The present writer refuses to be entangled in the maze and lost in the labyrinth of the dynastic lists. It may be there was no monarch of the name of Mena. No contemporary monument of his time has ever been found, nothing inscribed with his name, nothing is known to have been made, in the time of Mena.[1] Tradition ascribes to him a dyke which still exists in the neighbourhood of Cairo. Personally, however, the writer does not doubt the existence of Mena, only if the man were removed the era would still remain. That era has been made out to be nearly coincident with the time when the colure of the spring equinox entered the sign of the Bull. The name of Mena is identical with that of the Bull Mnevis or Men-Apis, of Heliopolis, the celestial birthplace of the Sun.

This apparently perilous subject of the mythological astronomy is only introduced here with the view of making Mena, the admitted head of a dynasty, a connecting link with the divine dynasties which also are astronomically dated, but have been assumed to be fabulous. It is certain that the dynasty of Mena coincides with the establishment of an Osirian myth that is for ever connected with the sun in the sign of the Bull by means of the legend relating to the death of Osiris when the sun was in Scorpio, and the bull and the scorpion

[1] Birch, *History of Egypt*, pp. 24, 25.

were the two equinoctial signs, showing that the bull of Mena and Osiris are one and the same. Now in the Egyptian zodiac[1] the crocodile is in Scorpio (so to say) that is, it occupies the whole three decans of that sign, and it was under the Osirian dynasty of deities that the crocodile was transformed into an image of evil and a symbol of Typho to be trampled under foot. This does not in the least destroy the personality of Mena, because the Pharaohs of Egypt were assimilated to the divinity, and monuments were raised and temples built to their own "name" in honour of the deity whose name they bore. The divine, not the human, was the object of the worship. In the present instance it is the divine, the solar Mena or bull that makes it possible to account for the various Menas in different countries, and gives us a common starting-point for the Manu who was the law-giver in India, Minos, who was law-giver in Greece, Menw, who was the law-giver to the Kymry, and Mena, who was law-giver to Egypt. The law first given in each case was that of time and period. Mena, as the bull who in Egyptian mythology (as Khem) is the bull of the mother, is one with the Minos whose wife was Pasiphaë, and this gives a natural interpretation to her passion for the bull. Mena is the head of the Thinite dynasty reigning at Memphis, and Thinis is in Upper Egypt. May not Tini and the Thinites derive their name in the same way from the new point of departure in the bull? The highest functionaries of the blood-royal were distinguished by the style of Princes of Tini.

Ten is the division, the place of the Tennu or equinoctial eclipses, and also means to fill up, complete, terminate, and determine.

The district in which Memphis stood was called Sekhet-Ra, the field of the sun. Mena is accredited with having built Memphis, and one of its names is Makha-ta, the land of the scales ; or as Makha equally applies to the equinox, from which the zodiacal scales were named, this may be equinoctial, and as the name of Mena is that of the bull Mnevis, there is nothing incredible in supposing that Mena of Makha-ta represents the sun (or vernal colure) in the sign of the Bull. The Arab name of Tel-Monf for Memphis points to its being Men-pa, the House (or sign) of the Bull. King Teta, who follows the name of Mena, built the Pyramid of the Bull at Ka-khema, the town or shrine of the Bull. It is probable that this step-pyramid of Sakkarah was the common grave or shrine of the dead Bulls, to judge from its bones of bulls and the inscriptions relating to the royal Apis, the Bull of that time being the representative of the sun instead of the later Pharaoh. The builder of the Bull Pyramid would enable us, if need were, to do without the human Mena. In Egyptian symbolry the celestial is primordial and continually contains the clue to the terrestrial ; the earthly is but the image of the heavenly. Thus the time of Mena is none the less real, and is all the more verifiable if astronomically dated. In Egypt the only fixed or definite

---

[1] Drummond, *Ædipus Judaicus*, pl. 2.

era was astronomical. All the reports show that prior to the period identified with the name of Mena the Egyptians reckoned vast lengths of time as "reigns" of some kind, sacred to monarchs left without human name, because representative of the divinities, and beyond these was the direct dominion of the gods. These last reigns are mythical, but not therefore fabulous in the modern sense. The truth is that a great deal of history and mythology have to change places with each other; the history has to be resolved into myth, whilst the myths will be found to contain the only history. The ancient fables were veiled facts, and when we can get no further records on the earth, it is in the heavens we must seek for the Egyptian chronology. It is the astronomical mythology solely that will reveal to us what the Egyptians and other nations meant by dynasties of deities and the development of series and succession in their rule. It was by astronomical numbers that the Chaldeans reckoned the age of their sacred books.[1]

"In Egypt, if anywhere," says Diodorus,[2] "the most accurate observations of the positions and movements of the stars have been made. Of each of these they have records extending over an incredible series of years. They have also accurately observed the courses and positions of the planets, and can truly predict eclipses of the sun and moon." Diogenes Laërtius states that they possessed observations of 373 solar eclipses and 832 lunar; these were probably total or almost total.

The Egyptians spanned spaces so vast that nothing short of astronomical cycles could be the measure and record of time and period for them. Plato, who spent some thirteen years in Egypt trying to get into the penetralia of their knowledge, reports that they had divine hymns or songs worthy of the deity which were held in round numbers to be 10,000 years old. He tells us that he does not speak figuratively, but that they are real and credible figures.[3] The authorship of these was assigned to the great mother Isis. These figures are not to be utilised forthwith or straightway; that can only be done by going round to work, when we shall see that such hymns probably dated from the time when the sun was in the sign of Virgo (Isis) at the spring equinox. When Plato wrote, the colure of the vernal equinox was coincident with the sign of Aries, and the time was just over 10,000 years since it left the sign of Virgo.

Previous to the reign of Mena, the Papyrus of Turin, and other documents assign a period of 5,613 years to twenty-three reigns.[4] These of course are not mortal reigns. They are identified with the Shus-en-Har, the followers, servants, or worshippers of Horus. A period of 13,420 years is also assigned to the Shus-en-Har.

Nineteen Han are likewise mentioned. The Han is a cycle, and

---

[1] *Diod. Siculus,* ii. 113.   [2] *B.* i. p. 81.   [3] *Legg,* ii. 657.
[4] De Rougé, *Six Premières Dynasties,* p. 163.

these were probably cycles of Anup or Sothis, the dog-star, whose period was 1,461 years, nineteen of which make a total of 27,759 years. Two dynasties of gods and demigods were collected from the Temple Records and rectified by Lepsius[1] from the various Greek chronological writers; these begin with Ptah (Hephaistos) and end with "Bitus." Of these—

Ptah	reigned	. . .	9,000 years.
Helios	,,	. . .	1,000 ,,
Agathodaimon	,,	. . .	700 ,,
Kronus (Seb)	,,	. . .	500 ,,
Osiris	,,	. . .	400 ,,
Typhon	,,	. . .	350 ,,
Horus	,,	. . .	1,800 ,,
Bitus	,	. . .	70 ,,
			13,820

After this, says Eusebius,[2] "came a series of reigns down to Bytis during 13,900 years;" meaning the above list ending with Horus. It is clear that Bitus has no business to be in the list of divine rulers. Nor is it necessary to assume with Bunsen that 13,800 years of reigns imply hero-worship in the modern sense. These like the preceding may have been astronomical cycles, but measured also by the reigns of sacerdotal kings, as according to Jamblichus[3] Bytis was a prophet of Ammon at Thebes, instead of divine names. According to Eusebius, Manetho had computed a total period of 24,900 years. Such numbers need not be rejected because they do not offer the direct means of correlating and reading them. They are quoted merely as mental eye-openers in the hope that by and by we may see a little farther and more clearly.

Herodotus says "from the first king to this priest of Vulcan (Ptah) who last reigned (Sethon) were 341 generations of men; and during these generations, there were the same number of priests and kings. Now 300 generations are equal to 10,000 years, and the forty-one remaining generations make 1,340 years. Thus they said, in 11,340 years, no god had assumed the form of a man; neither had such a thing happened before or afterwards in the time of the remaining kings of Egypt. During this time they related that the sun had four times risen out of his usual quarter, and that he had twice risen where he now sets (ii. 142), and twice set where he now rises." Again, he remarks, "Hercules is one of the ancient gods of the Egyptians; and, as they say themselves, it was 17,000 years before the reign of

---

[1] *Kritik der Quellen*, p. 484. [2] *Armenian Version*, i. 19.
[3] *De Myster.* viii. 5, ix. 7.

Amasis, when the number of their gods was increased from eight to twelve, of whom Hercules was accounted one " (ii. 43). That is they reckoned 17,000 years of time during which the eight original great deities were at the head of the Egyptian religion. Then four others were added. Still earlier, before the creation of Taht, lord of the eighth region, were the seven great gods as reckoned by the Chaldeans, and the great mother of all who is synonymous with the number seven, and who was personified as the goddess of the Great Bear. Whatsoever truth there may be in the statement, there are no phenomenal data known to the present writer by which the assertion of Herodotus, can be interpreted except those of the circle of the equinoctial precession. In no other circle or cycle does the sun ever rise at one time in the quarter it sets in at another. He therefore holds that the Egyptian priests did verily claim to have made chronological observations during a period, in round numbers, of 52,000 years. The explanation that the priests were referring in any wise to the Sothiac cycle, a period of 1,461 years, must be rejected as not only inadequate but perfectly puerile.[1] The speaking stones, the pictured papyri and written rolls, are all antedated by the celestial chronology of the divine dynasties, and if the present conjecture should prove correct it will drop from above the keystone into the almost completed arch of Egyptology.

My suggestion is that the divine dynasties founded on the cycles of astronomical time, were continued by the era of Mena. And there is evidence to support it in the table of Abydus.

Tinis is the name of the great city of Abydus, and the name of Mena heads the first Thinite dynasty. Now the new table of Abydus, discovered eleven years ago, in a corridor of the temple of Seti I., at Harabat-el-Madfouneh, gives a succession of sixty-five kings from Mena, the founder of the line, down to the last reign of the twelfth dynasty. If we take the accepted average of human life as about three generations to the century, this succession of sixty-five monarchs will extend over a period of 2,166 years, leaving a fractional remainder. M. Brugsch thus assigns to them a period of $65 \times \frac{100}{3} = 2,166$ years.[2] This is as near to the length of time during which the equinox remains in a single sign as need be, that time being 2,155 years. And this is the table of Abydus, of the Thinites, and of Mena.

Also the table comes to an end with a break so abrupt, an interregnum of some kind so marked, that it leaves us staring into a chasm, which is at present without a bridge, and we have to leap or scramble from the twelfth dynasty to the eighteenth.

In further illustration of this interregnum, Mariette-Bey has pointed out that the old Egyptian proper names of persons in the eleventh and twelfth dynasties recur in the same forms on the monuments of

[1] *Bunsen*, vol. iii. p. 60.   [2] *History of Egypt*, Eng. Tr. vol. i. p. 33.

the early part of the eighteenth dynasty, and the forms of the coffins are so alike as to be undistinguishable. Now the eleventh and twelfth are Diospolitan dynasties, and so is the eighteenth. After an interregnum of five dynasties the Diospolitans resume, and continue, as it were, where they left off.

Meanwhile the thirteenth dynasty introduces Sebek-Ra.

It is noticeable that towards the end of the twelfth dynasty Egyptian names compounded with Sebek become increasingly frequent. Sut-Apet, *i.e.* Sut-Typhon, is the name of the Mother of Mentu-Si, on a funeral stele of the twelfth dynasty, whilst two of his sons are named Amenemha and Usertasen, and whereas these names appear on the monuments of this dynasty, names compounded with Amen and Osiris are never found in the thirteenth dynasty. Mariette-Bey concludes from this that the later of the two families must have been the enemy of the more ancient one, the memory even of which it proscribed.[1] It was religious enmity. Sebek had taken the place of Osiris and Amen. The thirteenth may be called the Sebek dynasty. It is that in which the Pharaohs are assimilated to the god Sebek-Ra.

In the list of scutcheons given by M. Brugsch in his *Histoire d'Égypte*, Plate 6, Fig. 109, and in Bunsen's list, the last monarch of the twelfth dynasty is Sebek-nefer-Ra. According to Brugsch-Bey she was the sister of the last king of the dynasty, Amenemhat IV., and an heiress through whom the succession went by marriage to a new race, the Sebekhepts or servants of Sebek. Possibly the name of this Queen conveys the information that she was the continuer (Nefer) of Sebek as the Ram-headed Sun-god Ra, whereas he had previously been the Crocodile-god of the Fayoom, or Country of the Lake. Whether she is out of place here and should be the first of the thirteenth dynasty matters little. She marks the end of the twelfth, and is the first known royal Sebek on all the monuments, and the next dynasty is full of them ; it is, in short, the Sebek dynasty.[2]

M. Brugsch gives the genealogical table[3] of a distinguished family related to some members of the thirteenth dynasty in which the name of Sebek, in the male and female titles, occurs eighteen times. And the first mother of the king is named Aaht-abu, the house (literally the womb) of the lamb. The ram when young is the lamb, and the ram of the zodiac was represented by the Persians as a lamb. Now the old Sebek or Khebek was the crocodile-headed god of darkness, whose name of Khebek modifies into Kek, the Suchos of the Greeks. And he was resuscitated in a new form as the ram-headed god, found at Ombos and Selseleh. The ram's horns identify him with the Ram of the zodiac,

---

[1] *Notice des Principaux Monts. à Boulaq*, p. 74.
[2] *Egypt's Place*, vol. 2, Scutcheons. Königsbuch, *Lepsius*.
[3] App. 1, Vol. 2, *History of Egypt*, Eng. Tr.

and it is in strict accordance with this new character of Kronus that he should be styled the "youngest of the gods."[1]   The ram-headed crocodile god then we take to be the divine head of the Sebek dynasties, and this will enable us to interpret the fabulous end of Mena, who, according to the tradition and in the language of symbolism, was said to have been seized and devoured by the Crocodile; or the hippopotamus, another type of Typhon.   The crocodile was Sebek, whose reign, as the ram-headed god, began when that of the celestial Bull (Mena) with its sixty-three monarchs of the table of Abydus, came to an end.

The introduction of Sebek as the ram-headed Ra implies a religious revolution.   The capital of Shat, called by the Greeks Crocodilopolis in the district of the lake Mœris, identifies the Typhonian nature of the old Sebek, with whom the Osirians were at enmity.   In the list of Nomes the province of Lake Mœris was struck out as being hostile to Osiris.   In the Manethonian lists the names of the kings of the thirteenth dynasty are passed over in silence.   Their Typhonian tendencies will account for this neglect by the scribe who was too strongly Osirian and anti-Typhonian to register the names of these servants of Sebek or Satan.   Still the Typhonians also kept the Chronology. Everything in Egypt being typical, the name of the temple of Harabat-el-Madfouneh, the set or sunken Harabat in which the table was found, will now yield up its meaning. It was the set or sunken abode of Har.   Whatever sign the equinoctial colure was in there was the place of manifestation of the son, Har, on the horizon, during 2,155 years.   When this left the Bull of Mena for the Ram of Sebek the abode and birthplace was changed in heaven, whilst on earth Thebes took the place of Abydus, and the record of the celestial reign was left in the sunken abode.   The Egyptians built on celestial foundations.   They made the temple the centre of their city.   The dead were held in the heart of the living, and their place of preservation was the earliest sanctuary.   The tomb of one life was the womb of the next.   The meskhen, or place of new birth for the sun in heaven, furnished the type of burial and rebirth below.   Thus, when the sun was annually reborn in the Bull, the Bull city of Memphis was built as the representative place of resurrection; Abtu having become the sunken and superseded place of birth and rebirth.   The sign of the Twins is possibly implied in the name of Abydus.   Apt is a name of the pool of the twin truths, abtu is the double holy house of Anubis. It is the name of a dual place of beginnings which the Twins were when the colure was in that sign.   Indeed, in an ancient Egyptian zodiac the Twins are represented as a double Anubis;[2] this identifies one Abtu with that sign.   Moreover, the god Shu (Anhar), who, in his dual character, forms the Twins of the zodiac, was the divinity presiding over Abydus.

[1] *Wilk. Mat. Hierog*. pl. 27, f. 2.   [2] Drummond, *Ædipus Judaicus*, pl. 4.

Nor is this all. It is especially manifest that there is a parallel representation of the ram-headed deity, adopted about the same time, as the Amen-ra of Thebes, in whom Num and Khem, the old gods, were merged in setting up an image of the one god, the Amen of the hymn in which he is celebrated as the " one in his works, single among the gods," the " one maker of existences," the "one alone with many hands," the "one alone without peer." [1] Moreover, Amenemhat I., 2466 B.C. (12th dynasty), was the founder of the Temple of Amen at Thebes, as the especial shrine prepared for the Ram-headed God of the orthodox caste.

Osiris was the great god of the dynasty of the celestial Bull, who was succeeded by the ram-headed Amen of one cult and Sebek of the other. Brugsch-Bey has called attention to the remarkable zeal with which the kings of the twelfth dynasty, one of whom, Usertasen I. had himself represented standing as Osiris,[2] set themselves to maintain the cult of Osiris and his temple at Abydus.

The fullest testimony to this fact is supplied by a monumental stone found but the other day in the cemetery of Abydus, and now in the museum at Boulak, containing 163 inscriptions. This shows how a certain Sehept-ab-ra, who lived under the reigns of Usertasen III. and Amenemhat III., was commissioner, and commanded to attend to the service of the mysteries, the secret things and places in the temple of Abydus ; to regulate the feasts of the god and to build or rebuild the holy temple-bark and cover it with painted figures.

The next king Amenemhat IV. was the last of the twelfth dynasty. Thirteen years only are assigned by Brugsch-Bey to both reigns of the brother and sister, Amenemhat IV. and Sebek-nefer-Ra. May we not see in these instructions the preparation for the end of the divine reign, the regulation and adjustment of astronomical time and the change of the double-seated solar bark from the sign of the Bull to that of the Ram ? This suggestion is strengthened by the words assigned to him. He says, " I say a great thing ; listen ! I will teach you the nature of eternity." The Egyptian eternity was Æonian, and he had been the time-keeper of its cycles and master of its mysteries, the builder of the ark of the Eternal ; who was now to re-build it and cover it with fresh figures.

Nothing can be more probable than that the particular ram-headed representations of the deity were especially adopted as Amen and as Sebek-Ra about the time that the vernal equinox entered the sign of the Ram, and that this time coincides with the end of the twelfth dynasty, giving the exact date of 2410 B.C., and the date of 4565 B.C. for the commencement of the era of Mena.

In his list of the Pharaohs and their epochs, founded on the list of kings in the table of Abydus and on the regnal years actually proved,

[1] *Records of the Past*, vol. ii. p. 129.
[2] Mariette-Bey, *Notice des Principaux Monuments d Boulaq*, p. 298.

Brugsch-Bey gives the date of 4400 B.C. for Mena and 2266 for Amenemhat IV.,[1] the last king of the twelfth dynasty, or a total of 2,134 years for the twelve dynasties ; within twenty years of the time required in the celestial reckoning !

The monuments do not come down to the time of the entrance of the vernal colure into the sign of Pisces, but the Gnostics brought on the imagery, and on one of the Greco-Egyptian Gnostic seals in the British Museum[2] there is a figure of the young sun-god Horus, with the solar disk on his head carrying the fish as his latest type. He stands on the crocodile, and this illustration of the manifester as Ichthus, with the fish above and crocodile beneath, corroborates the view that the crocodile-god, with the ram's head, had represented the sun in the sign of the Ram.

The same sequence is illustrated by the types of sacrifice. The fish is now the sacrificial type, and has been ever since the equinox occurred in the Fishes. Before that the type was the Ram or the Lamb. Earlier still it was the Bull ; amongst the primitive races we can get back to the Twins as the typical sacrifice, and each of these types corresponds to the solar sign, and to time kept in the astronomical chronicles.

Also, the Egyptian month Choiak begins in the Alexandrian year, on November 27th ; and in the calendar of lucky and unlucky days in the 4th Sallier papyrus it is said to be unlucky to eat fish on the 28th day of the month Choiak, because on that date —our Christmas Day—the Gods of Tattu assumed the form of a fish, or in other words the sun entered the sign of Pisces, at which time the equinoctial colure must accordingly have been in the sign of the Twins.

The importance of this sequence and of the identification of Mena's era with the divine dynasties, and the consequent link established with the backward past, will become more apparent when we come to consider the cycle of the equinoctial precession, or the great year of the world which began when the vernal colure left the sign of Aries for that of Pisces nearly 28,000 years ago, and ended when it re-entered the sign of the fishes 255 B.C.[3]

In this sketch of Egypt the outlines are drawn in accordance with the intended filling in. The treatment will serve to show the ex-tended and inclusive sense in which the name of " Egypt " has often to be interpreted in these pages as the outlet from the African centre. We have now to turn and follow the track of the migrations into the north, called by the Hebrew writer (Gen. x. 5) the Isles of the Gevi (נג) or Gevim.

---

[1] *History of Egypt,* vol. ii., Appendix.
[2] No. 231, Dr. Birch's department.
[3] This is the date given by Cassini and Sir William Drummond, and adopted by the present writer on data kindly furnished by the Astronomer-Royal and the

In Hebrew, Gev (גו) is the back or hinder part, identical with the
Egyptian Khef; and the Children of Khef, the Æthiopic Genitrix,
are designated the Gentiles who went northward and carried with
them the primordial name of the Birthplace in the Celestial North.
The race of Japheth (יפת) are none other than the race of Kheft,
whom we shall find in Britain as the Great Mother Kêd.

calculations of an eminent mathematician. The following is the official reply to my
question as to when the vernal equinox coincided with the fixed point supplied by
the first star (the last in the backward movement) in the Ram constellation :—

" ROYAL OBSERVATORY, GREENWICH, LONDON, S.E.,
" *July 23rd*, 1877.

" DEAR SIR,

" IT appears from our computation, that the vernal equinox
passed through the star γ Arietis about B.C. 400, subject to an uncertainty of three
or four years, or perhaps more. The uncertainty of observations at that epoch
might easily produce an apparent error of thirty or forty years in the observed date
of such a conjunction.

" I am, dear Sir,
" Yours faithfully,

" GERALD MASSEY, ESQ.   " W. H. M. CHRISTIE."

SECTION II.

COMPARATIVE VOCABULARY

OF

ENGLISH AND EGYPTIAN WORDS.

THE following list of words, extant in the British Isles, compared with Egyptian words—itself a work of years—needs but brief introduction. The title is not meant to be taken literally; it is adopted simply for the sake of classifying the words. The chief authorities for the Egyptian are the *Wörterbuch* of Brugsch-Bey, the *Egyptian Dictionary* of Dr. Samuel Birch (Vol. 5, "Egypt's Place"), and the *Vocabulaire Hiéroglyphique* of M. Paul Pierret. In these the references to the texts are given for each word.

On the English side the *Provincial Dictionary* of Thomas Wright stands first, but numberless volumes have been ransacked for the result. Keltic, Kymraig, or Gaelic have not been especially drawn upon.

It may be stated generally, that the Hieroglyphics contain no phonetic C, D, E, G, J, O, Q, V, W, X, Y, or Z, but that certain of these letters are introduced as equivalents by later Egyptologists. One sign is rendered by L as well as by R.

Paralleled as the words are, they themselves sufficiently explain the laws of permutation or interchange of the equivalent letters.

ENGLISH.	EGYPTIAN.

A

ENGLISH.	EGYPTIAN.
a, one.	**a,** one.
aak, the oak-tree.	**akh,** how great, tall, green, magnificent.
acorn.	**aak** (oak), **ren,** the young, shoot, offspring, to renew.
ab, sap of a tree.	**ap,** liquid essence.
abode.	**abut,** abode.
aboo, Irish war-cry, expressing thirst, desire for battle, and delight in the onset.	**abu,** to dance, thirst, delight, brandish.
aby it, stand the brunt of it, stand against; **by,** against (1 Cor. iv. 4).	**abl,** against, in opposition to.
act.	**akh,** verb of action; **t,** participial terminal.
adragoul or **addergoul** (Irish), a place between two river-prongs.	**atr,** river-measure, limit, region.
aft, stern, place of helm.	**apt,** hold of a vessel, a corner or end; **apt,** guide; **aft,** hinder part.
agog, on the jog, on the start.	**akhekh,** fly, on the wing.
ahoy, sailor's cry of hailing.	**ahaui,** cry of joy or salutation.
ainted, anointed.	**ant,** anoint.
air.	**karh,** air.
air, or **aer,** appearance.	**her,** appearance.

ENGLISH.	EGYPTIAN.
ait or oat.	att, grain of some kind.
ait, island; ey, island.	aa, isle.
allies.	ali, companions.
am, them, both.	am, together.
amain (Irish), infernal deep.	amen, secret place, hidden region of th Abyss.
amakly, in some conscientious way or fashion.	amakh, mature, do justice to, fidelity.
ame, to guess, find out, tell.	am, to find, discover.
amell, between, passage between.	am, between, passage between.
amene, pleasing, consenting.	am, pleasing, in, with.
ames, a plural noose, round a horse collar.	aam, a noose.
an, of.	n, of.
an, hair.	anhu, eyebrow or bristle.
anaf (Gaelic), breath.	nef, breath, spirit of the firmament.
anaks, provincial name for a fine kind of oaten bread.	ankh, some kind of food.
ane, beard of corn.	an, horn or thorn.
ane, one.	an, one cycle.
anede, united, made one, oned.	un, t, an hour, the complete cycle, oned.
anker, clasp of a buckle; inkling, getting a partial hold or grasp of.	ank, clasp.
anne, to give, yield; annett, firstfruits.	annt or anent, tributes of Nile.
"anon, sir."	han-han, command me, at your command an, coming.
anotta, yellow colour, a chemical used for adulterating milk.	ant, yellow colour.
anshum-scranchum, to scramble after food in a wolfish manner.	ansh or unsh, the wolf.
anti, opposed to.	anti, go back, turn back.
aog (Gaelic), death, ghost.	akh, the dead, a spirit.
ap (Welsh), son of.	ap, ancestor.
ape or yape, a monkey.	kaf, ape.
appear, pour.	per, to pour out, appear.
apple-terre, apple orchard.	ter, garden; ter, a limit, the extent.
apt, to fit to, adapt.	apt, guide, lead, judgment.
arach (Gaelic), tie, bond, collar.	ark, a noose, a tie, a binding.
ard, height.	arr, t, staircase, height.
ardour, fiery fervour.	artaur, flames of God.
are, plural of "to be."	ar, to be (are).
are, to plough; ear, plough and sow seed.	aur, enter, go between, beget.
arra, either.	ari, one another.
ars, science; arstable, an astrolabe.	urshu, applied to astronomical observation.
Art (Irish), name for the Great Bear.	ta-Urt, Typhonian, Great Bear; ta is "the."
art or airt, quarter.	urt, a quarter; amurt, western quarter.
art.	ar-t, deed, form; rut, to retain the form, renew, sculpture.
arte, to constrain, urge, compel; whence to milk.	art, milk.
arted (Chaucer), constrained.	art, milk, neck-chain.
ask, applied to damp weather, a lizard or water-newt; "Askes and other worms fell" (MS. Med. 14th century.)	asskh, hurtful.
asise, a term of chess.	asb, chess.
ask or ash, stalk of corn.	askh, stalk of corn.
askings, publication of marriage by banns.	ask, delay, then.
assembly,	sam, assemble.
ate, eat, to eat.	aut, food of some kind.
ath (Irish), ford.	khat, ford.
athel, noble.	ati, a noble.
athene, to stretch out; athening, extension, heightening.	aten, extending in a circle.
athyt (Tusser.), condition of housing corn.	aat, house; at, type or condition; hit, corn.
atrick, usher of a hall.	aat, house, abode, hall; rekh, to speak, announce, declare.
attach.	tek, attach.
attack.	atakh, trample.
attend.	aten, to hear.

ENGLISH.

auch (Keltic), a field.
au, all.
aud, old.
audience, hearing (Chaucer).
aught, property, possession, anything.
"auh woot" direction to horses.
auk, invert.
aukard, backward.
auld, first, best, great.
autour, ancestor.
autum, slang term for banging.
autumn.

ave, to have.
avit, weight.
asize.

awen, the Druidic knowledge, science, gift, genius, inspiration.
awn (Welsh), the Word.
awyr (Welsh), the sky or heaven, (Eng.) air.
ax, to question.
axe.
ay, yea, yes, also used for "I have."
aye, ever.

EGYPTIAN.

uakh, a meadow.
au, being, was, is, and to be.
au, elders.
aten, to listen.
akht, thing or things, substance ; akat, claw.
ahi, denotes a forward movement.
akh, turn over.
akar, the hindward region.
urt, first, chief, great.
ata, father.
ath, drag, draw ; am, a noose.
atum, red autumnal sun of the west and the lower world.
afa, to be filled and satisfied.
apt, measure, quantity.
au, dignity, age, the old one, to chastise ; aut, the crook sign of divinity.
an or aun, priest, scribe, speech, decree, show.
an, speech of, speech to ; un, reveal.
aaru or aar, the sky or heaven, Elysium.
akh, how, why, wherefore.
aksu, an axe.
aia, I have ; ia, yes, certainly.
heh, ever.

B.

baa, lamb.
babs (Scotch), loops in garters.
back.

bad.
bad.
baffled, corn blown down by a gust.
baggie (Scotch), belly.
baide (Scotch), did stay, i.e. be'd.
bait, of corn.
bait, refreshment, luncheon.
bait, food, luncheon.
baith (Gaelic), grave.
ballow, cry of goal in a game.
balow, spirit.
ban and fen, "fen" such a thing.
bane, poison.
bane, destruction.
baptism.
bar, a horseway up a hill.
barley.
barley-bygge, corn for beer or strong drink.
barr (Gaelic), a height.
bass, fish.

basen, extended.
bases, aprons, also an embroidered mantle which was worn from the waist to the knee.
basin.
ba-sket, woven.
bast, a bastard.
baudrons (Scotch), cat.
baudy.
bauk, a cross-beam.
bay.
bay-salt, rock-salt.
bay-tree.

ba, ram or sheep.
beb, hole, circle, round, around.
pe'h and pekh, rump ; akh, spine (p, the article).
bat, bad.
but, abominable.
paif, gust of wind.
buk, belly.
ba, to be ; ba-t, made to be.
bet, corn.
ba-t, rations, food.
ppat, kind of food.
baut, hole of the tomb ; baita, house.
beru, cap, tip, goal.
beru, force, fervour.
ben, no, not, under ban.
ben, pollute, no, not.
ban, no, not, unclean.
âb (rab), pure, priest, wash, baptise.
baru, cap, tip, top.
peru, some kind of grain (? barley).
beka, palm-wine.
bur, cap, top, roof.
bes, transfer, pass, determinative, "bass-fish."
besa, dilate.
basu, an apron or tunic.

bashn, glass enamel, porcelain.
s'khet, to weave.
besh-t, revolt, hostile.
pesht (Buto), cat-headed goddess.
buta, infamy, abominable.
puka, a plank or log.
bau, a vase or container of water.
baâ, earth, stone, rock, or salt.
ba, wood, leaves.

bayete, procreate.
be.
beacon, hill.
beano, born (Eng. Gipsy).
bear.
beas, cows, cattle ; bu, buw, buwch (Welsh), cow.
beast.
beat.
beck, to bend the knee.
bed.
bed, uterus.
bede, bend to the right, command to horses.
bee.
beer, brew.
beggar, beggary, poverty, full of weeds.
being.
bekenn, to give birth to.
bekke, to beg.
belle, to swell ; brew, boil.
ben, good, well.
ben, inner room.
ben, a mountain.
ben, a figure set on the top of the last harvest load, dressed up in ribbons.
bend, a bond.
berry, a flood.
berry, an edible fruit, traced by Bopp to the Sanskrit "bhakjam," *i.e.* bhag-s-ja-m.
bert, perspire, bright ; berth, byrht, manifest ; purt, to pour out.
berewham, a horse-collar.
bese, to see, to behold.
besh, to sit (Eng. Gipsy).
besmear, harm, and besmear, in raiding.
bewe, obey ; beh, inclined.
bewe, drink liquor.
bewly, shining, lustrous.
bey, ox, bee, boy.
bib, drink, to bubble, to well forth.
bib, which goes round a child's neck.
bid, to command.
big.
bight, the bend or loop of a folded rope.
bill, a promontory.
bind.
bird.
bit (Scotch), place.
bite, feminine privities.
boban, pride, vanity, boasting.
bock (Scotch), to gush intermittently, to vomit.
body.

boggams, certain masters of ceremonies who wore red jackets.
boging, evacuating ; boke, to vomit.
boil.
bonnie, bears the palm.

boot.
boon-master, a road-surveyor.
border.
bore or eagre of a tidal river.
Boreland Hills, Scotland.
bosh (Eng. Gipsy), a fiddle.
bosom, also besom.

ba-t, inspire, give breath, beget.
pa, beings ; pu, to be.
bekhn, tower, fort, magazine.
benn, engender ; bennu, sons.
peru, to bear off.
behs, calf.

bes-t, skin of a beast.
pet, strike, beat.
beka, to pray.
pet, a crib.
bu-t, belly or womb.
pet, bend.
ba, a bee.
per, a liquid made from grain.
hekar, starve, famished (p, the article).
pu, it is ; pu, to be.
beka, to bring forth.
bak, servant; beka, to pray.
ber or beru, boil.
ban, enviable ; bent, excellent.
ben, place.
ben-ben, cap, tip-top, splendid, palm-branch

baent, bind.
peru, to pour out.
perru, food, appear, grow, manifest.

pert, emanate, proceed ; pert, put or pour out, manifest.
peru, to surround, go round.
bes, exhibited, proclaimed.
besh, to squat down.
bes-mer, bind and carry off.
beha, incline.
ba, drink, water (? bua).
ba, to illumine, radiate, diffuse light.
beh, creature.
beb, well, exhale.
beb, a collar.
pet, sceptre, sign of command.
pekh, extend ; bekh, enfanter.
betau, called holes for ropes.
ber, top, cap, roof.
baent, to bind.
urt, bird ; p, the.
bet, place.
bu-t, belly, womb.
bâ-bâ, boast.
bakh, to bring forth, void.

ba, soul ; ti, abode ; pauti, form, figure, body.
bak, hawk-sign of rule and lordship.

baka, to squat, bring forth.
ber, boil.
beni, the palm branch ; ben, tip-top, splendid.
bu, leg.
benr, road, outside.
per, go round ; ter, extreme limit.
ber, well, boil up.
beru, cap, tip, roof.
bes, god of dancing.
bes, dilate, pass to and fro.

boss, a head or reservoir of water.
bote, amends ; to **boot.**
bote, material to mend with.
both.
bottom.
bough.
bout, a round.
brag (Welsh), malt ; **brew** (Eng.).
bramble.
bread.

breath ; burr, halo.

breathe.
bright.

brittene, to divide into fragments.
broad.
broo, the top of anything.
brose.
bub, to throw out bubbles.
bucket, a bent stick to stretch the legs of a slaughtered pig by which it is hung up.
Budd-Ner, British god of victory.
bug or **puck,** a goblin ; **bagan,** devil ; **buggy-bane,** children's game, played in the dark.
bury, a castle or great house.
bush.
busk, inner support of stays.

busy, spelt **besy.**
but-shot, a bow shot.
butt, a measure ; **beat,** a measure.
butty-shop, where wages were paid in food.
bwrdd, a Welsh name for Arthur's table.
by, a place, **bi, by,** or **bye,** a town or village.
bysyschyppe, activity.

bes, inundater, transfer.
baat, recompense.
baaiuti, substance, material.
pehti, pauti, pet, both.
pet, foundation ; **am,** belonging to.
bu, a branch.
put, a round.
per corn ; **akh,** spirit.
bram, the snatem wood or thorny acacia.
prut, to appear, proceed, emanate from grain ; **perrt,** food of corn.
per, to surround, appear, emanate, proceed from.
prut, to void.
per, to appear, shed ; **akt,** light, fire, splendour.
prt, show, appear; **tna,** separated.
prut, manifest, proceed, spread out.
buru, cap, tip, roof-top.
pers, food.
beb, exhale.
pekht, a crooked stick, to stretch out.

pput, god ; **ner,** victory.
pukha, infernal locality ; **ha,** a dwelling.

buru, the lofty roof.
pesh, branch, flower, or fruit.
besk, viscera, heart, inner organ, mainstay of life.
besi, pass from one place to another.
put, a bow.
baat a measure.
pati, feed, food.
perr-t, food, appear, store.
bi, a place.

besi, pass, go, bear, from one place to another
skhep, transfer ; **skheb,** goad on.

C.

caad, cold.
cabbage, applied to the horns of a deer, and to come to a head or horn.
cabin.
cabobble, to puzzle.

cackle.
caft, intimidated.
cag-mag (Eng.), **cag-magu** (Welsh), a tough old goose.
cage.

cagg, to make a vow, firmly carry out a resolution.
caird (Scotch), carver of horn spoons.
calf, first form of the cow.
call.
cam, crooked.
camel.
can.
canal.
cannie, knowing.
canoe.

khat, a corpse.
kaba, horn.

kabin, vessel, ship.
ka, say, talk, type, cultus ; **beb,** go round and round in a whirl.
kaka, to cackle ; **Kakur,** the great cackler.
kaf seize, hunt, make desolate.
Kak-ur, the old Kak, a name of Seb, who carries the goose.
khekh, collar for prisoners ; **kak,** a shrine, shut place.
khaka, be obstinate, stupid, mad ; **ka's,** bind.
kart, sculptor, mason.
kherp, first form or formation.
khar, speech, speak, word, cry with the mouth.
kam, a crook, to bend.
kamaru, camel.
kan, service, power, ability, courage, valour.
khanru, a canal.
khennui, intelligence.
khenna, a boat ; **ma-khenna,** the boat of the dead.

ENGLISH.	EGYPTIAN.
cant, a secret language.	khent, inner, interior, concealed.
cant (Cor. Eng.), to contain.	khent, to contain.
cantlet, a little corner.	kan, hieroglyphic corner.
canty, festive, happy.	khent, joy, delight, circumstance of a festival
capes, ears of corn.	khepu, crop of corn; khepi, harvest.
capon and capul, fowls.	khepen, geese; khep, kind of duck.
capper, to coagulate.	khepr, the transformer, to transform.
car, low-lying land.	kar, under, nether, below.
carcern, a prison; carchar (W.), a prison.	kar, a prison.
carcharu (Cor. Eng.), to confine, imprison.	ker-ker, seize, hold, imprison.
care-cloth, marriage canopy.	ker, circle, round, zone, sphere.
carp, speech.	kher, word, to speak; p, the.
carry.	kar, carry, support.
carse (Scotch), fertile flat.	kars (Heb.), a vineyard or field that grew the finest produce; kar (Egyp.), power, property.
cart.	kart, to carry.
carve, a "carve of pasture."	kherp, sufficient supply.
carve.	kherp or kherb, to form, figure, model.
case.	kahs, habit, custom, state.
cat, to vomit.	ka, to vomit.
cat, pudendum f.	kat, womb.
caw, rot in sheep.	khau, malady.
caw (Welsh), enclosure.	kahu, enclosure, cover, or noose.
caw (Welsh), band, associates.	kaui, a herd.
caward, backward.	ka, tail, behind.
cayvar, some kind of ship.	khepf, hold of a boat.
ceannaigh (Irish), pedlar.	khennu, to carry, convey, traffic.
cease.	khes, to stop, turn back.
celi (Welsh), the mysterious or secret one.	kherui, the Word, logos.
cell.	ker, cell.
cent, one hundred.	shent, orbit, circle, million.
cerdd (Welsh), utterance, word, songs of Keridwen.	kher, voice, speech, word.
ces, measure of compatibility.	ses, the measure of compatibility.
chache (Scotch), blind man.	kak, darkness, god of darkness.
chair.	kar, under.
chapel.	kep, a sanctuary.
chapel, a printing-house.	khepui, types.
chaps, double, a pair of tongs.	kab, double.
charm.	kr, to lay hold of, seize, possess; am, charming, pleasing.
chase.	shas, to follow; khakh, to chase.
chat, child.	khat, child, race.
chates, gallows.	kaiti, punisher.
Chatham, where the waters of the Medway are landlocked.	khatam, a fortress shut and sealed; khat, shut, seal, lock.
check, stone chest, kistvaen.	kek, a sanctuary.
"chech-chech," cry to pigs.	khekh, whip, follow, chase.
chech, church.	khekha, an altar; kak, name of a sanctuary.
check or choke.	khekh, a neck-chain or collar for tying up with.
check, in reckoning.	khekha, reckoning.
checked, caught.	khekht, repulsed, collared, checked, caught.
cheens (Cor. Eng.), loins.	kena, loins.
cheper, an exchanger, a seller.	khepr, the changer.
cherry (Devon.), ruddy.	tsheru, red or red-wood.
cherven, to writhe, twist, turn.	ref, the snake and worm.
chest.	khest, some part of the body.
chet, slang for a letter.	khet, shut, seal.
chete, a ring.	kat, to go round, be round; khet, a seal ring.
chew.	khu-khu, beat, strike.
Child, one who bore arms; childing, bearing.	kart, to bear, have, carry.
chimney.	kam, black.
chin-chin (Gipsy), a carried child.	khen to carry; khen, child.
chirp, a first form of note.	kherp first form, formation.
chivy, to pursue.	kfa hunt, seize.
chop, to exchange, "chop and change."	khep, transform, change.

choppine, a quart measure.
chout, a performance, an entertainment.

chummy, a chimney-sweep.
chwed (Welsh), speech.
ci (Welsh), a dog.
ciric-sceat, church tithe paid in corn.
city.
clef, the initial mark for the key.
cleith-ras, a covered temple.
clowze (Cor.), tomb or enclosure.
cnight, a hard boy.
cnight or knight, pubescent.
cnuch (Welsh), to copulate.
coach-horse, the dragon-fly.
cock, the needle of a balance.
cock-brained, foolhardy.
cock roach.
cof or cove, an inner recess.
coff (Welsh), belly.
coff, to change.
coffau (Welsh), to gather.
coid (Welsh), wood.
comb.
coney, beehive (cf. the cony).
cooch (Scotch), a dog-kennel.
coof (Scotch), an ape.
cooked, to make the balance fraudulently.
cooser Scotch), stallion.
corder, a troop of soldiers.
corfe castle, once a royal residence.
cori (Eng. Gip.), mem. virile.
cote, a wood.
cote, to coast.
count, a title.
cover.
cow.
coward.
cows (Welsh), a wood.
cradle.
craibhdhigh (Irish), people who mortify the flesh.
crape, for mourning.
crawfish.
creed.
crefydd (Welsh), religion.
cro (Irish), fold, hut, hovel.
croot, a puny, feeble child, also colt.
crut, a dwarf; child (khart), by permutation.
crop, the top, produce, fulness, plenty, over-full.
crow, cry.
cuch (Welsh), to contract, a contraction.
cuckoo.
cuddle, to clasp in the arms.
Cun, chief, sovereign personage.
cup.
curr, to sit on the houghs or hams.
curve.
cut, a cut, a go.
cut.
cut or kutte, pudendum f.
cutty, short; get (Scotch), a child, a little one; kitty wren.
cwch (Welsh), boat.
cyphel, the houseleek.

khepeni, a measure of liquids.
khu, glorious actions, a ceremony, a represen-tation.
kam, black.
kheft, called the living word.
khaui, dogs.
akhet, corn.
ketui, orbit, circle of abode.
kherp, the first figure.
resh, a temple.
karas, place of embalmment, a tomb.
nakht, hard land.
nakht, phallic power.
nak, fornicate.
khekh or akhekh, the dragon.
khekh, a balance.
khekh, fool, obstinate, mad-headed
kaka, darkness, night, black.
kep, a sanctuary.
kefau, the navel.
khep, to change.
kaf, seize, claw hold with the hand.
khaut, wood.
kam, hair.
khennu, interior, cover.
khakh, a collar.
kauf, an ape, monkey.
khekh, a balance.
ka, typical male; ser, chief.
kar, war; ter, people.
kherp, first, chief, consecrated, the sceptre.
karu, testes.
kaut, wood.
khet, to navigate, ford, port.
kannt, a title.
kepher (Heb.), to cover.
kaui, cow.
urt, peaceable, mild, meek.
khaiu, a wood.
khart, child.
kherp, to consecrate.

kherp, consecrated, to offer, pay homage to.
krau, claw.
kher-t, the word, speech.
kherp, to consecrate.
khru, cell.
khart, child.
Harpocrates, the child.

kherp, first, principal supply, produce, excel, surpass.
kheru, voice, speech, word; karru, cackle.
khekh, a collar; keks, to bind.
ka-ka, to cackle.
khet, to enclose.
ken, titles, kan, able, victor.
kab, libation, refresh, enjoy, liquid, quantity.
kar, carry, bear, under.
kar, curve.
khet, to cut, go.
kt, sign of a blade, to cut.
khet, womb.
ket, little.

kaka or kek, boat, caique.
kefa, seize, lay hold, claw hold of.

ENGLISH.	EGYPTIAN.
cypher, a joiner's term—to cypher off a square edge; to make two edges in place of one.	kab, to double.
cyve, a sieve.	khi, a sieve.

D.

ENGLISH.	EGYPTIAN.
dad, a piece in the hand.	tat, a handful.
daddle, the fist.	tat, the hand.
daddy, father.	tat, father.
daft, to be put off, set aside; daft, mentally.	taft, desolate.
dag, day.	takh, to see, to behold; t, the; akh, light;
daimen (Scotch), rare.	tem, perfect; tameh, precious stones.
daive, to soothe.	tef, fragrance, pay attention.
daker, corn.	teka, corn.
dam, to stop up.	tem, no, not.
dandy, one hand.	tan-t, one half.
darfe, hard, cruel, stern.	taru (tarf), to afflict, bruise, drive.
daubing, making a house of clay.	teb, brick of clay.
daw (Welsh), a tie or bond.	tâ, knot, tie.
dawn, on the horizon, midway.	tan, half, half-way.
day.	tuai, time, morning, morrow.
deak or deke a ditch.	takh, a frontier.
dearly, extremely.	ter, extremity.
death.	tet, death, block, decapitate.
deem, judge.	tem, judge.
deme, to judge, condemn.	tma, make just, show, distribute justice.
den, evening, sundown.	ten, the inverted half (as the moon).
dent, a cut; dunt, a hard blow.	tent, cut in two.
dewskitch, a good pummelling.	skhet, to beat, a blow, knead bread.
dibstones, huckle bones used for gambling by guessing.	taba, some kind of game; tep, to guess, divine, announce.
dicker of hides, 10, a quantity, a "dicker of wit," Pembr. Arc.	tekai, a measure; tekh, weight, supply.
dickey, all over, an end.	tekai, a measure.
dickey-bird.	tekai, bird.
diet, a body.	tet, a body.
dight, to clean corn from chaff.	tekat, has possibly the same meaning (see Champn. N. D. 373).
digle, secret.	tekau, to lie hidden.
dike, a boundary.	tekh, a frontier.
dil, penis.	ter, penis.
dim (Welsh), no, none.	tem, no, not.
din (Welsh), hill.	ten, the high seat.
dinner.	tennu, ration; ten, reckon each, every; naru, aliment.
disen, to bedizen.	tes, ornament of dress.
dod, to cut or lop off.	tet, to decapitate.
doit, to stupefy.	tuha, to be drunk.
dole, to lay out, grief, mourning.	ter, layer out, mourner.
dose, sleep awhile.	tes, suspend, separate, leave, transport self.
dose, a given quantity; dossel, a bundle of hay or straw.	tesh, so much land bounded in a district.
doup (Scotch), backside.	tep, keel of a boat, the bottom, hinder part.
dout, extinguish.	tet, death.
dove, to thaw.	tef, to drip and drop.
down, a company of hares.	tun, to complete, fill up, unity, total.
draw or drew.	teru, drawing
drink.	ter, libation; ankh, to sustain life.
druid-heachd (Gael.), enchantment.	hekt, charm, magic.
dry, a sorcerer, "Try the spirits."	tri, invoke.
dub, to clothe, ornament, equip.	teb, to clothe, equip, be clad.
duck; dig, a duck.	tekh, the Ibis, a water-bird.
dude, the moon, (Eng. Gip.)	Tet, moon god.
dule, goal, doole, a boundary, heap.	ter, limit, extremity, frontier.
dumb-wife, a fortune-teller.	tema, dumb, announce, tell.

ENGLISH.	EGYPTIAN.

wfr (Welsh), water.
wyfol (Welsh), divine.
yke, a wall or ditch.

tef, drip, drop.
tef, divine father.
tek, a boundary.

E.

a, water.
arth.

ai, ia and a, water and a stream.
urt, car, that which bears ; ta-urt, the bearing mother.

ast.
aver, one quarter of the heavens.
dan (Ir.), hill brow of the rampart.
ft.
gg (on), to overrule.
ke, a final tumbler of toddy, a dominus.
lf, to entangle in knots.
ll and elbow.
Emma, a woman's name.
nd.

aat, a period of time, light.
aft, quarter, four corners.
aten, taking a circular form.
hef, viper, snake or worm ; hefnr, lizard.
hek or ak, rule.
hek, drink, rule, ruler.
arp, to tie up in a knot.
al, a measure of length. ¶
hema, the woman, wife, lady.
antu, division or limit of land ; unn-t, hour, end of a time.

nef (Cor. Eng.), the soul.
nough.
ntity.
ntire.
ric (Irish), fine for homicide.
sking, the penthouse.
ther, bindings for hedges.
ve, to become damp.
ver.
ayre, to go, move, haste, speed ; Justices in eyre, on circuit.

nef, breath, spirit.
henuü, fulness, riches.
enti, being, existence.
ter, entire, complete, all.
rek, culpable, criminal.
uskh, plan of a hall.
atr, limit, boundary.
aft, exudation.
ap-ar, type of totality.
ar, go along, make the circle.

F.

fad, fashioned.
fade, decayed, dirty, disgusting.
faff, to blow, move violently.
fag, paunch.
fag, coarse, reedy grass, meadow.
fag, to beat, thrash.
fagot, tie up, bundle together.
fair, manifestly, evidently.
fambles, hands.
famish.
fan, found.
fang, to catch, grasp, clasp, clench.
fantail, for reversing the sails of a mill.
far.
fare, to resemble or act like another.
fare, game.
fare, to go, cause to go.
fap, tipsy, drunk.
fang, to clasp, clench, bind, strangle.
fasguntide, ash-tide festival.
fash, tops of turnips, fibres of roots.
fash, trouble, anxiety, be troublesome.
fat.
fat, abundance, plenty, piece over.
fathom, a measure.
faud, a fold ; fawd, a bundle.
faugh.
faukun, falcon.
faut, to find out.
fauty, decayed.
feather, to fly with.

ât (fat), to build, form.
ât (fat), outcast, unclean, filthy.
paif, a gust of wind.
fekh, fulness.
âkh (fakh), reed, meadow.
âkh (fakh), to let fly, shoot.
fakat, the gathered result.
per, manifestly, apparently.
âm (fam), fist.
âm (fam), consume.
funa, sure, real.
ânkh (fankh), to clasp.
pena, to reverse, turn round.
âr (far), extremity.
ar, likeness, correspondence.
âr, game.
âr, to bear off, go along.
âp (fap), mount up, become tall, be elevated.
penka, to capture, squeeze, staunch.
âsha (fasha), applied to a festival.
âsh (fash), seed pods.
âsh (fash), cry, plaint.
âtu, fat, grease.
fat, a load.
fat, a measure ; am, belonging to.
paut, a circle, a company.
âuau (fauau), reproach.
âkhem (fakhem), eagle.
âtt, detect.
âut (faut), dead matter.
ppat, to fly.

ENGLISH,

EGYPTIAN.

fee, property.
fee, reward.
fek (Scotch), quantity, a number.
fekh, poke and other variants.
fet, fetched, borne, carried.
fetch, an apparition.

fie, a term of disgust, to reproach.
flie, bard or interpreter.
fine (Anc. Ir.), a tribal unit.
fingermell, a finger's breadth.
fire.
fob, froth.
fobble, quadruple.
fobedays, holy days.
fode, a youth.
farn, dirt.
fog and fogo.
fog, aftermath; feek, plenty, fulness; feckless, negative of.
food.
foot.
fote, a refrain.
fou (Scotch), to be elevated with ardent
  spirits.
fount.
foust, fogo, foul, fouty, a fool, roots in fu,
  pu and bu.
fouth, plenty.
foutnart, foumart, a polecat.
froth.
fuff (Scotch), to blow intermittently.
faff or five, as in the hand.

fi, his.
fek, reward, plenty, fulness.
fek, fulness, abundance.
fekh, to capture, denude, despoil.
fa or fe, to carry, bear, direction.
fetk, to sink, disperse, exterminate (make a
  ghost of).
fi, to disgust, repel.
pra, to interpret, make manifest.
fennu, an unit of number, a million.
meru, limit.
afr, fire, to burn.
ab (fab), to pour out.
aft (faft), four.
âb (fab), holy, pure.
at, lad.
fennu, dirt.
ákh (fakh), incense, censer.
fek, fulness.

put, food.
fut, separate, divided.
fut, a measure.
fu, dilation, ardour, large, extended, elevated.

fent, sign of the inundation.
fu, vice, fault, sullying; fu, dilation, large,
  be extended.
fut, load, measure.
futi, impurity, ordure.
pert, pour out, liquid, appear.
paif, wind, breath, gust.
f, hand.

G.

gaf, a sort of hook for catching eels.
gag, exaggeration.
gag, to hinder motion by tightness.
gaggles, nine-pins.
gair (Welsh), words.
gammer, the wife.
gammon of bacon.
gammon, to deceive, and many forms of pre-
  tending in a lowly posture; kham, to crook
  or bend; gammy, to "do the gammy" in
  begging.
gan (Irish), the little or young one.
gan, gons, or cons (Cor. Eng.), f. pudendum.
gant, a village wake; gantly, frolicsome.
gaoth (Irish), wind.
garden, enclosure.
garre, to chatter; garre, to chirp; gargate,
  the throat; garry-ho, loose language; ho,
  out of bounds.
garth, enclosure; garter, girdle, girth.
gash, cut.
gash (Scotch), talkative.
gat, a goat.
gat, a narrow passage; gut.
gate, road, difficult of ascent, cliff and coast
  roads.
gate, for shutting.
gauve, to stare.
gavel, a sheaf of corn before it is tied up.

kaf, hunt, seize.
kaka, to boast.
khekh, a prisoner's collar having nine points.

kher, word, voice, speech.
khem, rectum, rectrix, domus mulierum.
kamh, a joint of meat.
kamui, the lowly posture of the conquered,
  soliciting, imploring, adulating, adoring,
  bending.

han, the young one.
knau, f. pudendum.

khant, circumstance of a festival, joy, delight.
khet, to navigate, sail.
kerrt, zone, enclosure, en, to be.
kher, speech, to speak.

karr-t, orbit; kar-t, in-dwelling.
kas or kasha, cut.
kasau, tongue.
kah-t, a she goat.
khet, to enclose, ford, port.
katt, bad road.

khet, to shut.
khef, to look intently.
khef or gef, corn.

ENGLISH. | EGYPTIAN.

gavelet, a seizure of land. — **kafa,** to seize.

gay, joy. — **khaui,** joy.

ge, ye. — **ke,** thou, thee.

geasa (Irish), spells. — **khes,** a religious rite.

geb, to turn up the eyes in derision. — **kef,** some kind of a look.

geboned, polished. — **kabni, habni,** ebony.

geck or **gouk,** a fool, April fool. — **khak,** a fool; **keh-keh,** an old man.

ged, dead. — **khat,** dead body.

geese, a horse's girth. — **kes,** bind, tie.

gelt, deprived of testes. — **karut,** testes.

gereve or **greve,** a governor. — **kherp,** principal, chief; or **kherf,** his majesty.

gerse, grass. — **khersh,** a truss.

get, to catch. — **khet,** to net.

gib-fork, two-pronged harvest fork. — **kab,** double.

gig, cock-boat. — **kaka,** boat; **kek,** a boat.

gig, whipping-top. — **khekh,** whip.

gig, "the gig's up." — **khikhi,** to extend, enlarge, elongate.

gig, machine formerly used for winnowing corn. — **khekh,** fan.

gipseys, eruptions of water which break out suddenly on the downs of the East Riding, York, after great rains. — **kep,** the inundation; **si,** passing.

gis, an oath, "by gis." — **khes,** a religious rite; **kes,** to bind, covenant.

gise, to recline. — **kes,** to bend or lie down, be abject.

giss, the girth of a saddle. — **kes,** to envelope with a band.

git, yet, still, time, when. — **khet,** stop, when.

glamour. — **kra,** to lay hold, take, sieze, possess; **mur,** love.

glebe, sacred land. — **kherp,** consecrated.

globe, a sphere, model of the earth or skies, a symbol of imperial power. — **khreb,** the first form modelled as a sphere, the egg of Ptah; **kerp,** his majesty's sceptre.

gnar, as a dog. — **nar** (Heb.), to roar.

gnat, fly, anything small, worthless. — **nat,** small, enemy.

go, a measure. — **kha,** measure.

go. — **khu,** spirit, go, with whip sign.

goa, earth-goddess. — **ki,** land, earth, inner region.

goal, end of the course; **char,** a course. — **kar,** a course.

goat. — **kaht,** she-goat.

goats, stepping-stones. — **khet,** ford.

God, also **god's-good,** yeast. — **khut,** a spirit.

goge, throat. — **khekh,** throat.

goigh, very merry. — **khak,** to rejoice.

gome, a man; **gomman,** the father, game. — **khem,** he who has potency, virile power personified.

good, a measure of length. — **khat,** a dry measure.

gorsed (Welsh). — **kar,** course, circle; **keru,** word, voice; **set,** seat.

goud, the plant woad. — **kata,** young plants.

graf, first spadeful of earth. — **kherp,** first.

grape. — **arpe,** grape.

gruagach (Gaelic), Apollo. — **kheru,** word, Logos; **khekh,** light.

guare or **guary** (Cor.), a spoken play. — **kher** or **kheru,** to speak, utter a speech.

gue, quick to catch; **cue.** — **khau,** to seize.

guess, applied to barren cows or ewes. — **khes,** stop, stay.

guest, one who stops and stays. —

guide. — **khet,** navigate, steer, when, or when to stop.

gwes-par (Cornish), vespers. — **khes,** a religious rite; **per,** manifestation, go round.

gwiddion, British god; **cad,** a familiar spirit; **caddy,** a ghost. — **kata,** name of an Egyptian god; **khut,** spirit.

gwy (Welsh), sinuosity of shape. — **kai,** shape, figure.

H.

haap (Devon.), go back. — **happ,** to go backwards.

haben, ebony. — **habn,** ebony.

hack, place on which bricks are arranged to dry. — **aka,** dry up.

ENGLISH.	EGYPTIAN.
had, rank, quality.	**haut,** go first, precede.
hag, an enclosure.	**hak,** a fillet.
hag, witch.	**hek,** charm, magic.
ha-ha, of laughter in full tide of triumph.	**haa,** rejoice; **au,** long-continued triumph; **haa,** jubilation.
ha-ha, sign of rejoicing.	**haa,** rejoice.
hain, to spare, preserve, save.	**han,** for mercy's sake.
hak, serpent, snake.	**hauk,** reptile.
halo, a round of light.	**haru,** the circle of day, or round of light.
ham, a covering or enclosure; **home.**	**am,** tent, house, enclosure; **hem,** seat, place.
hammock.	**aam,** noose or sling; **akh,** up.
hanap, the San Greal cup was first called Hanup.	**han,** a vessel; **ap,** liquid, essence.
hand, the member.	**hunt,** the matrix or creative hand.
hand, a workman, performance.	**hanuti,** labourers.
hang, a tie, to stick to.	**ankh,** a tie, a noose, to clasp.
hang, a crop of fruit.	**hank,** an offering of vegetables.
hange, the pluck, heart, liver and lights.	**ankh,** symbol of life.
hank, knitted loop.	**ank,** clasp, a loop.
hansel, the money first appearing.	**han** or **an,** to appear; first sight of.
hap, to cover up.	**hap,** to hide; **hep-hep,** hide, screen.
har, light-ascending mist.	**hair,** fly up.
hard.	**kart,** stone; **ert,** the hard, that which retains the form.
har-har, shout of pride and pleasure used at witches' Sabbath (Henry More).	**har,** pride, pleasure.
harp.	**uarp,** to be joyful, charmed, delights.
harry, plunder.	**hura,** plundered.
haste.	**as-t,** hasten.
hasten.	**ustennu,** stride.
hasty.	**hasheta,** hasten.
hasty pudding.	**hastu,** some kind of drink.
hat, futuere.	**hat,** to espouse.
hat and ought, circles.	**at,** circle.
hat, to take off the; "**hat,**" to salute.	**hatte,** reverence, salute.
hathe, matted together.	**hut,** bundle; **hata,** mat.
hau, to the left.	**hai,** direction, go back.
have.	**haf,** to seize, possess, have; **afa,** to to be filled and satisfied.
havel, slough of a snake; **hoof,** goes on the ground; **havil,** a young crab; **ivy** or **ale-hoof; hafru** (Welsh), sluggish; **hafarch** (Welsh), listless, crawling along.	**hef,** snake; **hefu,** crawl; **hef,** crawl, squat, go on the ground as a snake, viper, or caterpillar.
hawbuck, a country clown.	**hau,** rustics; **bak,** labourer, servant.
hawid, hallowed.	**hatt,** reverence, salute.
haws.	**hau,** first-fruits.
hay, hail, stop.	**hai,** hail, stand.
hayne, park, enclosure.	**hain,** join, touch, near.
he.	**ui,** he.
head.	**het,** upper crown.
heaped.	**hept,** heaped.
hearse.	**hurs,** pillow or head-rest.
heart.	**artt,** milk, has the heart-sign; **aurt** or **haurt,** some substance.
heat.	**Heht,** goddess of fire; **uat,** mistress of heat; **ut,** fire.
heat, one course.	**hut,** hour or course.
heath, cover.	**hut,** shrine, secret enclosure.
heave, wave.	**hefa,** heave, crawl as the snake or caterpillar.
heaven, open.	**uben,** sunrise, shine, light.
Hebber-man, fisherman on the Thames below London Bridge.	**heb,** ibis, fishing, fisher.
heeze, to raise.	**hehs,** to raise.
hegar (Cor. Eng.), captive.	**hak,** captive.
"**heigh-ho,**" sigh of longing.	**uha-uha,** desire, long for, sigh.
height.	**akht,** height.

ENGLISH.	EGYPTIAN.

height.
height, to right, driving horses; **woot.**
heir, eldest son.
hell, a prison, Hades.
hem.
hemp.
hen.
henkam, henbane.
her, she.
heraude, herald.
herb.
here.
hero.

hal-t, ceiling.
ha-t, sudden; **aut,** go along.
ur or **her,** eldest child.
kar, a prison, under, Hades.
hem, border, frontier.
hema, hemp.
an, hen.
ankham, some flower, bud.
er or **ru,** she.
uarut, go, fly, carry on foot.
kherp, a first, chief, principal thing.
her, here.
ma-haru, the typical warrior, the true (**ma**) hero.

her-out or **erout,** for **without.**
Hertoga, an over-lord.
hessen, coarse flat woof.
heste, a command.
hete, to be named or called.
hetherims, rods twisted to keep the stakes of a hedge together.
hewt, high.
hey, hail, stand.
heyho, the green woodpecker, the long-tongued bird.
hey-lolli, a refrain.
hie, in haste.
higre (Welsh), boar, **pig.**
hingers, ears.
hingy, said of beer when it begins to work.
hip, hip, hurrah !
hipping-stones, for crossing a brook.
hips, seed of the wild rose.
hit.
hithe, an enclosed haven.
hobby-horse, ridden in the old mythical representations.
hob-nob, together, touch glasses, sign of conjunction.
hob-nob, a figure carried in the festival processions at Salisbury.
hocktide.
hocus-pocus.
hoddy, a net.
hoden, beaten.
hoer, she.
hoer (Cornish), sister.

her, with; **ut,** out.
her, over; **teka,** a fixed frontier.
ushen, net.
hes-t, order.
het, to consecrate.
hetar, to compel; **atr,** limit.

ati, hills.
hai, hail, stand.
hu, tongue.

heloli, mad, frantic.
hih, seek, search.
hekau, a pig, boar.
ankh, ear.
ankh, life, living, alive.
hep, unite, join together.
hepti, an ark or cabin for crossing the waters.
hebs, corn seed.
hit, strike.
hut, shrine, secret enclosure.
hebi, to celebrate the triumph of return; **heba,** joy, pastime.
heb, a festival; **nahp,** conjunction; **neb,** all together.
neb, lord of.

hak, a festival.
huka puka, magic and conjuring.
aat, a net.
hutn, strike, beat, smite.
ur, she.
hari, the name of Isis and Nephthys as the sisters.
heka, sow.

hog.
hoit, to indulge in riotous mirth.

Hathor, goddess of the dance and merry-making.
hanti, the returner that goes to and fro.

hondey, Lancashire name for an omnibus; **handy-dandy,** swinging to and fro on the **hands.**
honour.
honour, obeisance.
honor, a title, Norman fief noble called an "honor."
hoo, hunting cry; **woo, how? hey?**
hood, hut, hide.
hood, the raised crust of a pie.
hook.
hoop, to hide.
hooro (Irish), cry.
hop, plant.
hope.
hoppe, dance.

hon, majesty, sanctity, royal; **ar,** being.
han, to adore.
anr, hail; **anr,** a title, door-opener.
huh, to seek after.
huth, table cover.
hut, upper, height, the upper crown.
hek, hook.
hep, to hide; **haup,** spy, deserter.
hru, to cry, call.
ap, mount, rise, fly on high, climb up, tall.
ap or **khepr,** the beetle, sign of hope.
hep, festival.

ENGLISH.	EGYPTIAN.

hoppo, almost anything relating to the custom house laws in Anglo-Chinese.
hapu, laws.

horse and **arse**.
ur, principal; **ur**, to carry or bear; **ar**, fundament; **s**, he, she, it.

hot.
ut, glow; **haut**, fire.

hot, ordered; **odd-fellows**, an order
hut, order.

hot waters, spirits.
hut, spirit, good demon.

hot.
hut, onion; **hti**, consume.

hoy, a boat.
hau, transport boat.

huath (Cor. Eng.), fresh, anew.
uat, green, fresh, colour.

hucksy-bub, a name for the female breast.
akh, sustenance; **si**, child; **bub**, well or source.

hûd, mystery.
hat, terrify, fear.

hud (Welsh), charm, spell.
ut, magic.

huey, tramps' term for a town or village.
hui, a limit, boundary.

hufen (Welsh), cream.
hebnn, honey, conserves, &c.

huff, bully, hector, swagger, scold.
ufa, grasp, chastise.

hukni, (Eng. Gip.) art of fortune-telling.
heknu, supplicate, invoke, discourse, address.

hum.
hem, locust.

humstrum, female emblem.
hem, female emblem.

hunch of bread, lump.
uns of bread.

hunger.
hun, food; **kher**, take, seize, claw, reach after, claw hold of, fight for, have, have food.

hurdle.
kart, orbit, enclosure.

hurrah.
hurahu, courage, prevail over.

hurricane, terrific wind.
heru, terrific, terrifying; **khen**, blow, puff away, typhoon.

hurry.
uaru, go, fly.

hurt, blue in heraldry.
hert, above, heaven.

hush.
usha, nightfall.

hush, to wash ore.
ash, to wash.

hut, abode.
hat, abode, habitation.

hymn, to hymn.
ham, to invoke with religious clamour.

hyps, blue devils; **hypo**.
habu, haunt; **habau**, infest.

I.

inch, an island; **ing**, a field, enclosure; **ingan**, onion (clasped round and round); **inkle**, tie.
ank, clasp, a tie.

ink.
nâ or **nak**, ink.

ior, circle of the sun (Welsh).
har, day.

is.
as, is, it is.

ith (Irish), corn; **yd** (Welsh).
hit, corn.

ivin, north.
kheb, north.

J.

jads or **jouds**, rags.
uat, rags.

jock, to enjoy.
hak, a festival.

jowser, diviner for water with a rod.
user, a sceptre, the rod of authority.

joy.
khaui, joy; **ahaui**, cries of joy.

K.

ka, say.
ka, say.

kach, evacuated.
ka, evacuate, foul.

kae or **kye**, cow.
kaui, cow.

kaeawk (Welsh), Druidic wreath of beads.
khakr, adorn; **khakri**, necklace.

keb, name for children's crying.
kab, to be weak, feeble, wretched.

ENGLISH.

keb, a villain (in the modern sense); **kheppen**, to hoodwink.

keeche, to catch; **keck**, to choke; **keckcorn**, wind-pipe.

Kéd, British goddess.

keen (Irish), to recite the virtues of the dead.

kell, a welling water.

kep, to lie in wait.

keeper, a clasp; **kep**, to catch; **cap**, a shepherd's dog.

kemp, champion.

ken (Cor. Eng.), hollow.

ken, to know, to be acquainted with.

kensh, to shut up close.

Kent, called the garden of England.

kep, to catch or enclose.

kepe, to meet.

ker, business, occasion.
   " An hundred knights good of **ker**,
      Her better no man wepen ber."
            Guy of Warwick, p. 68.

kesh (Irish), kind of causeway made of wickerwork.

kex or **kecksie**, the dry stalk of withered hemlock.

kecky, for kecksie.

kest, a twist or knot; **kish** (Ir.), wickerwork.

ket, filth.

ketter, to diminish in size.

kevin, lower part of the round of beef.

kewte, to kitten.

ki, quoth.

kick, to sting.

kick the bucket. " The pitcher broken at the fountain."

kick ; kick, rebound of a gun.

kick-up, sort of balance used for weighing half-pence in the 18th century.

kid, of pease or beans.

kiddier, a butcher.

kiddle, a weir in a river with a cut to catch fish.

kill, to strike, to slay, to knock on the head.

kiln.

kime, a silly fellow.

king.

kisting, a funeral; **kist-vaen**.

kitchen.

kite, belly.

kith, knowledge, also a region.

kitten, young of the cat.

knap, to snap, to talk snappily, to browse.

ku-ku (Ir.), sacrifice.

kuf (Cor. Eng.), wife.

kuf (Cor. Eng.), the wife; **chavi** (Eng. Gyp.), girl, daughter.

ky or **chy** (Cornish), house.

kye, cows.

kyle, a vassal or serf.

kylle, to strike, strike off, or cole.

kyphor, to copulate.

EGYPTIAN.

kheb, hypocrisy, deceit, disguise, violate, change.

keks, to bind; **khekh**, collar, throat, gullet.

Kit-mut, a goddess; **Khept**, goddess.

khen, to tell, convey, be agitated, act of offering.

karua, water welling.

kepu, to be hidden, lie in wait.

kep, to seize, catch.

khem, to prevail, be master of.

khen, interior, hollow.

ken, accompany.

kensh, snap, extort, hunt.

khent, garden.

kep, closed hand, fist; **kep**, to receive.

kar, business, to bear battle-harness; **kar**, power, property.

khus, found, lay foundation, construct, pound, *i.e.* in road making, with a man pounding.

kehkh, the old (man); **si**, pass away, corrupt, decay.

kah-k, reeds.

kes, to bend, to bind.

shet, sin or crime of some kind.

ket, little.

khept, thigh, hind quarter.

khut, sign of maternity.

ka, say.

khekh, to sting.

kek, to break.

khekh, repulse.

khekh, balance.

khet, enclosure, shut, sealed.

ket-t, a butcher.

khet, to net.

kher, an animal going to be killed. ; **kar**, to strike.

kar, a furnace.

kemh, to stare.

ank, the king.

kes, embalmment, a funeral.

khet, small ; **khet**, building ; **khen**, interior, inner part.

khat, belly.

kaat, wisdom ; **khet**, circuit, enclosure.

katen, image, likeness.

nehp, to seize.

khu, ceremony, benefit, spirit ; **khuu**, sin.

kef, hinder, feminine half.

kefa, the genetrix.

ki, an abode.

kaui, cows.

kheri, a bound victim.

kar, to do battle with, sign of striking.

khepr, to generate.

ENGLISH.
　　　　　　　　　　　　　　　　EGYPTIAN.

L.

ENGLISH.	EGYPTIAN.
la, lah or **lack**, formula of exclamation, verily.	**râ**, formula, verily.
ladder.	**rat**, steps.
lage, to wash ; also **lye**, to wash.	**rekh**, to full, wash, purify.
lara, a round piece of wood turned by a turner ; **larabell**, the sun-flower that turns round.	**rer**, to turn round, be round.
latch, to invite, entreat.	**rtu**, urge.
lathe, invite.	**reti**, beseech, ask.
lawter, 13 eggs to set a hen.	**retar**, entire, a total ; **ret**, set.
leach-brine, purified brine.	**rekh**, to wash and purify.
leash, three dogs or partridges.	**resh**, written with three feathers.
leche, a physician, to heal.	**rekhi**, mages, knowers, doctors.
lede, lude, lithe, leet, people.	**ret**, race, mankind.
leek, onion.	**rekh**, heat, to be hot.
leiths, joints in coal.	**retha**, to quarry.
leits, footsteps.	**rat**, feet, steps.
lesse, to teach, lesson ; **llais** (Welsh), voice.	**res**, tongue.
let, causation of any action.	**rta**, cause to do.
letter.	**ret**, engrave, figure, write ; **teruu**, papyrus roll.
liege, applied to a king.	**rek**, to rule.
linn (Keltic), a deep still pool.	**renn**, virgin-pure.
liss, pleasure, joy.	**resh**, joy.
llan (Welsh), an enclosure or circle.	**ren**, an enclosing ring.
lloer (Welsh), moon.	**rer**, to make the circuit, grow round.
llyr (Welsh), sea beach.	**rer**, go round, surround.
llyther (Welsh), letter.	**ruit**, engrave, figure ; **rat**, engrave, letter.
lly-thrau (Welsh), Druidic signs cut in wood.	**rui**, scribe's palette with pen and ink ; **teruu**, a roll of papyrus.
loll, to fondle.	**rur**, to dandle.
look.	**ukha**, to seek.
loon, a boy.	**renn**, a boy, nursling.
loop.	**arf** or **arp**, to bind, tie round with noose.
lope, a **dog-lope** is a boundary between two houses and belonging to both.	**rupu**, either, or ; **takh**, frontier.
lull, to dandle a child.	**rer**, to nurse and dandle a child.
lum (Scotch), chimney.	**rem**, the erect ; **ram**, throat.
lycced tea, tea with spirits in it.	**rekhi**, spirits.

M.

ENGLISH.	EGYPTIAN.
ma (Irish), mother.	**ma**, mother.
mache, to match.	**mak**, match.
macks, sorts ; **make**, to mix ; **makke**, a mixed dish.	**maku**, mixed.
make.	**mak**, to control.
make.	**mâk**, work, inlay, composition.
make, mate.	**makh**, pair of scales, balance.
make, to rhyme.	**mak**, regulate, think, balance, measure.
Makhir, the divinity of dreams.	**ma**, true ; **khir**, word, speech, revealing, imaging, picturing.
mamma, the mother.	**mâmâ**, to bear.
man, man-ful.	**men**, to be resolute.
man, to man, the man.	**men**, the fecundator.
man-cowe, baboon.	**kaf**, a monkey.
mane.	**mennu**, some form of hair.
Manx arms, three legs revolving, counter poise.	**maank**, a counterpoise.
marge, margin.	**meri**, a border, a margin.
marish, marygold.	**maresh**, yellow, being yellow.
married.	**mer-t**, attach, attached.
marrow, mate.	**meru**, beloved friend.
mart, cow-fair.	**mer**, cow.

mash, to steep.
mason.

mast, nut of the beech.
mat, for the feet.
mate.
matere, womb; **mid, middle.**

matins.
matly, equal, true, like.
mattachin, ancient sword dance.
matted.
matter.
matty, equal, alike, corresponding, **mate.**
May, the month; you "**may.**"

mayor.
maze.
mead.
mean, resolve, intend.
meare, a boundary; **mur** (Welsh), circle, Stonehenge; **mill**, a round.
measure.
meat.
meer, a measure, (Peak of Derbyshire).
mehaun, the devil, the old **mehaun.**
mell, to wheel round.
mell, futuere.
mell, a mallet, hammer.
meni, stone monument.
mer (Eng. Gipsy), to die.
merkin, pudendum f.
meskins, is an exclamation, "By the mass"; **mass**, wafer.

mich, to act by stealth, to steal.
miching, false show.
middle.
might.
min (Cor. Eng.), brink, border, or boundary.
mine, to establish, to penetrate.
ming, to mix, to knead; **minginator**, one who makes fretwork.
minne, to remember.
minny, mother.
minnying, periodic memorizing.
minster.

minute, sixty seconds.

mirror.
mist.
mix.
moarg, to burn without flame, dead fire.
moat, water enclosure; **mota** (Irish), earthwork enclosure.
mob, a slattern, a sloven; **mobile**, the mob.
mock-shadow, twilight.
mog, to enjoy one's self.
mohyn (Welsh), bull.
moke, the mesh of a net.

mommy, all of a mass.
monument.
moor, to void blood.
moor, farm-bailiff.
moray, coast-country.

mas-mas, to steep.
ma, true; **sen**, brother, to fraternize, found, build, establish.
mast, born, produced; **mast**, seed.
mat, soles of the feet.
maht, to agree.
mat, mother; **mat**, middle, **matr**, centre, to centre.
mate, sing, praise.
matt, true, right, like, according to.
matai, soldiers; **ken**, to dance.
matet, unfold, unwind.
matrut, soil, stain; **matr**, a marsh.
mati, ankles or feet (a pair).
mâi, seed, germ, growth, renewal; **mai**, come, invitation.
mer, prefect, governor.
mes-mes, confusion.
meh-t, some kind of libation.
men, resolute.
mer, a boundary, a circle.

maser, an unknown measure
mat, some kind of food, dead.
mahar, a quantity; **mer**; a limit.
mehaun, serpent.
mer, circle, encircle.
mer, to love, attach, kiss.
merh, club.
men, an obelisk
mer, to die.
mer, circle; **khen**, inner.
meskin, place of new birth; **mes**, kind of cake, i.e. the mass-wafer or mass, the cake of new birth or birth cake.
makhau, to kidnap.
maka, an artifice or artificial.
mat, middle.
mat, granite; **mata**, backbone, phallus.
men, go round, that which goes round.
men, to establish; **min**, plodding penetration.
menkh, to fabricate, form, work, create, make, a mallet.
men, memorial.
mena, typical nurse or mother.
men, to go round, perambulate.
mena, the dead; **ster**, couch of the laid out dead.
min, a measure; **at**, a second of time; **menut**, a measure.
mher, mirror.
mes or **mest**, product of a river, river-born.
mak, mix.
merau, die.
mehat, enclosure, sepulchre.

mhubi, humble.
mak, mixed.
mak, to dance.
men or **min**, bull.
maka, a reel, some implement used with a net.
mem, pitch, wax, **mummy.**
men, to fix, monument.
mer, to die, end.
mer, superintendant.
mera, land, limit; **meri**, bank, shore.

ENGLISH.	EGYPTIAN.
mormaers, pictish governors.	mer, prefect, superintendant, monk; mer, circle, limit.
	mera, region, limit, boundary.
morrow.	mer-t, beloved person, attached to; mar-t, a female relationship or office.
mort (Gipsy), female mate, doxy.	
motion.	mat, to go: maten, to go, move away, give way to.
month.	ment, a total of two halves, as a lunation; men to go round.
mue, to change, to mue the feathers.	mu, to end, die.
mugget, a crispy ruffled shirt.	maku, some linen object, (t, is a terminal).
mul-berry.	mer, mulberry-tree.
mull, cow.	mer, cow.
mum, silent.	mum, dead.
mumm, to make up a character, transform.	mum, mummy, an image of transformation.
munch.	munka, to work, form, make.
munte, to measure out.	mant, a corn-bin; men, a liquid measure.
mur (Welsh), precinct.	mer, precinct, enclosure.
mush, anything mashed, as wetted meal.	mussh, mud
must, new wine.	mustum, an intoxicating drink.
mute.	mut, silent.
"Oh, my," wonder, surprise.	mahui, wonder, admiration.
my.	maaui, in the power of.

N.

nabbed, caught.	nabt, tie, plait, noose.
nac-an (A. S.), to slay.	nakan, slaughter.
nadredd, Welsh Druids.	nater, divine; at, father.
naf, pudendum.	Nep, goddess of seed.
nake, to make naked, known.	nakhem, know.
nape, to cut a hedge, for renewing or forming anew.	nahp, to form.
nart, a birth cake.	naar (Heb.), a newborn child.
nase, drunken.	nasp, delude, stupify, numb, i.e. drunk.
nash, firm, hard, gnash.	nash, strong.
nasty.	nah, foul; sti, stink, stench, smell, offend.
nasty, spiteful.	nashti, plague, torment.
nat, a kind of mat.	natt, shuttle.
nat or net (Ir.), little.	nats and netiu, little.
Nature.	Natr, goddess, time, season.
navel, first breathing place of the embryo.	nef, breath.
navigate, navy.	nef, sailor, to sail.
neart, night.	narutf (for anrutf), the sterile, infertile region.
neat, cow.	Neith, cow-headed goddess.
neb, to kiss.	nahap, to kiss.
nedder, an adder.	neter, a serpent symbol.
nedder, inferior.	neter, servants of priests.
need, anciently note.	neht, wish, request, vow; net, address, save, help.
neigh or nicker.	neka, provoke, incite, sign of male power.
neist, nearest, soonest.	nas, near, close, after.
neit (Irish), god of war.	netr, gods.
neme, care, to take care.	nemm, to take.
ueme, mate, associate, be near.	nem, to join, accompany, engender.
nemly, quickly.	nem, some kind of motion.
nen, an English river.	nen, water, inundation.
nene, neither, none.	nen, no, not, without.
nenet, will not.	nen-t, no, not.
nesh, applied to strong cheese, "choice of change" (1585), too nesh.	nash, strong.
ness, a height, jutting, or promontory.	nas, a tongue, out of, pedestal, upper crown.
net, to make water.	natra, maker; netur, water of the west.
net, total, profit.	net, all, total, limit.
nettle, a rope's end.	netur, to pull a rope; nnutt, a rope.

ENGLISH.

EGYPTIAN.

neuf, blaze.
"neus the matter," something near or like the thing.
nev (Welsh), heaven.

nevydd (Welsh), celestial lord.
newed, changed, renewed.
nice.
nick, to deceive, cheat.
nicka, a siren; nicka-nan (Cornish), night, when boys play tricks on the unwary; nickhem (Scotch), applied to a child.
nidde, to compel.
nide, of pheasants.
nig or nig-nag, coitus.
nigging or knocking shop.
niggle, dawdle, trifle.
night.
nim, to snatch; nimming, stealing.
ninny-watch, a vain hope.
ninted, perverse.
nirt, cut, gelt.
no.
noble, nab, or nob, head: nub, a husband.
nod.
nog, strong ale; nogged, strong-limbed.
nome, taken.
nominy, a public speech in "rough musick-ing."
none, early form nen.
noose or knot.
north.
not.
not, knot, naite, to deny.
not, a name for bandy; "out of notch," out of bounds (Eng. saying).
note, the time during which a cow is in milk.

now (1 Corinth. xiii. 13).
nowed, in heraldry for twisted or tied.
nowte, black cattle.

nub, nape of neck; nur, head; nare, nostrils of the hawk; norie, the nurse.

nuirt, an Irish amulet ring.

number.
nunt, to be obstinate or sullen.
nurt, Scotch birth cake.
nut, a small vase.
nyflo (Welsh), to snow.
nytte, to require.

nefer, heat, fire.
nas, near.

nef, spirit of the firmament; nu or nupe, heaven.
nef or neb, lord; it, heaven.
nnut, sweet, fresh.
nas, proclaim.
neka, delude, be false.
neka, delude, play false, provoke: nan, little boy, a ninny; nekhen, child; nakhem, knowing.
nat, submission.
net, a quantity, total, a collection.
nak, coitus.
nak, fornicate.
nikau, idle, lazy, dawdling.
akht, light; n, no, not.
naham, to take away; nam, forced.
nen, no, not, negative.
nent, ignorant, fools.
narutf, barren, sterile.
nu, no.
neb, lord, consort.
net, incline, bow, salute.
nakh, strong, strength; nakta, a giant.
nam, forced.
num, speech, utterance, tongue.

nen, no, not, without.
nuut, a rope; nuuh, knot or to twist.
narutf, name of the north.
neti, no, not.
natt, impediment.
nnu, go hither and thither; natsh, tie, attach.

net, all, total; neith, the cow-headed Isis as the nursing mother.
nnu, time appointed, this time, continually.
nuut, rope, tied.
neh, black; neit, goddess, the black Isis, cow-headed.
nahb, neck; nar, head of the vulture; nrau, nostrils of the vulture; nar, the vulture, emblem of Mut, the nursing mother.
nar, victory; ut, magic, inscription (with a tie); nnuh, knot, twist; ret, engraved.
num, to see, perceive, repeat, again, twice.
nent, ignorant, fools.
nehar, some kind of bread.
nu, the vase or jar sign of water.
nef, breath.
net, tribute, collection.

O.

ocub, cockchafer.
oddy, the snail that drags its abode also called the oddy.
of.
off, the line from which boys begin in playing a game of marbles.
ogha (Gaelic), prefix to names; Irish O or H.
ogham, monuments.
oh!
oil (by permutation).

khop, beetle.
athu, drag, draw, an abode.

af, born of.
af, born of.

akhu, illustrious, honorable.
aukhem, indestructible.
uob, very much, increase, augment.
ur, oil.

ENGLISH.	EGYPTIAN.

old (by permutation). — **urt**, old, oldest.
oned. — **unnut**, the double crown, twinned or oned ; **unn-t**, the two crowns in one.

oned, united, made one, one hour. — **unnt**, one hour, marked by the circle.
oo, one. — **ua**, one.
oont, want ; **oonty**, empty. — **un**, wanting, defect, open.
oose, mud. — **ush**, mud.
ope, open. — **ap**, opener ; **uben**, sunrise, shine, light, the opener.

orb. — **arf**, to bind, bundle.
orchard. — **arr**, fruit ; **karr-t**, orbit or enclosure.
orp, cattle. — **rep**, the beast.
ouche, a clasp. — **uskh**, a collar.
ought, should. — **haut**, ought, should.
ouris (Irish), management of cattle, flax, &c. ; a gathering of girls at one house to card wool or spin flax. — **ursh**, watch, rigid, observe, be attentive.

out, also **ut.** — **ut**, put forth, out.
oven. — **kafn**, an oven.
over. — **apheru** or **aper**, the place of crossing over ; **ap-her**, up over.

P.

pabo (Welsh), producer of life ; **papa**, father. — **pep**, to engender ; **papa**, to produce.
pad, to make a new path in walking. — **pet-pet**, trample.
pad, foot, pettitoes. — **pet**, foot, claw of animal.
page, one side of a leaf. — **pkhkha**, divide and divisions.
page, a boy-servant. — **bak**, servant, labourer.
paint. — **pant**, "all the colours of pant" (F. R. ch. 15).
pair-Keridwen (Welsh), Keridwen's vessel or vase. — **par**, a **pail** (pair interchanges with pail).

papa, pope, the father. — **apa**, ancestor, head, God ; **p**, the.
par, to enclose, as in a house or **paryard**, or with a **parapet.** — **par**, to surround, enclose, as in a house.
parish. — **par**, to go round, surround ; **sh**, space, or measure of land.

parson. — **par**, show, explain ; **sen**, a brother.
pasgadwr (Welsh), the feeder ; **cater-cousin**, one that is fed. — **pes**, food ; **ketr**, occasional.

pass. — **bes**, pass.
passe, extend. — **pesh**, stretch, extent.
past. — **past**, back, behind.
pat, a blow. — **pet**, strike.
path. — **pat**, a course, path of the sun.
patter, to talk, discourse, a judge's summing up, a broad sheet. — **ptar**, show, explain, a slip of papyrus ; **pth**, open-mouthed ; **ar**, to be.
patticake. — **pat**, cake, kind of food.
patty. — **ppatie**, cake, kind of food.
paup (Welsh), everybody, the whole race. — **på (pap)**, men, the race.
paw, fie ; **pah**, a monitory exclamation. — **pa**, an exclamation.
pay, satisfaction. — **peh**, function, arrive, attain, reach.
pay, with pitch. — **peh**, to follow up, penetrate.
peace. — **Pash** or **Pekh**, is the bringer of peace.
peach, a cloven fruit. — **pesh**, a fruit, cloven.
peck, victuals. — **beksu**, something eaten.
peck, a mattock. — **pekh**, to divide.
peck, food, victuals. — **pekha**, kind of food.
peck, a measure. — **peka**, peck-measure.
peg, a division, "take down a peg," clothes peg. — **pekkha**, divide, division, divided ; **peka**, gap.
peg. — **uakha**, peg ; **peka**, to divide ; **p**, mas. art.
pega (Cor. Eng.), to sting or bite. — **peshu**, sting, bite ; **pesh** permutes with **peka**.
pegma, bill of advertisement fixed up at ancient pageants. — **p-ka-ma**, the call to come, see.

pogma, a moving pageant.

pes. "My Gammer set her down on her pes" (Gam Gurton's needle, 2, 12).

pie.

piece.

pink.

pit, pat, pot, patty.

pobi (Welsh), to bake; pobs (Craven).

poet.

poop, let fly.

por (Irish), seed.

pore, cram with food, pour.

port, appearance, portment, porting the ensign.

port, harbour, storehouses, warehouses.

posh (Eng. Gyp.), half; pease.

poss, a waterfall.

posse, a number of people, a following.

post, stay, support.

pour.

power.

precept.

pree (Scotch), to taste.

pref (Cor. Eng.), a worm.

pren (Welsh), plant, tree; p is the Egpt. art.

puck, hobgoblin.

"pucker up."

puff or fuf, to blow.

puffs, light pastry.

pup, to produce.

put or puttoch, the frog.

put-pin and push-pin, one and the same in Eng.

---

peh, glory; khema, shrine.

pest, back, pelvis.

ppaui, a circular cake, or bread.

pekh, a division.

penkau, to bleed.

pet, a circle.

på (paf), to bake; pefs, to bake.

paut, type, image, figure; the Artist.

pepe, fly.

per, seed.

per, food, pour out.

per or pert, to show, sight, see, explain, appear.

per-t, granary, storehouse, grain, proceed, emanate, come forth.

pesh, to separate, halve, in two halves.

pesh, water.

pesut, followers, behind.

pest, back, spine.

per, to pour out.

ber-ber, fervour, ebullition.

sept, precept.

pre, show, see, perceive, explain.

ref, a worm; with the article p, pref.

renpu, plant, branch.

puka, magic, infernal locality.

puka, to divide.

pef, breath.

pefss, cook, light.

pep, to engender; pa-pa, produce, deliver, give birth.

Put or Putha, the frog-headed god.

put and pesh, both mean to stretch.

Q.

quash, old form cass, to stop, make null.

quat, a diminutive person, a pimple or rising-up spot.

queate, peace, quiet.

quede, evil, the devil

queek, to squeeze, pinch.

quiche, to move, kick.

quick, go fast.

quick, living.

quim, a feminine name.

quoy, enclosed land.

---

khes, stop, turn back.

kett, little; khut, the horizon of the resurrection, i.e. place of rising up.

kaut, rest; khat, corpse.

kheft, evil, the devil.

keks, to bind, entreat.

khekh, whip, repulse.

khi-khi, move with rapidity, be quick.

chich (Heb.), life.

kim, a female name.

ki, land enclosed.

R.

race.

raise.

rake, to deviate from a straight line, also ruck up.

raking coal, to keep in the heat.

ram, to lose by throwing out of reach; rame, to stretch.

ran, a noose or hank of cord.

rate, ratified.

rathelled, fixed, rooted.

ray of sunlight; ra, roe-deer.

ray-grass or rye-grass, called "ever."

---

rekh, race of people.

res, to raise up.

rukata, curve.

rekh, brazier, heat.

remn, extent, as far as, up to, extending to.

ren, an enclosing ring.

rate, tie, bind, make fast.

rat, fixed, rooted.

ra, sun, also means go swiftly.

ra, time.

ENGLISH.	EGYPTIAN.
re, again, to be repeated.	**re**, to, and to be.
rea (Gaelic), rapid.	**rua**, rush, go swiftly.
reach, the extent, measure, reckoning.	**rekh**, to reckon, know.
reap, a bundle, corn for binding.	**arp**, bundle.
rear.	**rer**, nurse a child.
rear, to expectorate.	**rir** (Heb.), spittle.
reck, to calculate.	**rekh**, know, reckon.
red, warn, advise, counsel.	**ret**, urge, beseech vigorously.
red, stark, entirely.	**retr**, entire.
reed, a water plant.	**ret**, a plant.
reek, smoke.	**rekh**, brazier, heat, emitting smoke.
reme, cry, moan.	**rem**, to weep.
reme, to froth up.	**rem**, to rise, surge up.
rene, to tie up, rein.	**ren**, noose for an animal's foot.
rene, a water-course.	**ruan**, a water-course, the gorge of a valley.
rene, to comfort.	**renn**, nurse, fondle, dandle a child.
rere, egg boiled but not set.	**rir** (Heb.), white of egg, spittle.
rere-mouse, the bat that flits round and round.	**rer**, circuit, go round.
resh, fresh, recent.	**resh**, joy, feathers.
rest, last.	**res**, absolutely, entirely.
rest.	**urs**, pillow or head rest ; **urst**, is pillowed ; **r-hest**, in place, seat, or throne.
retain.	**tenn**, hold.
rhiw (Welsh), a cliff.	**rru**, steps, ascent.
ric, kingdom.	**rekh**, people of a district, mankind.
ricky, masterly.	**rekh**, the wise man, the magi.
riddle, the ring to which the neckrope of an animal is made fast.	**ret**, a tie with a ring and stake for fastening animals.
rids, shoots or rays of sunshine.	**rat**, germs, shoots.
rie, to sieve corn.	**rii**, powder or dust with determinative of corn.
rike, to govern, to rule.	**rek**, to rule.
ring.	**ren**, a ring for enclosing names.
ring, rink, round, rounded.	**ren**, an enclosure or ring.
riot, a tumult, to be in a whirl.	**rruit**, whirl.
riote, a company of men.	**rrut**, those round.
rip, a lean animal.	**rep**, a beast.
rivo, drinking shout, Bacchanalian exclamation, Marston's " What you Will " (Act IV.).	**ravah** (רוה), Heb. made drunk, soaked or drunk, drunkenness, abundantly satisfied (Is. xxxiv. 7 : Prov. vii. 18) ; **repi**, of the Nile inundation.
riw, set, both.	**rehiu**, twin.
rixy, quarrelsome.	**rekai**, rebellious, scornful, culpable.
road, route.	**rut**, footstool, steps, feet, go out, gate.
road, at the mouth of a river.	**ruhaut**, a river mouth.
roar.	**ruhar**, superior mouth, authority.
roath (Cor. Eng.), form, figure.	**rut**, engrave, figure, form.
roke or **reac**, to cleanse armour by rolling it in a barrel of sand.	**rekh**, to wash, purify, cleanse.
root.	**rat**, plant, grow.
rote.	**rut**, to repeat.
rote, wheel.	**rrut**, whirl, wheel round.
rother, strong manure for forcing plants.	**rut**, to renew.
rother, a sailor.	**rut**, the Egyptians ; **urt**, the water-horse.
rout, green or yard for strayed beasts.	**rut**, tether of a beast whilst grazing.
routen, put to rout.	**ruten**, attack.
rowth, plenty, lush.	**rut**, prosper, vigorous.
ruck, to crouch out of sight.	**ruka**, to hide.
run or **rune**, name of alphabet, means of naming.	**ren**, name ; **rennu**, to name.
runt, an ox or cow given over breeding.	**renn**, cattle.
rush, a merrymaking.	**resh**, joy.
rut, dashing of waves.	**rut**, repeated, several.
rut, tracks cut by wheels, also to cut those tracks.	**rut**, to engrave ; **rut**, carved stone, to retain the form.
rutting stags	**rut**, to sow, germinate, renew, root.
Rye (Eng. Gip.), a lord.	**Ra**, royal, Pharaoh.

S.

sab (Irish Kelt.), counsellor of state.
sabbed, saturated.
sabre.
sack.
sack, be discharged, get the sack.
sack, a drink.
sad, heavy bread.
sad.
sage.
sage, saghe, speech ; **sager**, lawyer ; **sathe**, reason ; **say, seghe**, saw ; **segge**, to say ; **sacrament, sacred.**
sag-ledge, diagonal cross-bar of a gate.
sain, to make the sign of the cross.
saint, a cincture or girdle.
saite (Ir.), knowledge.

sam, to stand "sam," pay the reckoning.
samach (Ir.), happy.
same, likeness.
sand.

sandy, red ; **sang**, blood (**sang** is it—true it is, the probable origin of the oath "bloody ").
sane, sanation, curing, healing.
sanitary.
sap (Eng. Gip.), snake or serpent.
sap.
sapy, tainted.
sard, futeure.
sas (Eng. Gip.), a nest.
sash.
sasin, a reaping hook.
sasse, a river lock, a flood-gate.
sawed.
sax, a knife; **sock**, a ploughshare ; **sikis**, a scythe ; **seghe**, saw ; **sickle**, for cutting ; **seg**, a castrated bull ; **saugh**, a trench or channel ; **sice**, a gutter or drain.
say, the **say** of it, knowledge.
say, saghe, speech.
scabby.
scam, to stain ; **scummer**, to daub.
scambler, a parasite.

scan, scrutinize.
scap, a snipe.
scar, cliff, precipice, inaccessible.
scar, a piece or shred.
scare.
scarifying, called "sacrificing" (Brand, ii. 362).
sceat, a quantity or share.
scent, a descent ; **scent.**

school.

scimminger, base money rubbed over with silver.
scion, a shoot.
scochon, an escutcheon ; **scog**, to brag.

scote, to plough up.
scout, to field at cricket, a corner of a field.

sab, a counsellor.
sab, drink.
sapara (Ass.), sabre ; **khepsh** (Eg.), scimitar.
saka, sack, to sack a town.
suakh, cease, stop.
sakabi, a drink.
sut, a kind of bread.
saat, grief.
skh or **skha**, a scribe.
saakh, scribe, write, letter, influence ; illuminate, depict, symbolic eye.

sah, a cross strap.
senn, to found with sign of the cross.
shent, a circular apron.
sa, genius of wisdom, epithet of Taht ; **sai**, to know.
sam, to eat, drink, enjoy ; **sam**, total.
smakh, to bless, blessed, rejoice.
sem, emblem, similitude, image.
shant-bu, sand of the desert ; **bu** is the place.
sen, blood, red.

san, physician, healer ; **san**, to heal.
sannut, bath, medicament, or healing.
sep, the snake ; **sab**, serpent.
sefa, humidity.
sep, corrupt.
sart, to sow seed.
sesh, nests.
sesh, ring, roll, something twined round.
shesh, harvest.
sesh, draw bolts, open, a gate, pass.
ust, sawed.
sekh, to incise, cut, cut out, divide, sever.

sai, know.
ska, write, order, letter, depict.
skabui, eject, weaken.
skhmau, to paint.
skam, to stay, pass a time, dwell, remain, hang on.
skhan, recognise.
skab, to double.
skaru, a fort.
skar, cut in pieces.
s'har, scare.
skar, a sacrifice.

skhat, a quantity.
senti, to found ; **sent**, incense ; **senta**, to smell the earth, *i.e.* make obeisance to.
skher, instruct, plan, design, counsel, picture.
skhemau, to paint, to represent.

skennu, to multiply.
skhkha, to write ; **skhakr**, to emblazon, embellish.
ska, a plough.
skhut, field, a region.

scrag, to hang.
scrit, a writing, a deed, writ, act or deed.

scrow, a writing (in Dives and Pauper, 1493) ; scrowe, a scroll, a charm.
scut, a hare, also hare or rabbit's tail.
scutch, whip, strike ; scutcht, not killed.
scutlin, a small tart.
scuts, sort of barge used in carrying bread to London from the country.
Se, Druidic priestess.
se, assembly of the Seon.
seat.
seed.
seek.
seem.
seethe.
sefhte, seventh.
segd, a small ship.
sel', self ; sole.
semen.
sen, a name of Michaelmas.

senage, fines and payments levied in the sene court.
senche, to offer, place before.
sene, an ecclesiastical court for correcting the neglect or omissions of the church reeves.
sent.
sept, an enclosure by railing.
septical, causing putrefaction ; sepulchre.
ser, sure, safe.
serre, to join closely, pressed together.
set, to plant with dibble.
set, to settle, bind, "set" a broken limb."
set, to plant or sow ; sates, quickset.
set-sponge, cake it.
seth (Cor. Eng.), an arrow.
seth or sethen, both names of set, since, afterwards.
settle, for resting on.
seve, seven.
sew.
shaad, meadow.
shab, abscond, slink.
shad, divided.
shade.
shadow, shut of day.
shaft, handle.
shaft, anything created, creation.
shamnel, a masculine woman.
shamrock.
shan, to turn out the toes ; shins, two ; shoon, a pair of shoes ; shandy-gaff, a doubled drink ; shandery-dan, a two-sided car.
shanny, wild, half-idiotic, see Gaelic "Shoney," sea-deity. (Brand, i. 391.)
Shap, name of a place where there are Druidic remains.
shape.
shape, a dress of disguise.
shaped.
shaps, name of a prude.
sharp, scarped, parched, burnt up.
shaw, wood, small shady wood in a valley.
sheat, a young pig.

skarau, cut in pieces, destroy.
skher, plan, design, counsel, picture, instruct ; skha, write, letter, scribe.
skheru, writing, picture, scheme, plan, idea, doctrine, charm.
skhât, hare.
skhet, blow, wound, deprive, hinder.
sket, kind of cake.
skat, tow, conduct ; skhet, make bread.

Su, a royal scribe ; Sua, Egyptian priestess.
sehem, an assembly.
set, hinder part, seat ; asebt, seat.
set, corn, seed.
seakh, pray, adore.
sam, myths, representative likeness.
shet, effervescing wine.
seft or sefkh, seven.
skhet, an ark or boat.
ser, the chief one, private, reserved.
s'men, the erecting, preparing, establishing.
sen, division ; shen, a festival, one of two.
senhal, bind, conscribe, review, levy.

sensh, open, unclose.
shent, ancient name of magistrates, papyrus roll containing the laws.
sen, pass, traverse, extend.
sept, shore, lip, bank, margin.
sep, corrupt, corruption.
ser, the rock, a sanctuary.
serr, dispose, arrange, organize, augment.
setem-t, dibble.
sett, to catch, noose ; suta, heal, make sound.
set, impregnate, sow.
set, cake.
seti, an arrow.
sate, female set, called "yesterday."

seter, rest.
sef (kh), seven.
sa, Coptic so, to weave.
sha, field.
shap, hide, conceal.
shat, cut.
shut, shade.
shata, shade ; shuit, shade.
shep, hand.
sheps, conceive, create, bring forth.
shem, woman.
sam, a clover, with triple sign.
shena, turn away ; shen, two.

shannu, diviner.

shapt, solar disk, men belonging to religious houses.
sap, form ; sheb, shape.
shep, hidden, unperceived.
shept, figured.
shapsa, conceal, hide.
serf or serb, fire, flame.
shau, wood.
sha-t, sow.

shed, to separate. — shet, separate.

shed, for covering over. — shet, a small shed, a shrine.

shed, applied to light. — shut, illuminate.

shedar, female sheep. — shtar, betrothed wife.

shedle, a channel of water. — shet, a pool or ditch.

sheep, male. — shef, a ram.

sheen, play of light and shade. — shen, indicated by a storm cloud (shen, two-fold).

sheen-net, a drag-net. — shen, to encircle, and enclose, a crowd.

sheffe, 30 gads of steel. — shâ, 30.

shende, ruin, spoil, destroy. — shenti, rob, blaspheme.

shent, scolded, blamed, abused. — shenti, abuse.

shere, an egg with no tread in it. — sher, a junior, youth as non-pubescent.

shet, to join. — shat, to join.

shet, running water, water-shed. — shet, pool, ditch ; shetu, water-skin.

sheth, a division of a field ; sheading, division, water-shed. — sheth, ditch.

shield. — shlt (Heb.), shield.

shift, part allotted. — sheft, a section.

shin. — shenb-t, leg.

shindy or shinny, a game—also called bandy and hockey. — shenti, return, stop ; shenti-ta, go along, return ; shena, twist, turn back ; shenbtu, bandy.

shingle, on the shore, round the sea ; shingles, an eruption round the body. — shen, circuit, go round.

ship, an ancient piece of figured family plate. — sheps, figured ; shebu, traditions.

shit. — shat, noisome.

shittles, buns given to children. — shatt, food, sacred.

shod, covered. — shet, clothed.

shoe, for the feet. — shu, feet.

shoon, pair of shoes. — shen, two, a pair.

shoot, a. — s'hut, transmit ; shat, shoot.

shoot, an arrow. — shat, an arrow.

shop, a place where goods are sold by measure. — shep, an Egyptian measure.

shope, made, did shape. — sheb, shef or shep, to fashion, figure forth.

shore. — tser, to sit upon the waters, the rock.

shot, light and shade, shot-silk. — shu-t, light and shade.

shot, a reckoning at an inn. — sha, drink ; shat, book.

shote, a young pig. — shu, pig, swine.

shout, small, flat-bottomed boat. — shetu, a water-skin, straps used in a boat.

shout. — shetu, shout.

show. — sha, make naked, unveil, discover.

shrew, to curse. — sriu, curse, insult.

shrove, to be merry. — sheru, joy, to rejoice.

shun, to avoid. — shen, avert, turn away.

shun, to save. — san, to save.

shunt, turn off one line and back on another, put off, delay. — shenti-ta, go along, return ; shentt, return, stop ; shenti, turn down, shent, stop.

shuppare, the Creator. — skheper, to make live, creator.

shut. — shet, closed ; khut, to shut.

shuttle-cock. — shut, plumes.

shy, shady or shy. — shui, shade.

Sib (A. S.), Goddess. — Sef, a goddess.

sibrit, the banns of matrimony. — sop, time, turn, verify ; seb, priest ; rut, repeated several times.

sice, a sizing, college allowance of food or drink. — ses, measure of compatibility.

sich, a gutter. — sekh, to cut, incise, liquid.

sieve. — sef, determinative a sieve ; sef, to purify, refine.

sighed (Ir.), to weave. — skhet, to weave.

sighi (Ir.), a noose, or a score, twenty tied up. — sak, to bind.

sight, a great quantity, a "sight" of people. — sekht, a quantity.

sign. — sekhen, the sign of sustaining ; sekh, to represent, signify.

ENGLISH.

sign-tree, the support of the roof.
siker, sure, safe ; sikere, to assure.
sil, a bed of rock containing lead.
sill, step, ascent, throne.
sin, to stand, "don't sin talking, but go to work," *i.e.* don't stand like a statue (Norfolk).
sin.
sineways, sundry ways.
siney, bladder-nut tree.
sin-grene, evergreen.
sinoper, ruddle.
sinter, cincture, circle.
sis, the cast of six on the dice ; suse, six.
siss, a pet name for a girl.
sister.
sizer, a servitor.
sizing, yeast.
skaith, harm, damage, injure.
skate, fish prepared by squeezing or crimping.
skeen, sword and dagger.
skeke, contest or harryings.
    " With skekes and with fight,
    The ways look well aplight."
         —*Arthour and Merlin.*
sken, to squint.

sket, a part, a region.
sketch.
skew-bald, pied.
skill, reason, to know, signify, understand.
skink, to ladle out.
skink, to drink, pour out, fill the glass.
skittle, to hack.
skute, small tow-boat.
sky.
sky, to toss up.
slasher, name of the fighter.
slew, to turn round, re-arrange.
slewed, to be drunk.
slick or silk.
slon, a name of the sloe.
smack, sound of a kiss ; smacker, to kiss.
smeeth, to rub with soot ; smut, blacks.
smentini, (Eng. Gipsy) name of cream.

smeth, depilatory ointment.
smi, a fish, "Apua, a smi, which if kept long will turn to water."—Nomencla.
snaich, thief in the candle.
snap, take quickly, hastily.
snash (Scotch), blackguardly, foul abuse.
snatch, a hasp ; snitch, to confine by tying.
sneeze, with the custom of invoking a blessing.
snib, to shut up, fasten.
snitchel, used in measuring oats.
snoo, noose.
snood, to bind the hair.
snout, an organ of breathing.
snow, snew, old spelling ; sna (Scotch), snow ; snift, to snuff up with the breath, also a slight snow.
snuggle, hold close to the breast.
snuskin, a little delicacy.
so, pregnant (Gloucester).
soap.

EGYPTIAN.

sekhen, a prop, to sustain the heaven.
skarh, to soothe ; sakr, perfect.
ser, the rock.
ser, to extend, rise up.
senn, a statue, also to found, establish, make stand.

san, evil ; senn, to break open.
shent, crowd, many, various.
shenni, oak, acacia ; shennu, tree, oak.
san, preserve, keep long ; shen, ever.
sen, blood.
shen, enceinte, circle.
sas, six.
sest, she, her, the female.
sest, she, her.
shes, service or servitor.
ses, to breathe ; sesh, movement in general.
skhet, wound, blow, deprive.
skhet, to squeeze.
scheen (Heb.), knife.
skhekh, to destroy (*Rosel. Mon.* d. C. 72).

shen, bend, deflect, turn in, awry ; shena, turn away, twist awry.
skhet, a field, a region.
skhet, to paint or draw a plan.
skab, double.
skher, counsel, instruction.
skhenkt, a ladle.
skann, imbibe, give liquor to, pour out.
skhet, wound, blow.
skat, to tow ; skhet, a boat.
skhi, image of heaven.
ski, to elevate.
sersh, name of a military standard.
sru, to dispose at pleasure, place, arrange.
sru, to drink.
serk or selk, smooth, polish.
slon, a thorn, in Hebrew.
smakh-kh, rejoice.
smat, to black the eyebrows with stibium.
smen, constitute, fix, make durable ; tena, divided, apart.
smeh, anoint.
smi, name of Typhon the Apophis or "Appu" of the waters.
seniu, thieves, thievish.
snhap, take.
shanash, stink, foul, impure.
snett, to tie, tie up, to found by tying up.
snes, to invoke, to reveal, or discover itself.

snab, wall, case, enclose.
snith, measure.
senhu, bind, prison.
sen, to tie, bind ; aat, a net.
ssnut, breathe, with nose sign.
ssnu, breath, images, or to image ; sna, breathe.

snk, to suckle a child.
snus, nourish.
saa, belly, set up.
sef, to purge and purify.

ENGLISH.	EGYPTIAN.

ome, a quantity.	sam, a flock.
ome, (in blithesome).	sam, image, like, similitude.
on.	sun, to be made, to become.
oot, black.	sut, stibium, black.
oothe.	suta, to please.
op, soup.	sefa, make humid, moisten ; sefi, dissolve.
ort, to sort.	sert, arrange, distribute.
ough, wind.	suh, wind.
ound.	sunnt, to make a foundation, open the ground.
outh.	sut, south.
pace.	sheps, to be conceived, figured, born, spaced.
peed.	spet, pace.
pere, to ask, inquire.	sper-t, to ask, supplicate.
pew, a fourth swarming of bees.	spâu, make to fly.
pit (a) and a stride, very short distance.	spithams, span.
pit, to lay eggs, to dig ; spittling, preparing soil ; spitter or spit, a staff for weeding ; spade.	spet, create, prepare ; spet, a pole or staff (use unknown).
pit, with lips.	spt, lips.
pole, shoulder ; sprit, to split ; spur, of bacon.	sper, one side.
pot.	
pouse.	spet or setp, to select, choose.
punk.	spes, wife.
pur, to spur : sper, to prop up.	spu, creator, preparer ; ankh, life.
	sper, to conduct, lead, cause to approach ; spr, to come to the side.
s-collar.	ss-t, a collar.
srue, son, child.	sheru, child, son, junior.
ta, state, stay.	sta, entwine, reel thread.
tab, stop.	staibu, stop the ears.
tair or story.	st, to extend ; arru, ascend upward by steps.
take, to shut up, fasten.	steka, hide, lie hid, inside.
tall or stool for casks.	ster, couch, to lay out.
tan, to reckon ; stone.	sten, to reckon.
tathe, a wharf.	sta, conduct, tow.
tave, break down.	stafu, to melt down.
teek (Scotch), to shut, a ditch.	stek, to lie hidden, crawl, escape notice.
teg-month, the month of a woman's confinement ; stag, to watch, keep a look out.	stk, to lie hid, escape notice of.
tell, to fix.	ster, lay out.
ten (Scotch), to rear up, rise suddenly.	sten or stenh, turn back, fly.
tep.	step, walk, circulate.
tert, the point of anything.	steru, pierce.
teven, appointed.	step, chosen, appointed.
tew.	stu, pickle.
tick it in.	stekh, make not to see, hide.
till.	ster, to lie on the back, supine, stretched out dead.
tithe, firmly fixed.	stat, tied up, bound, noose, cord.
tool.	stur, couch.
tooth, to lath and plaster.	stut, to preserve, embalm.
totaye, to stagger.	stut, to tremble.
tow, hinder, stop.	stuha, repel.
toyte (Scotch), the walk of a drunken man.	stet, tremble, flutter, fearful ; stet, liquor.
traight, street.	stert, laid out.
try, spoil, waste, destroy.	stri, laid out, killed, destroyed.
tutter, stotor, a settler, a fearful blow.	steter, frighten ; stet, flutter, tremble.
uck.	sák, draw.
ucke, juice ; sukken, moisture.	sekh, liquid.
ue, to follow.	sui, behind.
ue, to entreat.	sua, come along.
ue, to issue in small quantities.	suh, the egg.
uet.	sat, grease.
uirt, to smoothe off the sharp edge of a hewn stone.	sert, sculpture.
uist, an egotist.	su, the personal pronoun.
uit.	suta, please.

ENGLISH.	EGYPTIAN.

sum and **some**, a total.
summit, highest point.
sur, over, beyond, supra.
sus, a nest (Eng. Gip.).
suskin, a very small coin ; **siss**, a great fat woman.
suss, an interjection of invitation signifying free to go ; applied to hogs, dogs, &c.
Sussex, divided into six rapes.
suse (Lanc.), six.
swack, a blow, to throw violently, a whack.
sweet.
Sywed, priests of Kêd.

sem, total.
smat, end of a time.
ser, to extend.
sesh, nests.
suskh, enlarge, stretch out.

suskh, free to go.

sus, No. 6 ; **sekh**, enclose, divide.
ses, six.
suakh, molest, cut up, harm, destroy.
set, aroma.
sha-t, men kept to enforce the sacred laws.

T.

ta, to take.
ta, the one.
tabby, pied, grey ("an old **tabby**").
tabn, a piece of bread and butter.
tabor, small drum.
tabs (Cor. Eng.), twitch harrowed out of the ground in cleaning it ; to be burnt.
tabularia, Druidic amulet shaped like an egg.

tache, clasp, tie, fasten, tacked together.
tack (ship), crossing.
tack, bad ale.
tade, to take.
taff, a Welsh cake shaped like a man riding on a goat.
taffy, a Welshman (" Taffy was a thief ").
tah, to drop, deposit excrement.
taht, given.
take, lease.
take-on, associate with.
takene, to declare, to show, token.
tall.
tally or **tolly**, male organ.
tam, to cut or divide.
tan, to pull, stretch out, beat out.
tane, the one.
tanist, heir-apparent.
tanist-stone (Keltic), coronation-stone.
tannin, a preserving substance.
tant, disproportionately tall.
tantivy, **toifonn** (Ir.), coursing with dogs.

tap, to call attention.
tap, of drum.
tar, to urge on.
tarse, mentula.
tart, a cake.
tasker, the reaper.
tass or **tassie**, a wine cup or measure.
tat (Ir.), first day of harvest.
ta-ta, leave-taking.
tatel, to stammer.
tath (pres. t.), taketh.
tatting, crossing by knitting.
tauntle, to toss the head ; **taunt**.
tavas (Cor. Eng.), a tongue, token.
taw, to stretch ; **tease**, stretch out.
tay, to take.
Tay, chief river in Scotland.
tazzle, entangle ; **tissue**.
te (Gael.), a woman.

ta, take.
tai, the.
ab, pied ; **t**, the ; **tabi**, a bear.
tep, bread.
tupar, tambourine.
teb, to purify by fire.

teb, circle, circumference, something round, sacred, an ark.
tekh, join, bind, unite, adhere.
tek, crossing.
tekh, supply with drink, wine, liquid.
tat, take.
tept, a cake ; **teb**, a goat.

tefi, he, male ; **tañi** (tafui), to steal.
ta, to drop, deposit excrement, earth.
ta, to give ; **tat**, take, assume, given.
teka, fix, attach.
teken, to accompany, frequent.
tekhn, an obelisk, memorial.
ter, extreme, extremity.
ter, male organ.
tem, blade, cut to pieces.
tun, to extend, spread, lengthen out.
tan, this.
tehan, heir-apparent ; **ast**, great, noble, ruler.
ten, seat, throne ; **ast**, sign of rule.
tahnen, a preserving substance.
tent, pride, rise up.
tanh, fly, flee away ; **tefi**, send away, dance, or, as we might say, " go it."
taf, attention.
tab, a drum.
tar, urge, require.
tar, mentula.
tert, a cake.
askh, mow ; **t**, the.
tes, a pint.
tat, beginning ; **tet**, seed, fruit.
tat, take ; **ta**, leave ; **tata**, go.
titi, to stammer.
tat, take, assume.
tat, cross, to establish.
tun, rise up, revolt.
tep, a tongue, typhon.
tehs, stretch.
ta, to take.
tai, chief, go in a boat, navigable.
tes, tie, coil.
t, feminine article and sign of gender.

ENGLISH.	EGYPTIAN.

te, to go, draw, pull, tug.
tea or **tay.**
team.
teata, too much, surpassing, continual ; **too-too,** exceedingly.
teb, fundament.
teche, teach.
tectly, secretly, covertly.
tee-hee, laughter.
teet of day, peep of day.
teme, race, progeny, to beget.

teme, make empty ; **tame,** to broach liquor.
tempre, to rule ; **tame,** to subdue.
ten, number in reckoning ; a half-score.

tench, "the tench's mouth," subanatomic term.
tende, to stretch forth ; **tense.**
tent.
tent, intent, design, "tak tent," take care of, reckoning.
tep, a draught of liquor.
term or **turn,** a time.
test, testing.
tet, cow-dung.
tether, bind.
tetta, "shall we?"
teyne, a thin plate of metal.
the, the article ; **taa,** one.
theat, firm, close, united.
thede, a country.
thedom, prosperity.
thekt, thought.
thepes, gooseberries.
thief.
thin.
thou.
thought, to image mentally.

thought, rower's seat across a boat.
thraw, a turn, a turn of time, to twist round.
threp, torture.
through, thorough.
throw, a space of time.
thunder.

thunk, a lace of white leather.

tidings.
tie.
tig, a game of crossing the frontier and touching.
tight, quick, instantaneous.
tight, fixed.
timarrany, two poor things.
time, to call.
tin, money.
tine, bind a hedge, enclose.
tined, divided.
tinestocks, the two scythe-handles.
ting, to split ; **tine,** fork.
tinse, to **tinse,** cover a ball with worsted work.
tint, gone, fled, lost.
tint, half-bushel of corn.
tiny, a moth.

ta, to go, to go in a boat.
tua, some kind of liquid.
tum, altogether.
teta, eternal.

teph, the abyss, the lower heaven.
Tekh (Taht, Divine Word), the teacher.
teka, lie hid, escape notice, unseen.
tehhi, rejoice ; **tehhut,** rejoiced.
tuaut, morning, dawn.
tam, male, to make again, issue ; **tamu,** generation ; **tam,** race.
teham, visit, waste.
tam, sceptre.
ten, reckon, amount ; **ten,** highest, Egyptian troy weight ; **tna,** a half.
tensh or **kensh,** snap at, extort, hunt, delight, determinative a fish, perhaps the tench.
tens, stretcher.
tenn-t, throne, cabinet.
tent, reckoning ; **tent,** having charge of.

tep, to taste.
ter, a time.
tes, self, substance, verdict.
teta, defile.
tethu, contain, imprison.
tet, speak, tell, speech.
tahen, tin.
ta or **te,** article **the.**
tut, to unite.
tata, the double land.
tet, to establish, sign of being established.
hek, thought, charm, magic (t, the article).
tep, kind of goose.
taf (tâ), to steal.
tun, extend, spread, lengthen out.
tu, thou.
tut, image, type ; **Taht,** the god of thought, is also named **Tekh.**
tat, to cross.
teru, a turn, a turn of time.
taru, afflict, bruise, rub.
teru, extreme limit.
tru, a space of time.
tun, to extend, spread, lengthen out ; **ter,** the extreme limit.
unkh, strap, sash, noose, support, dress ; (t, the.
tet, speak, tell.
tâ, not ; determinative, tie of a book or roll.
taka, a frontier, to cross.

teka, a spark.
teka, fixed ; **teka,** adhere.
tem, a total of two.
tema, announce.
ten, account.
tenh, to bind, to hold.
tent, separated.
tena, two halves ; **tenh,** to hold.
tena, cut in two ; **tenh,** divided.
tens, part of net, a mesh.

tenh, flee.
tent, half-measure.
tenhu, wings, fly.

tip, the tip or forecast as to the winner of a race.

tip, head.

tip-teerers, Christmas mummers.

tipe, globe; **tips,** faggots; **tippet.**

tipped, headed; **top,** head.

tiret, leather straps for hawks; **thwart,** strap across.

tirr (Scotch), work.

tit for **tat.**

tit, a galloping goer; **teite,** quick.

tith, fiery.

titmose, pudendum, f., Rel. Ant. 2. 28.

ti-top, name of a garland.

tist, tied.

toast.

toddy, the two-one, as a drink; **titty,** ditto, as breast; **ditto,** doubled; **teeth,** double-rowed.

toes, the five.

token (early, **teken**), to mark.

token.

ton, unity of weight.

tongs.

toot, to blow a horn for telling.

top, a-top, head (the Irish "top of the morning to ye").

top, which spins round.

torete, a ring; **turret.**

tosie (Scotch), ruddy with warmth.

tot, generative member.

tot, a hill.

total.

tot-hill, head of cross-ways.

tou, snares for game.

tourte, bread made of unbolted meal.

tow, tow a boat.

toy, an ancient Scotch female head-dress.

trape, the wing, said of the lover in wooing.

tree.

tremble.

trip.

trip, new, soft cheese, made of milk; **tripe.**

tro (Cor. Eng.), turn, circuit.

tropic, two circles parallel to the equator between which the sun's annual path is contained.

true, the **true.**

try, a sieve, to sieve.

tu, "Where hast **tu** been" (Lancashire).

tu, work hard.

tuadh (Ir.), North pole.

tuagh (Gaelic), battle-axe.

tuar (Ir.), a bleaching green.

tub, for washing in, an old ship.

tuen, to go.

tuff, to spit like a cat.

tuff, a tassel.

tuir (Gael.), to mourn.

tunic.

tup, to bow to a person before drinking.

tep, to guess.

tep, head.

tef, dance; **ter-rer,** go round at certain times or seasons; **tep-ter,** commencement of a season.

teb, crown, a jug, jar, chaplet, a round, clothe, clad.

tep, head.

tehor, leather; **tehert,** leather buckler.

ter, work, fabricate.

tut, to give; **tat,** to take.

tat, go galloping.

tet, fire, jet of flame.

mes or **mest,** the place of birth, sexual part.

ti, honour; **teb,** chaplet, a crown of flowers.

test, tie.

tes, hard, dense.

teti, the two-one, as lunar deity.

tu, No. 5.

tekhen, obelisk, monument.

tekhen, to wink.

tennu, unity of weight

ankh, a pair, and to clasp; **t,** article.

tet, to tell.

tep, head, heaven; **tep,** the point of commencement.

tep, potter's clay, spun round.

taru, encircle, a cell, a college.

tesh, red.

tut, generative member.

tu-t, rock, mountain.

tot, the hand; **totu,** both hands; **L (ru),** mark of division.

tata, head of roads; **tat,** pillar, landmark, cross.

tuia, net, catch, return.

tert, kind of food, a cake.

tau, navigation.

taui, a cap, kind of linen, linen object.

trup, receive favourably.

tru, a kind of tree; **teru,** roots, stems.

trem, cause to weep.

tref, dance, sport, lively.

trep, food.

tru, turn, time, season.

tro, time, a turn: **pekh,** division, dividing time.

tru, time.

tar, a sieve; **ter,** to rub, drive away, question.

tu, thou.

tu, go.

tat, image of the Eternal.

tu, slaughter, kill; **aka,** an axe.

tur, wash, purify, whiten.

teb, box, jar, chest, or ark.

tu, go along.

tef, to spit.

"tuf," papyrus reeds, which were tasselled.

tur-t, mourners.

unkh, dress; **t,** the.

teb-teb, humble, fall low.

ENGLISH.	EGYPTIAN.

turret. — **ter**, kind of crown ; **ret**, steps up.

tut-sen (Cor. Eng.), healing-plant, supposed to be the French **tout-saine.** — **tut**, type ; **san**, to heal.

tutsen (Cor. Eng.), sweet-leaf plant. — **tut-san**, health-giving

tux, knot of wool or hair. — **tes**, tie, coil, joint, a tied-up roll.

twat, pudendum, f. — **tuaut**, strumpet, a receptacle.

twig, to perceive. — **tek**, to see, hidden.

twist. — **tust**, twist.

twitty, cross. — **tat**, cross.

twy, two. — **ti**, two.

"tyb of the buttery," a goose ; **thep**, a gooseberry. — **tep**, a kind of goose.

typh-wheat, corn like rye. — **tef**, grain

U.

ubbly bread, sacramental cakes. — **apru**, consecrated.

ucaire (Ir.), a fuller. — **akh**, white.

uchab (Welsh), supreme. — **kef**, force, puisance, potency ; **khep**, to create.

uchel (Welsh), lofty ; **ucht** (Irish), hill. — **ukh**, a column ; **akha**, elevate.

ucht (Gael.), breast, bosom, lap, womb. — **kat**, womb.

ugh ! house ! ugge ! — **ukha**, night.

ughten-tide, morning.

uist, western isle, English west. — **uast**, the Thebiad, West Thebes.

um, "I see," or "let me see." — **um**, perceive.

un (Cor. Eng.), a while. — **un**, a time, an hour.

up up. — **ap-ap**, to mount swiftly on high.

uppen, to disclose, open, **heaven.** — **uben-t**, informant ; **ubn**, sunrise, opening, light.

ure, use, custom, practice. — **ur**, that which is first, chief, oldest.

urf, a stunted elf-like child. — **erp**, the Repa-child.

urne, to run. — **urn**, a name of the inundation.

usere, a usurer. — **user**, power, force, rule, sustain, possession, prevail.

utensil. — **utensu**, some utensil, a box.

utter, publish. — **utau**, speak out, give out voice.

V.

vaga, to wander about idly, whence vagabond. — **uka**, idle, idleness, rob.

vane, weathercock. — **pena**, to reverse, return.

vat and **faud**, a fold. — **a-t**, place, cabin, house, shrine, enclosure ; **aut**, shepherd's crook.

vese, to hurry up and down. — **as** (fas), haste.

victor. — **fekh**, to capture, take captives ; **teru**, all, wholly.

W.

wack, sufficient quantity of drink. — **uika**, a week, a full measure of time.

wad or **woad**, colour for staining, also black-lead (see Pliny). — **uat**, colour for eyes, collyrium ; **uat**, a blue cosmetic.

"wad," a mark to guide men in ploughing. — **uat**, order, transmit ; **uti**, word ; **uah**, ploughman.

wad, word. — **uti**, the word, Taht.

wahts, greens. — **uat**, different kinds of green.

wain, carriage. — **khen**, carriage.

wake or **week.** — **uaka**, a festival and a week.

wakes, rows of damp green grass ; **wax**, to grow, to thrive. — **uakh**, marsh, meadow.

wane. — **annu**, recoil, look back, turn back ; **un**, show, appear wanting, defective.

wap, futuere. — **khep**, to generate.

waps, wasp. — **kheb**, wasp.

warp. — **kherp**, a first form, figure.

warre, wary, aware, war ; **harr**, dog's snarl. — **uhar**, dog.

wart. — **uart**, a knob at the top of the Atef crown.

was.

wash,
wassail, call health, wish.
waste.
wat (Scotch), a man's upper dress, a wrap.
watchet, blue.
water.
wath, a ford.
wattle-jaw, a long jaw.
waught, a deep draught.
wawe, woe.
way, modified from the harder form.
wb-wb (Ir.), cry of distress, **weep.**
weaky, moist.
wearied.
weave.
wed, a pledge.
weem, Pict's house.
weigh.
weights, waite, to know; **weight,** means of; **what? weet, wit.**
Wessex.
wet.
weyd-month, June.

weye, to go; **ye,** to go.
wharf or **warp,** a shore, boundary, bank.
wharre, crabs.
whaten, what like.
whatten, what kind; **wita-gemote,** place of deciding.
wheam, womb.

whean, a worthless woman.
wheant, quainte.
wheat, corn; **ith** (Ir.), corn; **yd** (Welsh), corn; **had** (Welsh), seed.
wheden, fool.
whee, a cow.
wheel (by permutation).
wheen-cat, a female cat.
whenny, make haste, be nimble.
whent, terrible.
wherry, a light rowing boat.
whet, to cut.
whiff, a glimpse.
whimling, childish, weakly; **whim.**
whip.
whip the cat, get drunk.
whip-jack, a beggar who pretends to be a distressed sailor.
whipper, a lusty wencher.
white.
" **white it !** " "devil take it."
why ?
wi, a man.
wicca, magic, witchcraft.
wick, entrance, bay, inlet, narrow passage.
wicke, wickedness.
widdie (Scotch), a rope or binding made of withys or osiers; **with,** for binding; **with-bine.**
widow.
wig or **wicche,** an idol or spirit.
wine, the wind.
winnow.

uhas, lose, forget, neglect; **uh,** escape from; **as,** depart.
ash, wet, moisture; **kash,** wash.
uash, call.
ust, waste, ruin, blot.
uat, wraps.
uat, a blue cosmetic, paint.
uat-ur, water.
khet, a ford; **uat,** water.
uat, distance, length.
nah, very much; **uat,** transmit, liquid.
uha, sack, lay waste, want, long for, desire.
uakh, road, way.
heb-heb, weep, wail.
uakha, marsh, meadow.
urrut, peaceable, meek, figure squatting.
ââ (faf), knit.
khet, a seal, to seal.
khem, shrine, box, or small house.
khekh, balance.
uts, try, examine, suspend, weigh, stillyard, stand, weights of a balance.
uas, western.
uat, water.
uat, green, fresh plants, herbs, grass; **Uat,** goddess of green things.
ui, go, go along.
arp, to bind or bound, a bundle.
kar, to claw, seize with claw.
kheten, similitude.
uta, examine, question, verify, decide; **khe-mat,** for securing truth and justice.
khem, shrine, feminine abode, snug, very close.
khennu, concubine.
khent, feminine interior.

uahit, corn, wheat.
khat, fool.
kai, cow.
urr, wheel.
khen, female.
khena, fly.
khen-t, typhonian, adverse, disaster, calamity.
kheru, an oar sign; **urri,** chariot.
khet, to cut.
khef, a look.
khem, small, weak.
uaf or **auf,** to chastise.
kabh, libation.
kheb, hypocrisy, deceit, be in disguise.

khepr, generate, beget.
hut, white.
hut, demon.
uaui, meditate, discourse, to reason why.
ui, he, him.
huka, magic.
uakh, entrance, road.
khakh, obstinate, coward, mad, fool.
huit, bundle, bind up reeds.

uta, solitary, separated, divorced.
ukh, spirit.
khen, to blow, impel a boat.
khena, to blow, puff away, avert.

ENGLISH.	EGYPTIAN.

vish.
vish-rod, must be forked.
visp, bundle of straw.
vit, witch, wite.
vit, to know.
vitch.
vite, to depart, go out.
voh, to horses, be still, stop.
von, to dwell, inhabit.
voot, call to horses.
vord.

vorth.
vorth, a nook of land.
worth-ship.
vrap.
vrit, a legal process.
vrite.
vysse, to direct, conduct, command.

uash, invoke, wish ; **usha,** to aspire.
uas, a sceptre, forked.
usb, stack.
ut, magic.
uta, to examine, verify.
uit, magic ; **ka,** person or function.
uti or **utui,** journey, expedition.
uoh, abide, be quiet.
un, to dwell ; **uni,** inhabitants.
aut, go along.
uart, indicated by foot, leg, go, fly, that is, to carry word.
her-t, superior, above.
hert, a park, garden.
hert, superior, high above ; **shep,** type of.
arp, to bind, bundle.
rut, the judge—**Rutamenti,** judge in amentes.
ruit, engrave, figure.
hess, to order, will, command.

X.

xenia, New Year's offering.

khen, act of offering ; **kennu,** plenty, abundance, riches.

Y.

yacht.
yanks, labourers' leggings.
yape, futuere.
yat, heifer.
yaud, an old horse.
yede, went.
yep-sintle, two handfuls.
yeth-hounds, spirit hounds of clear shining white.
yetholm, the Gipsy locality near Kelso.
yore.
young, junior.
youth.
ysse, in thy seat.

iti, a boat.
ankh, strap, dress.
hap, unite, couple, marriage.
Athor, heifer, Goddess ; **kat** or **hat,** cow.
aat, old, outcast.
khet, to go, went.
sen, two.
hut, demon, spirit, light, white.

uat, a name of Lower Egypt.
ur, old, oldest.
hun, youth.
uth, youth.
hess, seat, or throne.

NOTE.—The compiler is, of course, aware that a few of these words may be claimed to have been directly derived from the Latin or Greek, but they are printed here on purpose to raise the question of an independent derivation from a common source. **Sane,** for example, and **sanitary,** believed to come from the Latin, may be directly derived from **san** (Eg.), to heal (p. 71); and he holds that in this and numerous other instances the Egyptian underlies both.

SECTION III.

HIEROGLYPHICS IN BRITAIN.

IN his Treatise of the "*New Manneris and the Auld of Scottis*," the ancient chronicler Boece says the old inhabitants "*used the rites and manners of the Egyptians, from whom they took their first beginning. In all their secret business they did not write with common letters used among other people, but with cyphers and figures of beasts made in manner of letters.*"

He is unable to tell how the secret of this crafty method of writing was lost, but affirms that it existed and has perished. We know the principle of Ideographic writing was extant in what are termed the Tree-Alphabets, in which the sprigs of trees formed the signs. Taliesin alludes to this kind of writing when he says, " I know every reed or twig in the cave of the chief Diviner," and "I love the tops of trees with the points well connected ; " these were the symbolical sprigs used for divination and as ideographs before they were reduced to phonetic value. The language of flowers is a form of the same writing.

In the time of Beli the Great, say the Welsh traditions, there were only sixteen "AWGRYMS" or letter-signs, and these were afterwards increased to twenty and finally to twenty-four. One account states that in the first period of the race of the Kymry the letters were called "YSTORRYNAU." Before the time of Beli-ap-Manogan there were ten primary YSTORRYN or YSTORRYNAU, which had been a secret from everlasting with the bards of the Isle of Britain. Beli called them letters, and added six more to the earlier ten. The sixteen were made public but the original ten YSTORRYNAU were left under the seal of secrecy. It will be suggested hereafter that Beli is the Sabean Baal, the first son of the mother who in Egypt was Bar-Sutekh (Sut-Anubis), the earliest form of Mercury, who became the British Gwydion called the inventor of letters ; that Gwŷd is Khet or Sut, and that the same original supplied the Greeks with their Kadmus, who is also accredited with introducing the sixteen letters into Greece.

But at first there were only ten primary letters or YSTORRYNAU. Now in Egyptian "TERU" is a type name for drawing, writings, papyrus rolls, stems, roots, literature, the "rites" of Taht, the divine scribe. Nau (Nu) denotes the divine or typical. Ys is the well-known Welsh prefix which augments and intensifies. There were ten of these branches on the first tree of knowledge. Kat (Eg.) is the name of the tree of knowledge. That is our British Kêd, who is the tree, and Kat or Kêd reappears as the Gwyd or wood of the Druids. The typical tree of Kêd or Ogyrven, one of her two chief characters, was an apple-tree, on which the mistletoe, the divine branch, is often found growing, and this gave the type name to the tree of knowledge with the British Barddas. Taliesin says seven score Ogyrvens pertain to the British Muse.[1] The brindled ox of Hu had seven score knobs on his collar. The number of stones in the complete temple of Stonehenge has been computed at seven score. The "Avallenau," or apple-trees, were the wood of the tree of knowledge, and these are represented as being 147 in number. From a poem written by Merddin we gather that there was a garden or orchard containing 147 apple-trees or sprigs, which could be carried about by him in all his wanderings. The bard bemoans that the tree of knowledge and the shoots have now to be concealed in the secresy of a Caledonian wood. The tree still grows at "the confluence of streams" the two waters, but has no longer the "raised circle" and the protective surroundings of old. The Druids and their lore are being hunted to death by the Christians, the "Men in black." Merddin and a faithful few still guard the tree of knowledge although their persecutors are now more numerous than their disciples. This tree of knowledge has seven score and seven shoots or sprigs, composing the whole book, and these may now be claimed as ideographs and hieroglyphics which deposited their phonetic values in the tree alphabets. Thus the tree of knowledge the KAT (Eg.), the Welsh GWYD is the representative of the mother Kêd, who is identified by Taliesin with Ogyrven.[2]

Ogyrven has an earlier form in Gogyrven, and Khekr (Eg.) means to adorn, a collar, or necklace, which in the lunar reckoning had ten points or branches as is implied in the name of Menat. Afterwards the collar worn by the mother Isis had nine points or beads according to the solar reckoning. Ogyrven is one of two characters of Kêd and Keridwen the other. When interpreted by the doctrine of the Two Truths, these are identical with the divine sisters, Neith and Nephthys. KHARIT (Eg.) means the widow. Keridwen is the widow lady in the mystical sense. She was the one alone. Ogyrven is called Amhat, she like Nephthys is the goddess of seed, or the seeded (Neft). It is in keeping that Gogyrven (Ogyrven) means some kind of spirit.

---

[1] Preiddeu Annwn. 5.   [2] Skene, *Four Ancient Books of Wales*, ii. 154.

Breath was the first spirit and this was the mother of breath, the other of the water.

In the chapters on the Typology of Time and Number, it will be shown how the tree of knowledge put forth its ten branches. It was at a time when the number ten was reckoned on both hands. In Egyptian Kabti is two arms. Khep is the hand and ti is two, thus Khepti or Khep, which becomes Kat and Kêd, is equivalent to both hands or ten digits. The Ogam alphabet is digital, and five of its digits read QV (Welsh), that is Khef (Eg.) one hand. Two hands or ten digits then represent the tree of Kêd, or Kat called knowledge. And as the ten digits were a primary limit it may be conjectured that the ten original YSTORRYNAU were represented by the ten first signs of the Ogam alphabet, the ciphers spoken of by Boece.

A document on Bardism cited by Silvan Evans,[1] says the three so-called "beams" were the three elements of a letter. These three consist of the right hand, left hand, and middle, and that from these were formed the sixteen Ogyrvens or letters. If we reduce the sixteen to the earlier ten we have the ten digits. And these mystical strokes include a right-hand one and a left-hand one. They are a figure of the Triad consisting of the one, who was the Great Mother, with two manifestations, whether these two were the two sisters, the twin brothers, or the dual IU, the young God. The Barddas assert that the three strokes, beams, or "shouts" rendered the name of the deity as IAU the younger. These emanated from the hieroglyphic Eye, to form the name that is one with the AU or IU of Egypt. They represented the "CYFRIU" name of the Trinity, or the thrice-functioned HU, says Myfyr Morganwg.[2] Khpr-iu as Egyptian would denote the transformation of the one into the Duad, and this meaning has been preserved in the "Cyfriu" sign of the Druids. Also Gafl in Welsh is the fork; GEVEL (Breton) is dual. The two hands, the ten branches of the one tree of knowledge then, were a dual symbol of KAT, KÊD, or KHEFT.

The number seven is the base of the seven-score and seven branches, and in Egyptian the word Khefti or Hepti also means number seven. Thus the seven and ten meet in one word, because the ti may be read as two or as twice. Khep is the hand and Seb is five, thus Seb-ti—five-two—is seven, and Kepti, or Khefti, is two hands, or number ten. When this tree of knowledge is recovered it will be seen that the Druids possessed it, root and branch.

TERUU (Eg.) denotes branches and stems, we may say, of the tree of knowledge, and it is also the name of drawings and of papyrus rolls. The first Ystorrynau were the branches of the hands, the ten digits; then the branches of the tree were sprigs of trees. Now in the old Egyptian stage of speech the words are isolated, and there is no distinction between the root and stem. The word stands for the thing

---

[1] Skene's *Four Ancient Books of Wales*, ii. 324. [2] *Vide* Letter.

rather than the thought because the thinkers were the THINGERS, and they who used the word as bare root had to obtain their variants by pointing to still other things. These variants were not represented by kindred words but by ideographic determinatives. It is here we have to look for the cognates, the branches, so to say, of the root. These were expressed in pictures, in gestures, and in tones, before the verbal variants were evolved. So the Grebos of Africa are still accustomed to indicate the persons and tenses of the verb by gesture signs. The Druidic sprigs belong to this ideographic stage. They were both the typical and the literal branches of speech in British hieroglyphics.

It is noticeable that the Damara tribes of Africa are subdivided into a number of EUNDAS (Clans or Ings) as the Sun-Children, Moon-Children, Rain-Children, and other Totemic types, and each Eunda has a sprig of some tree for its emblem. The Druidic sprigs, based on the number seven and extending to seven-score seven, doubtless originated in the same primary kind of heraldry, and their phonetic deposits are, as said, the tree alphabets.

In connection with the British TORRYNAU, it may be remarked *en passant* that, according to Eliphas Levi (*nom de plume* of the Abbé Constant), in his *History of Magic*, there is a hieroglyphic alphabet in the TAROT cards; these were distinguished from cards by the license of the French dealer who sold "Cartes et TAROTS." There is a Kabalist tradition that the TAROT was invented by Taht the originator of types. These cards are still preferred by fortune-tellers. The name, if Egyptian, connects them with magic and divination. Ter or tri is to interrogate, invoke, question. Ut is magic, TAR-UT therefore means magical evocation; magic applied to "TRYING the spirits" or evoking them. Also TERUT signifies the teru (Eg.) or coloured drawings, with the determinative of the hieroglyphist's pen and paints. The Tarot cards when interpreted by the Kabalist tradition were Terut or hieroglyphical drawings.

The Druids were in possession of the symbolic branch for the type of the youthful Sun-god, who was annually reborn as the offshoot from the tree. The mistletoe was their branch that symbolized the new birth at the time of the winter solstice. All its meaning is carefully wrapt up in its name. Mes (Eg.) is birth, born, child. Ter is time, and a shoot, which was the sign of a time. Ta is a type, also to register, the Mis-tel-toe is the branch typical of another birth of time personified as the child, the prince, the branch; prince and branch being identical, a form of the branch of the Panygeries on which Taht the Registrar registered the new birth of the Renpu. The branch in Welch is PREN corresponding to RENPU (Eg.) the shoot sign of youth and renewal. The branch of mistletoe was called PREN PURAUR the branch of pure gold, and PREN UCHELVAR. It had five names derived from Uchel the Lofty. The word will yield more meaning.

The Lofty was the tree, the aak or oak, and Al (Eg.) is the child or son of; the mistletoe was at times born of the oak.

The hieroglyphic shoot is well preserved in the Mayers' song in celebrating their form of the branch—

> "It is but a sprout,
> Yet it's well-budded out." [1]

The shoot or RENPU is carried in the hands of Taht, the god of speech, of numbering and naming, who is the divine Word in person. From the branch the Druids derived their Colbren, the wood of credibility, the staves on which their runes were cut. BREN is a later form of PREN. PREN is the REN, just as PREF, the snake, is the REF. The REN is the branch, and ren (Eg.) means to call by name. Coel answers to Kher (Eg.) the word, to speak, utterance, speech, voice. Thus the Colbren is the branch of the word, the wood of speech, identical with the REN (renpu) of Taht, and the emblem of that branch which was the word or Logos impersonated as the British DOVYDD.

The palm is the typical tree of time and of letters, a form of course of the tree of life and knowledge; it is the symbol of reckoning time. In Egyptian it is named BUK or BEKA, the original of BÔKA (Gothic), and the English BEECH. In Persian the oak is named BÛK. The trees may be various, the name is one on account of the type. The palm-shoot was the book of the scribe, and the beech-tree is the book-tree, because its bark was used for inscribing upon. The Buka or palm-branch on which time was reckoned finally yields us the name of the book.

One of the types of Palm Sunday is the cross made of palm-branches. In the northern counties Palm Sunday is a day of great diversion, old and young amuse themselves in making crosses to be stuck up or suspended in houses. In the latter part of the day the young of both sexes sally forth and assault all unprotected females whom they meet with, seizing their shoes, which have to be redeemed with money. On Monday it is the turn of the men, who are treated in the same manner. [2]

The palm cross primarily represents the Crossing, not the crucifixion, and is the same symbol of the equinoctial year as it was in the hands of Taht or An, as the sign of a time.

In the unavoidable quotation from Pliny we are told "the Druids (so they call the wise men) hold nothing in greater reverence than the MISTLETOE, and the tree on which it grows, so that it be an oak. They choose forests of oaks, for the sake of the tree itself, and perform no sacred rites without oak leaves, so that one might fancy they had even been called for this reason, turning the word

---

[1] Brand, *May-Day Customs.*   [2] Dyer, 133.

into Greek, Druids. But whatever grows upon these trees, they hold to have been sent from heaven, and to be a sign that the deity himself had chosen the tree for his own. The thing, however, is very rarely found, and when found is gathered with much ceremony; and above all, on the sixth day of the moon, by which these men reckon the beginnings of their month and years, and of their cycle of thirty years (the Egyptian Sut-Heb), because the moon has then sufficient power, yet has not reached half its size. Addressing it in their own language by the epithet of all-healing, after duly preparing sacrifices and banquets under the tree, they bring to the spot two White Bulls, the horns of which are then for the first time garlanded. The priest, clothed in a white dress, ascends the tree, and cuts the mistletoe with a golden knife; it is caught in a white cloak. Thereupon they slay the victims, with a prayer that the deity may prosper his own gift to them, to whom he had given it. They fancy that by drinking it, fertility is given to any barren animal, and that it is a remedy against all poisons." [1]

He also says, " Like to the Sabine herb is that called SELAGO. It is gathered, without using a knife, with the right hand wrapped in a tunic, the left being uncovered, as though the man were stealing it; the gatherer being clothed in a white dress, and with bare feet washed clean, after performing sacrifice before gathering it—with bread and wine. It is to be carried in a new napkin. According to the tradition of the Gaulish Druids, it is to be kept as a remedy against all evil, and the smoke of it is good for diseases of the eyes. The same Druids have given the name of SAMOLUS to a plant that grows in wet places; and this they say must be gathered with the left hand by one who is fasting, as a remedy for diseases of swine and cattle, and that he who gathers it, must keep his head turned away, and must not lay it down anywhere except in a channel through which water runs, and there must bruise it for them who are to drink it." [2]

In this account the Branch is identified with that of time. The number six is the measure of compatibility, and in the representation of the coronation of Rameses II., at Medinet-Habu, the king is offering to the god Amen-Khem, in presence of the god Sut and the WHITE BULL, six ears of grain which he cuts with a golden sickle.[3] The juice of SAMOLUS is the equivalent of the SAMA juice of the Hindu ritual and the HOMA of the Avesta. SMA means to invoke, and LUS (rus) denotes the rising or resurrection. Also SEMHI is a name for the left hand in Egyptian, SAMILI in Assyrian, Sema in Fijian.

The reed in the hieroglyphics is the symbol of the scribe. RUI is the Reed-Pen, whence RUIT to figure, engrave, our word write, and the reed was the sign of writing. Taliesin boasts that he knows

[1] Pliny, *Hist. Nat.* lib. 15, s. 95. [2] Pliny, lib. 24, s. 62-3.
[3] Wilkinson, *Thebes*, p. 62.

" every reed in the cave of the chief diviner," a holy sanctuary there is; the small reeds, with joined points, declare its praise. The Egyptians made their pens and paper from the Papyrus reed, and the reed is celebrated as composing the mystical characters of the Druidic writings. The rush and reed may be hieroglyphically read.

The name of the Son SU is written with the reed shoot. The shoot, whether of the reed or branch, is the emblem of renewal, the symbol of the Solar Son who was reborn at the time of the Vernal Equinox.

As late as the end of last century, at Tenby in Wales, young people would meet together to " make Christ's bed " on Good Friday. They gathered a quantity of reed-leaves (the shoots of the stem) from the river and wove them into the shape of a man; they then laid the figure on a wooden cross and left it in some retired part of a garden or field.[1]

The reed symbol was practically extant in our rush-bearings and carrying of reeds at certain seasons, as the emblem of a time and period such as the Feast of Dedication and various other festivals. Armfuls of reeds were cut and tied up in bundles, decorated with ribbons and placed in churches.

In the hieroglyphics a bundle of reeds is the sign of a time (ter) and indicator of a season. It also reads RET, to repeat, several.

At Weybridge, near Maldon, when the rushes were placed in the church, small twigs were stuck in holes round the pews.[2] The twig is the other and chief hieroglyphic of a time, held in the hand of Taht, the recorder of time. These tend to identify the symbolic nature of the Man of Reeds stretched on a cross at the time of the crossing.

The Man of Reeds was placed on the cross at Tenby. BI (Eg.) is the place. Ten is the division, and the word means to extend and spread, to fill up, terminate, and determine. TENNU are lunar eclipses, which take place equinoctially. Tenby then is the place of the crossing.

The leek or onion, the head-crest worn by the Welsh as a national symbol, is one of the hieroglyphics. Leeks and onions were identified with the young sun-god Adon, at Byblos. They were exhibited in pots, with other vegetables, called the Gardens of the Deity. The Welsh wore the leek in honour of Hu, one of whose names was Aeddon. The onion with its heat and its circles was a symbol of the sun-god Hu, in Egypt. It was named after him the HUT. Hut the onion is also Hut the Hat, and Hat the Mace. The hieroglyphic Mace or Hat is onion-headed.

One sign of Hu in the hieroglyphics is the Tebhut or Winged Disk, sign of the Great God, Lord of Heaven and Giver of Life. It

---

[1] Mason's *Tales and Traditions of Tenby*, p. 19.
[2] *Notes and Queries*, vol. i. p. 471.

is the solar disk spread out. The leek or sprouting onion (Hut) of Taffy is equally a Tebhut and a type of the solar god and source of life.

The adder-stone of the Druids, said by Pliny to have been produced by serpents, was called the GLAIN, the Welsh GLAINIAU NADREDD. Glain appears to have been a primitive kind of glass, not yet transparent but glassy. Some of the specimens are composed of mere earth glazed over. But they were all polished and reflected light. In the ANGAR CYVYNDAWD, the question is asked, "What brings forth the clear GLAIN from the working of stones?" obviously referring to the polishing of the surface. From this polish or glaze it had a mirror-like power. Hence the glain was a type of renovation and the resurrection. Mielyr, a bard of the twelfth century, sings of the holy island of the GLAIN, to which pertains a splendid representation of "re-exaltation," or resurrection. My object here, however, is only to point out that an Egyptian reviewer says of a work, "Thy piece of writing has too much GLANE in it," meaning glitter, as if he had said it was like the glazed earth, not like the transparent glass. The Egyptians of that time—Rameses II. —were making literary GLANE.[1] A variant of the Glane is to be found in CLOME, a Cornish name for a glazed earthenware cup. Also *Gloin*, in Welsh, is a name of coal.

There can be nothing incredible in supposing that the Druids also made pictographs, and drew the "figures of beasts." In fact we know they did. Near Glamis there is, *or was*, a stone, on which was engraved a man with the head of a crocodile. Sefekh the Capturer was the crocodile-headed god of Egypt, who was depicted as a man with the head of a crocodile. Now the name of Sefekh also signifies No. 7, the crocodile identifies the figure with Sut, and the Druids employed what they termed their seven-stone, which was also called the SAID or SYTH-stone. Thus hieroglyphically Sefekh identifies the seven-stone with Sut, whose emblem was especially the stone of the seven.

Plutarch, on the authority of Manetho, tells us that "SMY" was one of the names of Typhon the Ap, Appu, or Apophis of the Waters, the Dragon of the Deep. SMI, as we learn from the monuments, means the conspirator, the lurking, deceitful one of the waters. SMI has given his name to a small kind of fish, "APPUA, a SMIE" (Nomencl). "In Essex is a fish called a SMIE, which, if he be long kept, will turn to water." (Elyot.) Here we have the AP and SMIE in one, based on the treacherous transformation which was Typhonian. SMUI (Eg.) also means to pass, traverse; and this enters into the name of the rabbit's passage, the (A.S.) SMIE-gela, or coney-hole.

The hinder thigh, called the Khef, Khepsh, or Kheft, *is* an ideograph of the Great Bear, and its goddess, the Good Typhon. In

[1] Brugsch-Bey, *History of Egypt*, vol. ii. p. 104. Eng. Tr.

Herefordshire, a particular part of the round of beef is the KEVIN, that is the hinder thigh of the animal; and the joint constitutes one of the hieroglyphics in England. Khep, the hinder part, is likewise extant in the end of a fox's tail, called a CHAPE. Khef is the gestator, the image of brooding life, and covey means to sit and brood as a bird, hence a brood is the covey; and an Irish name for pregnancy is Kobaille.

Other Typhonian relics might be collected, but none more curious than this. Sut was depicted as the ass-headed god. One of the heads assigned to him as Sut-Nahsi is a black ass-headed bird, the Neh, a foul night bird. This hieroglyphic in England would seem to have been the bittern. And in the Arms of "ASS-BITTER" there is a bird called the "ASS-BITTERN," which is a Chimera with no likeness in nature,[1] but it is the very image of the ass-headed bird assigned to Sut-Typhon.

The Cockatrice, a fabulous animal supposed to be hatched from the eggs of a viper by a cock, is represented in heraldry as a cock with the tail of a dragon. This is obviously a form of the Akhekh Dragon of the hieroglyphics, a Chimera or griffin compounded of beast, bird, and reptile. The Cockatrice is either the threefold Akhekh, or a triple type of time.

Heraldry is as full of hieroglyphics as Derbyshire stone is of fossils. The Unicorn of heraldry is identical with the type of Sut, to be found in Champollion's *Dictionaire.*[2] Also Suti, in the hard form is Khuti, and the ass type of Sut is in Scotch the CUDDY.

The British Triads celebrate the great event of the bursting forth of the Lake of Llion, caused by the AVANC, a monster that had to be conquered and drawn to land by the three oxen of Hu-Gadern, so that the lake should burst no more.[3] The monster is identical with the Egyptian Aphophis or Akhekh, overcome by the Solar God, depicted with sword-blades as the cutter, the destroyer of bounds, and the analogue of the burster through the banks of the containing lake. The Egyptian monster is called the Apap, Ap or Af, Hef, Khef, Baba, in agreement with the Av–ank. Neka (Eg.) which also reads N K, is an epithet of the Ap or Av, meaning the evil enemy, the false, deluding, impious criminal one; the Crooked Serpent set with blades. Therefore it is inferred that the Druidic AVANC is none other than the APP–NK or Evil Aphophis that dwelt in the Pool of Pant, the place of destruction and dissolution, in the bend of the great void where the break might occur in the annual circle. The App or Apap is also called Baba the Beast, and the animal identified with the mythical Avanc is the BEAV-ER; this includes the name of BABA and Apap.

In Dselana, an African dialect, the AFANK is a pig, an animal

---

[1] Lower's *Curiosities of Heraldry*, p. 104.   [2] No. 115.
[3] *Archæology of Wales*, vol. ii. pp. 59-71.

that also represented the Apap in later Egypt, or was turned into a type of the evil Typhon.

The pig, like the beaver, is an animal that routs in the earth. In Malay, Salayer, and Menudu, the pig has the Typhonian name of Babi, and in Fulah, Babba is the Ass.

A form of the Aphophis monster or Akhekh dragon chained to the bottom of the water, usually of a lake, is common to the legends over all Ireland. The monster is supposed to rear its head aloft once in seven years. The number here corresponds to that of its heads in the Babylonian legends. The old Irish name of the dragon is the "BEIST," and Typhon is called "Baba the Beast." St. Patrick has now taken the place of Horus and St. George as the slayer or conqueror of the dragon.

The title of this chapter however is used figuratively, although the writer is not sure that even the phonetic hieroglyphics were not once used in Britain. Among the stones at Maes How is one that has some sort of inscription cut upon it. As no sense could be made of the letters, the inscription was thrown aside, a process never-ceasingly applied to our rude stone monuments. The inscription is on the inner edge of the stone, and would be hidden when the stone was *in situ*, therefore it was not for public reading. It has the look of a mason's mark, which would direct the workmen in placing it. So far as one can judge from the drawing made by Farrer, and copied by Ferguson, it is quite possible the inscription consists of four badly executed hieroglyphics.[1] The y-shaped figure is a perfect likeness of the Prop sign, the ideograph of SKHEN, to prop, and UTS, to sustain and support. The whole, from right to left may be read ABTA UTS-TA or ABTA SKHEN-TA. In either case it would intimate that the particular stone was intended as one of the supports. Ab-ta will also read to be carried to the front passage. But the Egyptian ABTA was the double holy house of Anubis, the place of embalming, the abode of birth and re-birth ; the womb and tomb in one. The stone was discovered at Maes How, the name of which may be found to have bearings on the meaning of the word "abta." The hieroglyphic symbols however are extant all over the world, in immemorial customs, in the mind's chambers of imagery, and in every language. But nowhere do they abound more than in our own land. Any number of these exist. We use them constantly, without dreaming how intimately we are acquainted with those strange-seeming hieroglyphics. We clothe our minds and our bodies in them, and include them in every domain of Art. The present writer has a very common bronze lamp for most ordinary use, and of the cheapest kind. It is ornamented by two herons fishing. The heron, or Ibis, the fisher, was the type of Taht (Tahuti), the Lunar God who carried the lamp of Ra through

---

[1] Orkney inscriptions. Ferguson, *Rude Stone Monuments.*

the nocturnal heaven, and the lamp still carries the emblematic bird of the reduplicator of light. That is the Lunar light; another is the Solar. In this the lamp issues from the lotus just as did the youthful sun-god Horus in the symbolism of Egypt.

In the Neolithic Age the axe made of jade or nephrite was most highly prized for its rarity, its hardness, and especially for its beautiful deep-green hue. According to Mr. Dawkins, the only places where nephrite is known to exist in the Old World are Turkestan and China.[1] He therefore infers that it could only have been brought into Britain from the East, along with all the superstitions attending it.

This mineral stone is found also in New Zealand, where the natives hold it in the same regard as do the Chinese, and as did the Cave-dwellers of the Neolithic Age, who made charms and amulets of it as well as axes.[2] It is possible that the name of JADE, or the green stone, represents the Egyptian Uat, from an earlier KHUAT, which is extant in an unexplained title. JOUDS or JADS, in Devon, are RAGS, and UAT (Eg.) is the name of rags. To JOUDER is to chatter or speak rudely, and UTA (Eg.) is to speak out, give out voice. JUT and JET also render UT, to put forth. Uat, the name of the Goddess of the North, likewise corresponds to that of KÈD. Uat is the type-name for Green, and for the hard green stone, the emerald, and the green felspar, and if so, the different UATU of Egypt are tolerably certain to have included JADE among the green stones. It will be important to ascertain whether jade is or was a product of Africa. The green felspar was, and that is so hard, a good knife will scarcely scratch it. The Uat Stone was used for tablets, and the word means to transmit. The green colour had all the significance for the Egyptians that it had for the Aztecs, Chinese, and the Neolithic men. The Uat (green) was a vestment used in certain religious ceremonies. The Uat sceptre was carried by the goddesses. A variant of the word, Utu, (tablet) also signifies to wish, command, direct, texts, inscriptions, and is a name for magic and embalmment. The green stones were formed into amulets worn by the living and buried with the dead. Moreover, it is manifest that the green axes were often worn suspended as charms and ornaments, and not put to common use. Now, although we cannot prove the axe of the hiero-glyphic Nutr sign was of Green jade, or felspar, yet it is a stone axe or adze of a most primitive type of the Kelt stone, and it is the ideo-graph of the Divinity, God or Goddess. One meaning of the word Nutr is to cut, work, plane, make. One Nutr is a carpenter, the Kar-Nutr is a stone-mason, i.e., the stone-polisher. Thus to plane wood and polish stone were once divine, and the stone-adze type of

---

[1] *Early Man in Britain*, p. 281.
[2] There is also an Assyrian cylinder, in the British Museum, made of jade, incribed with the word "ABKIN."

working and making is the ideograph of the maker as the Divine
Creator, the Goddess and the God. The axe was buried as an amulet
in Egyptian tombs. When the the coffin of Queen Aahhept, the
ancestress of the 18th dynasty, was dug up not long since, an axe of
gold and others of bronze were found to have been buried with it.

In a paper read at the Anthropological Institute, May 25, 1880,
Professor J. Milne states that axes, generally of a GREENISH STONE
which appears to be a trachytic porphyry or andesite, have been
found in the mounds or middens of Japan, ascribed to the Ainos.

The axe in the British burial-place had the same symbolical
value, and its colour was like the evergreen, a type of the eternal.
UAT itself was of a bluish-green like the dual tint of water or the
holly—green with a bluish reflection. It was the colour of reproduction from the underworld. The Gods Num and Ptah were painted
with green flesh in this sense. The Egyptians made the symbolic
Eye with the Uat colour, another figure of reproduction. The ring
of jade placed in the tomb was an ideograph of the resurrection as
much as the seal-ring (Khet) or the Eye in Egypt.

Mr. Renouf, the excellent grammarian, has lately made a special
study of all the passages in the texts in which the word NUTR
occurs, and his conclusion is that the one dominant sense of the word
is power, potency, might.[1] But this endeavour to arrive at a
general sense in which the particulars are swallowed up, and to
attain the abstract of all that preceded the later stage of thought is
fatal to an understanding of the origines. We have to do with the
Thingers as the earlier thinkers, and must keep to their region of
things if we are to follow them instead of making them follow us.
We must abide by their ideographic types and variants, and find the
way to the origin of words by means of things.

The word Nutr has an earlier form in NUN-TER; Nu and Nun
are interchangeable, because both are written Nunu. Nun, like Nu,
means the type, image, figure, likeness, the portrait, and statue. Ter
is time, a whole, a repeating period of time. These are fundamental.
And Nutr or Nunter signifies a mode or means of portraying time,
i.e., duration and continuity. Hence the stone NUTR is an emblem
of duration, naturally an early type of time, a double type, because
it was permanent in time and sustaining in power. One form of the
Nutr was the eyeball or pupil of the eye, the mirror in which an
image was reproduced. The eye, Ar, was synonymous with conception, making the likeness, repeating, hence it was an ideograph of
the Child, and the Year as the Eye of Horus, and of the repeating
period as the Uta, the symbolic Eye of Taht which indicated resurrection and renewal, like that of the new moon, for which reason the
word UTA means salvation, going-out, to be whole for ever. The
serpent is a determinative of Nutr, and that is a type of the

[1] *Hibbert Lectures*, 1887, pp. 94-99.

renewing period. The gestator (with the corn measure) is a determinative of Nutr, as the reproducer. The star is also a determinative of Nutr, and that likewise is a type of repetition in time. The Phœnix and the Circle are both Nuteru. The root-idea found by the present writer is that of renewal, reproduction and rebirth in time, and continuity. TER is time, and NATR is time and of time; NETR-TUAU is time, and which form of the period (or TER) must be determined by the type (Nu or Nun). One Nun-ter is the time of the inundation, as Nun is the inundation. Another is a day (tuaa or the morrow), and the type is the star. A third is that of gestation or of menstruation, hence the serpent type and the two goddesses.

The Nutr' must convey the two truths which are not to be found in the single notion of might, or power, or potency alone and unexplained. The bluish-green stone will tell us more than that. Green was the colour of renewal; the earth is re-clothed in green, and Ptah, who re-embodies the soul, is himself depicted with green flesh. Blue is the hue of heaven, hence the typical colour for the soul. The Coptic equivalent for Nuter, Nomti, is based on the Egyptian NEM to repeat, again, a second time, to renew. The blue and green were both combined in the jade Netr, or axe, just as they were in the Genitrix Uati, whose name indicates the dual one of the Two Truths.

The bluish-green jade-axe in the burial-place was primarily a type of the Two Truths, applied eschatologically. But it was a multiform image, one of the things that represented various accumulated ideas. As stone, it was a type of strength. As jade, the hardest stone, it was of great strength and everlasting power. As blue-green it mingled earth and heaven in its mirror. It was a sceptre of the Genitrix Uati, a sign of the Twin Goddesses, who brooded over the mummy and brought to rebirth, by which the deceased arose from the green earth and reached blue heaven. Hence the Nutr in the Demotic text of the tablet of Canopus is rendered by Khu, another sign of power, protection, and reproduction, for the Khu fan signifies breathing, spirit, power. As an axe or adze the implement of cutting, planing, polishing in the hands of the carpenter Natr, and the stone-polisher or Kelt-maker called the Nutr-Kelt (Kart), it was the artificial type of making, shaping, creating, hence of the maker, creator, god, and goddess, and lastly the idea of protecting power was added to the Nutr by its being a weapon of defence. To reduce all this to the one notion of power is to whittle away the substance of things to that fine vanishing-point at which the past of Egypt becomes invisible.

Wheresoever the jade came from to make the axes and ring-amulets, it would seem that our stone-men were able to cut and polish it. The jade was the hardest stone known, and to this day the stone-cutter in Gloucestershire is known by the name of a

Jadeer or Jadder, jade being the typical and divine stone on account of its hardness; the stone-cutter is thus designated as the jade-cutter.

The seat Hes is a symbol of the Great Mother with which her name is written, and the seat was an emblem of Kêd or Keridwen, who was likewise represented by the Chair of the Bards. The prize for poetry given to the minstrels or singers in the Eisteddfod was a medal marked with the figure of a chair, the chair of Eseye (a name of Kêd) of the seat in the Great Stone Sanctuary. The Hes or As seat identifies the Goddess called Eseye with Hes or Isis.

The Hesi in Egypt were the bards of the Gods attached to the divine service, primarily of the Goddess Hes. Hes means to sing, celebrate, applaud, hence the Hesi. A hieroglyphic S, the seat, is written as the child's first pothook, and the pothook of the chimney is called the Es-hook, whilst the "Ester" is the back of the fireplace where the Es-hook hangs. The Hes (Hest) seat or throne was also continued by name in the British war-chariots, called by Cæsar Esseda, 4,000 of which, belonging to a corps of observation, were left by Cassivelaunus to watch the Roman movements after the landing.[1]

"Hess" is the name of the so-called Treaty-stone of Limerick. The seat was a feminine symbol. The Coronation-stone of England is still placed under the seat, and has the same significance as the Hess of Limerick, and the Hes of Egypt personified as the bearing Great Mother. The Hes seat and throne has yet an extant but lowly form in the hassock. The use of it for kneeling in churches is a reminder that Sukh (Eg.) is the shrine, and Sakh means to adore and pray. When houses were built the dwelling takes the name of the rude stone seat, and the Genitrix is represented by the House.

A pair of foot-soles are common hieroglyphic figures on the rock sculptures of the North. They have been found in Ireland on a stone which was sacredly considered to have been an inauguration stone of the ancient Irish kings or chieftains.[2]

In Samoa the natives show the pair of footprints in the rock, consisting of two hollows nearly six feet long. These they say are the footprints of Tii-tii, marking the spot on which he stood when he pushed up the heaven and divided it from the earth.[3]

This pair of footsoles appear in the Ritual where the astronomical imagery becomes eschatological. In the chapter of the "Hall of Two Truths" (125), the Osirian who has crossed by the passage from night to day says, "I have crossed by the northern fields of the palm-tree." He is asked to explain what he has seen there. His reply is, "It is the FOOTSTEP and the SOLE." A footstep and a sole of the foot are equal to the pair of footsoles. The "FOOT AND SOLE" are mentioned (Ch. 144) as the "foot and the sole of the foot of the

---

[1] Cæsar, *Commentaries*, lib. v. 15.   [2] Simpson, *Archaic Sculpturings*, p. 183.
[3] Turner, *Nineteen Years in Polynesia*, p. 245.

Lion Gods," one of which is MA'TET. " Hail to ye, Feet!" (Ch. 130) is addressed to them.

"The Chemmitæ," observes Herodotus, "affirm that Perseus has frequently appeared to them on earth, within the Temple, and that a sandal worn by him is sometimes found which is two cubits in length." [1] Two cubits in Egyptian would be represented by Mati, the name of a pair of feet, and therefore the sandal of Perseus is equivalent to the "footstep and the sole." The same writer mentions another footprint, that of Hercules, upon a rock, near the river Tyras. This was also two cubits (Mati) in length. [2] Mat is the ancient name of An, the boundary and division in the celestial birthplace, that of the two truths, and of the two footprints. This was the seat of Atum, who is the original of both Adam and the mythical Thomas. Atum was rendered Tomos by the Greeks. The so-called footstep on Adam's Peak in Ceylon is assigned to Thomas, who has also left his memorial at Bahia, on the American continent, where the footsteps are exhibited in proof of the "Saint's" visit to that shore. The pair of soles then are typical of the two truths, the basis of everything in Egyptian thought. MATI, the two soles, has the HES THRONE for determinative, which suggests that the inauguration stone with this dual emblem on it was a primitive shape of the SEAT or THRONE of the Two Truths, and the king seated on it would be assimilated to the great "Lord of Truth."

The stone also appears in the same passage of the Ritual as a sceptre if not as a throne. The Osirian has found or made that "Sceptre of Stone; its name is PLACER," so rendered by Dr. Birch. But in MATI, to give, to place, we further claim the two soles of the feet, and in the stone of MATI the stone of the "Footstep and the Sole." This duality of truth called the Two Truths of Egypt is identified with the two feet by the English proverb, "A lie stands on one leg, but the truth upon two."

The HES is the stone seat, chair or throne, a type and namesake of the great mother. This stone seat is also the Kat as in our Cat-stone; the Kat is an abraded Khft, the primal seat or hinder part whose image was the Typhonian seat of stone. Now it is the same stone wherever found. The stone that was brought from Egypt by Scota, the stone of destiny, the Lia Fail, the stone that sounded when the true king sat on it, the stone that lies at last beneath the Coronation chair in Westminster Abbey. That is, the typical stone is one; the copies may be many. The literalists have been befooled by the legends because they knew nothing whatever of the ancient typology. The stone can only be identified by what it represents. It is no mere question of a bit of red sandstone, even if it were, the stone of Scone had the true Typhonian complexion, and the colour of the lower crown of Isis. It is found at Scone, and Sekhn (Eg.) is the seat; at Beregonium, and Kani (Eg.) is also the seat; at Tara, and Ta (Eg.) is the seat. Even the tradition of the sounding stone may be traced. HES is the stone-seat, and HES means to celebrate, proclaim, sound forth. Another form of the HES is the vase symbol of the genitrix, and the vase, or cauldron, the sacred Pair of Keridwen is a famous symbol in the Druidic mysteries. It is often referred to in the writings of the Barddas, and may be seen figured on the back of the Mare on the ancient coins or talismans.[1]

The hieroglyphic vase (hes) has neither spout nor handle, and in Devonshire they still use a beer-jug named a " HESTER " which has neither spout nor handle. In Egyptian this may be read HES, liquid, TER, limit or measure. Through the HESTER we can recover a lost link with the Goddess EOSTRE, who was AS or HES in Egypt, Ishtar in Babylon, Astarte in Phenicia, and Eseye at Stonehenge.

The Ankh sign is one of the hieroglyphics in Britain. Indeed, we have several forms of the Ankh. It is extant in the great seals of England, in a reversed position, as the token of power and authority. The Ankh is a circle and cross, and in Kent a " WENC " is the centre of cross roads. Our winch for winding up is a kind of Ankh. So is the anchor. " Ankh " is to pair, to clasp, a sign of covenant, and to this corresponds our " WINK " and to " WENCH." Another form of the Ankh is a hank or noose. Putting a patient through a hesp or hank of yarn, and then cutting the flax into nine portions as a means of cure, was practised in Scotland. The pieces were buried in the lands of three owners.[2] Here the hank was like the Ankh, a symbol of living. In the " Hange " we have a threefold symbol of life. This is the name of an animal's pluck, consisting of heart, liver, and lights, three seats and organs of life. The Ankh emblem of life and sign of Ankh, to pair, couple, a pair, is extant, minus the cross-piece, in the English DIBBER, for planting seed. A connecting link between the two is supplied by the Maori " HANGO,", a dibber, for planting potatoes. In this the emblem has kept its name. Teb, the Egyptian for Dib, is movement in a circle. This is Dib-ling. Dib-ling is a form of doubling, this the Ankh images. Doubling or dibling is seed-setting, reproducing. So read, the Ankh makes the dibber, a sign of Deity and a type of creation which commenced with " movement in a circle."

The Ankh, as an Egyptian symbol of oath and covenant, is represented by our HANK, noose, or a knot. This may be at the root of the superstition not yet extinct in England, that one man may lawfully sell his wife to another provided he places a noose or halter round her neck. This is not only a belief, it is yet a practice. The noose in the hieroglyphics is a form of the Ankh, and on the wife's neck it is the token of a covenant in the transaction. A much earlier ideograph of marriage (by capture) than the gold ring round the finger! Also,

---

[1] Gibson's *Camden* (Tab. one) and Davies' *Mythology*.
[2] Perth, K. S., 22-26, May 1623.

as the halter is a Hank, and the Ankh is an ideograph of life as well as of covenanting, this may explain the origin of the superstition respecting the curative virtue of the hangman's noose when applied to ailing and diseased persons.[1]

The French have a form of the Ankh-ring in the JONC, a wedding-ring; in one shape this was a ring of rush. Those who were married by compulsion at Ste. Marine were wedded with a ring of JONC or rush. The custom prevailed at one time in England of marrying with a rush ring as a sign of junction. The Jonc-ring is the most ancient form of the Ankh, going back to the closest nearness to nature, an the time when metals where not wrought into rings, or flax into nooses. Marriage with the JONC, and divorce or re-marriage with the halter, are correlated by means of the Ankh symbol of pairing, clasping, covenanting. The custom of strewing rushes was at one time called JUNCARE in Gloucestershire. The practice was periodical. As the rushes were also bound up in an Ankh image of life, like the hieroglyphic doll, and dressed in the living likeness, we may look upon the JUNCARE as the ceremony (arui, Eg.) of the JUNC, rushes, or ANKH, and the meaning of ANKH, to covenant, an oath, explains the rest in relation to the rite repeated annually.

Ankh, Unkh, and Nak are interchangeable. A "curious kind of figure," described by Brand,[2] used to be made of the ears of the last corn harvested and brought home by the farmer when he finished his reaping; it was hung up over his table and sacredly preserved until the next harvest. The image consisted of ears of corn twisted and tied together, and it was called "A KNACK." This KNACK answers to the Ankh. One form of the Egyptian Ankh is a nosegay, another denotes ears, therefore the ankh may have been a handful of ears as in England. The KNACK of corn is a knot of ears, and the Ankh permutes with the hieroglyphic of Life, signified by tying up or binding in a noose or hank. This tying up in a knot applies to the harvest, as well as to the more recondite meaning of the Ankh ideograph. Marriage, however, is still figured as tying the knot or Ankh. The harvesters shout, "A knack! a knack! Well cut! Well bound! Well shocked," of which tying and binding the Ankh is an ideograph. Another form of the hieroglyphic Ankh, or image of life, is a doll or baby, and the Harvest-Knack is also known as the doll or Kern baby, the seed or child-symbol of future harvests. The nature of the Knack may be determined by the shape of the Maiden in Perthshire, in which the handful of ears was tied up in the form of a cross, and hung up in the same way as the Knack. The Ankh was the Crux Ansata. In Herefordshire the knack is called a Mare. The sign of this is a bunch of wheat from the last load. In Hertfordshire there is a custom of harvest-home called "Crying the Mare." The reapers tie up the last ears of corn that are cut, which is the Mare,

[1] Brand, on *Physical Charms*.   [2] On *Harvest Home*.

and standing at some distance, each throws his sickle at it, and he who cuts the knot wins the prize. After the knot is cut and the shout is raised "I have her," the others ask "What have you?" The answer is "A Mare, a Mare, a Mare." "Whose is she?" The owner's name is then announced, and it is asked "Whither will you send her?" To so and so, naming some one whose corn is not all cut. The Mare, also found as the Mell, is the symbol of harvest-home, the end of the harvest; the figure of the Mare is passed on in token of the termination. Mer (Eg.) signifies limit, boundary, swathe or tie up, with the noose determinative. This is the Mare of our harvesters.

The Mell-Supper, at which the employers and employed feasted together or *pêle-mêle*, is not derived from MEHL, farina, nor from MÊLÉE, mixed, but from Mell, extant as a company. Men who coöperate in heaving and hauling constitute a Mell, and this is the Egyptian Mer, as a company or circle of people joined together and co-attached, who are the Mer, *i.e.*, Mer-t. The Mer, as circle, is illustrated by the expression, when a horse is last in a race, "he has got the Mell." He being last of the lot completes the Mer.

Also in English the mill is the round.

The "MAIDEN" is another form of the "KNACK," "BABY," "MARE," or Harvest QUEEN. The last handful of corn reaped in the field was named the Maiden. The Maiden Feast was given when the harvest was finished. In Egyptian "Meh" has the meaning of wreath, crown, girth, and ten is to fill up, complete measure, terminate, and determine. The word Maiden, as the girl arrived at the age of puberty, may have the same derivation. The harvesters in Kent form a figure of some of the best corn the field produces, and make it as like the human shape as their art will admit. It is curiously dressed by the women, and ornamented with paper trimmings cut to resemble a cap, ruffles, handkerchief, and lace. It is then brought home with the last load of corn, and this is supposed to entitle them to a supper at the farmer's expense.[1] It was a form of the corn-symbol variously designated the Maiden, the Mell, the Mare, Knack, the Corn-doll, Corn-baby, and the Harvest Girl. In this instance it was called the "Ivy Girl."

Ivy the plant is out of the question for any explanation. The likelihood is that the word has been worn down. Our Goddess of Corn is KÊD, the Egyptian Kheft or Khep. And KHEPI (Khefi) is the name of harvest. The KHEFI Girl is the Harvest Girl. Ivy has earlier English forms in hove and hoof, and the Khefi Girl has become our provincial "Ivy Girl." Khepi (Eg.), for harvest, is also represented in Cornish English by HAV for corn, and in English Gipsy Giv for wheat.

The original meaning of the word THING was to thing, to make

[1] Brand, *Harvest Home.*

terms. Chaucer's Serjeant of Law, as a good conveyancer, could well endite and make a thing, that is a legal contract. Thingian or Thinging was to make a covenant, hence the German Bedinging, contract, terms of agreement, and the Norse Thing, a place where terms were covenanted. This comes from the Egyptian Ankh, an oath or covenant with the article T prefixed : T-ankh, the Ankh (T-ankh) was in the Noose form the visible sign of binding on oath or thinging. The Angnail is also thangnail in English, a very bad form of binding hard ; it being a bunion on the toe. Our Tank, as the encloser and container of water, is a good illustration of the Ankh and Thing. This word is as universal in its use as was the Ankh sign in Egypt, and just as purely symbolic in its values. Where the sign of the Cross is made at the end of an agreement in token of the covenant the Ankh ideograph is visibly presented and used with its true typical power. To negative this image of the Ankh is to express the meaning " No-thing." This gave the power to the name of a "NITHING," one with whom no covenant was kept, as he was not within the social pact or bond figured by the Ankh ; was no longer one of the "hank," or "Ing," a body of people confederated. We have the word without the Egyptian article as Hank, to fasten, a hold on anything, and hang, to tie, and stick to. Shaking hands over a bargain is a form of making the sign of the Ankh, the Cross of Covenant, as Ankh means to clasp as well as to covenant, and to thank is a form of the ankh-ing. The Ankh (Hank) as symbol of Thing crosses curiously in Welsh and Egyptian, one name of the Hank and Noose emblem is TAMI, and in Welsh, DIM is etymologically the type-word for thing. Thus two things are two of the hieroglyphics which have one and the same meaning as the Ankh.

The Ankh symbol of life appears in the form of Kankh in the Cangen or branch carried by the divining bard. Cang-en, or Ankh-un, is the repeater of life, *i.e.* the branch. Likewise in the kink, or kneck, a coil, to twist, to entangle ; curly hair is said to kink ; also a rope when it does not come out freely ; kink is used in binding a load of hay or corn. Our "knack" then is finally the kink, and whereas Ankh (Eg.) is life and living, kink, in the Eastern counties, signifies to revive, and to be ANXious is to be very much alive.

An ancient British origin has been claimed for "the Feathers" by Randall Holmes. The Rev. H. Longueville asserts that the arms of Roderick Mawe, prior to the division of Wales into principalities, was thus blazoned, " Argent, three Lions passant regardant with their tails passing between their legs, and curling over their backs in a feathery form." [1]   And in the parent-language MAU is the name of the lion, and the variant of Shu for the feathers is MAU or MA. The three tails of the three lions (Mau) curling in the form of

---

[1] Brewer, *Phrase and Fable*, p. 291.

feathers (MAU) in the arms of Mawe is one of the most perfect of the hieroglyphics in the Islands.

A leash of dogs or of partridges is a triad, three leashed together. The three feathers of Wales are a leash attached by a band. The three feathers of the hieroglyphics are likewise a leash, in Egyptian, Resh. The three feathers are the determinative of Resh, which means joy, also res, is absolutely, entirely. With the terminal *t* we obtain the word rest, and the three feathers of the Prince, the Repa denote the joy of fulfilment in the Prince of Peace to whom the leash of plumes belongs. Very rarely we find a third feather added to the two. Father, Mother and Child are the Three Truths of the Trinity. And in the Solar Myth the Child was born as Horus every spring-tide at a part of the zodiac where the Egyptians located the Uskh Hall of the Two Truths, we might say of the two feathers. But here at the birthplace of the Son, the Repa, who is the heir-apparent and the prince—for these are his titles—we find the three ostrich feathers. The three have been found as an ideograph of Egypt.[1] They are probably very ancient though not common. It suffices that they are extant, and that they add a third to the Two Truths, as the sign of the Son, who is the Prince and heir-apparent. The three feathers are placed over the Cross-sign of the completed course at the crossing where the Solar Prince was born. Three is the Egyptian plural because the Trinity had to blend with, and come out of the duadic one who was female at first, then two females, then the male-female, and lastly, father, mother, and son. The mounting of three feathers shows the addition of the prince, the heir-apparent, who was considered as much a part of the ruling power as the Pharaoh. The three feathers are therefore the especial symbol of the prince, the heir-apparent, Repa, Har-em-heb, or Har-em-khebt. These are the three feathers of the Prince of Wales. The earlier form of Wales was Gales, that is in Egyptian Kars. The Repa, or Prince, was the completer of the solar course, and in him the trinity of father, mother, and son was fulfilled. The Prince of Gales, or the Kart, is independent of a land called Wales, because the imagery belongs to mythology. The three feathers are a sacred symbol in Egypt, and as such were brought into this country, before the English Prince of Wales could be a title. There is a coin of Cuno-Belinus in the British Museum which has on it a horse galloping to the left, and the symbol of a diadem with a plume of ostrich feathers.[2] The first Prince of Wales, the Repa, the heir-apparent was Prydhain, the Horus of the Bards, son of Acddon, or Hu. When the English Prince of Wales was in India the Three feathers of his insignia provoked much curiosity, for it was, as they thought, an indigenous emblem. It is well-known to the Buddhists.[3]

[1] *Pierret*, 754.   [2] POSTE, *Celtic Inscriptions*, p. 130.
[3] *Journal of the Royal Asiatic Society*, v. 18, p. 391.

Both must have been independently derived from the same Egyptian source, the one centre where all these things will be found to meet at last by many winding ways.

Hor-Apollo says the Egyptians indicate the rising of the Nile by depicting three waterpots, "neither more nor less, because according to them there is a triple cause of the inundation. They depict one for the Egyptian soil as being of itself productive of water, and another for the ocean, for at the period of the overflow, water ascends up from it into Egypt ; and the third to symbolize the rains which prevail in the southern part of Æthiopia at the time of the rising of the Nile."[1] And in the poems of Taliesin, Hor-Apollo's description of the three sources of water has a perfect parallel. The Bard teaches that there are three primary fountains in the Mountain of FUAWN ; three fountains of Deivr Donwy, the Giver of Water. Tep is the Egyptian Source ; Tephu is the gate, valve, hole, abyss of Source, and Tennu is to bring tribute in the form of water. The Three Waters are the increase of salt water, where it mounts aloft to replenish the rain which innocently descends, and the springs from the Veins of the Mountain. This "odd sort of philosophy" about the origin of salt water rains and springs is contained in an account of the Creation, and word for word it is the same as Hor-Apollo's rendering of the water symboled by the Three Vases. The three primary fountains in the mountain of Fuawn correspond to the Triple Vase, one of the names of which is the Fent or fount. The mystical rendering in either case does not cancel the suggestion of origin in the Three Great Lakes at the head of the Nile ; a type has manifold applications.

NEF in Egyptian is breath or soul. The Welsh NWYF is a subtle pervading element ; NWYVRE, the divine source of motion. ANAF is Gaelic for SPIRAVIT. ENEF in Cornish is the soul. To NIFFLE in English is to sniff. Khnef (Eg.) is the breath of those who are in the firmament. "From NAVE are God and every living Soul."[2] NEVOEDD in Welsh is the Heavens. "It" in Egyptian is Heaven, and NEF is breath. NEF-IT would be Breath of Heaven. NEVION is a Bardic name of God. NEF-UN (Eg.) is breathing being. NEVYDD NAV NAVION, the Celestial Lord Navion of the Welsh, is in Egyptian, the Breather in the firmament and life of all breathing being.

The Egyptian NEB (another form of Nef) is the Lord. The Welsh NAV is the Lord. NEB means the Supreme, the All. NAV was the Supreme, the Lord of all. NEF is Lord of the Inundation ; NAV was the British Neptune, ruler of the seas. Neb, the all, is synonymous with enough. NEAPENS is English for both hands full : a primitive measure of enough. Neb was Twin, both hands, right and left, the whole of being. In the hieroglyphics NEF is breath, a wind,

[1] B. i. 21.   [2] *Barddas*, vol. i. p. 381, Williams.

fan, inflation of a sail, the name for sailing and of the sailor. NEF also names an old goat. Khnef (Num) the sailor of the Argo was personified with the head of an old Goat. The he-goat is the type of the breath or soul, the Ba. Ba-t, a participial form of Ba, means to inspire, give breath. This the Wind, NEF, did to the Sail, and the Goddess Nef-t, and Ba-t as genitor, did to the child.

Now when Martin was in the Western Isles of Scotland he found it was a most ancient custom for the sailors, when becalmed and praying or whistling for a wind, to hang up a he-goat to the mast of the vessel as the symbol of their beseeching.[1] This symbolic custom signifies the worship of Nef in these islands, or at least amongst the sea-faring folk, whether the Divinity be personified in male or female form. The Irish have a tradition that 600 years after the Deluge NEVVY led a colony into Ireland. He came, say the Welsh Barddas, in the ship of Nevydd Nav Nevion. This is the Egyptian NEF, the Sailor.

One of the master works and great achievements of the Island of Britain was building the ship of Nevydd Nav Nevion. It was the vessel which safely carried the male and female of all species over the waters of the Deluge. Stonehenge was a vast hieroglyphic of this vessel, called, as it was, the "Ship of the World." The Stone-Ankh, or Temple of Life, likewise designated a Ship, was the Ark of Nevydd Nav Nevion. We shall find the use and meaning of all the old Deluge paraphernalia by and by; at present we are stating facts, and swearing in our witnesses. This Ark of Life, or Ship of the World, is known as the seat of NÖE and ESEYE, and is designated the great stone fence of their common sanctuary.[2] Nöe is a modified form of Nef, and Nevydd is the Welsh form of the Egyptian Neft, personified as the Goddess Nephthys. The first representative of the Breather is feminine, and the Ark is likewise a type of the female. This Ship of Nevydd has not been left without its witness in a mocking world. These old roots of the past went deep into the soil, and though treated as weeds wherever they cropped up, the roots lived and held on below reach, and could not be eradicated. Stars do not disappear, or seasons pause, though we may lose our almanacs.

There is a small island named Inniskea, off the coast of Mayo, whose few inhabitants are purely pagan. They have an image which they call NEEVOUGEE, a long cylindrical stone that is kept wrapped up in flannel in the charge of an old woman who acts as its priestess. The name of NEEVOUGEE is still identifiable as the plural of a word signifying a canoe.[3] Waka in Maori is a canoe, and Oko in the Aku language. But, far more to the point, the UKHA of the monuments is the Sacred Solar Bark, an Ark of the Gods, and NEF-UKHA is the

[1] *Western Isles*, p. 109.   [2] *The Gododin*, Song 15.
[3] Sir J. E. Tennant, *Notes and Queries*, 1852, vol v. p. 121.

symbolic Bark or the ark of the Sailor. It is of especial interest that, although the NEEVOUGEE is connected by name with the canoe, it is not ship-shape itself, but a mere cylinder or type of the circle, the arc round which the Divinities sailed in their Nef-ukha. Stonehenge was not in the shape of a boat, yet it was an Ark, a circle, and here is the UKHA of Nef represented by a round stone, the true symbol in its simplest shape. The first Ark was uterine.

The Hamiltons quarter a ship on their shield, reputed to be that of Nevydd. In the hieroglyphics the Hem or Ham is a paddle, a rudder, to steer, and fish, which connects their name with the Bark or Ark of Nevydd, the Divinity of Breath or Wind, and identifies them with the Hemu as sailors and fishers, the people of the Hem or Water Frontier. There was also an ancient stone-temple at Navestock in Essex, where we still find a family of the name of NEAVES.

In Yorkshire the country folk call the night-flying white moths, souls. Our moth is the Egyptian Mut or Mat. Mat is to pass; Mut to die; Matt unfold, unwind, open, as the chrysalis entered the winged state and passed. The winged thing was a symbol of the soul; it appears in the hieroglyphics as the Moth or Butterfly. The Butterfly has no direct relationship to Butter. In the one case Butter is probably derived from PUT (Eg.), food; and TER (Eg.), made, fabricated. Our PAT is Egyptian for the shape. The Butterfly may be the type PUT (Eg.), TER, complete, perfect. Thus in death (Mut) the soul passed, unfolded like the Moth, whose chrysalis, like that of the Beetle, showed, and was the type of the process, whence the Butterfly. Calling the moth a soul identifies the imagery as Egyptian. In Cornwall departed souls, moths, and fairies are called "PISKEYS." Piskey is the same word as PSYCHE, and both are derived from the Egyptian in which KHE is the Soul, and SU is She; hence the feminine nature of the Greek P-SU-KHE. Without the article, SAKHU is the understanding, the illuminator, the eye and soul of being, that which inspires. So in Fijian, Sika means to appear as spirits.

It was said at the British Association meeting held in Newcastle, 1863, so great was the ignorance of natural history that a short time ago, when a man in the North of England was remonstrated with for shooting a cuckoo, the defence was that it was well-known the bird was a sparrow-hawk in disguise, as sparrow-hawks turned into cuckoos in the summer.[1] This confusion was the result of symbolism. The sparrow-hawk in Egypt was the Bird of Horus and of Ra the Sun-God, who ascended once more at the time of the Spring Equinox to complete the circle of the year. This hovering, circling bird was the type of the circle. The cuckoo, likewise the typical bird of return, is the Bird of the cycle. In the emblematic language the hawk and cuckoo were two symbols of one fact, the return of spring, and the cuckoo had to suffer for it. On the other hand, we

[1] *Times*, September 3, 1863.

have the reversal in the popular German belief that after midsummer the cuckoo changes back again into a hawk.[1] As hieroglyphics, they were similar, only the cuckoo had retained something of its sacred character after the hawk had become secularized, and had to suffer for its synonymousness as a symbol. Also the feeling that prompted the remonstrance against killing the cuckoo was a relic of the same religion. Word for word Gec or Cuck(oo) and hawk are one.

The hawk is the bird of Breath or Soul, and the sail is an emblem of soul or breath. The hawk on the monuments carries the sail as the sign of the Second Breath. With us the hawk's wings are denominated sails, so that the two types meet again in one figure.

The Cherry Tree was a form of the Tree of Life in Britain. Children in Yorkshire used to invoke the Cuckoo in this tree, singing around it

> " Cuckoo, cherry-tree,
> Come down and tell to me
> How many years I have to live."

It is a popular saying that the Cuckoo never sings until he has eaten thrice of cherries. The Cuckoo is a bird of the period, and is here connected with the Cherry Tree as a teller of time among the modes and appliances of popular reckoning. Telling leads to divination or foretelling. Hence the appeal made to the time-teller to foretell.

Any Latinist would assert that the word NARE for a nose and the nostrils of a hawk was derived from the Latin NARIS, the nostril. Yet it is not. The NARE for the nostrils of the hawk is not only the Egyptian NAR or NARU, but the ideograph of the word is the head of the vulture used for the value of its nostrils or keen scent of blood. This head of the vulture, NAR, is in English NUR, the head. The vulture's head was the sign of the bearing Mother in Egypt, both royal and divine—that is, the Nursing Mother in mythology ; and our NORIE is to nurture, and the nurturer is the Nurse, the NORU or NORIE.

The Magpie is one of our sacred birds, a bird of omen and divination, like many others suffering for its symbolry ; nine Magpies together being reckoned equal to one Devil in an old Scotch rhyme. If you see one Magpie alone you should turn round thrice to avert sorrow, and for good luck's sake try to see two. Why two ? "One's a funeral; two's a wedd'ng," says the proverb. Hor-Apollo tells us that when the Egyptians would symbolize a man embracing his wife they depicted two crows, for these birds cohabit in human fashion.[2] He also says they depict two crows as the ideograph of a wedding,[3] our "Two's a wedding." But why should turning round and making the figure of a circle obviate the disastrous

---

[1] Grimm, *D. M.* 1222.   [2] B. i. 40.   [3] B. i. 9.

halfness of the single Magpie when you would have found fulfilment
had you seen two? The bird is obviously connected with duality
and with making a circle. His name of Pie signifies twinship.
The full and early name is MAGOTTY, or, in the West of England,
MAGATI-PIE.[1]

The fact is, the MAG-ATI-PIE, the black-and-white bird, was the
equivalent of the Ibis, whose black and white feathers were emblematic of the dual gibbousness of the Moon.[2] But the clue to the
nature of its twofold character in colour or piedness being lost, the
twinship of completion is sought for in Magpie No. 2, or the dual
circle completed by the act of turning round.

TAHTI or AAHTI, the bi-une lunar deity, was imaged with the Ibis
head. Tet signifies to speak. Mag means to chatter. The Magpie
can be taught to talk. The Ibis cried " AAH-ΛAH." The Magpie
is therefore a dual form of the Word, as was Aah-ti. And in Mag-
ATI-pie we have this plural Word identified by the name of " AATI,"
the Lunar Deity or Bi-une Word. This is the reason why the
Mag-ati-pie was once a Sacred Bird and is now looked on as uncanny.

We might note that the PYE has other corroborative names in
PYNOT and PYNU. In Egyptian both NET and NU signify time;
also NET is a total; and the Moon, the Ibis, the PYNOT were each
the representative of plural time or the twin manifestation of Time,
whether signified by the two halves of a lunation or by other forms
of the Two Truths of Egypt.

The Lark, our Bird of Light and up-rising—we speak of rising
with the Lark—is a kind of Phœnix, the bird of re-arising or the
resurrection. The Egyptians called the Lark Akha-ter; AKHA is a
bird of light, a type of the spirit, and the word denotes spirit, light,
up-rising, lively, joyful; TER is a time. The AKHATER is a type of
rising-up time, which is, eschatologically considered, the resurrection.
Akhater is literally rising-up time, or time to rise. In a world
without clocks or watches the Lark was a voice of morning calling
out of heaven. So the Bennu (Eg.) Phœnix is a type of rising up,
Ben being the cap, tip, top, roof, highest point. Another form of the
Phœnix is the determinative of the word REKH, which means the
pure wise spirit, the spirit of intelligence. This is the Arabian Roc,
and our Lark as the Laverock is the Rekh of the Lift or Sky, the
soaring intelligent spirit; and either Laverock modifies into Lark or
the latter name is formed of Rekh with the *l* prefixed. The sole
point is to identify the bird as one of the Phœnix type.

The Phœnix was an image of the Sothic year. This constellation
came to the meridian at the time of the rising of Sothis. A star of
the first magnitude, Acharnar, belongs to it, and this name tells a
story. Akar (Eg.) is a name of the underworld; Nar signifies victory;

<hr>

[1] Brand, " Magpie."   [2] Plutarch, " of Isis and Osiris."

thus Akarnar in Egyptian denotes the victory over Hades, symbolcd by the Phœnix, the bird of resurrection.

In Hindu tradition the Crow or Rook personified the shadow of a dead man, and food was given to these birds as if to the souls of the dead. The Egyptian Rekh and English Rook having the same names as the Spirit (Rckh) will enable us to understand the typology. The Rekh was the emblem of the pure wise spirits of the dead, and the living bird is as good an ideograph as one portrayed on papyrus or stamped in stone.

Of the Great Bennu it is said that it caused the divisions of time to arise. One form of the Phœnix of Egypt, called the Bennu, is a Nycticorax. It was an announcer of time and period. The English Nycticorax is an owl called the Night-jar or Night-crow, which an-nounces the time of sunset almost as truly as the almanac, as the present writer has often proved. This peculiar bird, says Gilbert White, can only be watched and observed during two hours in the twenty-four, and then in dubious twilight—an hour after sunset and an hour before sunrise. It is consequently a Phœnix. The Night-Raven is one of its names; and the Phœnix is a deter-minative of the REPA, a type of time ; Seb (Kronus) being a true Repa of the Gods. REP and RAV are interchangeable, and our Raven is a Repa, and a Rook or Rekh of the night.

The word Jar represents the Egyptian Kher for Voice. The Night-Jar is therefore the Voice of the Night, that announces at the time of sunset. The Lark is a Phœnix of Dawn, the Night-jar of Sundown. And the Night-Gale or Nightingale is likewise a Voice of the night. Gale is a song, to cry, scream ; garre is to chirp. The root of all is KHER (Eg.), voice, utterance, and the Night-gale, like the Night-Jar, is a Voice of the Night.

The Robin is a Repa by name, and the FINCH is a form of Phœnix by name.

The Cock is also a Phœnix, a bird of annunciation. He must have been so in Egypt, where the later sensitiveness to his well-known character caused him to be prohibited. The Cock was a type both of Mercury and Apollo. Cock-crow is the first time marked after midnight, and Cock-shut is a name of eve. The Cock is named from the Egyptian Khekh, which denotes Light, the Horizon, Equinox, cackling or crowing, also to turn and return. The Cock is the same to the night that the Gec (cuckoo) is to the year—the Phœnix of its cycle. As the Bird of Returning Light he was made a Sun-Bird in relation to the Equinox, and a victim of theology in the cock-throwing sports of Shrove Tuesday. As the Weather-Cock, he is the emblem of turning and returning, or the Khekh.

The BEAN Goose is a northern form of the Bennu, the bird of return that typified renewal and renovation. It is a bird of passage which is one of the first to arrive on the English coast about the end of August, and is known on the continent as the Harvest Goose.

The Bennu in Egypt was the symbol of Osiris in Annu, the risen god or soul of the deceased, and this eschatological character has been conferred on the Bean-Geese. As they fly by night they make a strange noise, and are called "Gabriel's hounds." The word hounds is possibly a corruption of HAN-SA. HAN (Eg.) means to return : SA is the goose. HANSA is the Sanskrit name of the goose. Our word goose may be derived from KHES (Eg.), to return, come back again. We have our Bennu too in English, as the BOON, an undistinguished fowl. Hor-Apollo tells us that when the Egyptians would denote a son they delineate a CHENALOPEX, a species of goose, because this bird is excessively fond of its offspring, and if it is pursued and in danger of being taken with its young, both the father and mother will voluntarily give themselves up to the pursuers so that their offspring may be saved. For this reason the Egyptians consecrated the goose.[1] Now if this foolish fondness of the returning (Han) goose (sa) be applied to the son, who is also SA (Eg.), we have the German goose of a son called HANS. HANSA, besides being the bird of passage, also reads the young (han) goose, (sa) Hansa, or Hans. The early reverence for the goose changed into derision at its simplicity, its SILLINESS in the later sense.

The BEAN was used in Egypt to throw upon graves. This signified the resurrection. The rising again thus typified by the bean was also symboled by the Ben or Phœnix. The Bean, which is synonymous with the Ben, is obviously the same by name, and the various uses to which it has been applied show its hieroglyphical nature in our land, as a type of transformation and renewal. The "Bean-feast" is especially celebrated by builders. The name has been erroneously derived from the BEAN-goose. When the employer gives his men an outing in the country it is called a "Bean-feast." But the true BEAN-feast of the builders is the one commonly known as the Roofing. When the building is reared and the roof is put on, the event is celebrated if in ever so small a way. We often see a red cotton handkerchief hung up as a symbol. The roof is the type that identifies the BEAN. BEN, or ben-ben (Eg.), means the cap, tip, top, supreme height, and is the name of the roof. The determinatives of this height are the obelisk and pyramid, and in the parish of Monswald, Dumfries, there were about twenty years since some large grey stones called "a BOON of shearers," said to represent a company of reapers, who were turned into stones on account of their kemping, i.e. striving.[2] The sole point here is the correspondence of the Boon to the stone BEN of the hieroglyphics, the mountain (Ben) raised in stone. The Bennu (Eg.) is a great stone of some kind. The

---

[1] B. i. 53.　　　　　[2] Brand, *Harvest Home.*

Boonwain was one that would carry the loftiest load, and ABOON means above, overhead.

Nor shall we find a more satisfactory origin for the name and signification of the BON-fire than this BEN, the roof, tip-top, the lofty and splendid. In English, Bin is a heap ; in Welsh, Ban is high, tall, lofty ; and the Bon-fire is a Ban-ffagl.

The Bon-fire belongs properly to the time of the Midsummer solstice, when the sun was at the summit and its light at the longest. The fires were kindled at the top of the highest hills, and the time of lighting them was at midnight. Everything was symbolical of the topmost, i.e., of BEN-BEN, the cap of the hill, the tip-top of time and roof of the house of heaven.

There was a form of the ben-ben or pyramidion of the solar god which was equinoctial in the worship of Atum ; but the Bon-fire was the Baal-fire, the fire of the Sabean, not Solar Baal ; the fire consecrated to the reappearing Sothis, the star whose rising crowned the summit of the year, as the star crowns the Pyramid in the hieroglyphic representations, when the Bennu came to meridian. The Bon-fire typified the fire in which the Phœnix (Ben) was fabled to transform.

No symbol in Egypt was more reverenced than the Beetle, in whose likeness the god Khepr was fashioned, as the Former and Transformer. He is represented as rolling the solar disk, and has the title of Khepr-Ra. But transformer of time, of one cycle into another, is the idea conveyed. Khepr was the type of transformation, the Egyptian mode of figuring immortality as continuity, and the Beetle (or Beetles) was stationed where the Crab is now. This point was the beginning and end of the solstitial year. Khepr clasped the zodiacal circle of the sun with one hand to each half of the whole. Here he received the sun, and passed it on in what was termed his boat. The beetle was made the great symbol on Egyptian rings and commemorative coins, as an image of Khepr, whose sign was the fibula of the starry round, Khepr being, so to say, the keeper of the solar wedding-ring.

Khepr was also identified with the sun itself, that went round for ever and ringed the world with the safety of light continually renewed. Khepr in his boat was the antithesis of the Deluge. Khepr-Ra is literally the sun-beetle, and this symbol of continuity, transformation, and resurrection was so profusely lavished in burial of the dead that the ancient scarabæi are plentiful in Egypt to this day. All that pertains to Khepr must have been as familiar to the British people as to the Egyptians, and the beetle was regarded with a feeling as religious as theirs. In English folk-lore, if you kill a beetle it will be sure to rain. The reader will not see the full symbolic force of that until we have mastered the Deluge myth. Khepr rolled up his ball and built his Ark to save the seed against the coming inundation. If you tread on the dark shimmering beetle called the sunshiner, the sun will suffer eclipse; as it is expressed, the "sun will go in." That is, because it was a symbol of the sun, and treading on its image was figuratively covering and eclipsing the sun. The beetle was the sign of the summer solstice, and our SCARABÆUS SOLSTITIALIS abounds at midsummer. PUTAH is a beetle-headed god, and the BETE in Devon is a black-beetle. Thus Put the opener and circle-maker keeps up his character as the Bete, whence the Beetle. In the monuments one name of Ptah, the Scarab-headed god, is Khepr-Ra or Sun-beetle, and Ptah was often painted of a green complexion. One particular sun-beetle with us has a head of gilded green, and is called a CHOVEE or CHOVY. Also we have the name of Khepr in the CHAFER or Dor-beetle, and CHAFER-DOR is a name doubly Egyptian, it shows that TER (now almost given up) was also a name of Khepr. Shevdilla, an Irish name of the beetle, also equates with the Dor-Chafer.

Our COOPER is by name and nature a form of Khepr; he rings or hoops round the staves of the cask as Khepr clasped the circle of the signs. In English keeper is a clasp, and to kep, an earlier form of hoop, is to enclose. From this comes the keeper-ring of marriage that encloses the plain gold ring. Khepr made the circle of Time as the Sun, and his image was placed at the juncture where one cycle was transformed into another, and the year renewed. In our childhood we were taught that if we found the beetle lying on his back it was a good deed to turn him over and set him on his way. This presented the image of pause and retardation, meaningless, except related to the creator of time and keeper of continuity, but the act was still performed when the consciousness was lost; the ideograph, no longer read, was interpreted by faith. With the Norsemen, this aid to the beetle was supposed to expiate seven sins. The beetle was called the Bug of Thor, Egyptian Ter. On the introduction of Christianity, the Thorbug was christened the Thor-Devil, to be kicked out of the way rather than helped upon it, yet the simple countryman, unthinking of Thor, will stop and turn over the poor beetle, "that we tread upon," who is a dark shadow on the earth of things heavenly.[1] These superstitions do not need to be damned; they want to be explained; and they were only damned for the purpose of foully discrediting them as witnesses to the religious origines.

How ancient, for example, is the order of the Sacred Heart! The thirtieth chapter of the Ritual was frequently inscribed on a scarabæus of hard stone, and placed inside the heart of the deceased, and the Rubric directs that these words are to be said over it with magic: "My heart is my Mother, my heart is my transformations." "My heart was my Mother—my heart was my Mother—my heart was my

---

[1] Thorpe, *North. Mythol.* ii. 53.

being on earth, placed within me, returned to me by the chief gods."
The transformation symboled by the beetle was the Egyptian " change
of heart," and renewal. The heart, mat or hat, as an abode of life, really
represented the hat or kat, the womb, hence the meaning of " my
heart is my Mother," and its relation to the re-birth by transformation.
The sacred denotes the secret heart, the same that became the type
of Cupid and the object of his shaft. The " Sacred Heart " of
Rome is a flaming heart, and, as may be seen by the Rosaries, it
represents the uterus of Mary.[1] The Deceased, lying at rest, in the
thirty-second chapter of the Ritual points to his beetle, and other
potent talismans, and says, " Back, Crocodile of the West, who livest
upon the Khemu who are at rest; what thou abhorrest is on me,"
or it was placed within him in the tomb which, like the heart, imaged
the mother as the womb of re-birth.

Moufet[2] says the beetle hath no female, but shapes its own from
itself. This did Joach. Camerarius elegantly express when he sent
to Pennius the shape of this insect out of the storehouse of natural
things of the Duke of Saxony with the lines :—

> "A bee begat me not, nor yet did I proceed
> From any female, but myself I breed."

For it dies once in a year, says Moufet, who thus enshrines Egyptian
mythology in a popular superstition, " and from its own corruption,
like a phœnix, it lives again, as Moninus witnesseth, by the heat of the
sun." According to P. Valerianus, there was a notion that the scarab
only rolled its ball from sunrise to sunset. The Singhalese show great
anxiety to expel the beetle that may be found in the house after
sunset, though they do not kill it.[3] Moufet repeats Plutarch, who
asserted that the beetle was male only in sex. But this is to mistake
the symbol for the thing signified. It was depicted as rolling the sun
through the heavens, and that course ended visibly with sunset. It
made the annual circle, and was thus the symbol of a year, or Ter,
hence said to die and be renewed once a year. There is a more
remarkable misunderstanding connected with the beetle, concerning
the "death-watch." Sir Thomas Browne observed that the man who
could cure this superstition and "eradicate this error from the minds
of the people, would save from many a cold sweat the meticulous
heads of nurses and grandmothers." It is easily explained. The
beetle was the type of Time, and associated with the end or renewal
of a period. The Beetle was that celestial sign in which the solar
year ended and a new year began.

The " death-watch " is a kind of beetle (*Scarabæus galeatus*

[1] Inman, *Ancient Faiths*, Figs. 47 and 48, v. ii.
[2] Moufet, *Theatrum Insectorum*, p. 149.
[3] Tennant, *Nat. Hist. of Ceylon*, p. 407.

pulsator). It is a helmeted beetle, and this identifies it with Khepr, for Kheprsh is a helmet, and the word denotes the horn of Khepr.[1]

Melchior Adams records the story of a man who had a clock-watch that had lain for years unused in a chest, which of itself struck eleven in the hearing of many before the man died. This indicated the nearness of the end of time, or twelve o'clock. So the death-watch denotes the end of time for some one belonging to the house, because it is still a symbol of Khepr. When the sun entered the sign of the Beetle, the clock of the year struck twelve : it was the end. This superstition shows the beetle to have been as sacred in Britain as it was in Egypt, about whose worship of insects and animals so many shallow things have been written.

The ancient Britons not only buried the beetle with their dead, but the same genus of it was chosen—the DERMESTES. In one of the stone coffins exhumed from the Links of Skail, which barrows are of the remotest antiquity, a bag of beetles was found, the bag having been apparently made of rushes.[2] They belonged to the genus DERMESTES, four species of which were found by Wilkinson in the head of a mummy brought by him from Thebes.[3] Obviously the beetle was buried in both instances, for one reason, it was the emblem of Time, ever-renewing, whence came the Eternal. The name DER-MESTES still tells the tale. " TER " was Time, the beetle-headed Khepr, MES is birth and to be born, TES "in turn." The scarab not only represented the circle of the sun, but the ever-turning time, the renewing cycles of the soul.

The beetle was buried with the mummy of the dead. This in Egyptian is the Mum image or type of the dead. And one name of the beetle, the type enclosed with the dead, is in English " MUM." There is a reason why the beetle insect and the beetle as Maul have the same name. The principle of this naming alike is to be found in the two-fold nature of each. The beetle, in making the circle, worked at both ends; so does the Maul, swung in a circle. In the Maori the maul is called " TA," which is an Egyptian name of the beetle. TA, in the hieroglyphics, is the head of a mallet, the wooden beetle, as well as the name of Khepr. So that this double meaning of the beetle, applied to both mallet and scarab, was Egyptian, as it is English. Another conjunction of this kind occurs in the person of Thor (Ter) and his beetle or mallet. These are types of the biune Khepr who made the two halves one. To Kep is to close two into one, make the copula. This Khepr did. We have the root meaning in cop and kyphor, whilst to kipper fish is a form of making two into one by taking out the backbone.

We still call half-and-half by the familiar name of COOPER, a drink

---

[1] *Ritual*, ch. 93.
[2] Gough's *Sepul. Monts.* vol. i. p. 12.
[3] *Anc. Egypt*, ii. (2nd scr.) 261.

composed of two in one. In Ireland the Mountain Kippure, from which the two rivers Liffey and Dodder run down to the Dublin plain, is named on the same principle as the beetle of Egypt, the one that was held to be bi-une, the one image of source in which the two factors met, or from whence the two sources issued.   The Mountain Kippure was the starting-point of the two rivers.

HEPT (Eg.) is to unite by an embrace. So the staves of the cask are united by the embrace of the rings, or as we say, it is HOOPED, in Egyptian, "HEPT." A quart pot used to be called a hoop ; it was bound by hoops like a barrel. "HOOP" also denotes a measure of liquid. Generally there were three hoops on the quart pot, so that three men drinking together took each his hoop. Jack Cade proclaimed that when he was king the three-hooped pot should have ten hoops, which would not have suited at all unless the pot had been greatly enlarged.[1]   Hoop is a measure of corn as well.   The word "Ap" (Eg.) is a quantity of liquid, to take account, reckon. "Ap-t" is measure and judgment. If we place the article first, we get the Tap.   The Tap was the tavern or Tabern, sometimes called the Tabard.   And "Tebu" (Eg.) is to draw liquid, that is to tap.   With us the place where it is drawn is the Tap, the instrument it is drawn with being a tap.

The soul of man, says Spenser, is of a circular form.   That is a hieroglyphic to be read by the hieroglyphics.   The circle is the symbol of a period, in this instance masculine.   The same sign, an Eaglet, says Hor-Apollo,[2] symbolises the seed of man, and a circular form.   The soul was the seed of man, a determiner of time and period in creation.   BA is the soul, a circle, a metal ring, and seed-corn.   In Chaldee a circle is zero ; our zero is still signified by a circle, and zero, Egyptian Ser, is the seed, and Ser is the same as soul. The minds of philologists have wandered the world round, always excepting Egypt, in search of the word "BODY."   Every sense of the word is found in Egyptian.   "PET" is foundation ; "PAUTI" is type, form, image, to figure forth or embody. PAUT is a company, the Paut the company of nine gods, the whole BODY of them. BA is to be a soul, AAT house : BAAT adds the feminine terminal, whence Beth, the abode of the Ba, or soul, that is the BA-T, BU-T, BETH, BOTHY, abode of soul or the BODY. Again, BA is the soul and TI is a boat ; in this combination the boat and body are identical, as are the abode and body. With the Polynesians a body of men or gods is a houseful or a BOAT-ful, a body, the boat, in Maori, being a POTI. The Ba (Eg.) is the soul, and HAT the heart, and the heart was con-sidered to be the shrine or body of the soul, so that BA-HAT, BA-TII, BA-T, the abode of the soul, is the house, or place, so named in Egyptian. But the word BODIG, Gaelic BODHAG, is an earlier form,

[1] ii. Hen. vi. 4, 2.   [2] ii. 2.

and it is suspected that the hieroglyphic TA had the force of TCH, going back to the click, just as body was the earlier Bodig.

The typical circle PUT is another name for heaven. So that the vulgar expression, "gone to POT," may not be so brutal as it sounds, for, in Egyptian, gone to PUT would mean gone to glory, to heaven; more literally, gone to join the divine circle of the nine gods. And this is our POT, as a circular form both in the cooking utensil or drinking measure, and as POT, the name of the circular black pudding, made of blood and groats. Going a-puddening is going round. PUT in Egyptian is to feed as well as food. Old English "POT days" were sacred to receiving and feeding of friends thrice a week.

"PAUTI" is a name of Osiris as the dual Creator; the biune Being. This is the full form of Put, and gives the plural of male and female, the circle of two halves, Osiris and Isis conjoined. They are, as we say in English, the two BUTTIES or mates. TI is two, reduplication, and a BUTTY is one of two mates who work together. The company of nine gods called a Pauti are equivalent to nine Butties, the number nine being the full Egyptian plural. A still more striking instance of descent from the divine to the dunghill occurs with this word PUT or PAUTI. In the hieroglyphics, as said, PUTI or PAUTI is the circle of heaven divided into two halves, upper and lower, north and south. And this image, as the initiated know, was sacredly perpetuated in the genuine English PETTY, with its upper and lower, larger and lesser halves of the whole. The PETTY-toes of the pig are likewise divided into upper and lower, larger and lesser, as their form of two-foldness. The petty-sessions again imply the same duality of being, as the lesser of two. And just as puti becomes put in Egyptian, so does petty become pet, hence the diminutive; also PUD is the hand or foot, one of two, as is the paddle and puddock (frog); the pod being a whole formed of two sides. PETI (Eg.), for two or both, is found in PAITA, Tariana; PAIHETIA, Brierly Island (Australia); BIT, Chinese; BAT, Basque; BOTEWA, Talamenca, and English BOTH.

The CARE-CLOTH was a kind of canopy used at one time during the marriage service. At Sarum when there was a marriage before mass, the parties kneeled together and had a fine linen cloth, called the care-cloth, laid over their heads during the time of the mass till they received the benediction, and then were dismissed. In the Hereford Missal it is directed that at a particular prayer the married couple shall prostrate themselves while four clerks hold the four cornered care-cloth over them. The care-cloth occupied the place of the Jewish canopy. The word Kar (Eg.) means a circle, sphere, zone, round, with the especial sense of being under and with. Kher has the meaning of being under and with, and Kher is the name of a shrine. Khar also means to enter, go between, beget. From which we may gather the care-cloth was symbolical of the marriage shrine, and all that is implied by marriage.

"SNATEM" (Eg.), rendered reposing, to be at rest, is applied to the bearing mother. It literally means the mother tied up. The GREAT mother is the mother great with child, and she was so represented as TA-URT with the tie or SNAT in front of her. SNAT or SENT means to found by tying. SNATH (Eg.) is a tie, and to tie. This we have in English. A SNOTCH is a knot. SNITCH means to confine by tying up. The SNOOD may be derived from SEN (Eg.), to bind, AAT, a net. Our SNOOD is the net-fillet for confining, that is SNOOD-ing up the hair; and the snooded maiden is our form of Mut snatem, the tied up or snooded mother. Snooding the hair was one of the various symbolical customs of tying up and knotting used at marriage, having the same significance as the true love-knot, the enfolding scarf, the garland, girdle, and the ring of gold, all of which were typical of the tying up of the female source by the male on which procreation depended. The hair of the woman in Egypt was not tied up or snooded until she was wedded. This, too, was a custom in our islands either at marriage or betrothal. It is alluded to in the song, " He promised to buy me a bunch of blue ribbon, To tie up my bonny brown hair." The "top-knot" of the bride is frequently mentioned. To " tyne her snood" was a synonym for loss of virginity; only because of being a mother but not a wife. She was unsnooded or not snooded. Camden, in his *Ancient and Modern Manners of the Irish*, says, they presented their lovers with bracelets of women's hair,[1] for which ornament the hair was cut off to form a typical ring. This is the equivalent of tying up and snooding the maiden's hair; but the symbolism goes still farther in converting the type of maidenhood into the tie of marriage.

The glove sent or thrown down with a challenge identifies it as a symbol. Gloves were ensigns of a bridal given away at weddings. White paper cut in the shape of women's gloves was hung up at the doors of houses at Wrexham in Flintshire as late as the year 1785, when the surgeon and apothecary of the place was married.[2]

It was at one time the custom in Sheffield to hang up paper garlands on the church pillars, enclosing gloves which bore the names and ages of all unmarried girls who had died in the parish. Another custom renders it imperative for the gentleman who may be caught sleeping and kissed by a lady, to present her with a pair of gloves. In the North of England white gloves used to be presented to the Judge at a maiden assize when no prisoner had been capitally convicted. These are still presented to the magistrate of the City of London when there is no "case." The glove is a hieroglyphic of the hand. The hieroglyphic hand is TUT, and the word signifies to give, image, typify, a type of honour, distinction, ceremonial.

Gough's *Camden*, iii. 658.   [2] Brand, *Gloves at Weddings.*

One naturally turns to the hieroglyphic symbols to see what help they will give in unriddling so universal a thing as the wearing of horns assigned to the man who has a wife untrue to him. Horns are generally taken to be symbolic of male potency. We forget that the cow has horns as well as the bull, and that the horn is not limited to sex.

The horn is a masculine symbol, but like many others, most ancient, not solely male. Cornutus, to be horned, is the Egyptian Kar-nat, the phallus placed in position as horns. Kar-nat is derived from karu—support, bear, carry; and nat, the tool or instrument. This applies to both sexes. The feminine Nat was the Goddess Neith, the cow-headed bearer and bringer forth of Helios.

The fact is that horns on the head are chiefly a feminine symbol. The cow and moon were the typical horn-wearers, and both were feminine signs. Cow and moon carried the orb between their horns, as bearers of the light. The cow in agriculture draws with its horns. The emblematic value of horn was in its hardness. This made it an image of sustaining power. Hence the horns sustained the solar orb. The horns belonged to the beast of burden; the bearer was by nature the female, thus the horned cow bore the burden and carried the sun, the type of masculine source. A curious application of this imagery is seen in the monuments. In the time of Tahtmes III. the subject race of the Uauat send tribute to Egypt, and amongst other tokens the horns and tufts of cattle are made use of to represent a negro with arms raised as if in supplication, whilst others carry their offerings between the horns. "I passed over on her fair neck," says the Solar God of Israel, speaking of the HEIFER of Ephraim.[1] She was my beast of burden is the sense.

The horns, then, are a symbol of bearing and sustaining. The Great Mother, the bearer, was not only horned like the cow and the moon, for Neith and Mut were also given the horn of male power. In the hieroglyphics the cow and the victim are synonymous, as the Kheri bound for the sacrifice. The horns of the Kheri, cow, victim, were wreathed and gilded for the sacrifice. And the horns figuratively applied to the cuckold have the same meaning; they are the hieroglyphic of the man who patiently bears, and who is the victim led to the sacrifice by his wife.

According to symbolism the husband of an adulterous woman is not only the common butt, but he is the pitiful beast of burden, willing to bear; willing to be the sacrificial victim, and as such he is crowned with horns. This reading is sustained by the custom of horn-fair, anciently held at Charlton, in Kent, on St. Luke's or Whip-Dog Day,[2] October 18th, to which it was the fashion for men to go in women's apparel.

Hor-Apollo[3] says a COW's horn when depicted signifies punishment.

---

[1] Hos. x. 11.　　　　[2] Brand.　　　　[3] ii. 18.

Doubtless the sign stood for a fact, and the custom of imposing horns, whether figuratively or not, would be Egyptian. The cow's horn, the horn of the victim of the sacrifice, the type of punishment, proclaims this to be a cow of a man, not a bull; hence a coward.

To be cuckold might be derived from the habit of the cuckoo in making use of another bird's nest for laying its eggs, but that would make the term cuckooed, and cuckoo is not a primary. Gec is the old name for the cuckoo, and this correlates with the Gouk as a fool. Cuckold read as Egyptian is the peaceable meek worm or the old man. KAK is a worm, KEH-KEH, the old man; URT is meek, feeble, inactive, bearing. This is probably the terminal syllable of coward —the one who is meek and peaceful of bearing as the cow. The cuckold is the coward, hence the horns. There was a subsidiary sense, which contains the postscript, the sting in the tail. The cow is horned at the head, but that does not make it a bull. Greene[1] says the cuckold was as soundly armed for the head as Capricorn. The cow-horned man is a sort of fellow-figure to the woman who wears the breeches; he takes her place as the BEARER!

The sacred origin of the bishop's apron can be illustrated hieroglyphically; it is an extant form of the figleaf or skin with which the primal parent clothed herself, and of the loin-cloth of the naked nations. The apron of the goose or the duck is the fat skinny covering of the belly. The apron is a Base, a garment worn from the loins to the knee in the mythical representations, in which six Moors danced after the ancient Æthiopian manner, with their upper parts naked, their nether, from the waist to the knee, covered with bases of blue.[2] Butler, in *Hudibras*, calls the butcher's apron a Base. The BASU was worn by Egyptians as an apron or kind of tunic. It is found on the rectangular sarcophagus in the British Museum. The Basau is also a sash with ends behind. The name relates the garment to the Genitrix Bast, and to the feminine period, BESH in Egyptian, PUSH-(pa) in Sanskrit, BOSH in Hebrew, PISH and BISI in Assyrian, BAZIA in Arabic, and to Bes the beast. After its first use the Basu became a type of the second feminine phase, the covered condition of the gestator. Hence Bes to bear, dilate; Bes, protection, the amulet (of the true voice), the candle (cf. AR, the candle, and to conceive). The Basu was made of the skin of the tiger or spotted hyena, the beast of blood. It was worn by the sacrificer and the later butcher.

The one who hunted and slew the beast and wore the skin for his Basu was an early hero. Hence it was worn in the form of an embroidered tunic by the knights of chivalry. "All heroic persons are pictured in BASES."[3] Bes (Eg.) means to transfer, and the Bes

---

[1] *Conceipt*, p. 33 (1598).   [2] Jasper Maine, *Amorous Warre*, iii. 2.
[3] Gayton, *Festivous Notes on Don Quixote*, 1654, p. 218.

skin of the beast was transferred to the conqueror. This was typical of another conquest, and of the Basu, whether as apron or tunic worn by the male. To cover and to cure are synonymous. The Basu as loin-cloth was emblematic of both. It was then transferred as a trophy to the male, and was promoted from the domain of the physical to that of the spiritual cure. The Egyptian king wore a kind of apron in certain ceremonies, and it was a part of the rite for him to furtively take and conceal some object beneath his apron. This act was typical of " MEN," to conceal, which is an euphemism for fecundating, used in the expression, " O creator of his father, who has concealed his mother "—the literal meaning being, " who has fecundated his mother." [1] The king's apron was a form of one worn by Khem as the sower of seed. The seed in Egyptian is NAPRA, and the English Apron is the NAPRON. If the wearers of these relics of the primitive past did but know their typical nature, they would hasten to deposit them in the nearest museum of antiquities, and never again wear them in the presence of men and women. They belong to the " Mysteries " that have not borne explanation.

In the Semitic languages a skull-cap is named TAKIYYA. In Chinese a helmet is THUKIU. In Cornish-English TOC is a cap or hat. TAJ (Arabic and Persian) is a modified form, meaning a skull-cap. TYU, in Zulu-Kaffir, is the cap or cover. The cap or cover includes THEAK (Eng.), to thatch, and the bed-TICK, TEKE (Maori), the pudendum mulebriæ; TAKARI (Sans.), a particular part of the same; Degy (Cor. Eng.) to inclose and shut in. In Egyptian the original form is worn down to TAAUI, a cap with a tie, that is, close-fitting like the skull-cap or helmet. TEKA (Eg.) yields the idea of all as to fix, fit close, cleave to, adhere. The Cornish TAKKIA, to fix, TACHE (Eng.), to clasp and tie, TACK, to make fast, whence TACKED is tight, and tight is tied close, TEKA (Eg.), close-fitting, fixed. Once the root is run down and detected in Egypt, it may be followed on the surface the world over.

In the Welsh writings we meet with three crowned princes, whether mythical when called Mervin, Cadelh, and Anarawt,[2] does not matter; each one wore upon his bonnet or helmet a kind of coronet of gold or head-dress made of lace and set with precious stones : this in Welsh or the ancient Kymry, was called the TALAETH, the crown, diadem, or band, a name given by nurses to the band or natural crown that determined its being a hero. In relation to this it is common amongst the English peasantry for the nurse to examine the child's head for the double crown, and if it be there, the child, who of old was to become a hero, is now to " eat his bread in two countries."

The TALAETH is the tiara, but with the Egyptian terminal T.

---

[1] Birch, *Records of the Past*, v. 10, p. 143, note.
[2] *Girald. Camb. Description.* cap. ii.

The "TARUTU" is found on the monuments[1] as a band of lace or net-work, with the determinative of hair. Taru is the name of the hero. Tu is a tie, ribbon, or band, hence the TARUTU, the head-band of the hero, is the TALAETH worn by the Welsh princes. Many conjectures have been made respecting the origin and meaning of the S-collar worn by our Lords, Chief Justices, the Lord Chief Baron of the Exchequer, the Lord Mayor of London, the Heralds and Sergeants-at-Arms. The collar consists of a blue and white ribbon lettered with S's in gold. Gold, to begin with, signifies a Lord (Neb) in the hieroglyphics, and the collar or tie is a syllabic SA and phonetic S. Another symbolic SA is an ornament with ten loops. These signs are ideographs of rank, kind of officer, virtue, efficacy, protection, amulets, and of an order. The ten-looped SA is also the determinative of a court, which may be royal, or a Court of Justice. The SHABU is a collar of nine points, the name of which appears to identify it as worn by judges, and this same collar is also called the USKH. S is our representative of both Sa and Us. SA or Us then is the Tie, the sign of rank. SA for instance is the Genius of Wisdom, and KH denotes the title; hence the USKH, S-KH, or, as we say, the S-collar. We know the Egyptian judges wore the USKH or SA collar, with the amulet and sacred symbol of Ma, the Goddess of Justice, attached; and the cause was not opened in the Civil Court till the collar was put on.

The USKH collar was the Judge's collar, as the Uskh (Hall) was the "Hall of the double Justice." The English S-collar, then, is a form of the SA-collar or USKH. The S-collar read as Egyptian is the SA-collar, a symbol of wisdom, the sign of an order, or rank, or judgeship. USKH, the collar, also means broad, corresponding to the broad riband.

In addition to the S-collar and the broad riband the USKH had a third form in the ESCU, the knight's shield and sign of service. From this comes the title of esquire, one who was a shield-bearer, and he who had carried the ESCU had the right to be called esquire. The shield ESCU, like the USKK collar, was the symbol of an order.

A piece of rag is a hieroglyphic which to some extent can be read. It was a common practice for those who visited the Holy Wells, and drank of their waters of healing and purifying, to leave upon the tree or bushes near some shred of their clothing, or bit of rag. The rag was apparently offered on the principle of "gif-gaf." It was a token of exchange, the rag representing the disease deposited in return for the healing. But the word RAG recovers an ideograph. It is identical with REKH (Eg.), to full, wash, whiten, purify. And when a well is found named RAG-Well, between Benton and Jesmond, near Newcastle, and it is a famous well of healing, we may suppose it was not named from the RAGS left there, but from REKH,

---
[1] *Select Pap.* xlvi. 4.

to purify, make white, and heal. The rag becomes hieroglyphic as the symbol of sores. The rag, as chalk, is named from its whiteness.

"He's off his CAKE" is a provincial phrase, explained as meaning he's off his head. It signifies he's loose-witted, out of bounds, "out of all Ho." Caker is to bind with iron, and this connects the cake with a boundary. The hot cross-bun of Easter is a cake-symbol of the equinoctial boundary. There can be no doubt about this hiero-glyphic having the same value as it had in Egypt, where it was the sign of boundary, orbit, and circumference, which the daft man is out of who is off his cake.

One symbolical phrase for dying is to "KICK the bucket." Our kick is the Egyptian KHEKH, to recoil, return, send back, return. To KHEKH the bucket in that sense would be to return it empty; the one of the two that returns for the water. It has, however, been suggested that the image of kicking the bucket was drawn from the mode of hanging up a dead pig by the hind legs; the crooked stick which stretches the legs being called a bucket. This is the PUCKHAT (Eg.), a stick (crooked) or rod, with the meaning also of stretching out.

Chaucer in the "Pardoner's Tale" speaks of Dice as "bicched bones," and offers a *crux* to the philologists, which the Egyptian word "PEK" solves in a moment. PEK is a gap, hole, shape. And with the terminal T "PEKT" is "BICCHED;" that is, pitted with gaps or holes. It is a variant of PECKED and POCKED. Pight is pitched, placed. To peck is to make the gap or hole, and the instru-ment is a pick-axe. The pig is the animal that routs or pokes. Peckled is speckled, spotted with pecks. The cowslip is called a paigle. The Egyptians had the "PECK," measure, and this, like the pock in the dice, is a form of the pek or bicch. Next we have the biggin. The PEKHA is an Egyptian rod, English PEG, and PKH-KHA means divisions. Both peg and division met in one word in the custom of drinking from a tankard marked inside by pegs graduated for the purpose of dividing the liquor into equal shares; whence the phrase of taking one down a peg.

Plutarch informs us that the Egyptians called the loadstone "the Bone of Horus." And Martin tells us[1] that in the little island of Quedam, in front of the Rock of Quedam, there was a vein of ada-mant, the loadstone: and some of the natives told him that the rock on the east side of Harries had a vacuity near the front in which was a stone called the Lunar Stone, and this advanced and retired according to the increase and decrease of the moon. This is the legend of the loadstone found in Quedam. The Bone of Horus was found in Quedam, where there is a place named Harries; this may read the Temple of Har. Now the loadstone is adamant (earlier kadamant), *ergo*, Quedem-ant, which is Egyptian for the

[1] *Western Isles*, pp. 41-50.

stone of Quedam. This will enable us to recover the name of the loadstone as Khetam (Eg.). Khetam means to be shut, sealed, a lock, a fortress, all that we term adamantine. Khetam was the name of a seal-ring, the sign of shutting, sealing, stopping, locking. And it is now suggested that the Bone of Horus was Khatam-ant (Adamant) the Stone of Khetam. Khet means to cut, to reverse and overcome; am or ma is with. Khetem. is a place named on the Monuments; locality unknown. Possibly the mineral called Khetem may not always have been gold, the Harris Papyrus mentions a mythical monkey having an eye of Khetam. Why not of the loadstone, as a type of attraction?

A Devonshire talisman in possession of a Miss Soaper, of Thurshelton, North Devon, a *bluish-green* kind of stone, is called a KENNING-Stone. This was not a knowing stone, but a charm against disease, which it averted or sent away. It was held to be particularly potent for sore eyes when they were rubbed with it. An operation for the cure of the bite of a mad dog is called KINSING.[1]

The sense missing is found in KHENA (Eg.), to avert, blow away, puff away, repel, carry off, with the determinative of Typhon the Adversary.

The Lee-stone is a curious talisman belonging to the family of Lee in Scotland. When tried by a lapidary it was found to be a stone, but of what kind he could not determine. It is dark red in colour and triangular in shape; and is used as a charm against disease and infection, by dipping the stone in the water and giving the water to the cattle for drink. In the case of a bite from a mad dog the wound is washed with the water. It is said that Lady Baird, of Sauchton Hall, near Edinburgh, was bitten by a mad dog, and that after she had shown signs of hydrophobia, she was cured by drinking and bathing in water which the Lee-stone had been dipped in.[2]

Now the earlier form of the name of Lee is Leigh, and this is the Egyptian Lekh, or Rekh, a name of the Mage, the Wise Man, the English leech, and healer. The Leighs were probably Leeches (or Rekhi). Rekh also means to wash, whiten, and purify, and do that which is attributed to the Lee-stone.

An or Un, in the hieroglyphics, is the name of an hour; English one, Welsh, un. UNNT is oned. The sign is a five-rayed star which also reads number five. Here the hour or period is denoted by a figure of five, as with us it is signalled by the hour-hand and the number five. The English hand answers to the Egyptian ANNT. We have the AN for number five in AN-berry, or five fingers, the name of a wart on horses and a disease of turnips.

Shâ (Eg.) is number. It signifies the first and stands for thirty. Thirty days, of course, made one month. Hence thirty (Shâ) made one sheaf of days; our English sheaf is the first binding up of corn. But

[1] Hall's *Epigr. against Marston.*   [2] Brand, on the "Lee-Penny."

we have the word signifying thirty. A SHAFFE is thirty gads of steel. Our sheaf of corn is a measure, an armful. And SHA (Eg.) is the arm and measure ; " F " signifies to carry ; hence the armful carried is a sheaf.

In English the " CRIB " is both a manger and the bed of a child, and in the hieroglyphics the PET (bed) is the same image as is the APT or manger; PET and APT permute, and the crib and manger are identical as in English.

A " SPARE-RIB " of pork is usually explained as meaning a thin or lean rib. But why does spare signify lean ? The hieroglyphics answer because it is the rib. SPER in Egyptian is the rib, and one side. SPER also was a measure, which was called a side, as we have it in English, a spur of bacon for one side. Sper as measure denotes the thin lean part found in the ribs or side named a " SPER."

Champollion gives a hieroglyphic BAKAN as some unknown kind of altar. It is a framework with what may be four pieces of meat suspended within it.[1] " BA " is food, and " KANN " is smoke. " Kanf" is a baker. And the Indians of Brazil were found in 1557 to be in possession of a kind of wooden grating set up on four forked posts on which they prepared food with a slow fire beneath, for preserving it. This in their language was called a BOUCAN.[2] By aid of which we may identify the altar as an instrument or framework for preparing or smoking meat. Boucanning, the art of smoking meat to preserve it, is found in Africa, the Pelew Islands, Kamkatka, the Eastern Archipelago, and it gives us the name of our smoked pork or " BACON " in England.

" The honest miller has a golden thumb" is a proverbial phrase. Chaucer says his miller " had a thumb of gold pardie." Brand suggests that this typical thumb may have been the strickle with which corn is made level and struck off in measuring. It was, but that does not explain the origin. The thumb is a measure still, and TUM (Eg.) is a total of measure; the word means to cut, strike, announce. TEMA is to make both true and just. It also means complete, perfect, perfected, to satisfy. The thumb of gold is probably a symbol of measure, typical of truth and justice, the twofold truth of Egypt, therefore of Tumu or Tum, the Great Judge. It may have relation to the Two Truths, that the bushel used to consist of two strikes, and the strickle called the Thumb would be the analogue of the TAM, sceptre and sign of just rule; the thumb of gold would correspond to the TAM of gold, Tam being the golden.

An ancient piece of family plate used to be set on the tables of the old nobility, called the SHIP, although it was not always ship-shape. SHIP is a name for a censer. However, the name SHIP is supposed to denote its origin. But in Egyptian we meet with the SHIP, image,

---

[1] *Mon.* iv. 322.
[2] Levy, *Hist. d'un Voy.* 1600, p. 153, quoted by Tylor, *Early Hist.* 261, 3rd ed. 262.

which is not the ship, vessel. SHEB is a clepsydra. SHEB is a figure and to figure; SHABAU, a figure belonging to the heraldry of death. And SHEBU is the name of traditions. Possibly this and not merely a ship was the name that rendered the meaning of the heirloom piece of plate. If so, it was an image of descent, a type of transference from generation to generation.

The BESOM is an emblem of passing and crossing. The besom or broom is used by witches in passing to and fro. In Hamburg they have a nautical tradition that if you have had an adverse wind at sea, and you meet with another ship, if you throw a broom before it the wind will change, and the bad luck pass to the other ship.[1] Here the broom is a symbol of passing and crossing. So is it in the laying a broom across the inner side of the threshold for the nurse to step over, when the child is taken to be christened, and the making of a besom during "the twelve days" to lay on the threshold for the cattle to step over when they are first driven out to pasture in the spring, which was intended to protect them against witches. It is a belief in England and Germany that no witch can step over a besom laid across the threshold. She must push it aside if she would enter as she cannot cross it.[2] In both cases whether it be the broom which the witch does stride or the besom that she cannot pass, it is a symbol of crossing, and as a symbol has divers applications. The besom attached to the mast-head in token that the ship was for sale was emblematic of this passing by transfer from one owner to another. The burning of besoms was a part of the sports in the fire-festivals at the summer solstice. In the Harz the fires of St. John were accompanied with burning besoms which were whirled round in the air. The Czechs of Bohemia do the same thing, and all the old worn-out besoms that can be begged or stolen are collected for weeks beforehand to make the *feu de joie* on this occasion.[3] In the churchwardens' accounts of St. Martin Outwich (1524), we have "Payde for byrche and BROMES at Midsommr, ijd." "1525. Payde for byrch and BROMES at Midsomr iijd."[4] These brooms doubtless ended as torches. The burning broom was still the hieroglyphic of the passage, that of the sun now culminating at the point of the solstice, the worn-out broom being a symbol of the passed circle of the year, utilized in feeding the fire which typified the renewal of another annual passage through the heavens.

In Egyptian the word "BES" signifies to transfer, to pass from one person, thing, or place, to another: am (Eg.) is belonging to. BES-AM is our besom, an ideograph of transfer and passage. The BES is also an amulet for protection. The bush as a sign of sale and transfer used by vintners and also by horse-dealers in the shape of green boughs worn by cattle, is a variant of the besom ideograph.

---

[1] Kelly, *Curiosities*, p. 226.
[2] *Ib.* 226-7.
[3] *Kelly*, p. 228.
[4] Brand, *Midsummer Eve*.

Bes (Eg.) also means to be exhibited and proclaimed, as was done by means of the bush and broom.

The Skimmington was a kind of representative and burlesque procession, employed, for one thing, to ridicule a man who suffered himself to be beaten by his wife. In Dr. King's Miscellany Poems[1] are the following lines :—

> " When the young people ride the Skimmington,
> There is a general trembling in the Town,
> Not only he for whom the person rides
> Suffers, but they *sweep other doors* besides,
> And by that hieroglyphic does appear
> That the good woman is the master here."

This shows that the besom, true to its name, from Bes, to transfer, was an emblem of the transfer of power. SKHEMA (Eg.), to accuse, drag forth, represent, figure, offers an explanation of the name of this ideographic ceremony.

The Druidic speakers constantly talk in hieroglyphics, which may be understood when we have collected and massed the original matter. We meet with the horse or mare, CEIDIO, named CETHIN, which has the horn of Avarn. It is also called KARN GAFFON, and the hoof or foot was guarded at the end with a band or ring. It is likewise described as being cut off at the haunches. The symbolic mare of the Druids is representative of KÊD. In Egypt the water-horse was her type, the Kheb or hippopotamus form of the genitrix, who became the later Hippos of Italy, the Mare-Mother of Greece, and the Dobbin of our nursery stories. KAT (Eg.) signifies to go round in a circle; IU the two houses or halves of heaven. Keten (Eg.) is an image or likeness of the goer-round. The hinder quarters cut off form a hieroglyphic determinative of Kefa. Khefiu (Eg.) means tethered, and Gaffon was tethered with a band or ring. This tether is also a hieroglyphic, a cord or noose for an animal's foot called the REN.[2] Ka-ren (karn) is Egyptian for the type of tethering: and Karn Gaffon was the horse tethered by the foot. The "REN" tether was the sign of binding within a circle, an orbit, and the symbolic horse of the Druids and the British coins was so bound. The mare has the horn of AVREN. This may name the Typhonian type of animal, the mythical unicorn sometimes represented by the rhinoceros, and REM or REN. REN is an animal, Ap is a hieroglyph'c horn. Ap and Af are names of the old Genitrix, who is possibly identified as AVARN. She was depicted as the pregnant water-horse. Afa (Eg.) means filled, satisfied, and AFA-REN would answer to AVARN. The animal is called the hideous. KEFA was the hideous. Strabo mentions the CEPUS, sacred at Babylon, near Memph's, with a face like a satyr, and the body a combination of Dog and Bear.

---

[1] *Works*, 1776, vol. iii. p. 256.
[2] Bunsen, *Egypt's Place*, vol. i. p. 551, No. 149.

The Unicorn, an express symbol of Sut-Typhon, was deposited at last in the arms of England as one of the supports of the crown; that is Typhon as the beneficent, not the dark demon of later times. The mare of Kêd and the conventionalised animal, sometimes called an elephant on the Scottish stones, may be explicated in this way. There being no hippopotamus in the country, the horse or cow of the waters would be more naturally represented by those of the land, and this would lead to enigmas of allusiveness in compounding the symbolical type. Tef, for example, is the water-fowl, duck, or goose, and this is identified by name with the goddess of the Great Bear. Now if the sculptor wanted to indicate the animal of the waters he would or might give it the head of a waterfowl. This was done. The duck or swan is found as the head of the enigmatical animal on the Scottish Stones. This identifies the old Genitrix Tef, Kheft, or Kêd just as well as the hippopotamus. Another mode of denoting the horse of the waters would be by giving it a boat-shaped body. This too was done, as may be seen on the coins, where the Chimera is found as a monstrous horse, having the body of a boat and the head of a bird. Bird, ship, and mare are compounded in the portrait of Kêd, or Keridwen, who carried the seed of life across the Deluge waters, and the emblem is equivalent to the old genitrix, who included the hippopotamus, crocodile, lioness, and kaf. The mare cut off at the haunches corresponds to the lioness divided in two, the hinder half of which represents the north or west, and is the type of force and attainment. Possibly because in lower latitudes the hinder part of the Great Bear, the Khepsh, dipped below the horizon in crossing the quarter of the north!

The name of the water-horse Khep is found in the word CAPPLE, a horse in provincial English and in Keltic. A proverb has it, "'Tis time to yoke when the cart comes to the Capples." Another proverb says, "The grey mare is the better horse," and the typical grey mare is the old DOBBIN of our nursery lore, who still retains the name of TEB, like the star DUBHE in the Great Bear.

In the British Mythology we have the solar bull and the solar birthplace identified with the sign of the Bull. The birthplace is where the sun rises at the time of the vernal equinox, and this in the Druidic cult is continually identified with the bull, which must have been over four thousand years ago, as the equinox entered that sign 6190 years since (dating from the year 1880), and left it 4035 years ago.

In the Mysteries we find the priest exclaiming after the manner of the Osirian in the Egyptian ritual, "I am the cell, I am the chasm, I am the bull, 'BECR-LLED.'"[1] The cell was the womb of Kêd; the chasm, the equinoctial division. The title of the bull, says Davies, has no meaning in the British language. It has in

[1] Davies' *Mythology*, p. 137.

Egyptian. Lled is of course, people, the race, one with the Rut (Eg.). Bekh (Eg.) means to fecundate, to engender, beget. The Bekh was the birthplace of the sun in the mount of the horizon, or sign of the equinox. Bekh-r (Eg.) is to be the begetter. The sense is purely Egyptian like the words. "I am the bull, Becr-Lled," is "I am the bull of men, the fertiliser of the race; I am the procreator in the image of the bull," as was Khem, Mentu, and Mnevis.

The Ape as a sign of station was solstitial as Kafi (Shu) and equinoctial as An. The "mouth of the ape" and the "mouth of the star" are names applied to outlets of the Nile. The Druids also had the symbolical ape called Eppa. "Without Eppa or the cowstall or the rampart, the protecting circle," says the Bard, no time can be kept.[1] The imagery can be read as Egyptian of the earliest time. The egg also remains as an ideograph of the circle, as it has been ever since it was shaped and named by Num, or laid by the Goose. You ought never to take eggs out of or into the house after sunset. Why? because the cycle is completed of which the egg was an image. For the same reason an egg was considered the luckiest gift for a newborn child. For the same reason originally but now the symbol remains and passes current without the sense as people keep on talking after their reason has gone.

The Egyptian goddess Hathor or Athor is the feminine abode, the habitation of Har the child. The abode Hat, earlier Kat, is the womb, and in Cornish English the belly or womb is called ATHOR, the goddess being thus reduced to her primitive condition.

The white cow was especially the symbol of Hathor, the Egyptian Venus, whose title is the nurse of the child. She is depicted suckling the child, and her type as the nurse is the white cow. "Hat" is both cow and white, "Har" is the child. In Wiltshire the superstition is still extant that the white cow gives the motherly milk. There is a symbolical saying, "A child that sucks a white cow will thrive better.[2] Hathor, the divine nurse, still survives in the image and ideograph of the white cow that nursed the divine child. The white cow that rises from the lake is a familiar figure in the Irish legends. In the time of Kufu there was a priest of the white bull and sacred heifer of Athor. And it is to this sacred symbolry that the present writer would look for the remote origin of the wild white cattle of Great Britain. The Bulmer crest was a white bull, and the primeval Bulmer may have been a priest of the white bull or cow, as Mer (Eg.) is not only the cow but a form of Hathor, the goddess of the white cow, and the English Mart was a cow fair. Bul-mer (or Bar-mer) is the son or bull of the white cow.

The Ponsonby crest is a serpent issuing from a crown that is pierced by three arrows.[3] This heraldic device may be seen as mythological

<hr />

[1] Gwawd Llud y Mawr.   [2] *Choice Notes*, p. 244.
[3] Letter in *Notes and Queries*, October 20, 1871.

symbolry in the *Antiquities of Egypt*, the French work,[1] where arrows are entering and the serpent is issuing from the crown of the great mother, who wears the feather of Ma. The arrow is a symbol of Seti, the wearer of the white crown. The serpent represents the lower crown of Neith. The feather shows that the Two Truths were signified. Sen or Shennu is the circle of the Two Truths; these were imaged in the white and red double crown called the Shent.

SEN is also the Egyptian name of the temple of Esne, as the house of the circle. Pen is an emphatic the, and PINU, a name of the double-Crown; bi (or by) is the place. PEN-SEN-BI reads the place or circle of the Two Truths. Thus the House of Ponsonby would seem to be an English form of the mythical hall of the Two Truths, localised in this instance at SEN in Egypt.

An oar is the ideograph of Kher. "It" (Eg.) means to figure forth. Khart is the child. The oar is the symbol of Makheru, the divine child and true Word. The oar as a means of crossing the waters is thus the synomyn of the solar child who crosses the waters. In the constitution of the boat of the sun,[2] the paddles are said to be "the fingers of the elder Horus." The boat itself is primarily the feminine abode. This boat is personified in KÊD, the great mother of British mythology. One of her names is KERID-WEN. Wen, like Ven, Ken, Gwen, is the lady, the queen, Oine, Venus. She is represented as a sailing vessel, that is, as the boat of breath, but the paddle is before the sail, and the paddle is also her hierogly-phic. Her name KERID might be read KHER-it, the figurer of the oar or of the child. She is called the modeller or figurer of the young. And the oar is her symbol. When Gwion the Little let his cauldron boil over she seized the oar and struck the blind Morda on the head.[3] Morda is called the demon of the sea. MERTA (Eg.) is both the sea and the person attached to it. The action is equivalent to crossing the water by means of the oar. This will suffice to show the hieroglyphic Oar is the same in Britain as in Egypt.

An Oar is also a name of the Waterman. This is in the hieroglyphical tongue. An oar is the sign of Har or Khar, the Sun of the Cross:ng, whether as Horus or Makheru. Oar and Har are identical; the oar or paddle being a type of crossing the waters in the passage through the underworld. Horus or Har, as the oar of the Boat of Souls,[4] is the Waterman; the Child that crossed the Waters first of all in Womb-world; secondly, in the Planisphere, and, lastly, in the Eschatological "Boat of Souls."

In *Hudibras* Butler says:—

> "Tell me but what's the natural cause
> Why on a sign no painter draws
> The full moon ever, but the half?"

---

[1] *Description de l'Égypte* (1809).   [2] *Rit.* ch. 99.
[3] *Hanes Taliesin*, ch. 2.   [4] *Rit.* ch. 9.

The answer according to Egyptian symbolism is that the moon was masculine up to the fifteenth day and then entered its secondary phase. The half-moon was personated by Taht, the male lunar deity. Taht signifies a sign, image, type, and the half-moon was the sign outside of the house. The Inn itself was the feminine sign, the abode invited to by the outside sign. The half-moon, as a sign in England, is a synonym of Taht, the word, the tongue, the proclaimer and manifester in Egypt. This may explain the origin of divination or forecasting by means of the ominous swinging of sign-boards mentioned by Gay in *Trivia*. The board itself was a sign, a symbol, part of a system of symbolism. Set in motion by the wind, a living voice was given to this sign, which to the decaying sense of the symbolical uttered portentous, but undefined meanings, to be shaped for the listeners by their ignorance.

An ivy bush was at one time a vintner's sign : an ivy bush is a tod of ivy. Our tod is the Egyptian Tet, the type, image, mouth, tongue, and to speak, manifest, tell, proclaim, or make the sign. Taht carried his TOD or branch of the panegyrics not in ivy but as a shoot of palm. This branch has the meaning of showing, explaining, as did the Tod or bush, hence the saying, "Good wine needs no bush." Taht, however, was not outside only. The full name of this deity, "TAHUTI," signifies the double one, the double gibbousness of the moon or light. This duality of the Divinity is sacredly preserved in the dual drink named TODDY.

Lluellin in his poems [1] wonders—

> "By what hap
> The fat harlot of the tappe
> Writes at night and at noon,
> For a tester half a moon,
> And a great round O for a shilling."

There was no hap, as chance, in the matter. The tester was then sixpence, or one-half of a whole, earlier it had been twelve pence. The coinage was changed, but not the symbol of one-half of a total, that lived on in the half-moon, the hieroglyphic of one-half or TNA, the fortnight, as one-half of a month. Our vagabonds still call a month a moon, and thus use the hieroglyphical mode of the Egyptians and Red Indians, with whom a month was a moon, the fortnight a half-moon. The word leg answers to the Egyptian REKH, to reckon, keep account, and the LEG is yet used as a sign of reckoning, a leg being one-half and two legs the whole game.

Various of our public-house signs are of Egyptian origin, and can only be read by the hieroglyphics. In a list of curious signs in the *British Apollo*,[2] there is the "Leg and Seven Stars." Now the "Leg and Seven Stars" is not known to English astronomy as a constellation. But it was to the Egyptians. The Seven Stars of "Ursa

---

[1] 1679, p. 40. [2] *London*, 1710, vol. iii. No. 34.

Major" was a constellation of theirs called the Thigh of the Northern Heaven.[1] The English "leg and seven stars" answers to the thigh and seven stars found in the Great Bear.

Drink and drinking were sacred customs long before they were profaned. In Egyptian the "KAB," libation, liquid, to refresh, enjoy, is our Cup. And as Kaba is a horn it shows they also drank by the horn. Our drinking-horn is a TOT—the name of Taht, again, who represented the horned moon, and wore its crescent on his head. Many pretended explanations of these signs are on a par with the English sailor's rendering of the name of a French vessel called *Don Quichote* as the "donkey shot," such as the "Bull and Mouth," rendered by Boulogne Mouth. In Egyptian the mouth and gate are one in the "Ru." Our "Bull and Mouth" alternates with the "Bull and Gate."

Bull and mouth (or gate) are male and female signs; they represent both sexes in one. The bull is personified as Khem; the mouth, as Mut the mother. "KAMUT," a title of the bull and mother, is literally our bull (Ka, Bull) and mouth. Bull and mouth is the sign of male and mother, or the male as mother. An ancient picture of the bull and mouth given by Hotten in his book on sign-boards places the mouth under the belly of the bull, which makes the bull an image of the creative Khem in the drawings at Denderah.

As Khem is our bull, it is probable that Num is our "green man." Num was represented in the Egyptian portraits of him as the green man, and his name signifies the winepress. The green man and winepress is equivalent to our "Green Man and Still." Num wore the ram or goat's horns on his head, and our green man also carries the horn. If it be said that the horn was to blow, and this was mere Robin Hood imagery, the answer is that such imagery is wholly mythical.

Another of our old signs is the "Axe and Bottle." These are two hieroglyphical types. The Egyptian axe is the Nuter, symbol of divinity. The bottle NU is just our common water-bottle. Out of the NU the goddess of the water of life pours the divine drink of immortality for thirsting souls. The NU is the sign of drink and within. The axe and bottle read as hieroglyphics proclaim to the passer-by that there is divine drink inside : they also denote "shelter within." It is certain that the axe had the hieroglyphic value of the Nuter type in the British Isles, as in the tales of the Irish Gobawn Saer, the goblin-builder, it is the same image of power and potency as in Egypt. When, once upon a time, he came to a place where the king's workmen had finished building a lofty palace all except the extreme part of the roof, the Gobawn completed the dangerous task by cutting some wooden pegs with his axe and throwing them

[1] *Egyp. Rit.* ch. 17.

up one by one into their places, and flinging the magic axe after them with such unerring aim that each peg was driven home.[1]

An English sign not yet extinct is sometimes called the " Good Woman," and in the neighbourhood of Rippenden, Yorkshire, there is one named the "Quiet Woman." It shows a woman without a head. The common notion is that it conveys a satire on woman's tongue. The headless woman, however, was an Egyptian goddess, Ma, who personified the truth itself. She is the original of the Greek Themis or Justice, whose eyes were bandaged. Diodorus Siculus[2] mentions a figure of Ma, the goddess of Truth and Justice, as being without a head, standing in the lower regions at the "gates of Truth," and this headless woman was found by Wilkinson in the judgment scenes attached to the funeral rituals on the Papyri of Thebes ; the true and just without eyes or even a head. In place of a head she has the stone (T) and feather (Ma) T-ma ; the true.[3] Another figure of this divinity may serve to explain why the headless woman is found on the sign-board as the type of genuineness. She is seen issuing from a mountain presenting to the deceased two emblems, which represent water or the drink of heaven, the true drink of life. The headless woman may be the cause of the expression, "to put a head on it."

In Scotland an "ALE-WISP," a bundle of straw on a pole, was a public-house sign.[4] This is very Egyptian. The straw was emblematic of the grain thrashed out to produce the drink. So women who have just produced are said to be in the straw. Our "WISP" is the Egyptian USB. US or USH is to mow, cut ; "USHM" is the corn, also the essence, decoction, or brewing. "USB" is to stack the corn. "USF" is leisure. "USHB," to consume. The USB, or Wisp of straw, says in the ancient language the corn is cut, the malt is brewed ; we have leisure now, come and consume it. The wisp is a sign of call for the house of call.

After all, these are but gleanings from a wide field of research. A lifetime might be spent in gathering and re-publishing this book of the hieroglyphics in Britain. We might have made a collection of sayings, proverbs, blazons, and legends, which can be interpreted by the ancient typology.

In Somersetshire it is a saying that a child born during CHIMETIME will see spirits, and in Egyptian one name for spirits is the Khemu.

Drayton in his *Poly-Olbion* (song 3) records the Druidic prophecy, that if the River Parret were to fail, they should be suppressed, the end would have come. This, of course, could not depend on the drying up of one stream among so many, but must be connected with

---

[1] *Choice Notes*, p. 107.   [2] B. i. 96.
[3] 2nd series, vol. ii. p. 30, plate 49.   [4] Dunbar's *Poems*.

its symbolic name. The Egyptian shows us how. The Parret was a "BORIAL stream." The word "PERUT," literally pour out, proceed, emanate, is not limited to water alone. PERRUT is to run away, bear off, carry away. PARRIT is food, grain, germinate, manifest, corn. The ceasing of the Parret therefore included the non-production of the corn on which the Hut and Cruitnich-men laid such stress, and the prophecy has a double meaning.

Again, it is said the Men of Kent are born with TAILS, and "Long tails and liberty" is a Kentish blazon. This the hieroglyphics will explain.

Khent is the hinder part, south; an Egyptian name of the south, and for going back from the north. Khent is the south both in Egypt and England. The south, as the quarter of the summer solstice, was the solar point of turning back and beginning of another year. Khennu means to go back, and Khent is the place of turning back in the circle. Sut-Typhon, with long tail erect, was an early type of the Turner back in Khent, the South, where Sothis (Sut) rose to announce the returning back of the Inundation at the turning-point of the year. One type of Khent is the Cynocephalus with its long tail irately stiffened, and another ideograph of Khen, therefore a Khent, is so definite an image of this hindward part that it is a decapitated animal; it is all tail and no head. Another sign is the animal's skin with the tail attached. The long-tailed men of Khent are imaged in the likeness of the long-tailed Kant or of Khent, the Cynocephalus of the south with its long tail erect, denoting the upward hinder part.

Khent, to go back, Khent, the place of going back, has another illustration in Kent's Cave, Devon. A local legend assigns the origin of its name to the circumstance that once upon a time a dog—in another version a hawk—entered the cavern and emerged in the county of Kent, which identifies the Egyptian meaning of the word. The dog went back to the typical south.[1] Khent means to go back, and names the place of going back as the south. Kent is not south from Torbay, but Lower and Upper Egypt as north and south were Khebt and Khent, and when the dogstar went down it descended into the celestial Khebt or cave of the lower world, and when it rose again it emerged in Khent.

This earliest mode of reckoning the year by the Great Bear and Dogstar has also left its imagery in legends about the caves of the Mendip Hills.

At Cheddar they still repeat the story of the dog that entered the cave at that place and came out again shorn of all its hair at the Wockey Hole. At the Wockey Hole it is said to enter the hill and to issue forth at the Cheddar cave. This is identical with the dog

---

[1] Pengelly, *Kent's Cavern: a Lecture*, p.

and the hawk entering the hole at Torbay and issuing forth in the county of Kent. The Basque proverb says, in flying from the wolf (or dog) he met the bear. The two constituted Sut-Typhon. The Great Bear was the type of the North ; the Dog of the South; the one belonged to Khept (was Khept), the other to Khent (was a Khent), the hinder part South, the place of the long-tailed Kant, and thence of the long-tailed Kentish men.

Time was when the Great Bear below the Pole represented the Great Mother in relation to water, hence the type of the Water-horse. Above it she represented the element of heat. And a writer in *Notes and Queries*[1] was informed by a countryman that the cause of continued drought was the Great Bear's being on this side the North Pole ; so long as it continued on this side the weather would keep dry, and it had been there these three last summers. If it could get to the other side, we should then have a wet one.

Lastly, in the Egyptian Mythology there are Seven Khnemu or Pigmies, called the Seven Sons of Ptah, who stand by his side as architects to help him. These are our seven goblin-builders. The seven, whether the first boatmen (Kabiri) or builders, are representatives of the Seven Stars first observed to bridge the void below the horizon. Our name of goblin and of the Irish Gobawn Saer identifies them with the name of Kheb, the Goddess of the Seven Stars. Goblin, as Kheb-renn, is the Child of Kheb.

Kheb or Khept is our Kêd, and in the Lancashire Traditions[2] the Chapel of St. Chadde was erected on the height by the Goblin Builders. The legend is, that Gamel, the Saxon Thane, Lord of Rached, now Rochdale, intended to build it on the bank of the Rache or Roach, in a level spot, and the foundations were laid three times, but on each occasion the Goblin Builders removed the materials to the more elevated situation, the high place, the Mount that is sacred to the Great Mother, whether called Khept, Kêd, or Chadde.

The dove was also a type of the Genitrix, and bears her name of Tef. The dove was sacred to Hathor, and there were seven Hathors. Seven Doves in the Christian Iconography represent the Seven Gifts of the Holy Ghost, or earlier, the Seven Stars of her who was termed the " Living Word," at Ombos, and in some of the legends of the Goblin Builders, the stones of the building intended to stand down in the dale are carried away to the height by doves. So was it with the church of Breedon, Leicestershire ; the foundations were dug and the work was begun, but all that was built by day was carried away by doves in the night to the top of the hill where the church now stands.[3] Another type of the Great Bear and its Goddess was Rerit, the sow, in Egypt, as in Britain it was Kêd, the sow, and in the case

---

[1] April 8, 1871.   [2] Roby, 1st series, vol. i. 23.
[3] *Choice Notes*, p. 1.

of Winwick Church, Lancashire, the pig was the cause of removal to the height. It was seen to take up one of the stones in its mouth and carry it to the spot said to be sanctified by the death of Oswald, and during the night it removed all the rest. There is a figure of the pig, or sow, sculptured on the tower just above the western entrance, in witness of the transaction.[1]

[1] *Choice Notes,* p. 2.

SECTION IV.

EGYPTIAN ORIGINES IN WORDS.

THE comparative vocabulary in the second section furnished evidence of something or other not yet taken into account or even dreamed of by the comparative philologist. But if it had stood alone, unsupported by further evidence, one might have felt inclined to suppress it for fear of Grimm's Levites. They have so worried us into believing that verbal likeness is no sign whatever of relationship. They have so incessantly insisted that, when we find a word spelt the same in one language (Greek) as in another (Sanskrit), we may be certain that it cannot be the same word.[1] Grimm's law forbids. Which, from my point of view, is somewhat like saying that, if two men have a strong family likeness, and bear the same surname, they cannot be brothers. It is positively asserted that "SOUND etymology has NOTHING to do with sound." Philologists, says one Sanskritist, who bring in Chinese, New Zealand, and Finnic analogies to explain Indo-European words, are thoroughly unsound. They need to reform their science from the foundation.[2] "To compare words of different languages together because they agree in sound is to contravene all the principles of scientific philology: agreement of sound is the best possible proof of their want of connection."[3] Of course, no one would compare them if they did not retain the same signification. Even then they are valueless for those with whom language begins a very long way on this side of Babel, and who assume that there was no unity of origin on the other. Grimm's law forbids that origin should ever be proved by likeness because it only shows difference. This has limited the comparative philologists to the narrowest possible area, and their verdicts are often as unsound as their generalisations are premature. It looks as if the discovery of Sanskrit were doomed to be a fatal find for the comparative philologists of our generation.

No foundation in ancient language is perhaps so late as Sanskrit, and dogmatism on a basis of Greek and Sanskrit is the most bankrupt

[1] Max Müller, *Introduction to the Science of Religion*, p. 307.
[2] Whitney, *Oriental and Linguistic Studies*, p. 213.
[3] Sayce, *Introduction to the Science of Language*, vol. i. p. 347.

business in the world of words. We have to dig and descend mine under mine beneath the surface scratched with such complacent twitterings over their findings by those who have taken absolute possession of this field, and proceeded to fence it in for themselves and put up a warning against everybody else as trespassers. We get volume after volume on the "science of language," which only make us wonder when the "science" is going to begin. At present it is an OPERA that is all overture. The comparative philologists have not gone deep enough, as yet, to see that there is a stage where likeness may afford guidance, because there was a common origin for the primordial stock of words. They assume that Grimm's law goes all the way back. They cling to their limits as the old Greek sailors hugged the shore, and continually insist upon imposing these on all other voyagers, by telling terrible tales of the unknown dangers beyond.

As the palimpsest of language is held up to the light and looked at more closely, it is found to be full of elder forms beneath the later writing. Again and again has the most ancient speech conformed to the new grammar until this becomes the merest surface test; it supplies only the latest likeness. Our mountains and rivers still talk in the primeval mother tongue, whilst the language of men is re-moulded by every passing wave of change. The language of mythology and typology is almost as permanent as the names of the hills and the streams.

" Care must be taken," says Professor Sayce, " to compare together only those myths which belong to the languages shown by comparative philology to be children of a common mother. Where language demonstrates identity of origin, there will be identity of myths; not otherwise." [1] But the identity of the myths and ideographic types is demonstrable, and can be demonstrated among various races of the world whose languages are supposed, by comparative philologists, to have no relationship whatever. This is their root relationship.

In proportion as we get back towards a beginning it becomes more and more apparent that *Comparative Philology* and *Comparative Mythology* have to make way, in a double sense, for *Comparative Typology*, as this only can show the stage of language in which unity is yet recoverable.

It is the especial province of the present writer to identify the myths and what he terms the Types, for in his view there is a typology anterior to what is known as mythology, and if their identity shows an identity of origin for language, we are surely on the way to the abode of the common parent of all.

The chief evidence of this origin will have to be brought forward in two volumes of *Comparative Typology* now ready for the press, and intended to follow these.

[1] *Introduction to the Science of Language*, vol. ii. p. 260.

The founders of philological science have worked without the most fundamental material of all, the Egyptian; this they neglected early and avoided late. From lack of the primaries to be found in that language, a vast number of their conclusions are necessarily false, and their theory of the Indo-European origin of languages and races is, in the present writer's opinion, the most spurious product of the century. This list of words at least will give no countenance to the theory; they point to Egypt, and not to India, as the place to look for the origines of the language that first came into the British Isles.

These words are neither Keltic, Kymric, Gaelic, nor Anglo-Saxon in the restricted sense; they belong chiefly to the provincial dialects among which we find the *débris* of the oldest language dissolved by the influence of time, and of which the Kymric, Gaelic, Manx, or others, are but localised after-drifts and developments.

If two words found in Sanskrit and Greek, when spelt the same, cannot be the same, what does comparative philology say to many hundred words being the same, generally spelt the same, and having the same meanings, being found in Egyptian and English? Of course with the difference that followed the evolution of sounds, as from *t* to *d* or *k* to *c* which may disguise but will not determine the origin of any word.

Now, supposing this old starting-point in Egypt be the true one, it is no longer necessary, for example, to derive HIMU (Sanskrit) from a root ZBE, to invoke, when HAMA in Egyptian means to invoke with religious clamour, and SEBA is to pray? Numbers of such equivalents meet in Pahlavi and Sanskrit which are distinct words in Egyptian.

Again, it is assumed that the Zend ANHU, life, is derived from ASU (Sans.), life. This, however, is unnecessary. ASH (Eg.) is life, the tree of life, and ANKH is also life. ANHU derives from ANKHU, and ASU represents ASH or AS. Asu denotes life as breath, and as-asni (Eg.) means to breathe. Ash (Eg.), emanation, emission, applies to both principles of life, the water and the breath.

Why should a Sanskrit root DA supply the Greek with the words δίδωμι, δαιτρός, and διδάσκω, as Max Müller asserts? He affirms that this DA, to give, supplies the Latin DO; Greek δίδωμι; Slavonic DA-MI; Lithuanian DU-MI; and various others; it also means to cut, and furnishes the Greek δαιτρός, a man who carves; and still another, DA, identical with these two, means to teach and to know, preserved in διδάσκω. Now, if we turn to the word TA (Eg.), we find it means to give, and a gift, and that it is an abraded TAT. TAT has every meaning of the Sanskrit DA; TAT is to give, to cut, the scribe, language, discourse, tell, the mountain, fire, and others; also the modification of TAT into TÂ is marked by the accented vowel as in Sanskrit. Moreover, the hieroglyphic Tat, the hand, is the pictograph of the TAT-ing or TA-ing, whether in giving, cutting, typing, or writing. In that we have the ideograph in which all the meanings meet. Professor Sayce affirms that "By tracing the Greek δῆμος to the root δα, 'to

divide,'" (the philologist) " can show that private property in Attica
originated in that allotment of land by the commune which still prevails
among the Slavs."[1] But TEM (Eg.) means to cut, divide, make separate, and relates to division of land ; the Tem was also a district, a
village, a fort, a community, and a total, as in the English TEAM.
So the Greek TEMNO signifies to cut, and δῆμος, for the people,
represents the Egyptian TEMU, the people, created persons, mankind.

Again, M. Lenormant is of opinion that the Akkadian word IT for
the hand and the Semite יד were not derived from each other, but
had an independent origin in Assyrian and Akkadian. They had,
according to the present hypothesis, a common origin in Egyptian,
which supplied this type-word for all the chief groups of languages.

The hand or fist is found as Khept and Kep, and Khept has a
worn down form in IT, to figure, paint, portray, with the hand
of the artist for determinative. This IT corresponds to the
Hebrew יד (Jad), the Akkadian IT, and the Assyrian IDU. Now
the Jad stands for No. 10, that is, for two hands, and Khept, the
fist, is the sign of the hand doubled, therefore, of two hands. Thus
Khept or Khepti is the Dual form of Kep, and Khept wears down
to Jad for number 10. Kabti (Eg.) is two arms or hands.

The following list of words contains the name of the hand ranging
from KHEPT to IT, and the same process of modification from
the one to the other which took place in Egyptian may be seen in
universal language under all the changes of phonetic law :

KHEPT, Egyptian.
KEPITEN, Micmac.
GAVAT, St. Matheo.
GÓT, Vayu.
KUTT, Chepang.
KUT, Kapwi.
KHUT, Khoibu.
CCUTA, Mokobi.
KHUIT, Tshetsh.
KIT, Tsheremis.
KET, Lap.
KET, Ostiak.
KAT, Assyrian.
KAT, Vogul.
TA-KHAT, Tengra.
TA-KHET, Khari.
AGGAIT, Labrador.
CHETABA, Bororo.
KUTANGA (handful), Maori.
SECUT, Adaihe.
HUT, Maring.
HATH, Shina.
HATHA, Bowri.
HATH, Gohuri.
HATH, Siraiki.
HATH, Hindustani.
HATH, Gujerati.
HATH, Kuswar.
HATH, Kooch.
HATH, Hindi.
HAT, Mahratta.

HÁT, Acam.
HAT, Ruinga.
HATO, Uriya.
HAT, Durahi.
HAT-KELA, Pakhya.
HAT-THO, Pali.
JAD, Hebrew.
JAD, Syriac.
JAD, Arabic.
JAYATHIN, Thaksya.
JATHENG, Garo.
ATTH, Lughman.
ATHA, Cashmir.
ATA, Singalese.
AITILA, Maldive.
ATHENG, Borro.
OTUN, Chutia.
OTOHO, Gunungtellu.
YUTU, Tawgi.
UDE, Upper Obi.
UTO, Tschulim.
UDA, Baika.
UDA, Karyas.
UDA, Yurak.
UDE, Samoyed.
EUTIJLE, Canichana.
EED, Tigré.
EED-gekind, Amharic.
IDA, Guato.
ID, Gindzhar.
IDU, Assyrian.

IT, Akkadian.
IT, Egyptian.
T, Egyptian.
GAP, Akkadian.
KAPH, Hebrew.
KEP, Egyptian.
CAB, Mexican.
CHOPA, Movima.
GAUPEN, English, a double
    handful.
KOPO, fingers, Lutuami.
CHU, Tibetan, No. 10.
TCAPAI, Pujuni.
NUCAPI, Isauna.
NUCABI, Barree.
WACAVI, Toma.
ERIKIAPI, Uaenambeu.
IN-KABE, Guinau.
CHABEN, Koreng.
CIPAN, Kusund.
TSHOPRE, Coroato.
ISIP, Vilela.
WOIPO, Mundrucu.
YOP, Mijhu.
IPOHA, S. Pedro.
IPAP (hands), Walla-walla.
EPIP, Cayus, also fingers.
APKA, Shasti.
IPSHUS (hands), Sahaptin.
UBIJU, Angami.
AFA, Enganho.

---

[1] Sayce, *Introd. to the Science of Language*, vol. i. p. 161.

Other modifications might be followed, as in the Kamkatka, SYTHI ; Gafat, TSATAN ; Chinese, SHEU ; and Gyami SYU. KHEPT (Eg.), the doubled hand, permutes with KHEMT (Eg.), the number 10, and in the Philippine and other languages we find both GAVAT and CAMAT for the hand :

GAVAT, St. Matheo.	CAMAY, Tagala.	TSEMUT, fingers, Upper
GUMUT, St. Miguel.	CAMOT, Bissayan.	Sacramento.
CUMOT, Umiray.	CAMAT, Pampango.	SHUMI, Zulu, 10.
KAMOT, Sulu.	YAMUTTI, Maiongkong.	QUIPU, Peruvian, knot of 10.

The Welsh LLYTHER and Latin LITERA have the same meaning, but the one was not derived from the other. They had a common origin, from which they were independently derived. The first lettering was done in stone; hence RET (Eg.), to engrave, cut in stone, denotes the earliest letter, the Akhamenian Ritu for writing. Ret means to figure and retain the form first incised in stone or bone. Ru or Er signifies the word, discourse, a chapter ; ar is a type. Thus Ret-er would be the word engraved, Ret-ar the retained type. The Rui (Eg.) is the reed-pen of the scribe, also the colour used for the hieroglyphics. Teru (Eg.) is a roll of papyrus, and the word means drawing in colours, or making hieroglyphics. Rui-teru is the equivalent of the Latin Litera, a scroll, a writing, or a letter, and of the Welsh Llyther. The engraved stone and hieroglyphic scroll were the letters. Hence we have leather for letter (in Leland), *i.e.* the Rui-teru or scroll of the scribe, the written parchment or leather; the Egyptians also used leather as well as papyrus.

It is assumed that the words WEB, WEAVE, WOOF, Greek ὔφος, are derived from a Sanskrit root VABH, to spin, whence UNAVABHI, the spider. And, of course, the V does pass into U, and VABH, VAP, and WEB meet in one meaning. But VABH and WEB may be and indeed have been derived on two distinct lines. The English WEB implies an earlier KEB. KAB (Eg.) yields the principle of weaving with a shuttle. KAB, to turn, double, turn corner, return, and redouble. The KA are the weavers, those who KAB. It is not necessary for our W to come from V. But V implies PH, F, and B, and VAB has an equivalent BAB. BAB (Eg.) is to turn, go round, circulate, revolve, a collar. The bobbin is still used in BABBIN or weaving. There is also Â Â (Eg.), to knit, and these accented A's (the arm sign) denote earlier F's. Thus to knit was FAFA, or FABA, as in fabric, worn down to â â. UAB, to spin, is an intermediate for both FAB and BAB. Now, if we drop both K and B, we have AB (Eg.), to weave. Ab is also to net and tie ; Abt is linen, the woven. BAB is AB with the article P (B or F) prefixed, whence VABH. And at the origin we have both KA, the weavers, and AB, the weavers, that is, on the principle of word-building enforced by Grimm's Levites. Any number, however, of words in Sanskrit, considered to be roots, are but the worn down forms of words. Further, KA becomes SA (with the signs of the tie and the crocodile's tail), and we have the name of sewing and the sewers, following the weavers from the same root-origin.

The Egyptian BAB signifies going and being round. BAB is a hole, a whirlpool, a whirlwind, a circle, to circle, revolving circularly, anything going in a round. Beads are known as BUBU. In English a BOB is round; the plum-BOB, the shilling, or the BAUBEE, are round. The Scotch BAB is the round, as a loop in a garter. The BIB is tucked round. The BAP is a round cake. BABBART is a name of the hare that doubles round. A BOBBIN is round, and in machinery it revolves. The BOBBIN, faggot, is a round bundle of sticks. BEBLED is covered all round. To BUBBLE is to bladder round. BOBY, a cheese, is made round. BOB is the name of a ball. BOB is a round in ringing bells. To BOB the hair is to twist it round. BUBBIES are round. The PIP is a round spot or seed; the PEBBLE, a round stone. The PIPE, a round tube or a cask; the POPE'S eye, a round of fat in the leg of mutton.

This original meaning of BIB is still applied to the Bible in the practice of divining with a key placed in it, the result depending on its turning or BIB-bing round.[1]   The Bible is the Book of revealing. The first revelation was that of time and period, that is, of revolution, and BEB is the name of both the revolution and revelation, also the Book of Revelation. The planets in Babylonian astronomy were the BIBBU, as the revolving stars, the revealers of time. The Seven Bobuns are revolving spheres. BABA is a name of Typhon, whose starry image was the Great Bear, with the seven turners round. Midnight is considered a good time in Bibliomancy, that is, at the turn of the night. Also—and this is very Typhonian when we bear in mind that Ursa Major was the Thigh Constellation—the proper thing is to bind a garter round the Bible, but it must be one that is by woman worn. For this Typhonian Thigh was the hinder thigh, that is, the feminine symbol, and in this image we may possibly see why it is the sacred usage for woman to garter above the knee! whereas the male wears the garter below; the male is foremost, as in the hieroglyphics "bah" means the male, and in front, whilst the hinder thigh is feminine. The explanation of this is that both Kabbing and Babbing are derived from turning round and crossing, whereby the figure of a loop and a knot were made. One name for this figuring is KAB, one is BAB. The stars turning round and crossing over and under were, on one line, the first AB-ers, HAB-ers, or KAB-ers, and, on the other line, with the article prefixed, P-ab-ers, BAB-ers, VABH-ERS, UAB-ers, Wab-ers, Weavers. The two origins passed separately into Sanskrit and English, and all that can be said is that the words are now equivalents. But, to speak of *derivation* implies knowledge of origin.

[1] "BIB." The word Bible, Greek Biblos, for the Book, may be traced to the Egyptian PAPU, for Papyrus. That meaning is not here in question.

Egyptian gives us a glimpse of language in an ideographic stage. For example, the child and the seed are synonymous. SU is the seed, the egg, the child. In an earlier form this is SIF. Now the SIEVE is a sign of corn, by reason of sifting. SIF, the child, is also SIF, to purge and purify, in the sense that Siva, the generator, is designated the purifier. The name of the sieve is KHI, and this, as shown by a form of the child, written with the KHI (sieve), must have been an earlier KHIF, whence the modified SIF, the child and seed, which further abrades into SU and SI. We have the KIPE, an osier basket, as a sort of water-sieve, for catching eels, answering perfectly to KIF, for the sieve. The Kif, for sieve, is also found in CYVE (Sieve). This retains the ideographic KHIF, worn down in the hieroglyphics to a phonetic Khi, the child (Sif) written with the sieve sign. The present point, however, is this. A word like KHIF is a primate which yields KHI or SEF, SU and FI. It is anterior to gender. SIF, the child, may mean the boy or the girl, without distinction. But SIF splits into SU for THE, HER, IT; and FU for HE, HIM, HIS, IT, when gender could be phonetically distinguished. As the sign of U denotes the earlier FU or KHU, according to its line of descent, the Egyptian UI is the deposit of a consonant as is our Y. It occupies the place of the Y, supplies its sound, as a participial terminal, and also means he or him. This may serve to show how it is that the letter Y in English comes to represent F and Z. In Scotch of the sixteenth century YEAR is written ZEIR. Chaucer writes JOLLY as JOLIF, and GUILTY as GUILTYF. DAY is the earlier DAG; YE, earlier GE; YES, earlier GESE. Taking the G as equivalent to the KH (Eg.), we find the scattered KHI, S, and F, which can be traced back to an ideographic khif, all meeting once more in the Y, and the Y we are taught is Greek, the F is Old French, and the G is Anglo-Saxon. The process of this dispersion of visible speech, so to say, into divers variants of language is more or less extant in Egyptian, and the whole matter has to be put together again before we know anything of origin.

Language must have emanated from the centre in the ideographic stage, and the primitive types are more or less extant to prove the unity of origin. For instance, there is an ancient Roman tradition of twelve vultures, or twelve ages, no clue to which has ever been found. This belongs to language in the ideographic stage. MU (Eg.) means a year, cycle, or age. In Akkadian, MU is a year, and a memorial name. In Chinese, the MU are the eyes of the four quarters. The MU (Eg.) ideograph is the vulture, a symbol of sight, and sign of the year. Thus the keen-visioned MU of Egypt denotes the eyes of the four quarters in China, and MU is the year in Egypt, China, and Akkad; the vulture representing a year or age in Rome. MU (Eg.) is also the mother, and the vulture was the type of the virgin motherhood of Neith, who came from herself, a type belonging to a time *before paternity was established*. Origen defends the Immaculate Conception on the ground that the vulture, as stated by Hor-Apollo (I. 11), procreated without the male. The monuments show the MU with the male member, which emblem was necessary to express the earliest ideas, when both truths were given to her who was the VIRGIN Mother. This will show that religious doctrines, founded on a typology misinterpreted, may be in a perilous predicament.

Certain ideographs are compound types. The hippopotamus, for instance, is KHEBMA, the earliest form of KAM. The goddess TA-URT or KHEBT is compounded of the hippopotamus, crocodile, lioness, and kaf. The tail of the crocodile is an ideograph of KAM, a syllabic KA, later SA. So that the goddess is both KHAB and KAM, and midway between the two we have KVM (Heb), for Kam, Kvm being a reduced form of KHEFMA, further abraded in the final KAM. This process deposits Khef (Kheb) and Kam as two distinct words, but the ligature of their twinship is visible in KVM and in the after permutation of B and M, of Kheb and Kham, Khebt and Kamit for Egypt, Neb and Num, Nebrod and Nimrod.

There is an ideographic TES (the bolt sign), that bifurcates and supplies a phonetic T and S. Hence the permutation of T and S in Hebrew and Chaldee.

Again the Basque type root for stone, and the stone-weapon, is AITZ, with the variant AIZ. In Egyptian, the typical stone as the seat or throne is found as the AS, AST-B, ASB, and HES. The Basque has retained the ideographic TES in AITZ, and only the phonetic in AIZ. This ideograph is represented by the phonetic in the Egyptian AS or HES. AITZ (Basque) is a stone; the AITZURRA is a pickaxe. In Egyptian this is extant in another form of the TES. This Tes is a weapon, and to cut, determined by the stone and a knife, therefore it is a Stone knife. By permutation of TA and AT, TES is Ats, the equivalent of the Basque AITZ and the English ADZE, which in the hieroglyphic NUTER is likewise of stone.

Words such as the English door, Greek θύρα, Mæso-Gothic DAUR, Lithuanic DURRIS, German THÜRE, Sanskrit DVÂR, N. H. D. TOR, Latin FORES, have never been traced to any root. Yet this will be found in the hieroglyphic Ru, the door, gate, mouth, outlet. With the feminine article TU (the) prefixed (or earlier TEF), the Ru (door) becomes our door. The RU (Rue) in French is the street. And in Egyptian the Ru is also the path, way, road, and going-place. TERUAA is a doorway. In Chinese TAU and LU joined together mean the road. TARU (Eg.) is to encircle, enclose, a cell, a college; this is the Dravidian TORU, a fold, and TRU, a vase; TRO, Cornish, a circuit; Irish, TORA, boundary, border; Hebrew, דור, a circle. And the reason why the RU is a door, a circle, a passage, a street, a cell, a fold, or a boundary, is revealed in the hieroglyphics by its being the mystical mouth, the RUE DES FEMMES, the ideograph of Her or She. From RU with the feminine terminal T comes RUT (Eg.) progeny, the race, the route, road, rota, and all the rest. RUT permutes with URT (Eg.), the chariot, in Latin the ROTA; the Great Mother as URT or RUT being the chariot of the child, the bearer and bringer forth by the RU.

Roots like RU are visible at last in hieroglyphics in which the idea is figured to sight, and only in these can we see bottom or get to it; they are the final determinatives of language, and no comparative philologist has ever yet touched bottom anywhere, because the primitive types have never been taken into account. These offer particular means of identification where we are otherwise left all at sea in discussing language in general.

The word WHARF has been hunted all Europe through in search of its origin. Philologists assure us that it can have no relationship to the word warp. But their science does not include a knowledge of the things which are the foundation of words, still visible to some extent in the hieroglyphic types. The wharf, to begin with, is not merely the modern landing-stage, but the bank or shore of a river. Shakespeare uses the word wharves for the banks.[1] A wharf then is a bank, a boundary, a binding round or alongside of the water. It is the Egyptian ARP, to bind round, engirdle, with the determinative of a skein of thread, a bundle, and of linen generally; that is of WARP or ARP. ARP as wharf is that which binds or bounds the AR (Eg.), river, and thus is identical with a RIPE, Latin RIPA, a bank. But ARP and RIPE have their earlier form in KHERP (Eg.), a first shape, a model figure of binding and boundary. It is here that we reach rootage. The WHARF or RIPE, as bank, is a boundary, a primal form of binding in relation to water. The KERB, however, is equally the WHARF of the street.

The WARP is also a first form in weaving, the boundary for the woof. ARP, the bundle, to bind, is the same as our WRAP, to wrap round, as do the KERB and the WHARF. A far earlier wharf is the WARP, the deposit of the river Trent after a flood; this is a first formation or KHERP. Also soil between the sea-bank and the sea is called a WARP; these preceded the artificial wharf. KHERP is the Kar, a circle or bound, to put a bound to anything, encircle, go round, contain. Kh or K denotes the type, which type is finally the hieroglyphic RU, a water-way, a water enclosure, the edge, border, round about the water, a mark of division. Here we have a visible and typical "root," in the same RU, with many radiants.

At this stage we are midway in the genesis of words. Beyond the hieroglyphic RU lies the range of symbolism and the origin of sounds, and on this side the etymology of formation with the different

---

[1] *Ant. and Cleop.* ii. 2.

prefixes, suffixes, and affixes. This belongs to a domain that philology has not yet entered, and can only enter by means of Things.

There has been as great quackery, on the part of "Scientific Philology," in the fortune-telling of words as in any kind that is punishable by statute law, because the original data have been left out of sight. On the assumption that the oyster is named from its shell, and a root identical with Os for bone, we are told who ate the first oysters. The oyster has a common name throughout Europe ; in the Welsh OESTREN, Latin OSTREA, Greek ὄστρεον, Russian USTERSII, Scandinavian OSTRA, Old French OISTRE, with which agrees the Armenian OSDRI. The Sanskrit PUSHTIKA will not correlate with the "OS-TER," and "the only inference from this fact is that the Western Aryans became familiar with the Caspian Sea, and therefore with oysters, long before their Eastern brethern, who, not meeting with them till they reached the shores of the Indian Ocean, hit upon another name for them, derived from an entirely different root." [1]

There is, however, a possible origin for all, a root in which the meanings meet. PUSH (Eg.) means to divide into two halves, and therefore supplies a name for the bivalve. TEKA (or tika) has the same significance as TER ; it denotes boundary, frontier, margin, and as the word PUSH is a form of FISH, the PUSH-TEKA is the fish of the margin or shore, the one that could be first caught, and named from the time when food was something ready to hand. TEKA (Eg.) denotes that which adheres, is attached and fixed. TEKAI is the adherer personified and PUSH-TEKAI is the bivalvular adherer. That is what Egyptian says where Sanskrit is silent.

Push-tika was the fish that could not swim away. PUSH in Egyptian modifies into USH. USH is to cut, to saw or divide in two. Both PUSH and USH are applied to opening, dividing. The USH, Os, or OYS, then, is the modified form of the earlier PUSH, and both PUSH-TIKA and USHTER read the Bivalve of the Shore. What then is the inference ? That the PUSHTIKA was known on the shores of the Indian Ocean, and named before the migration from Africa to India took place, and that PUSH was worn down to USH before the migration into Europe followed ? That is the first look of the facts, and, in the Xosa-Kaffir, IMBAZA, *i.e.* BAZA with the "im" prefix, is the name for oysters ; but these wide deductions from a single word are often critical indeed, especially where nothing is known of the origin. And this would not do, as we have our word FISH, and "FISSE" for the fist that is formed of the closed hand ; our PESH or PEASE are also named like PUSH-TIKA from their dividing in two. It is curious that TER and RET permute in Egyptian, and the Irish name for oyster is OISRIDH. Ret (Eg.) also means the one made fast, or, both hard and fast. Thus OIS-RET has the same name as the Ruti race, and is literally the bivalvular RUTI. The Ruti species

[1] Farrar, on *Language*.

first among fish as the Egyptians claim to be among the human species.

TEKA in a hard form becomes the Hebrew type name for the fish, as Dag (רג), and Egyptian will account for both on the ground that the oyster was the first fish caught because it could not escape. Also KHA (Eg.) is the fish and TE means to remain. TEKHA would be the fish that remained fixed, attached to the shore. TIO, the Maori name for the oyster, probably a modified TIKO, also means a landmark, and ice, a form of the WATER-FIXED like PUSHTIKA, whilst Ika is the fish, and T the article The.

Kadmus, again, is said to mean the East and can mean nothing else, and it has been argued that as Kadmus was the bringer of the Greek Letters, and as his name signifies the East, the letters must have been brought from the East. But Khetem or Khetmu is the Egyptian name of the Seal-ring, the type of lettering, and therefore of Letters. Khet means to cut, to stamp, or seal, and the Khetem ring is the type. Ketu was an Egyptian God of Things or Letters.[1] Kadmus is the Greek form of Khetmu, to seal or to letter, and the Divinity was extant in Khetu. Kadmus is the Phœnician name of Taht as the inventor of letters or types. Khet implies an earlier Khept, and Khepui (Eg.) is a plural equivalent, meaning Types.

And so it is with the names of mythical characters. Max Müller says:—"If the first man were called in Sanskrit Adima, and in Hebrew Adam, and if the two were really the same word, then Hebrew and Sanskrit could not be members of two different families of speech, or we should be driven to admit that Adam was borrowed by the Jews from the Hindus, for it is in Sanskrit only that Adima means the first, whereas in Hebrew it has no such meaning."[2] In this we have the two sides of an arch, impassable until the key-stone is dropped in. Adima appears as the first man in India, as well as in Jewry, because both are independently derived from the common source in Egypt, where Atum is not only the first created, but self-created. "I am Atum, Maker of the Heavens, Creator of Beings coming forth from the World, making all the generations of existences, Lord of life, supplying the Gods."[3] Also ATI (Eg.) means the King, the first, chief one. Adima has for consort an Eve in India, as in the Genesis, because the original of both is the Genitrix, the goddess of the Great Bear, and mother of flesh (Af), as Aft in Egypt. We shall find two or three Eves in Africa.

As another illustration of comparative mythology, the Fijian divinity KALOU-GATA, when juxtaposed with the Egyptian Har-Makheru, will shed a light on his name and nature. Ma is truth and true ; Kheru is the word or voice. Plutarch tells us that when Isis felt herself to be with child (*i.e.* when she quickened), on the 6th

[1] *Pierret*, p. 452.   [2] Max Müller, *Science of Religion*, p. 302.
[3] Birch, *Eg. Ritual*, ch. 79.

day of the month Paophi, about the time of the autumn equinox, nearly six months before the time of the vernal equinox, she hung an amulet or charm about her neck, which when interpreted signified a TRUE VOICE.[1] At that moment the one Horus was transformed into the other whose title is Makheru the True Voice. Now the Fijian KALOU is the God who is as good as his word, and fulfils what he promises. The first Horus was a dumb image of the Word, the flesh-type in embryo, the promise of life to come. The second —after the quickening—was the True Word, the word of promise made true, or as the Fijians have it, the God who fulfils what he promises, and is the Justified.[2]

But we must keep to our words.

According to Wedgwood, the word WAKE is the old Norse VAKA, Gothic WAKAN, Anglo-Saxon WACIAN, German WACHEN, to wake, O. H. G. WACHAL, A.-S. WACOL, Lat. VIGIL, waking. From O. N. WAKA, to wake, was formed VAKTA, to observe, watch, guard, tend. The corresponding forms are O. H. G. WAHTEN, to watch or keep awaké, to keep guard; G. WACHE, watch, look-out, guard; WACHT the guard; Du. WAECKE, WACHTE, watching, guard, and E. WATCH, N. Fris. WACHTGEN, exspectare, and from Northern Fr. descended E. WAIT and WAYTE, a spy, explorator.

These are all derived from the root represented in Egyptian by UAK, a festival, and UKHA, to seek and search after. In the funeral WAKE and the statute fair of that name, we have the festival. In the Irish WAKE of the dead we find the seeking, searching, and calling after. The Christmas Waits (or WAIKTS) are a form of the seekers for the winter sun, as in the Egyptian search after Osiris during seven days or nights, whose UAKA, or festival, is kept by us at Christmas, as the end of our year.

The Egyptians held an annual Wake or feast of the dead, called the Uak Festival, on the 18th and 19th of the month Taht, the first month of their year. And in Egyptian we find the primal KAKA, to rejoice, eat, feast; KAKA (Choiak), a festival; KHAKH, to follow, seek after, chase.

KAK (Eg.) is darkness, and all watching turns on that.

The stars were the earliest watchers on this account. The sun that watched through the darkness, was named KAK or HAK. The modified AKH is the name of the illustrious watchers, the stars, and the Akkadian moon-god. The earliest WAKE is the KAK, as in the Assyrian KAK-KARRIT, an anniversary. The first watching is KEKING, or KEEKING. KA-AKH (Eg.) denotes a calling for the light, or for the dead, manes, spirits. And in a magic papyrus at Berlin occurs this formula of KA-ing or KHA-AKH-ing, "KHAAKH! KHAKHAKH! KHARKHARAKAKIIA!" An invocation that probably preserves the language of calling used by

---

[1] Of Isis and Osiris.　　　　[2] Hazlewood, *Fijian Dictionary.*

the oldest watchers in the world, who besought the AKH by night and rejoiced at certain recurring periods, and held their KAKA, and uttered what the costermongers still designate a " KIHIKE," *i.e.* a kind of hurrah, a cry in praise of, to call attention to.[1] The Hebrew AKAK-AK (אך אנך) means to cry out AK, or AH, as a mode of invocation. KA-KA (Akkadian) is to confirm the word by repetition of KA, to speak. The AKKADIAN Amen (AMANU), is " KAK-AMA."

The word YES, we are told, is Anglo-Saxon, the same as the German JA, and it conveys the historical information that the " White masters of the American slaves who crossed the Atlantic after the time of Chaucer had crossed the Channel at an earlier period after leaving the continental fatherland of the Angles and Saxons."[2] But IA, the equivalent of YEA, means Yea, Yes, Certainly in Egyptian, and may have been in the island thousands of years before an Angle, Jute, or Saxon came. It is questionable, too, whether YEA is not distinct from Yes. Chaucer always distinguished between them, and in the original tongue IA, the sign of assent and assuring, can be paralleled by HES, to obey, be obedient, which goes still farther as KES, to bow, bend down, be abject. Here are three degrees of IA, HES, and KES. IA with a mere nod of assent, HES a bow of obedience, and KES to bend down abjectly and entreat. The English Y like the J is not a primitive, and was preceded by the G or K. YESTE is GEST, a history. Yes was the earlier GESE. The YES is an earthworm, and this has its prototype in KEK—later KES—(Eg.) a worm. YIFFE was GIVE; YESTE was GEST in Anglo-Saxon. So that YES may have been GES or KES (Eg.) meaning to bend down, bow ; in fact to enact our Yes in *ges*ture speech. This has the emphasis of Chaucer's Yes. KES (Eg.) means to lie down, and this we have as GISE, to recline. Our YES is also GES or GIS, as an oath. " By GIS, and by St. Charity " (*Ham.* iv. 5), and GIS corresponds to KHES (Eg.), a religious rite. GIS has a variant in GOSH.

When the word CURSEY is found in Cornish English signifying a friendly chat in the house of a neighbour, it is forthwith assumed that CURSEY is the French CAUSER. But KHER (Eg.) is speech, to speak a word, and SI, to pass, has the sense of *en passant*. KHER-SI would be a passing word. " KAU " (or Ka) (Eg.), to call, and say, is a distinct root, it exists in the Cornish COWS, to say, speak, tell, and " Ser " (Eg.) means privately. Perverted pronunciation is by no means such a factor in the development of language as it is assumed to have been.

There is no need for going to the Sanskrit SVETA, White, for our English wheat, when we have it extant as wheat in the Egyptian UAHIT, corn, and HUT, the white corn. Wheat, we are told, means the white corn. So it may, but not simply so. The Egyptian HUT is white, also Hut is the white corn or wheat. But white is not

---

[1] " *Chi-ike,*" *Slang Dictionary.* Hotten.
[2] Max Müller, *Lectures,* first series, p. 225.

the origin. Some of the most famous wheat in the world was the oldest RED corn. We may get at the original signification in another way. "UAHIT" is the Egyptian word for corn, extant as our English wheat. The terminal T is a suffix. HU is corn, aliment, white, and UAH, a name of corn. UAHIT without the IT (Hit) signifies to cultivate and increase, it is likewise the name of the ploughman, the cultivator. So that UAHIT, or wheat, is named as the cultivated corn. Thus the name of wheat in Egyptian signals the nature of this corn in contradistinction to the cereals that grew wild and uncultivated, and the UAH, to increase and augment, shows the joy of producing the HIT (Welsh Yd) by means of the Uah-Heb, the Ploughman. The root of "UAT," water, green, green things, shows that "UAHIT" was also named in relation to the need of water in cultivating it. Green rather than white is the colour primally associated with wheat. And here one of many byeways opens, which the present writer must not follow. The reader will perceive that the provincial pronunciation of the word "Wheat" without the *e, i.e.* "WAHIT," preserves the sound of UAHIT, and makes Egyptian still the spoken tongue.

One of the Irish names for wheat is CRUITH-NEACHT. In Egyptian, arable-land and a field sown with corn is "NAKHT;" "KAR" is food, "KHARU" is bread, therefore "CRUITH-NEACHT" in Egyptian denotes the bread-food of the ploughed land, and this is the Irish name for wheat. Here again the name, like that of wheat, tells us that it was the cultivated corn, therefore the corn first cultivated. The ancient Scotch were called CRUITNICH, rendered corn-men. This in Egyptian, KAR-UAHIT-NAKH, is the corn cultivators, which agrees with the tradition of the Welsh that the God Hu, whose name signifies corn, taught the first settlers in these islands the art of growing wheat, previous to their emigration from the land of Hav, that is, in Cornish, the land of corn. A kind of corn highly thought of and much grown in Sussex is called CHIDHAM white ; this in Egyptian, KHETAM-HUT, reads the golden corn. The English BERE for barley and BEARD, an ear of corn, are the Egyptian PAR and PERT for grain. Pliny says the oldest name for corn in all Latium was FAR. This is a form of the word PAR. The Maori PURU for seed, PAROA, flour and bread, and the Irish POR, seed, for race, are from the same root. PAR was the name of the seed-time in Egypt. Another type-name for grain is supplied by AB, corn, earlier KA-AB, food-corn, whence the Zend and Sanskrit YAVA, Lithuanian JAVAI, and English Gipsy GIV, for wheat. ZEA, also the oldest known Greek name of barley, derives from the Eygptian Sif (Su), corn, bread, seed, the boy as seed; earlier KHEFI, Harvest.

CREEING is a word used for steeping grain, whether rice for a pudding or wheat for making furmety, by putting it into an oven to become soft. CREED and CREEDED describe the process of oven-ing without baking. In Egyptian KRA is the name of the oven or furnace, and the form of the word for the thing CREED when done would be KRAT. It must have been used for malting or distilling, as the KARAU is the jar, a vessel from which steam is issuing. And this gives us the Welsh word for strong ale, whilst KRA, the furnace, with the Egyptian terminal T, yields the word GRATE.

Bopp works a long way round to derive the word BERRY from the Sanskrit BHAKSJAM (BHAG-S-JA-M); it is the Egyptian PERRIE or PERI, food appearing, with a branch sign of bearing; PERU, to put forth, manifest. BERRY (Eng.) is a flood, and in Egyptian PERU is to pour forth, flow out. PER (Eg.) also is grain, corn, PERRI is a granary. In English to BERRY is to thrash corn, and the thrasher is a BERRIER, which, according to Egyptian, is the person who makes food appear.

PEF (Eg.) is breath; PAIF a gust or puff of wind. In English PUFF is a name of the breath. To PUFF is to blow with the breath, to pant; to PEFF is to cough faintly. PEFS (Eg.) signifies to cook, bake light or lightly. POBI (Welsh) means to bake. English PUFFS are light tarts. From PEF, light food, we derive the word BEAVER, a very light intermediate meal. Another name for BEAVER is BAIT. BAAT in Egyptian is food, a kind of loaf called "BOTHS"; BAAT is especially food distinguished from flesh, and this is the character of our BAIT or beaver, which is a meal without meat.

In Egyptian USKH and SEKH are variants of one word. SEKH is a liquid, the same as suck and sack in English. USKH gives us our river-names of ESK and the Irish UISGE and SUCK (river), the ISCA and WHISKEY. UISGE combined with BAKH (Eg.) for beverage, forms the word USQUEBAGH. SHEKU (Eg.) is an intoxicating drink, and SEKHT is the Goddess of drinking and of fire. USKH passes into OX, as in Oxford, the waterford. But Egyptian shows that Oxford may mean much more than this. The USKH is also the Hall, the Temple, with the leash of feathers for determinative, and Oxford as the place of the Halls of Learning is the USKH-ford in the Hiero-glyphic sense. The Hall, the Abode, is determined also by the quadrangular sign. This likewise appears in the buildings and the four-cornered cap of USKH-ford.

Again USKH (Eg.) means broad, wide, to range, stretch out, extend, and there is an old English cry used in hunting, "ASYGGE," "ASYGGE," in the sense of making a broad and extended cast round about.

"Ye shall say 'ILLEOSQUE, ILLEOSQUE,' alway when they fynde wele of hym, and then ye shul keste out ASSYGGE al abowte the feld for to se where he be go out of the pasture, or ellis to his foorme."[1] This "ASSYGGE," never yet explained, is the Egyptian USKH, to stretch out, to extend, range out, and around. The instructions are to cast out broadly and ring round on a large scale to ensure the run. A collar is one type of the Uskh.

In Egyptian SEKTI is a bark; one name of the bark on which

---

[1] *Reliq. Antiq.* i. 153.

the sun made its upward passage from the lower signs.   The
divine bark of the Gods was the SEKTI or SEKT when contracted.
SKT may be read SKAT, and the SKATE survives with us as a
small boat or wherry.   The wherry, by the by, is the Egyptian
URRI, a form of carriage.   The wherry carried passengers.   The
SAKTU or mariners were the rowers of the boat, and in English to
SKUT is to stoop or crouch down, as the rowers do in pulling.   SKAT is
Egyptian, for towing and conducting a boat.   SKA is to cut, scrape,
play upon.   SKA is the plough that cuts the earth.   SKAT is to tow
or conduct a boat on the water.   Our form of SKAT in scraping,
cutting, ploughing is applied to SKATING on the ice ; like the original
SKAT it is still conducting on or over the water, whilst the SKATE
now represents the SKT, or divine Bark of the Gods.

KHEN (Eg.) means to go by water, to navigate, impel, convey.
The sailors are the KHENT, a means of navigating as rowers.   But
KHEN is also to impel, to blow, as a means of sailing, and in this case
the impeller, the KHENT, is the WIND.   The word KHENT becomes the
English WIND.   To winde is to go, to bring, and the wind was the means
of going and bringing by water, by which the boat WENT, and therefore was the WIND or KHENT, the impeller, the conveyer, the sailer
or cause of sailing and conveying.   It was not named merely for its
going, but as the means of going by water, and the antithesis to water
which was at first the natural opposite of going.   Answering to KHEN
(without the terminal), to blow, puff, impel, blow away, avert, we have
the form WINE (the wind) used in Somerset, and WINNY, to dry up.
KHENA (Eg.) also is to refuse, and WINNA (Scotch) signifies will
not.   KHENA is to be agitated, fearful, and WINNY means to be
frightened.   KHENA is to blow away, puff away, inspire, avert.   This
process we call WINNOWING.   KHENA is also to attain, alight, rest,
and WINNA (Eng.) is to attain, reach, gain, WIN.   These prove the
equivalence of WIN and KHEN, WIND and KHENT.

The root of ANIMA, wind, breath, air, soul, has to be traced back to
Khn (Eg.), to blow, puff, breathe, whence wind, in the form AN.
The spirit, or anima, founded on breath, is not so much the breath as
the breathing, the repetition of breath.   The root " AN " (Eg.) is
neither blowing nor breathing, but means to repeat, renew.   It was the
repetition and not the vapour on which the observers founded the being.
AN is being and repeating in one.   So with the word SPIRIT.   This is
not derived from the breath, but from the breathing.   SEP or SPI
(Eg.) is a time or turn, manifestation, spontaneous act.   RET (Eg.)
means repeated.   SPI-RET is the spontaneous manifestation repeated
in breathing.   This is shown by the SPIRT, for a short space of time
or a brief emission.

The Mum (Mummy) was the very self preserved, the self-sameness
kept in death.   The Egyptian MUMU means also, or likewise.
It is extant in the French MÊME, for self, which signifies likewise

and also. The MUM type of the self-likeness yields the Zend MAM for me; Lap, MON; Yakut, MIN; Mordvinian, MON; Akkadian, MU; Finnic, MÂ; Esthonian, MA; Proto-Median, MI; Zyrianian, ME; Etruscan, ME; Ostiac, MA; Welsh and Irish, MI; English, ME; and Latin, MEMET, for me, myself. MUM in English means be silent, hold your tongue; MEM, in Quiché, to be mute; IMAMU, Mpongwe, to be dumb; MAMU, Tahitian, to keep silence, and MUMU in the Vei language, because MUM in Egyptian means death, dead, silent, of which the Mummy is the ideograph. MEMN (Eg.) is a memorial, and the MUM was the memorial figure of the dead, by which they were kept in mind and memory and re-MEM-bered. MUM was the visible, not an abstract form of memory, and our mumming was a similar mode of representation. Thus Memory, from MEM, the mummy image of the dead, and REKII (Eg.), to know, or the Intelligence, was named as the faculty of keeping the dead in mind, and being able to reproduce their likeness, or figuratively make the mummy. To keep the Mummy (Mum) constituted the first Memory. The later phase is to call up an image mentally. The MUM type was continued in the MOMENE, an idol, and the MAMMET, a puppet, idol, the dolly, the Mammy, a Swiss doll; and in the image of Memory.

The internal organs of the body, the type of which was the heart, all that we call the viscera, the inner support and mainstay of life, were designated the "BESK" by the Egyptians, with the heart for determinative, denoting inward substance, the seat of life. With us the BESK survives as the BUSK, a piece of whalebone or steel, worn inside of STAYS to give support and keep them straight. To BUSK a lace is to put a stiff tag on the end called the busk point. The BUSK in the stays still images the BESK of the body.

In Egyptian hemp is named HUMA, and flax is HUMAMAUI, i.e. hemp made bright or beautiful. In India the HUMA for hemp becomes UMA for flax, and the Latin LINUM, for flax, adds the LIN of Linen to the UM, ÛMA, HUMÂ, of flax and hemp. RENN (Lin) in Egyptian signifies the unblemished, the pure, the virgin. If applied to blanching, this would be the bleached. RENN-HUMÂ would be the bleached or whitened hemp. The full word is RENEN, the virgin, pure, unblemished, hence the white. N often interchanges with M, but the Greek λίνον, the Welsh LLIN, Irish LIN, old Norse LIN, German LEIN, and English LINEN, only repeat the RENN or RENEN as the bleached, the virgin-white.

Flax is the prepared hemp. This we may derive from REKH, to full, purify, make white. P-REKH (Eg.) or F-LEKH is the thing whitened, purified, blanched, whence flax. Whilst the word bleached is the REKHT (P-REKHT), the fulled and whitened.

In Old English cloth is TUCK, the TUCKER was the weaver. Tucker Street, Bristol, was an abode of the weavers. TUCKING Mills are extant in Cornish village names. The name has been derived from

that of the river Toucques in Normandy. But it goes back to the
origin of weaving as crossing. TUKA (Eg.) means crossing, to cross,
twist, unite, attach, as in weaving. The crossing is still visibly preserved in the texture of DUCK, and the pattern of TICKING. Duck
was the especial wear of sailors, the crossers of the waters, a form of
the Ducks.

The HEARSE, we are assured, simply means a Harrow, because in
French the harrows used in Roman Catholic churches for holding
candles are called "HERSES." [1]   Our hearse is the Egyptian "HURS,'
a wooden head-rest, the Assyrian IRSA, a bed or couch. This
might be for the living or the dead, and it is supplemented by the
word "HERSA," signifying on it, and after. The "HEARSE" has both
meanings. Hearses were set up in churches after death. Hearse
is also an English name for a hind in its second year, the year after
the term of its being a hind calf. The custom of following after the
hearse likewise illustrates this meaning.   The same thing as the
Egyptian HURS, a head-rest, has been found in the crescent-shaped
objects discovered in the lake-dwellings conjectured to be head-rests.[2]

In English one form of REAM is to hold out the arm to receive. In
the hieroglyphics the "REMN" is an arm, to touch an arm, shoulder;
a hieroglyphic action with the arm. REM also means to rise up,
surge up, weep. In English REM is to cry, moan, froth up, stretch
forth. REM is cream, and to cream. This motion of surging up and
going forth is the act of yearning and grieving expressed by the word
ERME, used by Chaucer in the Pardoneres Prologue.

> "I cannot speak in *terme*,
> But well I wot thou dost my herte to *erme*."

ERME, to grieve, to lament, is one in English with REM, to moan,
cry, weep. And the Egyptian REM, to weep, also reads "erm," according to the placing of the vowels. Further, one hieroglyphic sign
of ERM or RMEN is the arm; that is ARM in English. To stretch
out the arm, to take, to desire to take, being a form of ERM-ing or
REAM-ing. Hence to stretch out the hand to take is synonymous
with grieving or weeping with desire, that is yearning. To ream in
English is also applied to stretching the legs. It is supposed that to
ERN or YEARN is a corruption of ERM. But the more we see of the
Egyptian origines, the more we shall doubt the breeding capacity of
corruption as a generator of language. It is true that ERN or RUN
has the same value as ERM or REAM.

Running is the contrary to standing still, whence the identity
of flowing and running. To RUN is equivalent to REAM-ing in
stretching the legs. REM, to weep, is the same as ERN, to RUN or
RENDER, to melt down. ERM and REAM, ERN and RUN, all meet in

---

[1] Brewer, *Dictionary of Phrase and Fable*, 393.
[2] Drawings by Dr. Keller.

this sense of flowing, and in that of a motion toward, as in yearning, rising up, stretching forth. RENN in Egyptian signifies to dandle, a nurseling, and may denote the mother yearning over her child ; such is the image. In the hieroglyphics the signs for *m* and *n* have at times the same value, Ma and Na both read of, from, to, by. Nu and Mu both denote water. The word REMN deposits a rem and a ren, our erm and run, hence the interchange of the one with the other.

URM (Eg.) is a name of the inundation, the overflow. In English URNE is to run. The URN is a synonym of RUN or to URNE. The prototype of our tea-urn is an Egyptian vase with a spout, the Kabh sign of refreshment and libation. Three of these joined together constituted the symbol of the inundation or URN, that is, with permutation of *m* and *n*, an URN. It may have a bearing on the subject that an URN measure contains twenty-eight pints. The URM or inundation, says Plutarch, at its highest rise (at Elephantine) is twenty-eight cubits, which is the number of its several lights. The Egyptians, he observes, consider the risings of the Nile to bear a certain proportion to the variations of light in the moon. The number twenty-eight is also a mystical measure of time.

The REM is one thirty-second of a measure of land ; one Rem is a span. The quantity may also be varied, as the word REMN means the extent, extending to, up to, thus far. The Irish ROM or ROME, from whence the ROME-feoh was a certain extent of land. In a report on the state of Ireland made to Henry the Eighth we read, "First of all, to make his Grace understand that there be more than sixty countries, called Regions in Ireland, inhabited with the King's Irish enemies; some regions as big as a shire, some more some less unto a little ; some as big as half a shire, and some a little less ; where reigneth more than sixty Chief Captains, whereof some calleth themselves Kings, some King's Peers in their language, some Princes, some Dukes, some Archdukes, that liveth only by the sword, and obeyeth to no other temporal person, but only to himself that is strong : and every of the said Captains maketh war and peace for himself, and holdeth by sword, and hath imperial jurisdiction within his ROME." This is the ROM or ERM of Egypt, the measure of land extending to, as far as, the border or limit determined, that is, the RIM, margin, the space or ROOM.

The hieroglyphic RENN for cattle is a noose or cord for the foot of the animal, to determine its Run or Remn when grazing. From this lowly origin it rose to become the ring-enclosure of the royal names.

The English RIM for margin, edge, circuit, answers to REMN (Eg.), extending to, so far as, up to. With the (Eg.) article prefixed, we get BRIM, as the RIM. And here the RIM (Eg.), to rise and surge up, has another correlate. The REM or BRIM, as sea-margin, is not only the edge, but the sea, the flood, the surging-up itself. Another BRIM or BARM is the uterus, the bosom ; BARM also being a name of yeast.

Rem (Eg.) the extent, boundary, at the place of, may be written Erm, or Elm, the name of the elm-tree in English, and thinking of our universal hedge-row Elms it seems likely that the Elm is named as the tree of the boundary, the extent, at the place of, the edge or hedge, the hedgerow elm. This tree of the terminus is also a sacred tree, used for the coffin, at the end of life. The elm or Ailm of the Bethluisnion Tree-Alphabet is A, which answers to the Runic A or Arm. In this case the boundary is at the beginning. In Devon, the elm is called Elemen, and in the Gaelic Leamhan or Leoman we have the equivalent of the Egyptian Remen. This "Remen," the type of boundary and extent, takes another form in the Gaelic Ruimne, a marsh land, as in Romney Marsh and Ramsey in the Fens. Marshes were limits. And still another in the port of Lymne, from which an ancient road runs across the Kentish hills to Canterbury called the Stone Street. Ermin Street, which joined London and Lincoln, is also a form of Remen (Eg.), extending to, and has no relation to Paupers. It is one of the four great roads ascribed to the Romans. The names, however, will show that the Romans did but follow and reface the track of our far earlier road-makers and boundary-namers. In Lambourne, Berks, the bourne repeats the Rem as the Limit and extent, and the Remn (Eg.) with the article the (P) prefixed passes into Perimeter for the circuit, the orbicular extent. The Rim, Akkadian, is a mound. This Rem (Eg.) at the place; Remn, the extent, enters into Sanskrit as Ram, to stop, stay, remain; the Gothic Rim-is, Lithuanic Ramas and Rem-ti; Latin Remaneo, to stay, continue (also Remano, to turn, flow back) and our English word remain, which is an equivalent of Remn (Eg.) at the place, and Remnant, for an end.

Uah, Uas and Uam are interchangeable words in Irish for the top and summit, the supreme height. In the hieroglyphics Uah is the crown, the word also means very much. The Uas is a sceptre, the sign of supremacy, and the Uam-ti is a rampart, a raised wall or height. Whilst the root Uâ signifies the one, the one alone, solitary, isolated, equivalent to the top and summit.

In the North of England the word "Leet" is used in the sense of a meeting of cross-roads. In Essex the fuller form is Re-leet. A two-releet is the meeting of two roads; a four-releet a meeting place of four roads. Our words road and leet both derive from the Egyptian Ru, a road and pathway; the French Rue, a street, with the terminal T suffixed gives Rut, our Route and Road. The point in releet is the meeting of the roads. This we get from ru, road, division, and ret, several, repeated, bound up together. We have the Leet as a meeting answering to Ret, repeated, several, and with Ru or ret for the road, and a division, we get Ru-ret or Re-leet the meeting of roads. The same word as Ru-ret (re-leet) is extant in the hieroglyphics, as ret-ru, with the meaning of entire, answering to the totality of the RUT or road in RE-LEET. There is an old word SART, applied to a piece of woodland, newly stubbed up and turned into arable land, which the Egyptian SART perfectly explains. SART means to sow seed and also to cut down, prepare, to dig, plant, sow, grow, renew, and augment. A form of the word exists in the Latin SARTURA, mending and weeding, but the English keeps all the senses of the Egyptian. "ASSART Rents" was the term employed for the payment made by those who recovered such lands, and SERETH was the earlier name of the territory of the Barony of Tir-hugh called Tir-Aedha, in Donegal, in the sixth century.

To grub or GRAFF land is to break up the surface for the first time. And in Ireland the peasantry use a sort of double axe for rooting in graffing the land, called a GRAFAN. This supplies a type-name for lands such as Graffa, Graffee, Graffy, Graffage, meaning the grubbed land. GRAFF is the Egyptian KHERP, the first cultivated land. A form of it is extant in the GLEBE land, sacred to the Church, the name of which points to a primordial form of tenure and cultivation. KHERF also signifies an offering of first-fruits, a supply, a CROP, sufficiency; to pay homage, consecrate, which meanings apply to the GLEBE land.

The CELT-Stones are the KART stones by permutation, and these are the stones cut by the mason who is the KART. Kar-Nater is one name of the Mason.[1] NATER means to cut, plane, work, make smooth. Kar is the stone, and *t* the participial, so that the KART is the cut and smoothed or polished stone. One form of the KAR stone is the teste, KAR-IU, and KARTI, the two Kars are the testes, the KART is the cut stone, and in English this cutting of the stone is to GELD. Our Kelt stones then are named in Egyptian as the KART stones, cut and polished by the KARTI or KAR-NATER, and the KELTÆ bear the same name as these cutters of the KELT stones who in Egypt are the KARTI, the masons, the stone-cutters and engravers on stone, the stone-polishers, in many lands; the men of the Neolithic Age and Art. Every way we take leads to Egypt, every word we hunt down runs to earth at last in the land of the Rut and the Karti, or the Kafruti, who were the Rut of the primordial race.

The term WHIN is applied by miners to certain hard rocks. "WHIN Sill" is a sheet of basalt, spread out between the carboniferous limestone strata. Our WHIN is the Egyptian Han; and Hanna is a quarry, to yield tribute. The Han-ser (whin-sill) is the rock that pays for quarrying. Our whinstone is a basalt used for whet-stones. This is the AN stone, also used by the Egyptians for whet-stones,[2] the AN that gave the name to their typical column. Ba-salt again is ba-sert (Eg.), gravingstone.

[1] *Denkmäler*, ii. 134, *a*.   [2] Rosellini, *M. C.* 33, 14.

"THREPE" is used for arguing, disputing, maintaining, saying, and has a meaning of urging, insistance in saying. The Egyptian deity Taht is the scribe of the gods, lord of the divine words. He is called SA, the clever, skilful. His name Taht denotes the word, mouth, tongue, speech, and his "rites" are called "TERP." A rite is a formal act, in this case of the utterer, Taht. TERP may be dissected, as P is the article. The word ter means interrogative, question, who? what? This includes discussing, arguing, and maintaining, that is THREPE or TERP, the RITES of Taht, the tongue or speech.

Burns speaks of "some devilish CANTRIP sleight." Cantrip-time is the season for secret magical wizardly practices. The Gaelic cantrip is a charm, spell, incantation, evocation of spirits. TERP (trp) signifies certain rites of Taht, the revealer or interpreter of the gods, and charmer of spirits. KHAN has the meaning of inward, interior, hidden, and is related to the dead. KHENF is the food offered to the dead. KHEN is a time. KEN-TERP would be the time for certain rites of revelation, known as CANTRIP-time.

The word Day is synonymous with the Egyptian TUAI, time, morning, the morrow, as we say, the morrow-day. TUAI also denotes two halves, and the day is one of these, the light half; the lower hemisphere is the TUAT as the other half. But day has an earlier form in DAG. TEK (Eg.) is crossing over, and TAKH a frontier, whence a crossing over from one side to the other is a DAG. This sense of day as boundary, so much cut off, gives a name to the instrument for cutting off, as the DAG, an axe, and the DAGGER. To DAG or DOCK is to cut off. A Dagon is a slice cut off. And to DAKER is to work for hire after the day or DAG is over, the boundary crossed. DUCKISH is twilight. TAKH, a frontier or cut limit, becomes DEKE, a ditch. It is said of a man whose life is cut off, " it's all DICKY with him." The DIG or DUCK is the crosser over the water. A DOG-LOPE is a boundary between houses that belongs to both. DAG is boundary; TAKH (Eg.) is frontier, and RUPU (lope) means either. When the dog turns round before lying down, he is said to be making his DOKE, his boundary. Time was cut or TICKED off by TEKHI, goddess of the months. TEKH is a name of Taht, the moon-god in his secondary character. This was represented by the dog-headed ape. TAG in English is one who assists another at work, in a secondary character, as the dog the shepherd; in the same way to dog is to follow after. Here our dog of the man in the moon, Taht, is TAG or TEKH by name, as the secondary and following character. Tekh (Eg.) is to fix, attach; the DAG was formerly the fixed day, and the dog is the animal attached to man.

TEACH, we are told, comes from the Anglo-Saxon TACCAN, to show. Taccan is a developed form of TAC or TEACH, no matter where it may be found. DICH (San.), to show, and the English TEACH are anterior to taccan, and the Greek δείκνυμι. TEKH is Egyptian, the name of

Taht, who was the teacher, the illuminator, revealer, shower, personified. Nor does TACCAN, to show, go to the root meaning. We
only get to that when and where the various meanings meet, as in
TEKA, to illumine; TEKA, to see, behold, fix, attach; and TEKH, the
divine teacher. The first teaching or TEKHING was by reckoning
the transits of days, moons, months, stars, as they crossed over, and
ticking them off on the digits by tens in the reckonings of time.
Hence Tekhi was Goddess of the Months, and Tekh, the lunar God,
the Teacher, as reckoner, calculator, measurer, the ticker-off of
periodic time.

The name of the duck answers to the Egyptian TEKH. TEKH or
TEKAI is the OTIS TETRAX, also the ibis, and is not known as the
name of a duck. But it is known that the wild duck covers up her
eggs with moss or grass every time she leaves them, and this is a
form of ducking or TEKH-ing. TEKH (Eg.) is to hide, to escape
notice of, to duck. Either would supply the name, and both senses
meet in the one word TEKH, our duck. The Assyrian Tukhu
means the descending, or more literally, the ducking Bird. This is
the same word with the same meaning as Duck, although applied
to the Dove, doubtless from its motion.

URT is the Egyptian word for Bird, and the B as representative
of the Article forms the word BURD. URT the old first genitrix is
the Brooder or Breeder, our Bird being named as the typical URT,
the brooder on the Nest. The Urt Crown, composed of the two
Serpents, was representative of both truths, of puberty and maternity,
showing that the Wearer was the Bearer or Breeder, an image of
Ta-Urt, or with the other Article, B-URT, the Bird; Scottish BURD,
as in "BURD Ellen," who was also the Brooder.

The word DATE, the name of the palm-fruit, still preserves the
name of TAHT, and of his reckonings or dates made upon the palm-
branch of the panygeries. TA is to write and register; Tat was the
registrar of dates, and his account was kept on a branch of the palm
that bore the DATES as its first fruits, the numbers (DATES) as
its last.

The origin of the word TREE may be traced thus : REP (Eg.) means
to grow, bud, blossom, shoot, take leaf. REP and leaf are equivalents. With the article the, we have TREP, the budding, leafing,
flowering, growing. This modifies into TERU (Eg.) with the meaning
of roots and stems, by aid of which we get pretty near the tree.
The *u* passes into *e*, and we have the word TREE. The family or
house-tree is the TREF, the TREV, TORP, and DORP, and the word
contains the roof—T-ROOF—hence the ROOF-tree as the TREF.

The REP, REF, or ERP, Greek ERPE, is an Egyptian temple, the
sacred house, ROOF, or, with the prefix TREF, as in our TREVS and
TRES, which will be separately described.

But the meeting-point of Tree and Three may be here noted in relation to the ERP, ARP, or REPA, who is the Branch, or Tree, and the representative of the three—as the manifester of the father and mother —he being the three in one, which is also typified by the leash of feathers.

The name of TAHN is found applied to such various things in the hieroglyphics as to be very perplexing. It is certainly applied to tin and other metals. It is found in the form of crystal, hyaline, rosin, and probably glass. Our English words TIN and THIN will give us the determinating idea that correlates the whole of the substances named. TIN and THIN are one word. Tin is that which can be beaten out thin. An old form used by Chaucer, TEYNE, is a thin plate of metal or tin; Latin, taenia, Greek, TAINIA. In Egyptian TUN signifies to extend, spread, or lengthen out; TAN (Zend) is the extent; and those substances which would do that or become THIN are the TAHN of Egypt. Glass is TAHN, or thin enough to let light through. This principle of thinning by spreading out and TEN, to the fullest extent, explains how THINDER and YONDER meet in one word that includes TINDER, as that which is spread or thinned out. It likewise illustrates how THUNDER and TINDER meet in the same word, with the same sense of spreading to the farthest extent and uttermost limit, the one audibly, the other visibly. TANNING by beating is a process of THINNING. In Numbers ten is the extent of that division, as ten (Eg.) is the total amount.

It is said that the name of the "Kit-Cat Club" was derived from Christopher Kat, who supplied the members with mutton pies. But KIT-CAT is the name of an ancient game played by boys. KIT or KITTY is little, small, as in kitty-wren; also a quantity. KIT-KAT, a small quantity, is the Egyptian Kit-KAT, a small number, a few. This makes the name of the KIT-CAT club signify a select few, whether or not it was known to the members. And this, it would seem, is the likelier origin of the KIT-KAT portrait, a three-quarter length, or the little picture.

The Druidic KOMOT, court or division, and later witanagemot is Egyptian by name. The KOMOT was the department in which a Druid was empowered to teach according to the grant and privilege of the lord of the territory.[1] One form of the Egyptian KHEMUT was a shrine, place, enclosure. UITA (Eg.) means to examine, specify, decide; UITAU, speak out, give out, voice. The WITA-GEMOT is thus the court or parliament invested with power to debate, examine, verify, and decide.

The Brehon Laws are probably named from HUN (Eg.), command, rule, to rule, flog, restrain, make to turn back, order; HUN, sanctity, majesty, royalty, divinity; and PRE, to show, manifest, explain. PRE-HON is to show, make manifest, explain the sacredness of rule and government, as set forth in the laws. BRE also answers

[1] *Barddas*, vol. ii. p. 31.

to the BAR (Eg. Par), BARRA (Gael.), the high place, BARRA (Gael.), a court of justice, the bar of justice, where the manifestation of the HUN took place and the culprit was hung.

AUTUM is a slang term for an execution by hanging, supposed to be connected with autumn by means of the drop or fall of the leaf. In Egyptian ATH is to drag, to draw, and AAM is a noose. ATH-AAM is to draw a noose, and hanging is a process of execution by means of drawing the noose. The slang autum is a derivative from Egyptian. ATAM (Eg.) also means to enclose and exterminate, as hanging does. This is related to Atum, the great judge, the avenger. TEMA is a title of the justiciar, who executes justice. AU to chastise, punish ; TEM, a criminal, offers another form of autum applied to the punishment of a criminal by drawing a noose.

In the "Babee's Book," an old tract for teaching children courtesy, entitled the " Lytylle Children's Lytel Boke," it is said of it, "THIS BOKE IS CALLED EDYLLYS BE." [1] " EDYLLYS BE " has baffled all investigators. Nor can it be read straight off by aid of Egyptian. The root, " IT," is Egyptian for to paint, to figure forth. This passes into the words idea, idyl, idol, and eidolon. An idea is a mental image, an idol a figure shaped, an idyl a picture painted. EDYL then might mean the portrait or portrayal. LYS or LOOS is (A. N.) for honour. BE, as we know, is an abraded BEIGH, a ring, an ornament, a sign of distinction, a jewel. In the hieroglyphics PEH and PEKH are synonymous like our BE and BEIGH. PEH is glory, glorious, glorified. This is our BE, a jewel, as the ornament of courtesy, the glory of knightliness. Spelling, quilting, husking, and other BEES derive their name in the same sense as a show, a mode of surpassing and excelling. But there is a still better reading perhaps for EDYL. ETTLE is to intend, contrive, attempt, prepare, set forth, deal out sparingly, as in teaching a child: that is our EDYL. LYS, or LESE is to pick, select, gather, glean. Thus, courtesy being the ornament of honour, " ETTLE-LYS-BE," is intended for the teaching and the learning of courtesy, as the " BEIGH" or jewel, an ornament desirable for the child. The Babees' Book is the babees' in a peculiar sense. The BABE is a " Child's Maumet " (Gouldman). The baby in the North of England is still used for a child's picture. A babby is both a baby and a sheet, or small book of prints for children.

A CAR-WHICHET, or Carra-whichet, is a retort, repartee, a witty word of quick return, or used, as we say, in giving "WHICKET fôr WHACKET." The English WH often represents the Egyptian kh as What does the old Quhat. Thus, WHICHET is the equivalent of KHEKT (Eg.), to follow, return, repulse. KHER (Eg.) is a word, speech, to say. KHER-KHEKHT, is the word that follows in return, the retort that repulses. KHEKH is check, and this is the check-word

[1] *Furnival*, p. 22, note 14.

of wit, or the CAR-WHICHET. The wicket in the game of cricket
is the place of repulse and return, or the KHEKH-T.

A story is synonymous with a lie, mildly described, but why
a story and lie should be identical, English does not show, and
Egyptian does. STERI is to lay out, be stretched out in death,
to lie on the back. That is STERI, to lie. To lie is to be stretched
out, a story is to stretch out, as in lying, *ergo*, a story, a stretcher,
is identical with a lie.

Mrs. Quickly's " TIRRITS "—

" Here's a goodly tumult ! I'll forswear keeping house, afore I'll be in these
tirrits and frights."—*Henry IV. Part II.*—

can be explained by the Egyptian TER and TERT. TER
or TERU means an extreme limit, extremity, to be hemmed in,
bounded, hindered. The *t* (ti) makes the plural ; so that Mrs.
Quickly means, by her TIRRITS, that she is driven to extremities.
Another form of the word TARUT (Eg.) signifies urgent require-
ments. In this sense she is not equal to the calls made upon her.

STUM is a name of strong new wine, used to fortify old weak
wine, and to STUM the wine means to strengthen it.' The Egyptian
STUM is used in both senses. To STUM is to paint the eyes, to
beautify the eye-brows, darkening the lashes to heighten the appearance of the eyes. The stibium, or kohl, employed for this purpose is
also called STUM.

> " If ever thou bist mine, Kate,
> I get thee with *scambling.*"

says Petruchio ; and, again, Shakespeare speaks of the " SCAMBLING
and ungenial time " (*Henry V.*, i. 1). The Mondays and Saturdays
in Lent, when no regular meals were provided, were called
SCAMBLING days. The word has been derived from the Greek
SKAMBOS, indirect, oblique. But SKAM (Eg.) contains the necessary meaning. SKAM is hurry-skurry. Shakspeare uses it in
that sense. Scamp, in Lancelot, is to run in a hurry. To scamp
work is to do it too hastily. And in Egyptian, SKAM means to
stay or pause only for a short time, thence to be in a hurry, or
hurry-skurry.

The word " TIT " is a philological perplexity. It is applied to
a small horse, and to a strumpet (a light TIT). The Egyptian TAT
will explain why. This signifies to gallop in going. The TIT is
the opposite of the jog-trotter ; it is the galloper or fast-goer. Hence
the allusion—

> " This good mettle
> Of a good stirring strain, too, and goes *tith.*"
> —*Loyal Subject.*

To go TIT is to gallop. " Tit-up a tit-up " is the mode of describing
the sound of the gallop, and a TIT, horse or woman, is the one that
goes with a gallop, the galloping goer.

There is an old English dance in which the suitor for the lady carries a cushion and presents it to her kneeling. It is called the cushion dance. KES (Eg.) means to dance, bend down low and entreat abjectly. The cushion is a type of kneeling down. KES (Eg.) is to kneel and to adore. This root KES enters into the name of another dance and tune found in Chappell's popular music of the olden time, called DARGISON. It is said to be, like the modern CAN-CAN, intended to provoke desire. DAIR (Gael.) means to rut, sexual intercourse. TER (Eg.), to engender, has the same significance. GEASAN (Gael.) means to charm, enchant. KES (Eg.) is to dance, allure, entreat ; and AN is to show, wanton, be wanton ; TER-KES-AN denotes the wanton dance as a mode of charming, inviting, entreating to sexual union, and is in perfect agreement with the asserted character of DARGISON. The Fijian Women dance the BOKOLE dance ; in this they expose their private parts in token of invitation to the returning warriors. One name of their dance is DELE, that is identical with TER and DAIR.

Many words not found in Egyptian were formed as English in the ancient mould. Light, for instance, is AKHT (Eg.), and this with the *l* prefix forms the word light. With the *n*, sign of no, negative, prefixed, we have NAKHT, or night, as the negation of light. The word SNOW, or as the Scotch have it SNA, is not applied to the phenomenon. But SNA means breath, to breathe, and is a first type of founding and shaping. SNA, then, is shapen breath. NYFIO in Welsh means "to snow," and Nef in Egyptian is breath. URS, again, is the name of a pillow or head-rest. URST would be the participial form of "to pillow." This is not known as Egyptian, but it forms our word REST. "ARK" in the hieroglyphics is the end of a time or thing completed. P is the masculine article The. At the end of its lifetime or completed period, the pig becomes PORK, which is, when read in Egyptian, the ended or completed pig. The period is represented by a circle, the symbol of enclosure, or Arking round. Thus Ark, with the article P, yields our word PARK. ARK (Eg.) with the Tie sign means the end of a period ; 30th of the month ; a binding, to swear, make a covenant. When the land was divided, the cotters were at one time bound to give the landlord a certain number of days' labour as rent. These were called DARG days. Hence a day's DARG for a day's work. DARG signified the amount covenanted for ; as Egyptian T-ARK is the covenant, the binding or bond, and end of the period. The same formation supplies the word DARK for the end of the day. So likewise with the word NARK. A Nark is a common informer, the French NARQUOIS is a thief. NARACH (Gael.) is shameful, disgraceful, as is the French NARQUE. These we may derive from ARK, with the N prefixed denoting No, not, negative, which makes N-ARK the un-bound, un-covenanted, an outlaw, one who is forsworn ; whence the epithet for the informer and the term of shame and disgrace.

GLAM is a northern name of the moon, therefore it is argued the word Glamour is derived from the moon. But the Glamourie of the moon is altogether figurative and unreal, or rather typical. On the lunar theory we can do nothing with GLAMS (Northumberland) for the hands, and GLAM, to snatch, seize hold of. Now KER (Claw) in the Egyptian means to seize, lay hold of, embrace, and AM is belonging to. KER-AM (Gl-am) is thus the hand. The seizure enters into glamour. KRA is to seize, embrace, and MER (Eg.) is love. That is one form of glamourie. But KR, to lay hold of, seize, possess, contain, and AM to be pleasing, charming ; AM, grace, charm, is more definite. KR-AM, then forms both GLAM and CHARM. GLAM as the lunar name, is derived from KR (or GL) the course maker, and some form of AM. AM (Eg.) means to wander, grace, favour, charm, visible, light. To charm with the voice, and a charm of birds may likewise be derived from KHAR, voice, utterance and AM, to be pleasing, charming.

As MER (Mel) is an Egyptian name for the cow and the genitrix this yields the root of the word milk. As the act of milking is named before the milk itself, and the word milk, Malg, Melgo, and Milchu, implies the sense of milking, we may find the act of milking the MEL (cow) expressed by KUA (Eg.) to compress, tighten, squeeze, as in milking. Khu (variant of Akhu) is white and spirit, and Mel Khu would be the white or spirit of the cow.

To bear is derived from the Egyptian PER to show, see, appear, appearing. The Egyptian has the *t* terminal in PER-T, to void. That is our BIRTH, or BEARED. And the *n* terminal of our past participle born, is the Egyptian un, to be. Born is to be seen, to appear, as the visible, BORNE, or BORN child. Our word CHILD is by permutation the same as the Egyptian KHART, a child. We have the same form with the *k*, *r*, and *t* sounds in "CROOT," a puny, feeble child. This corresponds to the maimed weakling child of the mother, Isis, Har-pe-KHART, who was born deformed, and who died prematurely.

The Egyptian BES is warmth, rising flame, to dilate, pass to and fro, transfer, and follow. AM, is pertaining to ; AMU, to desire, urge. Hence our word BOSOM. The old meaning of bosom is wish, desire. The bosom is that which dilates, moves to and fro, and transfers the breathing image of desire. BES, to dilate, AM, desire, is the Bosom in the oldest sense, and BES-AM is the dilating organ. Also from BES, to transfer, pass to and fro, from one place to another, and AM, belonging to, is derived the name of the besom.

The Cornish word for the wasp is SWAP, supposed to be an inverted word. But according to the laws of language, the word SWAP reproduces the Egyptian name of the wasp, which is KHEB, the symbol of Lower Egypt, or Kheb. WAPS is a vulgar name for the wasp, and the letter W often interchanges with the K. KHEP (Eg.) is to transform, become winged; Wap (Eng.) is to flutter, as the wings. WAP is the root, the *s* may be prefix or suffix, and WAP is the equivalent of KHEB. YS-WAP is the WAP or KHEB.

Puss and Pussy are world-wide names of the cat. Erse, "PUSAG;" Saxon, "PUS PUS;" Gaelic, PUIS;" Irish, "PUS;" Tamil, "PUSEI;" Afghan, "PUSHA;" Persian, "PUSHAK;" Brahui, "PISHI;" Chhintangya, "PUSU;" Yakha, "PUSUMA;" Santali, "PUSI;" North American Indian, "PWSH;" Hidatsa, "PUZIKE;" Mundala, "PUSI;" Nali (African), "MPUS;" "BOOSI" (Tonga Islands); "BES," Arabic, cat or lynx; and in Maori, which has no letter S, "POTI" (PEHTI).

The solution of this philological problem, as of a hundred others, is that the source is Egyptian, and the same word went out on each line of the radii from the common centre. The name of the cat-headed goddess is PASH, and PEKH. She who gives the names to our "PASCH," Pasag, or Easter festival. PASHT was the Great Mother, the cat said to have nine lives, the cat of the witches. An old black-letter book, called "Beware the Cat" (1584), says it was permitted to a witch to take on the body of a cat nine times.

That the origin of PUSSY, the cat's name, is Egyptian, is shown by the fact that "SHAU" is the cat.[1] The article "the" prefixed makes this "PSHAU," that is the cat. The cat-headed goddess is the cat personified, or P-SHAU-T, which welds into "PASHT." The French form CHAT, for the cat, is SHAT without the Egyptian article P, and with it P'SHAT is THE CHAT. The name "SHAU" was not limited to the cat, it also denotes the SOW and the BITCH, two other feminine representatives. PASHT was the she-lion as well as cat, and She is SHA. This she-lion was also a cat and a leopard (as mau) in short PASHT is BEAST, a typical name which various animals take. In Egyptian "BES" is the spotted skin, and BESSA was a form of Baba, the Beast. In the Chinook jargon cat and cougar are both called PUSS. The Cougar is great PUSS. This is the exact equivalent of Pasht, the Egyptian goddess, who is cat-headed, and as the she-lion stands for the lion-leoparded of heraldry, that was in Egyptian Mythology the shorn and maneless sun of the lower region. In this way the divine Beast or Pasht gave her name to the Beast as cat, Cougar, or other animals, symbolically named. In Malabar and Tamil, PUSSY, as "PASU," is the cow. The lion, panther, and cat, have all one name in Egyptian, as Maui. And the Chikasaw Indians called the panther the Cat of God, and this was their emblem for the Lion,[2] which was then unknown to them. The Great Mother was represented by various Beasts. Pasht implies an earlier form found in PAKHT, and this is THE cat in sound. We have CAT, SHAT, and

---

[1] "SHAU," not "MAU:" Barker, *Papyrus*, 217.
[2] Adair, *History of the American Indians*, p. 31.

CHAT. The blossoms of the hazel are called CAT-kins, and CHATS; gathering them is CHATTING. The French sound the Sh in CHAT, but our CAT is the Egyptian KHAT. The cat was adopted into the ancient symbolism that underlies Mythology and Language, as will be hereafter explained, as the CATCHER, the killer of the rat and mouse that ate the malt in Jack's House—the cat that had nine lives or a life of nine solar months during which no mouse or rat dared stir. KHAT in Egyptian is the womb, to be shut and sealed. To Khat is to net, and catch. There was a goddess named "MUT-KHAT," the catching Mother, the one of the female triad who kept the abode shut for nine months or during ten moons. Such is the nature of "PASHT," or "PKHT," the cat-headed divinity. "Animal-worship," as the foolish phrase runs, has no primary application to Egyptian Typology. These trivial little words like PUSS, are the oldest and most precious in language, they possess the magic power of making the hieroglyphic images live once more. And the effect of these buried dead and forgotten things being made alive again, will be like revivifying a formation, all fossils, and causing it to move off with all that has been built over it.

It may be that the name "SHAU," in Egyptian, being both that of cat and sow, is the origin of our "buying a pig in a poke," or making a blind bargain, for the Welsh and Bretons have it a "Cat in a poke," we also have the "Cat in a Bag." The trick is supposed to have been played by palming off a cat in a bag for a young pig, which was no more a practice possible in the past than it is now. It must have been a symbolical representation; a permutation of the ideographs. Also SHAU the Cat is She, and there is an English saying, "SHE is the Cat's Mother."

The name of Pasht is connected with our Pasch and Past; the time when the sun had passed the spring equinox in its yearly new-birth. "PESCH" in Egyptian is the extent; "PEST" is sunset, the extent of day; "PES" is the stretch, range, extent, or period; PEST is the No. 9, the number of gestation, the extent in solar months. Our Elizabethan dramatists denoted a certain extent of time as a "PESSING-while." A candle thrown in to make weight and turn the scale is named a "PESSING-candle," that is, reaching to the full extent. Here the scales were a hieroglyphic of the equinox. PASCH, Easter, is the extent of the year and turn of the scale. PESH passes into PITCH, the height or extent; to PITCH up is to fill up; a PITCHIN-net extends across the water. At a PITCHED-market the corn is not sold by sample but by the sack or full extent. This root PESH gives us a meeting-point in language for PEASE and VETCHES, as they had a point of departure in evolution. PEASE are also PESH and PESK. PESH permutes with PEK. P equates with F and V. Vetches in Chaucer's time were called VEKKE. Pease and Pasche are identical in name, and were associated in season at Easter-tide. Pease and

Vetches from one origin are an exact equivalent for Pash and Aft (Fet) the still earlier beast-goddess, the hippopotamus-headed, who in priority of origin is as vetches are to peas.

If we take three common names for a secret language, such as the English CANT, French ARGOT, and Italian Gergo, Egyptian explains them. KHENT denotes something interior, within, inward, secret, secretly intimated, as is the meaning conveyed by the Cant lingo. ARGOT and GERGO are two forms of the same word which apparently preserve the original between them. KHER (Eg.) is speech; KHR abrades into Ar. KHUT is shut, sealed; KHER-KHUT would modify into ARGOT in one tongue and GERGOT in the other. London thieves call their secret language ARGOT. GAR-GATE for the throat proves that we have the KHER for speech; also GARRE is to chirp and chatter whilst "GARRY-HO" is loose improper language, HO meaning out of bound, the GARRY then is the Egyptian KHERU, speech. In the north tramps and beggars still talk "the GAMMY," and Khamui (Eg.) is to bend and beg, cringe and sue; GAMMY, the beggars' language, is equivalent to KHAMUI.

The word "NIX" used in slang has come to be identified with nothing, or do-nothing. But "NIX" in the thieves Argot means more than that. NAKE is to strip, make naked, steal. And this is the Egyptian NAK. NAKE means to steal, NAKA (Eg.) is to cheat, play false, deceive. In English NICK is to deceive, cheat. NICK also means to take a thing àpropos, that is the thieves' NIX, to steal at the right moment, or in the nick of time. The do-nothing sense is found in Nikau (Eg.) to be idle, lazy. This meaning, together with the NICK of time, is found in the schoolboys' "NIX," a signal to the lazy "NIKAU" when the master is coming.

There is a curious practice still kept up in schools. When a boy is hard-pressed in any game and his antagonist is gaining ground upon him and he cries out "NIC'LAS" he is entitled to a suspension of the pursuit or play for a moment's grace. The cry of "NIC'LAS" always entitles him to this resting-space.[1] This is taken to be an appeal to St. Nicholas, and yet, absurd as it may look, NEK (Eg.) means to compel, and las (res) is to suspend; NEKA LAS reads compulsory suspension.

Amongst other origines given by Egypt to the world it would seem as if all the words supposed to be purely interjectional, which are commonly treated as spontaneous sounds rather than words, might be identified as proper words in the hieroglyphics with the meaning still more or less attached to them as it is found in other languages. It has been asserted that these interjections show a common tendency to utter the same sounds under the same circumstances as expressions of the same feelings. On this theory language must have had divers origins, and could not have met in the end to

[1] Brand, "St. Nicholas' Day."

be interpreted by a common alphabet. Waiving this, however, for the time being, it will suffice to show that this interjectional language of ours is not the original and spontaneous utterance of unthought-out meanings. The interjections are words extant in Egyptian, with the visible signs and ideographic value given to every sound. To this unity of origin we shall have to assign the common significance of the same sounds uttered with similar meaning in various languages, and not to spontaneous diversity or independent coincidence. An independent origin of the same utterances by different and far-divided people would not drift by any fortuitous concourse into oneness of meaning all the world over. The oneness was assigned in the beginning for us to find it in the end; this circle, like every other, had its centre. Meantime it is useless to speculate and theorize upon the origin of these exclamations, whether they were imitated from without or "divinely revealed" from within, until we have taken the evidence of the hieroglyphics into account, and are better acquainted with what they have to show us in their picture language concerning the nature of these primitive words and sounds.

The universal sound of sighing, longing, wanting, desiring, expressed in many languages, which is embodied fully in our "HEIGH-HO for a husband," has every meaning in the Egyptian "UAH" to be soft-hearted; "UHA," to long for, sigh; "UAH," very much; "UHA-UHA" desire, with "UHA" the voice of desire. This "UAH" is the nearest sound to the English "OH." In moments of intensest feeling we can find no utterance so expressive, or inclusive of all we mean as the "OH," but it is not a mere instinctive interjection. Every form of the "OH" used as the utterance of different emotions and the mode of freighting sound with feeling is extant in Egyptian. "Uah" is to be soft-hearted, whence the "Oh" of sighing, longing, yearning. "U" means to adore, hence the O of the vocative case. "Uh" is to increase, augment, intensify very much, hence the prolonged "OH" of poetry and prayer, the sign of magnifying. "UA-UA," to precipitate, cast oneself on, reduplicates the "OH" as an increased expression of emotion. U—U—U has the force and significance of No. 3 or thrice. All that we mean by the "OH" is explained by Egyptian in which "UAA" signifies transmission. Another form in "HEH" to seek, search after, wander in search of, gives us the "HEY" which with the Scotch still retains the place of the "OH." With the Maori "OHA" is called "dying speech," where we should probably say sighing. In the Wolof (African) dialect the sign is written HHIHHE, the same as HIH (Eg.), to seek, and search after.

The O prefixed to Irish family names, as O'Brien, was anciently the H. This affords double hold for its Egyptian origin. The equivalent for O, "UAU," or UA (Akkadian UA, sole one), means the chief, the one, one alone, unique, whilst HA is the leader, chief, duke, the one who goes first or precedes. It is the same prefix with the same origin as the Japanese O, the sign of distinction, and titular honour; the double OO being a hieroglyphic of greatness or chiefness. The Maori "OUOU" signifies the few. The Egyptian "UAU" is likewise the captain. The Gaelic form of this prefix is OGHA, which represents the Egyptian AKHA, the great, illustrious, honourable, pre-eminent, noble, or the HIGH.

The O is a modified HO, and these imply an earlier form with the K sound, as may be seen in the hieroglyphics. So in English the sounds of sighing in HA and HEY imply the more vigorous utterance in HEIGH, KA to call, and KYE to cry, and these also are Egyptian with the same meaning. So the O and H of the Irish and Japanese point to the KHU or AKH, an Egyptian title of the ruler and governor, with earlier forms in KHEKH the whip, and KAKA to be tall, high, and to boast. Thus the earliest OH is KHEKH.

The oriental "WAH-WAH," an utterance of open-mouthed wonder, is just the Egyptian "UAH," denoting great, very much; also the North American Indian "HWAH," and the Chinook "HWA-WA," expressive of astonishment, is the Egyptian "UAH," an indefinite very much, increasing, pouring forth. The meaning of UAH is to augment and increase the thing, as is done by the expression of wonder.

When the Camacan Indians want to convey the idea of many, or much, they hold out their fingers and say "HI," which is supposed to be a mere interjection[1] requiring the gesture as an ideograph. The full sound is "HIE," often repeated as "HIE-HIE." Now the two hands make the circle or total of ten. Ten is HYO in Nutka, HY-YU in Aht; that is, the circle or complete number. In the hieroglyphics, "HIH" is many days, an age, aion, a round number, or a cycle with a circle as determinative. And when the Camacans hold out their two arms and utter their "HIE," they make the hieroglyphic of ten, a typical sign of the total.

The Arab camel-driver urges on his camel with a "YAHH YAHH," the Basutos their oxen with "WAH, WAH!" and in Egyptian "HAA" means lift, carry away, go along; "UAH," very much, go it, increase. "HAI HAI" makes them proceed with caution. "HAI" is a cry used by the Lummi, and "HOI" by the Clallam Indians. AY or OY is a Quiché call. These agree with the Egyptian "HI, HI," to draw, drag, pull along. HI also means to strike, and is therefore a modified KHI, to strike with the Khi (a whip). The French drivers cautiously guide their horses with a humouring "HUE, HUE," the North German with a "YO." This is probably the sailor's "YEO" of the "HEAVE-O." In the hieroglyphics "HUA" is denoted by a rudder and to bear (we say bear a hand), therefore it means guidance. The Swiss driver slows and stops his horses with a long-drawn "HU—U—U—U." This brings us to our English "WOH," which is purely Egyptian. "WOH" is used not only for stopping, but signifies be quiet, be still. "UOH" in

[1] _Tylor_, vol. i. p. 171.

the hieroglyphics means abide, be quiet. And as we have the UOH in our WOH, we may infer that the "Dust O" and "Peas O" of our street cries are Egyptian too. With this O the arrival is announced, and U or UI means to arrive. In addition to this arrival the dust-man's cry may include the Egyptian "UAA," lift, carry away, kidnap, as his sound is generally identified with "Dust AWAY," and the exact sound is "Dust OO-WAY." The form "Geho," Italian "GIO," used by carters, is most likely the Egyptian "KHU," to govern, to whip; "KHU-KHU," to beat with the whip. "FAN 'EM ALONG" is said of driving horses, and it means whip them along, the fan and whip being identical. So in the hieroglyphics the whip and fan are one in "Khi," a whip and to fan. Thus the labourer with his whip (Khu) and his "GEHO" to the horses is a living repre-sentative of the Egyptian god who governed with this "KHU," whip in hand, as Osiris, Khem, or Ptah. The horses of the sun at one time required the "GEHO" or "KHU," and seem to have taken their cue from the whip.

In the Midrash Ekha rabba[1] the sun is made to complain that he will not go forth until he has been struck with sixty whips and received the command "Go out and let thy light shine." In an Arabic poem the sun is described as refusing to rise until he is whipped.[2] The whip would seem to be the antithesis of the noose sign.

A Galla orator is said to punctuate his speech with the aid of a whip which he holds in his hand and marks the pauses from comma to full stop, a flourish in the air denoting the sign of admiration.[3] The action was symbolic, and the whip meant the same thing as in the monuments; it was a hieroglyphic of rule, divine rule, the com-manding orator. The click of his whip probably represents the click, or "kh" sound that is older than verbal speech. The waggoner and his whip-cord vie with each other in "kh—kh," or cluck-clucking to the horses, and the sound means "go, go." The whip was the sign of "GEHO" and "GO." Lastly, the whip is the "KHEKH," and in its "KH," "KH," answering to the "cluck-cluck" of the clicks, we bottom all these exclamations in O, Oh, Hey, Ah, Akh, Khi-Khi, and the rest.

HUN (Eg.) is rule with the whip. And the Caribs described by Rochefort applauded the discourse of their chiefs with an approving "HUN-HUN." Hunt (Eg.) denotes the rite or act of consecrating. Hanu is to favour, nod, or cry; whilst HAN means to pay tribute, and assent, or express pleasure. HAN-HAN is equivalent to ENCORE.

The German drivers use the cry of "HÜF-HÜF." In the Wester-wald "huf" is the call in backing the horse, and "HAUFE" means

[1] *Introd.*, § 25.
[2] Goldziher, *Hebrew Mythology*. Martineau, p. 341.
[3] Sayce, *Principles of Philology*, p. 26, note.

go backward. The hauve (helve) of a thing in English is the hindward part. "Hüf" may be derived from the Egyptian "Hefu," which is applied to squatting down on the ground ; that is a form of going backward, hindward, backing. "F" (or Hef) also denotes direction in bearing and carrying. To HUFF in playing chess is to remove, put back a man. HAAP in Devon means go back. HAPP (Eg.) is to go backward. The Egyptian UFA signifies to chastise, whilst HUFF in English is to swagger, scold, hector; also AUF intimates what it will be, it is a threat of UAF-ing or whipping; OFA in the Galla language, means to drive. This UAF (Eg.), to chastise, give them a whipping, is the manifest root of the old German WAFENÂ, or call "to arms," and the English WAP to beat.

With us to hiss is an expression of disapproval and contempt, but there is plenty of proof that it was once a sign of applause. The Japanese express a feeling of reverence by a hiss commanding silence, identical with the English "HUSH." The "HUSH-SH" of the Sioux Indians, described by Catlin, has the same signification. With the Basutos, says Casalis, hisses are the most unequivocal marks of applause, and are as much courted in the African parliaments as they are dreaded by our candidates for popular favour.[1] According to Captain Cook the people of Mallicollo showed their admiration by hissing. Egyptian will show that hiss and hush are no mere inter-jectional sounds, but consciously compounded words. HES (Eg.) means to praise, applaud, celebrate, glorify. This answers to the Basuto hiss of encouragement. "HA-SA" (Eg.), reads the salu-tation or all-hail following; and HES, to celebrate, to sing, is a form of the after applause. HES is also a word of command, to will and order, which answers to the religious "hush." And in "HES," to repulse by a look, we find the meaning applied to the sound in hissing as a sign of reprobation.

The exact form of the town-crier's announcement known to the writer thirty or forty years since was "HOI-YEA-YES," in some cases "O YEA, YES." This formula is abbreviated in the Cornish "HOYZ," which has every element of HOI-YEA-YES. The English OH and AH are the Egyptian "HAI," and Hai is to hail, address, invoke, and means Oh, Hail ! "HEH" signifies search, seek, go in search of, wander about or look about and bring to light. HES is the order to be obedient ; HES, will, order, command. HAI-HEH-HES then, is Egyptian for a command announced with the Hail or OH, of the "Ha," who was the crier and proclaimer in Egypt, to go and seek for and find something "lost, stolen, or strayed" according to the mode of "crying" yet extant.

The Brazilians have an exclamation of wonder and reverence written "TEH-TEH." It is the same as the Egyptian Tehu, to beseech.

---

[1] *Basutos,* p. 234

Tua is to adore, Tua God of the morning, or day. "Tua" has the ideograph of worship and adoration.

A group of interjections which Mr. Tylor affirms[1] has not been proved to be in use outside the Aryan limits depends on the root and sound "ST," Latin "St," used by the French in stopping a person; Russian "ST," Welsh "UST," German "PST," English "HIST," Irish "WHIST," Italian "ZITTO" and many more, all having the meaning of stop, stay, or stand.

In all the languages of the Indo-European family, says Curtius, from the Ganges to the Atlantic, the same combination STA designates the phenomenon of standing, while the conception of flowing is as widely associated with the utterance "Plu" or in forms slightly modified. This, he observes, cannot be accidental. It is not. Nor is it because there was any general outbreak at various places and times of an universal consciousness which puts the one soul into the same sound. All these have it because in the hieroglyphics "ST" means to stand, sit or stay. One of the types of "SET" is the rock, an image of fixity itself: the stone is "ST." "SETT" has the sign of stopping and staying. "SIT" is the back of a chair; English SEAT, and SETTLE. "SAT" is the floor for standing on. "SETT" is to catch, lay hold, stop. "STAIBU" is to stop the ears. SUUT is to stand. This sound is not the symbol of any abstract idea of "to stand," that is a modern notion. It has for hieroglyphic the Phallus. The meaning of this standing, staying, stopping, was embodied in the ancient Deity SUT, who was afterwards dethroned in Egypt on account of his nearness to nature. The God Sut, Stander and Stayer was represented by the ass-headed Onocephalus, and this creature, according to Hor-Apollo[2] was adopted to symbolize the man who stopped at home and hugged his ignorance and had never been out of his own country. This certainly agrees with the meaning of Sut. The word SUT has still earlier forms in "SHET," true, real, and in KHET to stop, KHET to shut, KHET to catch hold, KHET the seat, KAT, a stone. The English "SHUT," and French "CHUT" are intermediates of SUT and KHET. Nor is the "t" terminal necessary, for in the hieroglyphics the various illustrations of staying, of "ST"-ing are expressed by KHA or KA, a sound with no audible relation to either STA or SUT. Yet "KA" is to call, cry, stand, stay, rest or be thrown to the earth. KA is the seat, throne, land, earth, stone, floor. KA is the Phallus and the God. Where we call to "STA" or "STAY" any one, the early Egyptians cried "Ka" and made the hieroglyphic with two uplifted arms, and as Ka means to figure, this was the figure of the full-stop, their earlier "ST."

In the Maori, which has no sound of S, "STA" is found in the same sense written with its equivalent "NG," NGATA, and gives the name to the leech, slug, and snail, as the stayers. What then becomes of the

[1] *Primitive Culture*, vol. i. p. 177.　　　　[2] B. I. 23.

"physiologic potency" of the sound "STA" when its sense is rendered by NGATA, and this takes us back to Egyptian where KT is the earlier form of Sut? The same word with the one meaning is "ST" in one language, "PST" in another, "UST" in another, SUT, SHET, HEST, KHAT and KA in Egyptian and NGATA in Maori.

With the Egyptian masculine article P prefixed to "ST" we obtain our words "POST" and "PAST." "SITH" is time past or since. In Irish SITH is to leave off. The post stands. With the *p* added to sith, we have the past, for time gone by. Sut in Egyptian is denoted by a cake, with the *p* prefixed we get our "PASTE." And a "PESTLE"-pie is a standing dish. "PASTE" is also hard preserves of fruit for keeping.

The PLU above mentioned as the pluvial symbol in sound is the Egyptian PRU to flow out, pour forth, emanate, run. There can be no "physiologic potency" in the sound of the L which was originally expressed by R, nothing can be more diverse to the modern ear, than the sounds of L and R, yet they are of equal value in language. This shows the Pluvial idea was not born of the sound "Plu." Comparative philology without Egypt has been trying to stand on one leg alone. But when the "Aryan" limits are proved to include Egypt, what will become of Aryan theories?

HEM is an exclamation, or so-called interjection, having the effect of stopping a person, or calling him back. HUM or HUMME in Low German is a cry to stop a horse, as is HUMME or HUMMA in Finnic and its kindred dialects. The Dutch HEM is explained by Weiland as an exclamation to make a person stand still. We call back or stop a person by crying "HEM" in a mystical manner, especially when addressed from the male to the female. The origin of this is traceable to the Egyptian HEM, the seat, abode, place of stopping, and dwelling. The HEM or HAM, as stopping-place, became the Hamlet, and other forms of the Ham. This is provable. Still earlier is Kam (Eg.) the staying place, to stop, and stay; Chinese, KIM, the hem and boundary; and with the causative prefix, SKAM (Eg.), stop, stay, pass a time, dwell, remain a while. These abrade into SAM. Thus we have SKAM, KAM, and SAM, with the same meaning of stopping, and staying. And because these, together with HEM, denoted the place of stopping and staying, the word HEM became the sign of calling to stop, and the German HEMMEN means to hinder, restrain, hold back, stay.

"We must not forget," says Max Müller, "that 'HUM,' 'UGH,' 'TUT,' 'POOH,' are as little to be called words as the expressive gestures which usually accompany these exclamations." [1] Whereas these our interjections are often the most secretly precious of ancient words in the world, most mystical and matterful in their meaning. "POOH" answers exactly to the Egyptian "FU," which means to interrupt,

---

[1] *Lectures on Language,* first series, p. 371.

and stop any one; "FU," vice, sully, fault. But the same signification is still more strongly expressed by "PAH"; and "PA" is an Egyptian exclamation, the meaning of which we are left to recover in our English PAH. "FOW" in English is foul, to cleanse out, ERUDERO, ALVUM, EXONERARE, it is used as a term of contempt. Some of these exclamatory words have too much meaning for fuller explanation. The beginnings were very lowly. The Egyptian "FI" means to disgust, be repulsive. Our English "FI" is a term of disgust at something foul and repulsive. "PSHAW" is an expression of disgust and rejection. It is applied likewise in repelling uttered foulness as an equivalent for dirty. In Egyptian "SHA-SHA," disgraceful, disgrace, renders it well. "SHANASH" is stink, putrid, impurity, "SHA" being a substance, "NASH," nasty. "SHA-TIRUTA" is foulness and dirt, "TIRUTA" is our word DIRT. This "SHA" is a word of mystic meaning, degraded to the dirt. "SHA" is the substance born of, that maternal source of which flesh is made. "SHA" is the feminine period and the name given to Cat, Bitch, and Sow. All words found in the mire were sacred words at first.

The interjectional "SHUT," or SHET, French "CHUT," twice or thrice repeated implies an immediate shutting or hushing up. It is used to children and grown-up babblers who talk what they should not. It is a sign of mystic significance had recourse to when plainer words do not sufficiently express the meaning, or may not be used. Then it is we employ our Egyptian. "SHET" in that language is the name of mystery itself. "SHET" is secret, close, shut, be closed, mystic, sacred, a sarcophagus, secret as death. In the form of KHET it means shut and sealed.

"HUM" is expressly made use of when we think "Let me see." In Egyptian, "UM" is to perceive. "UM-H" is to try, examine, or see. "HAM" is to conceive. "UGH" is to feel a repugnance, to be terrified. UKH (Eg.) is a spirit. UKKA, in the hieroglyphics, is the night, once a time of terror in a fireless, lightless world. Its earlier form is "KUK," for darkness. That this name of night is the original of our "UGH," may be inferred from the fact that UGHTEN-tide is a name for the morning. So far from "TUT" not being a word, it was in Egypt the Eternal Word itself, or Word of the Eternal. "TUT-TUT" we say, meaning don't tell me. "TUTTLE" is to tattle, or tell tales. In Egyptian TAT signifies to tell. TETI is to stammer. TET is to decapitate. Our "TUT-TUT" is to cut short, put an end to. TETHUT (Eg.) is to imprison; "TUT-TUT" is intended to shut up.

"LA" was at one time used as an emotional cry. "LA LEOF" was equivalent to O my Lord, or my very Liege; "LA" being a formula of reverence and obeisance. Slender says,

"Truely, *la*."

Mistress Quickly,

"This is all indeed—*la*."

LA was equal to verily, truly, indeed, and Shakspeare echoes this sense. It is the Egyptian RA. RA was a formula, probably of reverential address to the Râ; Râ (Eg.) likewise means VERILY. Râ has an earlier form in REK, and La in LACK.

We have a vulgar English exclamation in provincial use supposed to be "O MY!" It is an expression of astonishment or wonder, and, as all who ever heard it properly pronounced can testify, it is sounded "O MAUHI!" and this as "MAHUI" is the Egyptian word for wonder, and to be filled full of astonishment. Moreover, "O MAUHI" expresses the same mixture of wonder and admiration as the word "MAHUI." UAH (Eg.) is very great.

The frog in German is supposed to say "QUAK" and "KIK," but this is the hard form of the name of the frog, HEKA, in Egyptian. The dog is credited with saying "WAU" "WAU," and this is the Egyptian KHAU, or modified AU, the dog itself. Both "QUAK" and "PAK" are supposed to be uttered by the Duck. QUAK is the Egyptian KAK. Seb who carries the Duck, or Goose, on his head, is called the old KAK (Kakur); AK is a Duck or Goose, and KA denotes the caller, whence KAK or QUACK, the AK that calls (or the call of the Ak). With the article P (the) prefixed to AK, we have PAK, the Ak. AK permutes with KA and KAKA (Eg.) means to cackle or quack. In Chinese the wild-goose says "KAO-KAO," synonymous with the KAK of Seb, and the German "QACK" of the HEN. The hen, when laying eggs, says "GLU," "GLU," and that is the Egyptian KHLU-KHLU (Khru), to utter, give word, notice, cry. The "GLU-GLU" of the hen, the Mongolian "DCHOR-DCHOR" of the cock, the German "DECKEL-DECKEL," a call to sheep, and "KLIFF-KLAFF," ascribed to the Dog, are all based on this KHLU (Kheru), which includes many crying or calling forms of utterance; TEKHEL (Eg.) would be a call to remain; TEK-KHEL, a call to affix or attach. The Cock says "KIKERIKI," in German. But KIKE is its name as the COCK, KAK, or crower, and RIKHI (Eg.) means the intelligent, wise, knower. KAK-REKHI is the intelligent announcer of time. A crow-like bird in South America is named the CARACARA, and in Languedoc "CARACARA" is assigned to the crowing of the cock. In Polish the crow is a KRUK, so in the north of England it is a Crouk. The Egyptian KA-KA, to cackle, yields the various names of birds and their cries found in KAH-KAH (British Columbia) a Crow; KAKA (Sanskrit) Crow; KU-KUK, Malay, to crow: KUK-KO, Finnic, Sanskrit KUK-KUTA, Ibo AKOKA, Zulu KUKU, Yoruba KOKLO, to crow, KOKO-RATZ, Basque, clucking of a Hen, KHKUREKATI (Illyrian) to crow. Moreover, Egyptian shows the principle of naming the cackle and the cry. KHEKH (Eg.) is the throat, the gullet, and the QUACK, KAK, KAO-KAO, CACKLE, KOKORATZ, and others are guttural sounds. On the other hand the cry, crow, or KHERU, means to call with the mouth (Ru), hence Kheru is the name of speech, to speak, voice, the word, utterance with the mouth, distinguished from KUCKLING with the gullet.

In Suffolk, according to Moor, crow keeping, or rather keeping the corn from the crows, is called both "WAHA-HOW" and "KA-HA-HOO." As our English word wheat is Egyptian it seems likely these are Egyptian too. UAH is an agricultural labourer, and HU is corn. Or KA denotes function, person, or type; HA is to stand and shout; HU is corn. Both forms read by Egyptian will render the corn-keeper, or preserver. Also "HU" (Eg.) is to drive. In English "HOO" is a cry used in pig-driving and in hunting. The Suffolk people speak of a man who has no "HO" in him; GO and HO permute. In Egyptian we have the same modification of KHU into HU. KHU, the synonym of go, means spirit, with the whip sign of GO, and KHU with the whip permutes with "HU" to make go; HU also is spirit. Egyptian will show us how it is that "WAHAHOW" and "KAHAHOO" are two cries or names for the same thing.

The Spaniards drive away their cats with a "ZAPE-ZAPE," hurry away. This is the Egyptian SAPA, to hurry away, to make fly away, Arab Z'APH, English ZWOP, to drive away with a blow.

When a countryman sees a shepherd's dog astray without sheep or shepherd he shouts "SHIP-SHIP," to hurry it off. This is erroneously supposed to be a reminder of the sheep. SHAB is an old English word for absconding and slinking; the Egyptian SHAP is to hide, conceal; the dog is treated as a truant. Also in the hard form, SKHEB (Eg.) means to goad and urge on. The Indians of Brazil call their dogs with an interjectional "AA." In the hieroglyphics AU is a name for the dog.

The Bohemians call to their dogs when at work "PS-PS," PES being a name for the dog. Egyptian will tell us why in both cases. PESH means stretch out, extend, enclose, a shepherd would mean by it, range round. And PESHU (Eg.) is bite; so that two meanings of the call as "PS-PS" are found to be Egyptian words for calling to the dog at work.

The cry to the dog with the Portuguese is "TO-TO," said to be short for TOMA-TOMA, meaning Take, take. But in Egyptian TA-TA or TU-TU would signify to offer the food which the dog was called upon to eat, and at the same time say "take the food." It is probably the same at root as the German cry to the chickens "TIET-TIET." The Austrian "PI-PI" with which they call their chickens to be fed is rendered by the Egyptian PI and PI-PI, signifying come quickly or FLY-FLY. The word PI has wings for determinative, and one hiero-glyphic PI is a fowl flying with mouth wide open, it may be to be fed.

The Germans are accustomed to call chickens to be fed with the cry of "TIET-TIET"; and in the hieroglyphics TUT is a handful; TA is corn and to take or offer. TAT or TIET, therefore, would be to offer corn by the handful; so that we have the corn given, dropped (ta-t) by the handful, (tut) expressed in sound (tet, tell, speech, discourse) by the "TIET-TIET" of the caller, which says, as TA-TA would in Egyptian, take corn by the handful.

The English "COOP-COOP" instead of being an abbreviated "Come up," is more likely to be the Egyptian KAP-KAP, for hidden, concealed, as the fowls frequently are, hence the calling. For one reason this KAP permutes with HAP, to lie concealed, secret, screened, and in the child's play of hide and seek it is a law of the game to signify hidden by crying "HOOP" which has the same meaning as the Egyptian HAP or HEP. KAHAB, however, means to excite, incite, toss, as is done in calling "COOP-COOP" and throwing the seed.

The English cry for ducks and ducklings to come and be fed is DILL or DILLY, dilly being the diminutive for the young ones. The Bohemians call theirs with the same word, DLI-DLI. With the *r* instead of *l* this is the Egyptian, TERA, a young bird with the duck for determinative. Terpu is the name of some kind of duck. This suggests the American TERAPIN, a name that might read in Egyptian as the duck that smells or is fragrant. One of the hieroglyphic ducks is the type of fragrance.

TERA, the young one or little one, passes into our words DILL and DILLING. This is corroborated by TER (Eg.) for the male emblem. Another meaning of the cry may be found in TER (Eg.), all, the whole of the young brood.

In driving fowls from the door or out of the house our farmers' wives generally cry, "SHU, SHU, BIDDY, SHU," to make them go. SHEU is an interjection of disapproval, and this is one with SHA. In Egyptian SHA is to make go out; SHU-ing and SHA-ing are identical. BIDDY, moreover, has the most curious equivalent. The English BIDDY is applied to a chicken : in Egyptian PATI is a name for all clean fowl, and "PA-TI" is fly, "go along" with you!

Supposing the forbidden cat to be skulking in a bedroom, the English housewife will hunt her out with the cry of "SKHAT." This is an Egyptian word signifying an order, to make, drag, deprive. SKHAT is the order to come out, or be dragged out and deprived of its hiding-place. SKAT (Eg.) is to lie hid and escape notice. The English word of command expresses the Egyptian fact. The hare, which we call PUSSY, is SKHAT in Egyptian, the animal that hides and is hunted. In English also the hare is named SCUT.

The words "NAM NAM-NAM" have become a sort of baby language now, because they belong to the infancy of the race. The Chinese child uses the word "NAM" for eating nice things. "NAM" in the Negro languages is to eat. A negro proverb says "Buckra" man, "nam" crab, crab "nam" "buckra man." In Soosoo "NIMNIM" is to taste. In the Vei language "NIMI" is sweet, savoury, palatable. "GNAMO" in Bhutani Lhopa is sweet. "NAM-NAM" in Swedish is a tit-bit. NAMMET in English is a luncheon. NAMBITA, Zulu, is to smack the lips in eating, also in tasting something mentally pleasant. NEIMH in Irish is heaven, heavenly. NAM, in Sanskrit, is to worship. In Yakaana "NEM-NO-SHA" expresses the verb to love, as to make "N-M-N."[1] Again, there is one origin for all. Every value of the word quoted was assigned to it as current coin of language in Egypt. In the hieroglyphics NEM is sweet. NEM is delicious, delight. NAHM to wish, vow ; NAM, repeat, go again. NEM-NEM to engender. NAHAM is joy, rejoicing, to enjoy. NUM is speech, word, utterance. NEM is a religious festival. The hieroglyphics show that some of these words are in their second childhood, and not their first. NEM (Eg.), to be sweet and delicious, is NETEM or nem, according as the character is read as a syllabic Net or phonetic N. So with NAM, earlier KHNAM ; and as the KHA becomes SA, it will account for dialect difference, as in the Cantonese SAM for Nam, or *vice versâ*. KHNAM may thus yield KAM, HAM, NAM, SAM, in the process of derivation from the ideographs.

The Zulu pitches his song with a " HA " " HA." " HAYA" means to lead a song; " HAYO" a starting song, also a fee given to the singing-leader for the HAYA.[2] This is the " HA " and the " HAI" of the hieroglyphics. " HAI" is to stand and hail, invoke; and "HA" to go first, precede, be the leader. It is identical with our English " HEY " which leads off the refrain as in the old " HEY, Derry down." The earliest known form of this burden is " HEY-DERI-DAN." In Egyptian HAI is to hail (it may be howl) ; TERIU, is twice, TAN to complete, fill up, and finish. " HAI-TERIU-TAN " is what is meant by repeat in chorus : that is our HEY-DERI-DAN, or HEY DERRY DOWN.

We find our " HEY LOLY " also to be Egyptian given as " HÈLOLI " to be mad, frantic. The full chorus is " HEY LILLILU, AND A HOW LO LAN," in the apparently meaningless ballad burden: HOW is whole, or full, and LAN, the moon. This suggests the song and circular dance when the moon was at full.

It was the same " HA-LE-LU " heard by Adair[3] which the " Red Hebrews," as he called them, sang whilst encircling round the holy fire, and identical with the " ALLELU-JAH " of the Jews, the " ALALA " of the Tibetans, the " HALALA " of the Zulu Kaffirs, and the " ALALA " of the Greeks ; the Polynesian " LOLOLOA," meaning drawn out very long, the English HULLA-baloo (the yule or howling for Baal at the winter solstice), one with the " HI-LE-LI-LAH " used by the medicine-man of the Dacotahs who danced and shook his rattle and whirled himself round franticly in a state of nudity as a mode of charming away disease.[4] The same that Livingstone heard in Central

---

[1] Tylor, *Primitive Culture*, vol. i. p. 186.
[2] Tylor, *Primitive Culture*, vol. i. p. 171. Various other of these illustrations are taken from the same chapter.
[3] *History of the American Indians*, p. 97.
[4] *Schoolcraft Res.* ii. 199.

Africa when the natives kept him awake with their wild ceaseless "LULLILOOING" through the night. There is one origin for all. The root meaning is better rendered in Egyptian with the letter *r* instead of *l*, where "RU-RU" means to go, circuit, wheel, and whirl round. By aid of RERT or ERRT, the hippopotamus, we see the "RU-RU-ING" was once applied to the revolving stars of the great Bear in the Sabean ceremonies. Egyptian shows us how these primary sounds of the childhood of language were deposited as child-types. Thus RURU denotes the nurseling and the nurse, also to dandle and lull the child. LILLU in Coptic, LALA in Polish is the child. The English Lullaby is sung by the nurse in lulling the child. LALLE in Danish is to prattle. In Kymric LLOLIAW means to babble, prattle to a child. LYULIATI in Servian is to rock the child, and in Russian ULIOLIOKAT is to sing, rock, and lull the child to sleep. And the earliest nurse is RERIT, the Assyrian LILIT and Hebrew LILITH, who in the hieroglyphics is the genitrix, as the sow or the hippopotamus, the old Typhon of the beginning, who first reared the child in heaven. In the third Sallier Papyrus there is a vivid description of a battle in which the king Rameses II. is surrounded by the enemy. He calls on the god Amon-Ra for help, and suddenly hears the voice of Ra behind him shouting "Ma!" "HRU-HRU-HA-KA." This passage has perplexed Egyptologists. De Rougé renders it "I come quickly to thee." To me the "HRU-HRU" is a war-cry to be understood by its cognates. HURANU is a form of courage. HRUT means "Arm for war," and this "HURU" is the root of our "HURRAH." In the Maori "Aru-Aru" is a cry signifying "pursue relentlessly." "HIRI-HIRI" is also to rush with relief, and energetically assist, as does the god in the Egyptian poem of Pentaur. It is the cry of a god inspiring the warrior, and in Maori "HIRI-HIRI" is the word used in repeating charms over a person with the view of imparting energy and inspiring courage. Also "HORU" is a yell used in the war-dance. "HERU-HERU" (Eg.) signifies extension, dilatation with joy; and this is the connecting link of language between the ultimate "ALLELUJAH" and "HOORAY," or, in the Irish form, "HOOROO." The universal cheer this may be called, for it is the wild Irish "HUROO" of battle; the Norman "HARO," the shout "UR-RE" with which the Mahouts urge on their elephants, the Nepalese "HERO," Siamese "AURA," Arabic "AR-RA," the "HURRAR" of the Norsemen, the Armenian "HAURA," Æthiopic "HURHUR" (go along!), the Maori cheer of the rowers "HARI-HARI," and the French "HARER" for setting on a dog. By aid of our "Hip, hip, HOORAY" we may perhaps reach the root of the matter. The Irish form "HOOROO" answers to the Egyptian HURU, meaning additional, another, one more. Our "Hip, hip, hooray" is generally given three or nine times, often followed by the "one more" cheer.

The "Hip, hip, hooray" may be a salutation of the rising sun.

Heru (Eg.) is another day, or one more round, and the RURU and HERU both meet in the ROUND. HURI-HURI, Maori, is to revolve, whirl round. IRI-IRI, in Fijian, means to fan repeatedly. We may infer that the Druids used to salute the rising sun with loud rejoicings, for he is called, by Taliesin, the Lord or Leader of the din or hubbub. The dawn was a festival of his return. And in Egyptian the Triumph of Return is expressed by the word "HEB." Heb is the Festival, the triumph, and the return. The Leader of the Din is called by Taliesin and Aneurin "RHWYU TRYDAR," literally the Lord of the Rhwyu. And the Fugleman of the HIP-HIP, is the leader of the HOORAY. HEP (Eg.) however, means unite, join together. HEP-HEP-HURU (Eg.) is unite all together for one more round or cheer, no matter what may be celebrated.

Our interjectional "Marry," as in "MARRY, COME UP," in the combinations of "MARRY ON US," and "MARRY GIP" would be well explained by MERI (Eg.) a name for Heaven. MERI come up, would thus be an appeal in Heaven's name. "MERI KIP" is Heaven receive or keep, or clasp us. MARRY-COME-UP, and MARRY-GO-DOWN are allied to the Merry-go-round and the Merry-dancers of the Northern-lights which dance in heaven. A see-saw is a MERRY-totter, it goes up and down. And this is the up-and-down image named MERI. With the prefix Ta it is Tameri, the double land of Egypt, Upper and Lower, or if applied to MERI, the Heaven, the Egyptian Meri-come-up, and Merry-go-down, of the sunrise and sunset, and the Meri-go-round of the Solar Bark. The Merry-go-rounds of our country fairs go up and down and round and round, and are made of boats.

The Freemasons make use of a formula "So MOTE it be," instead of So be it, or Amen. This mote is purely Egyptian, a rare form for May it be. "MET" is to fix, establish. "Met" is an ejaculation. Met means to pronounce conservative formulas.[1] "So Mote it be," is the conservative formula of the Masons, as it was in Egypt of the Priests.

The present work is not intended to deal with the structure and formation of the various languages of the British Isles, which languages the writer looks upon as detritus and drift in new forms, of an older language common to them all. But it may be pointed out that our participial terminal ED or T is Egyptian.

TA is to cross; TA-T is the cross sign of crossed. TI is two; TI-T (Tat) crossed, TIED, Twoed. TNA is to separate; TNA-T is separated. TNA and TNAT are the same as our TINE and TINED, applied to the fork. TEHU is to rejoice; TEHUT to be rejoiced.

In the genders the feminine is formed by adding a T. The Egyptian explains the English. When we are assured that I loved is I did love, that tells us nothing of I am loved, I am proved. These latter did not originate in I love did, I prove did. They indicate the present condition as much as I love, and yet it is the

---

[1] Pierret, "Met."

second of two. The hieroglyphic *t* as *ti*, is a sign of reduplication and not merely the determinative of the secondary condition, a sort of figure of two which shows that beauty, ability, bounty, majesty are the reduplication of beau, able, boun, majes, not merely the secondary, but the doubled form, out of which comes the plural TI (Eg.) and TY in English. The hieroglyphic T in Ti indicates that the conditions are one of two in TA and TAT, love and loved, not limited to the present and past, or to the genders, and the second is the condition of being Twoed; a doubled condition or secondary stage of being. The word " DID " itself is in that second and dual condition as the past of " DO," just as in the words loved and proved we have the twoed condition of love and prove. In English " TIDDE," happened, is the equivalent of " DID." Two of the hieroglyphic T's are a hand and a half-sphere ; in each case one of two. The T was made the feminine article, as the secondary one of two. As specimens of this Twoed condition we have the ToD of wool, two stones ; the TOUT for two gallons ; TOUT, the posteriors ; also TITTY, the young cat ; the TADPOLE ; the Welsh TAD, our DAD, for the father, the second of two because the Son was acknowledged first.

Our common termination " EN," as in open, sweeten, ripen, craven, withouten, leaven, appears to be the Egyptian UN or EN, to be, being, condition of being. This " UN," being, is distinguished in English in the adjective as well as participial termination, as in wooden, golden, brazen, to be of wood, of gold, of brass. At this stage we see the Egyptian " UN," to be, modify into " N," of, and from ; OF gold, FROM gold, made OF gold, or golden, as the condition of its being golden. Also the Egyptian ER (Ru) has the force of the English ER, in greater, sweeter, happier. Er (Eg.) means more, more than ; the repetition in ER-ER denotes very much.

The sole object of the present quest, however, was such matter as retains the original likeness, and tends to prove identity in the beginning. Grammatical structure of languages is not of primary importance ; that belongs to the mode and means of dispersion and diversifying the one into the thousand languages which enable philologists to class them according to their later differences, and lose sight of the original unity. This, it has to be remembered, is only a Book of the Beginnings, hence the trivialities of the chapter now ended.

SECTION V.

EGYPTIAN WATER-NAMES.

THE author of *Rude Stone Monuments* appears to me to darken counsel merely and deepen the superficies of the subject only in his search after a theory to account for them. He sees that the architecture of Stonehenge is primeval, and that it ought, according to the laws of evolution, to belong to a time and an Art antecedent to that of the Great Temples of Egypt and India! Yet he does not dare to apply the law of evolution to these structures, and has no doubt that the rude erections are degenerate copies of the more perfect originals.

This is parallel with saying that the poetry of Cædmon is a lowly imitation of the work of Tennyson. There never was a crasser instance of not only putting the cart before the horse and riding backwards, but of flogging the cart to make it go THAT way. He admits, however, that the NAMES are everywhere the great difficulty. And as Max Müller says of Cornwall, " Where every village and field, every cottage and hill bear names that are neither English, nor Norman, nor Latin, it is difficult not to feel that the Keltic element has been something real and permanent in the history of the British Isles."[1] Gradually these names will yield up the dead past to live again as Egyptian. Left without likeness in the classical languages, our names of rivers, for example, have been felt to be unfathomable.

The Egyptian name of Water as an element is UAT, the same as that of the Goddess UATI. UAT is the English WET, and our word Water is the Egyptian UAT-UR, which is also applied to the ocean, as the Great UAT or Wet. Uat (Eg.) for Water shows the element in a dual aspect, necessary to note; the word signifies blue-green, and Water unites the two colours of earth and heaven in one, as blue-green.

We shall find however that all elementary naming from the Egyptian origin is divine, that is, mystical, that is, finally, physiological.

[1] *Chips*, vol. iii. p. 252.

For example, the Keltic UISGE is a type-name for water, and at the same time it supplies a title for the spirit-water, the water of life, as WHISKEY. This comes from a first origin. UAT means both heat and wet, both elements being represented by the Genitrix UATI; they are the Two Truths of the Water of Life from which the name of Whiskey is derived, in accordance with its spirituous nature. USKH or SEKH (Eg.), the liquid, has a still earlier form in KHEKH, a fluid determined in one aspect by the sign of bleeding. Blood was the primal suck, on which the child was nursed, the first water of life. Heat or spirit was a secondary element. These are represented by KHEP, UAT, and SEKHT. KHEKH has the meaning of blood and spirit, the Two Truths of being imaged by the spirit-water of life.

The Egyptian HES and USESH mean evacuation. UKA is the water of the inundation. We have the river ISE near Wellingborough, ISIS at Oxford, the ASH in Hertfordshire, the IZ in Bedfordshire, the USC in Buckinghamshire, GUASH in Rutland, OUSE in Bedfordshire, various USKS and ESKS, the ESKY in Sligo, the ESKLE in Herefordshire, ESTHWAITE and EASDALE waters; these, together with the AXE, OX, UX, and EX, the ISCA, the Welsh WYSG, a current, and Gaelic UISGE, for water, the WISK, the WASH, and other variants, are all derived from this root. The Gaelic and Erse UISGE, however, is a worn-down form of the Welsh WYSG and this again has a prior existence in GWYSG, at least the GWY is an earlier spelling of WY for water. WYSG is the name of a current. This points to a prefix corresponding to KHI (Eg.), to extend, expand, elongate, and run with great rapidity. KHI-SEKH (Eg.) would denote the current, as in WYSG, and the W constantly represents a K. KHI-KHI (Eg.) is the original of quick, and the KHI-SK, GWY-SG or WYSG is the rapid-running water.

The Welsh name of the river called ESK is the WYSG, and this points to a rapid or spreading water as the primary type of the rivers so named. The WYSG takes the English forms of GUASH (compare "gush") in Rutland, the WASHBURN in Yorkshire, and the WASH. WASHES are outlets in the sea-shore, and in the fen-country large spaces left at intervals between the river banks, for floods to expand in, are named WASHES. WASH, GWASH, GWYSG are represented in Egyptian by the word KASH, to water, spread, be in flood, inundate.

The French rivers in the high Alps, the GY, the GUIL, the GUISAVE, and the GUIERS, the GUER in Brittany, the GIRON and GERS are probably named from their movement; they are the goers, as our GO is a form of the KIIU or KHI, to go, and make go, quick.

The GEYSER is named from the same root KHI (Eg.), to rise up, elongate, spread with great rapidity. And as KHU or KHI is Spirit, Whiskey is the KHI-SEKH or Spirit-water, the spirit in this case being analogous to the motion and go of the rapid waters.

An old Irish Glossary says, "BIOR and AN and DOBAR (are) the three names of the Water of the World." This shows their naming goes back to the mystical one of water with two aspects found in the Pool of Two Truths, and in the name of Uat(i). The Duad may be expressed by BI-OR. AN represents the Egyptian HAN or NEN, the primordial water identified as the Bringer in the beginning. DOBAR, DOVAR, or DÛR answers to TEF-ER or TEP-ER, as the water of the commencement, the first water. TOBAR (Irish) is a Well, and the Teph (Eg.) was the Well of Source. Also TEBU (Eg.) means to draw water, and the Tober water is drawn from the well.

AN (Han) was the celestial water and the well-water was below. These were two aspects of the BI-OR or dual water, of the Pool of Two Truths, the Water of Life that made flesh in one form and fermented into spirit in the other. This will be illustrated in a chapter on the Typology of the Two Truths, both of which were at first assigned to the Motherhood.

In passing it may be noted that in two Egyptian names of water we find synonyms of Yes and No. IA is water, to wash, purify, whiten, and the word means Yes, and assuredly. IA is the positive of water, and with this agrees the English EA, and YEA. NU is water, and the word means not, negative, negation, the English No.

Also IA, the white, agrees with milk, the white, and NA, the red, with the blood. The Milk and Blood of the Mother were a primal form of the Twin-Water. The Two Truths were represented by White and Red, and both colours were combined in the "spotted cow" of Hathor.

The DEE springs from two fountains in the East of Merionethshire, and is probably named as the dual water, the twin water of the Egyptians. It was called DYVRDWY, the divine water of source in a dual character. Another of its names, Peryddon, read as Perit-ten (Eg.), yields the two-fold manifestation.

The river Deskie is formed of two streams, and in Egyptian Ti-sekh would signify the double water. D-eskie is the dual Eskie.

IVEL is a twin river: "Two rivers of one name," says Drayton. The IVEL was anciently called the YOO, and the IU (Eg.) is two or twin. The YEO is the same, and our YEO-math is the second mowing.

The river NEATH has a double head, and NET (Eg.) is a total, the All, which was composed of two.

If the Egyptians named our waters and rivers, it is tolerably certain those of Wales were the first-named, as the TAVES, TEFIS, DYVIS, DOVERS, and others. For this reason: AP or AF (Eg.) means the first, as a liquid, an essence of life, or essential life. This supplies another elemental water-name. With the article The pre-fixed, this is Tepi or Tefi, the Welsh Dyvi, English Tavy, and Dove, the first, primordial, ancestral source.

This Ap or Af, as first in the form of liquid, with the Egyptian T prefixed, furnishes all the primaries of water found in TOBOR and DOBAR, the well, the Kymric DWR (Dfr), water, and the river names, TEF, TAV, DYVI, and their kindred, the T and D, retaining at times the twofoldness of character. Thus, if, as will be maintained, the first landing was in Wales, one of the rivers named the first would be the DYVI, which flows into Cardigan Bay, and debouches through an estuary that divides North and South Wales.

The TAY, formerly TAVUS, is a first river, on account of its size.

The name of the Thames shows an example of compounding from Egyptian. If we take the ES for a reduced ESK, as it is found in SENAS, a name of the Shannon, the THAM may be accounted for in this way.

In Drayton's *Poly-olbion*,[1] a dozen rivers make up the Thames The Oxfordshire Colne, the Charnet, Charwell, Leech, Windrush Yenload,[2] join the Isis; the Ock and Ouse join the Thame, and these all unite with the Kennet, Loddon, Wey, and Hertfordshire Coln to form the Thames.

In Egyptian TEM is a total; the completed and perfected whole. TAM means to renew, make over again, the second time. It should also be observed that TEMI (Eg.) is a title of the Inundation in Egypt, and the Thames is a tidal river. Thus THAMES is the total and the tidal river. The same origin will account for the name of the river Tamar (Devon), which "sweeps along with such a lusty train" of attendant streams and rills as "fits so brave a flood two countries that divides."[3] Tamar takes in the Atre, Kensey, Enjan, Lyner, Car, Lid, Thrushel, Toovy, and others. As Thames means the collective or total water, so Tam-aru reads the collective river. In each case, Tem (Eg.) is that which divides the land into districts, a name of the Inundation, a total, and a created river.

We have another name for Thames, as the Tidal Water. According to Camden, this river was once known as the COCKNEY, and therefore a dweller on its banks is called a Cockney. The KHEKH-NUI (Eg.) means the tidal water; the to-and-fro of the water corresponding to the motion of the Khekh, as the Balance or Scales. This name, says Boileau, was likewise applied in Paris, where we find the water is the SEINE, and in Egyptian SEN or SHENA is the tidal-river of the Inundation.

Shuma (which permutes with Shuna) is the Pool of the Two Truths, in An; the water of the dual aspect here figured as the tidal river.

Ancient Paris stood on the island which divided the Seine into two halves, called the Isle of France. SHEN (Eg.) denotes the dual, twin-water. The names of the SEINE and the Cockney rivers are important. SHEN is to complete an orbit, it is the circuit, extent,

---

[1] *Songs,* 14 and 15.   [2] Drayton's spelling.   [3] *Poly-olbion*, Song 1.

perimeter; SHEN, to stop, bend, twist, turn away, turn back, turn down. It is the Egyptian name for the measure of the Inundation, the hieroglyphic of which is the SHENT. We have a form of the word in SHUNT, to turn back on another line. The game of SHINDY, also called SHINNY, is designated from its motion to and fro, like the ebb and flow of a tide, as Egyptian shows. Now, if we apply this to the name of the place on the river Thames called SHENE, we shall see that it was so named as the spot to which the flow of the tide reached, and where it ebbed again. That in Egyptian is SHEN.

This may supply a landmark for the geologist. Shandon on the River Lea (Ireland), on the Clyde (Gairloch), and in other localities, may be named on this principle of identifying the place where the tide once turned.

There is an old sacred place in the county of Durham called COCKEN Hall, round which the River Wear winds two ways, and KHEKH-NU (Eg.) means the water that goes to and fro. In this instance it is a tidal river, but the tide does not now ascend so far up; the name, however, appears to show that it did so in the past, and the Wear was a KHEKH-NUI, or tidal water.

The Cockney, as a person, is not named from the Thames. He is a form of the April fool called the Gouk. In the north of England April fools are April Gowks.' The Gouk, Gowk, or Goke, is sent on a foolish errand, repulsed, and sent back again, and Khekh (Eg.) signifies to repulse, and send back. Hence the KHEKH, or GOUK. The Cuckoo is the Gec, as' the bird of return, that goes to and fro. Thus we have the KHEKH, as a pair of scales, the tide, the cuckoo, and the fool. Now in the hieroglyphics the word for water or inundation is written nu, nnu (nenu) and nini (nni), and the same word is the name of the little boy, the NINI (Nui), who is our NINNY. The Ninny, or fool, who goes to and fro on fool's errands, is the analogue of the Khekh, river, and, like it, is the KHEKH-NINI, KHEKH-NUI, or Cockney. In relation to the First of April, Khekh (Eg.) is the balance, equinoctial level. COCKAIGNE, as place, was the mythical land of promise and plenty, that is the solar country lying eastward, where the waters were crossed and the manifestation to light had occurred, where the corn was seven cubits high, and the ears three cubits long.[1] This was attained at the time of the equinoctial level, the Khekh. London, on the COCKNEY river, as the land of Cockaigne, is connected with the mythological astronomy, as the Gate of Belin likewise shows. Uka (Eg.) means a festival and to be lazy. KHEKH-UKA-AN would answer to Cockaigne, the reputed land of laziness and luxury, a form of An (Heliopolis), placed at the Khekh, on the summit of the Equinox. At this point in the planisphere was the Pool of Persea, now represented by the double stream of the Waterman; the one water with two manifestations,

[1] Rit. Ch. 109.

which may be the two aspects of a tidal river, the water above and the water below, fresh and salt water, milk and blood, or, finally, male and female source.

The "Khekh" balance, the type of tidal motion, takes many forms in English. The sea-cockles are left on the sand by the turning tide. In Devon they are called cocks. Stairs that wind about are called cockle-stairs. The pilgrim and palmer wore the cockle-shell as a badge, not that they had been to sea, but because they were wanderers to and fro like the bird of passage, or the tidal water, or the cockle, a tidal shell-fish. They too were Gecs, Khekhs, or Cocks. Dampier speaks of a "cockling sea, as if it had been in a race where two tides meet;" the motion of contrary currents caused the "cockling." Shag is another variant. Wicliff translates "the boat was SHAGGED with waves," that was, in a cockling sea.

To cocker is to fondle, dandle, jog, or rock up and down. To joggle is to move this way and that. To juggle is based on rapidity of movement to and fro. One "Khekh" hieroglyphic of this motion to and fro, up and down, is the balance, as the figure of the equinoctial level, and the up and down of the two heavens. Khekh passes into our word Weigh, and in Bavarian Wäg is the Balance; Wage in Dutch; Waga, Russian; Vág, old Norse. To weigh is to balance, and all turns on the wagging up and down. Goggle, joggle, waggle, gaggles, quake, shiggle, gig, giggle, giglot, gigsy, and many other words are variants, having the same fundamental meaning. GICK-GACK is a name of the clock in nursery language, from the motion of the pendulum to and fro. A jigger in machinery goes to and fro. In giggling the body shakes up and down. A giglet is always on the go. The Gaelic gogach and English kickle denote a wavering and unsteady motion. Goggle-eyes roll to and fro. Nine-pins are called "gaggles," and they are set up to be knocked down, and thus illustrate the motion called Khekh-ing. In the children's' game, "COCKLE-bread" is made by wabbling the body up and down and to and fro,

> "Up with my heels and down with my head,
> And this is the way to make Cockledy-Bread." [1]

The goging or cucking-stool moved up and down in ducking the culprit. The cock on the vane turns to and fro. Hocking at Hock-tide is a custom of the male and female alternately lifting each other up and down. The game of hocky consists in driving the ball to and fro.

This derivation of the April fool from the KHEKH tends to prove that the so-called "All Fools' Day" is in reality the "Old Fools' Day," Scotch, "Auld Fool's Day." In an ancient Roman calendar, quoted by Brand, there was a feast of OLD Fools. The Khak in

[1] Brand, "Cockle-Bread."

Egyptian is the old fool, coward, nincompoop. Kehkeh (Eg.), like the Maori Koeke, and Kaffir XEGO, is the old man. Khekh-ing, hocking, and hoaxing are all connected with the Equinox. The Khekh represents the old "Kak," the god of darkness, who was derided and made sport of when the young sun-god had arisen, in the ascending scale or Khekh.

But to return. We might infer that a people coming from the land of the Nile would be sure to confer the names of the Inundation on our tidal rivers, and erect the tidal into a type-name. This we find they have done. TEMI, as before mentioned, was one name of the Nile inundation. ABHAIN is an Irish name for a river, and as ABH is the river, the suffix probably characterises it. HAN (Eg.), or An, is to bring, to come and go, turn and return. AB is the water, AB-HAN is the periodic tidal water, named after the typical inundation of the Ab, Hap, Kabh, or Nile.

The river DOVE, says Ray, is the Nile of Staffordshire when it overflows its banks in April. There is an old distich,

> " In April, Dove's flood
> Is worth a king's good."

DOVE rendered by Egyptian is TEF or TEB. TEF signifies dripping, flowing, and to evacuate. The TEPHT is the abyss of source, the Welsh DYFFED. DOVE in the West of England is the name of a Thaw. In Egyptian, TEP marks the point of commencement of thing, time, and place. This meaning combines the beginning of the overflow, or the Thaw, or the land, as in the names of Dyfved and Dover. The name of Staffordshire, we might suppose, would be based on the River DOVE or " Tef" with the S prefixed, as this is the causative prefix to Egyptian verbs. Accordingly we find that STEF is an Egyptian name of the Inundation of the Nile, and STAFFORD is the ford of the flooding river. Several of our river-names suggest this origin. NEN is a name and a type of the Inundation which, according to Hor-Apollo,[1] signified to the Egyptians the New or Renewer. NEN-UT is fresh and sweet. We have the river NEN in Northamptonshire, and one of its two sources springs near STAVERTON, which looks as if it also had been named from STEF, the Inundation.

The name of the river SHANNON is still more to the point of NEN being a name of the Inundation, for it is the Nile of Ireland in its overflowing and shedding of alluvial soil. SHEN-NEN (Eg.) reads the periodic type of renewal. NEN (han) also means the bringer, and SHANNON is the periodic bringer like the Nile. There are large tracts of marsh-land along the banks of the Shannon deposited when the river overflows its banks; these are called " CAUCASSES," and are famous for their fertility.[2] KAU, in Egyptian, is earth, and KHUS means to found, lay the basis. SHEN also is

---

[1] B. i. 21.   [2] *Enclyc. Brit.* 8th edition, "Limerick."

to turn away, and return; that is the tidal river; as SENAS also, it is the tidal or inundating water. LIMERICK appears to be named from the Inundation or tidal river. REM (Eg.) is to rise and surge up. REM likewise denotes the place of, REMN means extending up to, so far, and REKH is to wash and purify. Thus Limerick may be the place to which the tide ascended. Remi-rekh (Eg.) reads washed by the inundation or tidal water.

The SEVERN is a tidal river. Nennius (Ch. 68) calls it the HABREN. Hab (Eg.) means periodic, the type of return, tidal. The naming of SEVERN as the tidal river is also denoted by the two kindred divinities, SABRINA, and SEFA, who is the goddess of the tidal river in Egypt. The goddess of the Severn, and the inundation of the Nile, are one and the same at root. URNE and ORNE signify to run. RENE is a watercourse. Thus HAB-RENE would be the tidal watercourse. This RENE represents the RUAN (Eg.), the valley gorge, and outlet of water. In which case the HAB-REN, is the tidal river of the valley-gorge.

The HABREN or Severn has a different origin to the HAFREN and AVON. These are the crawling, sluggish waters. AVON is a Keltic type-name of the river. In Welsh HAFRU signifies the slow and sluggish. Hefu, Hef, and Af (Eg.), mean to squat, writhe, crawl along the ground like the caterpillar or snake. Nu or N denotes water. Thus Hef-n or Avon is the crawling or sluggish water, as the Avons are; the crawling, winding, serpentine water. The Gothic AHVA and Welsh ARAF, the gentle, include this meaning of AF and HEFN (Eg.) to crawl along the ground.

The AFF is found in Brittany, the IVE in Cumberland. There are a dozen AVONS in England, Wales, and Scotland. Besides Shakspeare's AVON there is one in Hampshire, one in Gloucestershire, one in Devon, in Lanark, Banff, Stirling, Monmouth, and other counties. The EVENENY, in Forfarshire, is a diminutive of the AVON. AVON abrades into the AON in Manx, the AUNE in Devonshire, also the AUNEY; the ANEY in Meath, and INNEY in Cornwall.

The LEVEN implies the al (or Ar) compounded with Avon. The earliest form of LEVEN is ALAFON, which modifies into Alauna. Leven also means the smooth, like AVON. So derived, these are the slow, smooth, crawling ALUS (Aru, Eg. River), unless we take Al to signify White.

But we have to include another type in LEN, as RENN (Eg.) is the pure unblemished virgin water. This is extant in the Linn, a deep, still pool. Drayton sings of the calm, clear Alen born of Cranborn Chase. Len (Renn) may enter into the name of the ELLEN in Cumberland, the ALLEN in Derbyshire and in Leitrim. Matthew Paris calls Alcaster on the ALN, ELLEN-Caster, so that the ALN, the LUNE, the ERYN in Sussex, LOIN in Banff, LINE in Cumberland, LYON in INVERNESS, LEANE, Kerry; LANE, Galloway; LAINE,

CORNWALL, have to be distinguished according to their character, whether they derive from ELLEN the pure, the virgin water, or from the LEVEN, ALAFON, the slow, sluggish, crawling, serpentining stream.

RUI (Eg.) is mud, muddy, red, or black-coloured. To this corresponds the ROY (the red) in Inverness-shire, and some of the other rivers of similar name, which include the RYE in Yorkshire, Ayrshire, and Kildare; RUE, Montgomery; RHEE, Cambridge; ROE, Derry; RAY in Oxfordshire and Lancashire; REY, Wilts; REA, Herts, Warwick, Shropshire. The Gaelic REA, the rapid, Welsh RHE and RAU (Eg.), the Swift, have to be taken into account. Still the rapid and the red or the deep-coloured are often likely to meet in the same stream. The Warwickshire REA and Hertford LEA are not the rapid but are the muddy (rui) rivers.

AUR, ARU, the Egyptian name of the river, has an earlier form of the water-name in KARUA, a lake, or some other water-source. These include such names as the URE, ARE, and AIRE, Yorkshire; AYR in Ayrshire and Cardiganshire; ARU, Cornwall; ARRO, Warwickshire; Arrow, Herefordshire and Sligo; ARAY, in Argyleshire; ARA-gadeen and ARA-glin in Cork; ARU, Monmouth; the Norfolk YARE, the YAIR and YARROW in Selkirkshire; the YARRO Lancashire, the GARRY in Perthshire.

The name of the NILE had to be derived from "ARU," the River as it was called. The River watered the Two Lands, and the plural definite article is NAI. From NAIRU or NAR comes the Nile.

In Ireland we find the River NURE is also THE OURE, that is by dropping the Egyptian article The. The River NURE or THE OURE is precisely the same as Aru and Naru (Nile) in Egyptian. Boate, in 1645, calls the Irish river "NURE or THE OURE." AN-FHEOIR is the full Irish name. So the Egyptian A is an earlier FA in more than one sign. The NURE, like NILE is a dual river.

The hieroglyphic sign of the Inundation is a triple vase, with two spouts, from which the water issues in two streams, one on each side of the sign. One name of this symbol is KHENTI; this abrades into KHENT. Khenti means an image. Khen is the waters, liquid, within. Ti is two. Khenti is the plural of water, say as red and blue Nile. Near Abury in Wiltshire a river rises in two heads, and realizes the Egyptian image of source called the KHENIT or KHENT. Our twin river there is the KENNET. This identifies the two-foldness of the ideographic KHENT. The hieroglyphic of the vases has been variously read KHENT, SHENT, and FENT.[1] Latterly "FENT" has been given up. Yet, the FENT, as the nose sign, is one of its determinatives. The English FOUNT makes it almost certain that FENT was one of its names. The nose, as FENT, is a fount of life, the dual organ of breath; and the FENT imaged by the vases symboled

---

[1] Birch, *Dictionary*, p. 548.

the water-fount of life. In English "FEND" signifies livelihood, means of living.

Diodorus says that when the Nile overflowed most parts of Egypt, and the waters were coming down full-sweep, the river, for its impetuosity and exceeding swiftness of its course, was then called the EAGLE. As the Egyptians rendered *l* by *r*, the Eagle corresponds to our EAGRE or ACKER, a peculiar vehemence of motion in the tide of some rivers. It is still applied to a dangerous surge and eddy in the river Trent called the "RUME" (Eg. REM, to rise and surge up.)

> "Well know they that the *Reume* yf it aryse,
> An *Aker* is it clept." [1]

The word AKER was explained by the early lexicographers by the Latin *impetus maris*, which they said preceded the flood or flow. In Egyptian the eagle as bird is Akhem or Akhmu, which likewise means an extinguishing wave of water. Akh is to rise up; mu, water. The eagle was of course a symbol of swiftness and ascending power. The BORE is the name of the "REUME" or "AKER," which occurs annually in the River Severn at Gloucester about the time of the Spring Equinox. BER (Eg.) signifies to be ebullient, and boil up to the topmost height. BERWI (Welsh) is to boil and bubble. Periodic manifestation was one of the first forms of phenomena observed and named, and the "Bore" of our rivers was an especial phenomenon. The Parret is a river with a "bore," and this apparently enters into its name. Per (Eg.) inter-changes with Ber, to well up, be ebullient, manifest periodically; ret (Eg.) means repeatedly. Thus Parret names the periodic high tide. In the Hoogly, one of the mouths of the Ganges, the Bore is known as the BORA, and in the Amazon the Indians call it PORO-ROCA. A reviewer in the *Athenæum* for July 3, 1880, remarks that under "Humber," a Shropshire name for the cockchafer, Miss Jackson, in her Shropshire word-book, gives additional currency to the old notion that the river Humber took its name from the humming noise made by its waters, and says "This is certainly wrong. The Humber does not hum more than other rivers, nor nearly so much as the Parret, the Ouse, the Trent, and other rivers on which the high tidal wave known as the bore-or eagre manifests itself. The origin of the name is at present an unsolved enigma." In Egyptian HUM means to return, to be tidal; it is a variant of Hun, to go to and fro. BER (Eg.) is to boil up; BER, the supreme height, cap, tip, top, roof; this names the "Bore," and the Hum-ber is doubly designated from Egyptian, as the tidal and Borial River. HUMMIE is a Scotch name of shindy or hockey, which has the tidal movement; and the HAM-mock is the swing bed. This sense of "HUM" passes into

[1] *MS. Cott. Titus A*, 23, f. 49.

"Hum-drum" and "Hum-strum," applied to recurring, and, in "Hum," to incessant motion, whence the Cock- (Khekh) chafer is also the Hum-ber.

The Keltic Dwr for water has earlier forms in Dovar and Dobar, which, as Tep-ar (Eg.) denotes a water from the point of commencement, as the well of water, like Tobermory, that is, water from the source. Durbeck in Nottinghamshire is thus the Tep-ar-bekh, or water from the beginning in the beck. The Beck is the Egyptian Bekh, for the place of birth, the begetting, birth itself. It is a delusion to look on this as a Teutonic addition to the Dwr. The Dowr-water in Yorkshire is also the water from the source. In this case the water is the Uat-ur (Eg.), after the Dour from the source has widened out. Many rivers were named from this origin. We have the Dour, in Fife, Aberdeen, and Kent; Duir in Lanark; Thur in Norfolk; Dore in Hereford; Durra in Cornwall; Doro in Queen's County; Durar-water, Argyle. But not all the Durs and Turs are to be derived from Dobar or Tepar. These have been lumped together as Dwrs, or waters, with no power of distinguishing them by any principle of naming.

The chief type-name in Egypt for River is Aru, with the variants, Aur, Ar, or Ur. But which Aru or Ar in a given case is the question. The Tep-aru is the first Aru, from the source. But we have Turs and Ders, which are only branch-rivers. And Teru (Eg.) means the branch of a river and a measure of land. This is included among our Turs or Ders, perhaps as Derwent. The town of Derby (or Deoraby) stands on a Derwent, which is an interior Ter to the Trent. Went possibly represents the Khent (Eg.) either as the inner or the dual water. Khent as the lake and interior water is certainly retained in Derwentwater, and probably in Windermere. Trent is absolutely a boundary water of the county, and as Trent is the lower Derwent, and Ent (Eg.) is the lower of two, and means "out of," it looks as if Trent were the Ter-ent, the lower of the two Ders or Drws. Nt (Eg.) also means limit.

The Tepar, Tobar, or Dru, as water from the source, becomes the Tur and Dur, as the name of a natural boundary, the land-limit, the first lines drawn on the topographical chart. Many rivers are Turs in this sense that are not named from the well-spring or fountain-head. They are named from Teru (Eg.), a measure of land, the boundary and margin of a shire or other district. The River Nidd in Yorkshire is still a land-limit and boundary for two different Hunts, those of York and Ainsty and Bramham Moor.

A good example of river-naming in Egyptian occurs at Duruthy Cave, near Sorde, in the Western Pyrenees. The name of Duruthy Cave actually describes its topographical situation in Egyptian. It overlooks the junction of the two rivers Gave—the Gave de Pau and the Gave d'Oléron—two tributaries of the Adour. The name of the

Gave corresponds to Kapu, the name of the Nile in the oldest form, and there are two Gaves whose branches blend at Duruthy. TURU (Eg.) is the branch of a river, and TI reduplicates. TERU-TI would indicate the double river-branch. They flow into the Adour, the name of which, as Atur (Eg.) signifies the river, limit, measure, and a region determined by the Propylon and house or temple. Duruthy is the place of the famous Bone-cave.

The TUR, as river-branch and boundary, has its earlier form in ATUR (Eg.), with the same meaning—the water, the river that constitutes a measure and limit of land. With this agree the ADURS in Berwickshire, the ADUR in Sussex and Wiltshire, the ADDAR in Mayo. This gives the name to several of our streams, as the Cornish ATRE, the Welsh ATRO, the ETHEROW in Derbyshire, the ADUR near Shoreham, which latter name echoes in modern phrase, the shore, coast, land-limit, signified by ATR in Egyptian.

The same thing occurs in THETFORD, the ford of the River THET. TET (Eg.) is to ford, cross over, pass through ; therefore the THET was the fordable river in Egyptian ; a thing of importance in early times. AT adds a different type to the AR or water, which has in each case to be distinguished before we know the nature of the particular name. It may go back to KAT, so as to include the CHEDDAR. In KAT or KHET we obtain the type of the navigable river. ATH (Eg.) is a canal. KHET is to navigate. Here it may be noted that Egyptian supplies a far better type-name for the Spanish rivers, the GUADA, GUADIANA, GUADARAMA, GUADALETE, GUADALIMAR and GUADALQUIVIR, than the Arabic WADI, the channel of a stream. The present writer would derive these names from KHATA (Eg.) to sail, go, navigate. These were first named as the navigable rivers, that is, KHATA supplies the navigable as its type.

But here again we find two types under one name, as KHAT also means a ford ; so GUADO in Italian is the ford. Thus the river named from KHAT may be the fordable, i.e., the WADE-able, or it may be, the navigable water, for which the water itself must be questioned. KHET, as the ford, is preserved in QUAT, near Bridgnorth in Shropshire, where we not only find the QUAT but QUAT-ford, and the ford repeats the QUAT, as in WATFORD. QUAT is an earlier form of WATH, a ford, therefore the KHAT.

Another good example of the primary nature of naming the rivers and flowing waters as the self-cut boundary lines may be found in the Irish SRUTH, for a stream, when interpreted by the Egyptian SRUT, to cut, dig, plant, as a means of arranging, distributing, organising, from Ser, to arrange, distribute, organize, make private and sacred. SRUTA is to cut out, engrave, as the stream did in its course, whence its adoption as the distributor and divider of lands and the establisher of frontiers and boundaries to be held sacred. The river formed the first shire and the SHERH (Eg.) is a river, a source. Rivers were the ready-made lines on the map of the land. This principle of naming them in Egyptian as water-boundaries of the district or region of land is very apparent. If we take a few of the border rivers and those found, or once extant, such as the river TEES in the north and the TEISE in the south, on the margin of counties this will be evident. TESH is an Egyptian name of the NOMES into which Egypt was divided; TESH is a frontier, and the TEES is a frontier river; TESH is a district, frontier, the Nome, made separate. TES is a liquid measure; so is the river Tees.

On the borders of Denbigh and Flint there are, or were in Drayton's time, the rivers HESPIN and RUTHIN. HESP is likewise an Egyptian name for a district, land measured off, from, or by water; a square inclosed. RUT means to cut, engrave, figure, girdle, tie, fasten, retain the form, separate. Both names agree with these meanings. This principle will account for the naming of our rivers STOUR. The OUR or AUR (Eg.) may be taken for the water-word as it is in various other names, the River, for example. RU (Eg.) is the path, channel, outlet, and with the *f* for "it," we have RUF or RIV, whence RIB and RIPE, the bank of the AR, or water, our river.

SAT is Egyptian for the NOME; AUR for the river. The Stours were the boundaries of districts. STER (Eg.), to lay out, agrees with the SAT-OUR or NOME-river, and this root enters into the name of scores of German streams, such as ALSTER and ULSTER, the STREN and STROO being akin to our STOUR and STREAM, the water-boundary of the NOME, or district laid out.

In the west of England STREAM means to draw out at length, to pass along in a set course actively. STER (Eg.) is to lay out length-wise and together; and as AM (Eg.) is belonging to, it seems likely that our word STREAM is SAT-AUR-AM, the water-boundary of the Nome, water and land being laid out together.

The ERIDANUS or IARUTANA (Eg.) of the Planisphere is the dividing river and the water that divides. The river PO is also known as the Eridanus, and in Egyptian the equivalent PU means to divide. TEN or TNA (Eg.) is to separate, divide in two, halve. This name of rivers, as the boundaries that divide the lands, is found in the DONS of Brittany and other parts of France, and also the DANUBE, DNIEPER, the ancient TANAIS, DONETZ, DNIESTER, DANASPER, ADONIS, TANARO, and others. This gives a name to our TYNES and DONS; DUN, Ayrshire and Lincolnshire; DEAN, Forfar and Notts; DANE, Cheshire, DEEN, Aberdeen; TONE in Somerset; EDEN in Yorkshire, Kent, Fife, and Cumberland; TEANE, Stafford; TEYN, Derbyshire; TIAN, Jura, whilst the TANOT in Montgomeryshire, and TYNET in Banffshire, show the participial form of TEN (Eg.) to divide, be TYNED or made separate. Some of these, no doubt, are modifications of TEIGN, a Devonshire river. TEKH

(Eg.) is the hard form of TESH, the Nome boundary, and denotes a frontier, a crossing. The type of this would be the TEKHNU ; and as nu is water, the TEIGN is the water-frontier, that which divides and makes separate in the Tun. A rivulet near Ambrose Hole, Hampshire, is called Danestream. Danesford also occurs in Shropshire. But these have no relation to the Danes except to show the perversion of the water-name. Tain (Gaelic), Don in Armorican, and Tonu in Sclavonic are the Danes meant. They are forms of the dividing stream. There is a river TOUCQUES in Normandy. If we take the UES (uskh) as the water, the TOUCQ answers to the Egyptian TEK (Tuk), a frontier fixed, and the TOUCQ-USKH is the river of a fixed frontier. The TAGUS agrees with the TOUCQUES, as the river of the TEK, frontier and land-limit. TESH and TEKA permute within Egypt, and the names of the TEES, TEISE, TAGUS, and TOUCQUES out of it. No better illustration could be given of the water being the first boundary than in the name of boundary, as the BOURNE. This was then applied to the BURN of water, and thus the BURN and BOURNE are one in the water-boundary.

> " Come o'er the bourne, Bessy, to me."
> *King Lear*, iii. 6.

alludes to the water-bourne. In Wiltshire we have the river BOURNE. In Herts the BULBOURNE. The ISBOURNE, the ASHBOURNE and OUSEBOURNE are all of them USKH-boundaries, not mere Scottish " BURNS."

The Keltic UISGE also contains the elements of U or UI (Eg) the line, limit, edge, canton, territory, so often applied to the country in Central Africa, as in U-GANDA, and Sekh is the liquid, the drink, English suck, the name also of a tributary river, and with the U prefix UISGE reads the Sekh of the limit, line, edge, direction ; the USKH, as the extent, range, boundary, equal to the USKH collar, a type of binding round, therefore of boundary for the land. In the earlier form of the UI we have KHI to govern, rule, protect, dominate. The same word SEKH or USKH means to cut out, incise, engrave, memorize. Thus the SEKH or USKH, as river, is the dividing line of the SOKES, as in Essex, with its six rivers and six sokes or divisions of the county. According to Drayton the river TEAM, which divides Shropshire and Hereford on the Cambrian side, is the furious.[1] Again, the TEAM (Eg. temi) reproduces the name of the Inundation. TEMA is also to swoop down, cut in two, announce, ticket. RU (Eg.) is a mark of division, a chapter, fraction ; RUA, to separate. This, which becomes the French RUE, a street, gives a title to several RUS and LLHUS, in Britain, as rivers. We have the river PENKA, and this word in Egyptian means to disjoin, separate. The

---

[1] *Poly-Olb.*, Song 8.

SPUT is a small river in Westmoreland, and SPUT (Eg.) is a district of country marked off, one form of this being the SEPT inclosure known by that name on the monuments.[1]

The river GYPPEN, now called the Orwell, the river on which Ipswich, that was once Gyppes-wick stands, repeats the Egyptian KHEPENI, a measure of liquids, a brimming measure of beer; our GYPPEN being a liquid-measure of land.

RU (Eg.) is the mouth, gate, way, gorge for water. AB (Eg.) signifies the water. From these come our ABERS. AP-RU, or AB-RU is the water-mouth, gorge, gate of passage for water, the outrance of a river. Aberdeen is AB-RU-DEEN, the outlet, gate, or mouth of the Deen-water.

From HI (Eg.), the water, stream, or more particularly the Canal, we obtain the HITHE as a landing-stage on the river, or seat on the sea-shore, as in Hithe, Rotherhithe, Queenhithe, Greenhithe. There are several Hithes on the banks of the Thames. The name has curious illustrations in Egyptian. HI (Eg.) is the water, the canal, or stream; TA signifies bearing, carriage, seat. Whence HIT is a boat, as a carriage on the water. The canal, as HIT, is determined by the hippopotamus, the bearer of the waters, who (as Taurt) was the seat. The HIT is also a seat, a station, a limited place on the canal, our Hithe. The Hithe, in the form of Hit or Hut, then, existed as the boat, the seat, the bearer of the water, and a water-region or inclosure. Still earlier is the HAT, our Hatch, or Dam, by which the waters were banked off and land created. This, the sign of Chaos and determinative of precommencement, so to say, of creation, is the earliest form of the Hithe as a landing-place, and its type is worn on the head of the goddess Egypt, as the sign of land obtained from the water. So ancient is the Hithe. In ERITH we have the ARU-HITHE, the landing-stage on the river. This will prove the origin and application of the principle of naming to be Egyptian.

Other names from the same source abound. At the beginning is the BECK, the infant stream. BEKH (Eg.), as before said, denotes the birthplace, and to *enfanter*. The BECK is the water from the place of birth, the river in embryo. In AVONSBECK the BECK crawls along like the child on all fours from the birthplace. The BURN may be derived from NU (N), water, and BUR, to well forth, or PER, to come out, appear, become a visible form of manifestation.

The MELTA is one of the "handmaids" of Neith. MERTA (Eg.) is a person attached to; also MERTA, a water attached to. HEPSEY is another affluent of Neith. HEB is the fountain of source; HEP, hidden; SI or SIF is a child.

The SMESTAL is a tributary of the STOUR; TAL being a permuted form of TER, the branch river, and SEMS (Eg.), the Minister, SMESTAL becomes the ministering branch.

---

[1] Leemans, *Mon. Egypt*, 2, 11, 45.

Those who know the nature of the rivers, if they have not lost their character since Drayton's time, will be able to identify the principle of naming the ART and WERRY in Cardiganshire. Both ART and URI are names of the Inundation. UARU represents our word hurry, and means to go swiftly, fly. But there is another form as HURU, the tranquil; URT is the gentle, peaceful, meek; and this may be the IRT in Cumberland, the " pearl-paved IRT," which, though the smallest of rivers, was, according to Drayton, the richest from the pearls found in it. Urt (Eg.) also means the crown, crowning; ARUT is spoil, and Art is Milk, the white.

There is the same difficulty or choice in the name of KART and its congeners. We have the KART in Scotland and TA-GRATH in Wales. In one case KART (Eg.) is the dark and silent; in another, KART means a cataract. So the Gaelic CLITH, the equivalent of KART, the cataract, means the strong, the typical Force, whence the CLYDE and CLUDAN in Scotland, the GLIDE in Ireland, the CREDDY in Devon, the CLWYD, CLOYD and CLYDACH in Wales.

AYSGARTH FORCE, on the river Ure, in Wensleydale, has a cataract when the Ure is in flood, which has been compared with the Nile. Its grandeur depends on the stream being swollen to the flood. One name of the flood in Egyptian is AASH,[1] and KART is a cataract. AASH-KART, or Aysgarth, means the cataract of the flood. URI, a name of the Inundation of Nile, is repeated in the name of the URE.

The ROTHAY runs fast; RAUTI (Eg.) is swift. RAUU, is swift; TI, go along; ROTHAY is the swift-running water.

The CALDER is in Keltic the winding water; in Egyptian KAR-TER would indicate the extremely winding.

The river BRY that rises in Selwood runs or glides along in such a winding course that in one part it almost encircles Glastonbury, Arthur's Avalon. PRI (Eg.) signifies to manifest, appear by sliding, slipping, and wrapping round, as does the river Bry. Our prying and peering come from this root.

The river MEDWAY is peculiar from its long wandering windings, covering some thirty square miles of the surface of the country with its trunk and branches. It is said to have been called the VAGA, on account of its wanderings.[2] The primitive of our word Way is the Egyptian UAKH, a road, or way, and " MATT " means to unfold, to unwind, round and round. MAT (Eg.) also denotes a surface of water, and MATR is a name of the marshes. At Tunbridge the river is separated into five different channels, just above the town, which join again into one below it. This names the town in Egyptian; TNA, or TUN, means to divide and turn away, or separate.

The Pool of PANT in the *Ritual* is the mythical Red Sea and lake of primordial matter, which was the place of dissolution for those who

---

[1] Lep. *Denk.* 3, 279, b.   [2] *Enclyc. Brit.*, " Kent."

were resolved again into the elements. PANTA was the ancient name of the English river now called the Black Water; black and red are permutable. PANT is PAINT, and the NA (Eg.) paint is red and black. Both are negative.

There is a river in the county of Limerick named the "MORNING STAR." The old name of it was SAMHAIR or SAMER; this in former times was a woman's name. SHEM and SEM in Egyptian are type-names for a woman, the woman who bears. SEM signifies breeding; SAM-HER is the delightful because fertile woman, a summer of a woman, for it has the same meaning applied to woman as to season. In Egyptian SHEM is summer, harvest, and the woman. AR or ARU is the river. SHEM-AR is also the tributary river, the fertilising or feeding river. The French and Belgic SAM-ARU, the SOMME, is equivalent to the Egyptian SHEM-ARU, the fertile river. The Persian name of a river, SAMAR, is the same. SHEM (Eg.), to recede and retrace, is likewise a tidal name. But the SAM-ARU is also known as the "MORNING STAR." Possibly in this way: SAMHAIR is sounded SAVVIR. Now the Morning Star, as Venus, was named ZIPPORAH and LUCIFER. This was the star of stars. SEP or SIF (Eg.) is the star; it is both star and morning, therefore the morning-star; the Lady of the Blush of Heaven, as the Akkadians called Venus of the Morning. This meaning is recoverable in SAV-VAR, the river of SAV, SEF, or SEB, the star of the gateway, the dawn, the one woman-star in heaven, as a planetary type, and the sole one as the feminine Seb (Sebt or Sothis), the star of SEFA, the Inundation, and therefore an equivalent of SHEM (Sam), the woman of rivers. Possibly the title of the Morning Star is not a mistake after all for a name of the Irish river SAM-HAIR.

The Great Mother, Teph or Typhon, gives the name to various streams called TAFF, TAW, TEIFY, and TEVIOT. TEPHT (Eg.) is the entrance, door, gate, valve, hole, cave of source, the mouth of the abyss. This was personified as the Great Mother, of whom we have an immense image in the Mendip Hills, famed of old for the caves in which it rained or drizzled; a type of Teph. A cave called by Drayton the WOCKEY-HOLE, is one of these. UAKH is Egyptian for wet and marshy. UKA is water of the Inundation (every name of the Inundation was reapplied in our Isles), and out of the Mendip Hills, says this writer, springs the FROME, that is, out of the place of the rainy or weeping caves. REM (Eg.) is to weep; F represents the article or pronoun. The REM (PREM, FROME) is the river that was wept. We have the REM without the article in the ARME in Devon, and again URM is a name of the Inundation. The FROME is said to rise near EVERSHOT. This name read by Egyptian UF-ER-SHET means to be shed secretly, squeezed out, expressed drop by drop. Shet is a name of mystery; secret, sacred, mystical; a pool of water; UF, drop, pressed out; R, to be. The name of MENDIP is apparently derived with the same signification. MENA is the wet-nurse, and to suckle. One of her names is TEF, and TEF (Eg.), the equivalent of DIP, means to drip, drip. Tef enters into the name of Tefnut, a Goddess of Wet. MONA was the nursing mother of the British Druids. The oldest name of St. David's in Wales is "HEN-MENEW," old mother; in Egyptian the divine wet-nurse, Teft, whence the name of David. And Men-dip, the modern name of the weeping cave, we take to be a personification of Mena-Tef, Teb, Tep, or Typhon. In Mendip are the Cefn caves, or caves of Khef, another name of the old bringer-forth.

The Yorkshire river Nidd is said to be designated from its course being for a considerable way subterranean. NET, or ENT (Eg.), means out of. Also NET is the lower region, the underworld of the goddess Neith. NIDD is thus a form of NEATH. The Nidd enters the Ouse at NUN-Monkton. NUN (Eg.) is the new water, a name of the Inundation.

The "DRIPPING WELL," on the banks of the river Nidd, in which a petrifying spring falls from above, is credited with being the birthplace and abode of Mother SHIPTON. Is Mother Shipton then a form of the primeval great mother, who personified source itself, and poured forth the water of life as NUPE, KEFA, and SEFA? SEFA, the earlier KEFA, is a goddess of the Inundation of Egypt. SEFA (Eg.) is to make humid, to dissolve and liquefy, hence to drop, as in the dropping well of the river Nidd, and the Mendip Cave.

SHEP (Eg.) means to exude, flow, evacuate periodically, like the Inundation. SHEP is interchangeable with KHEP. SHEPT and SHEPSH, the hinder-part, are one with KHEPT and KHEPSH. This makes it probable that Mother SHIPTON, the prophetess, is a form of KHEPT, the British KFD, as the Goddess of the North, the under-world of Neith or Nidd, where the Well of Source was placed or personified as the feminine emaner who wore the red crown, or poured forth from the Tree of Life, or fed with the breasts of the wet-nurse.

In AN, the place of commencement, was the Well, the Pool of the Two Truths. The well and the water are identified with our Easter customs.

The Aldermen of Nottingham and their wives from time beyond memory had, in 1751, been accustomed on Monday in Easter week, after morning prayer, to march from the town to St. ANNE'S Well, with the town waits playing before them.[1] The well of ANNE answers to the Pool of the Two Truths in AN. St. ANNE, as at Jerusalem and other places, has been adopted as patron of the Pool of AN. ANIT (Neit), the figurer in AN, who is represented with the shuttle or knitter, is perpetuated in the name of Nottingham.

[1] Dyer, p. 173.

ANIT was the weaver, and this became a great town of the weavers or NET-makers.

In the Keltic mythology the presiding spirit of the waters, in what is erroneously termed well-worship, was called NEITHE,[1] identical with the Egyptian ANIT or NEITH. The place of the well was that of the USKH Hall of the Two Truths, stationed in AN, the last of the three water signs.

At Whitby they had the formality of planting what was called the PENNY Hedge in the bed of the river ESK on Ascension Eve. "Nine stakes," "nine gedders," and "nine strout-stowers," were regularly planted, and a blast was blown on a horn by the bailiff of the Lord of the Manor, rendered, "Out on you! out on you! out on you!"[2] Our number 9 corresponds to the nine non-water signs, and relates the custom to the deluge-imagery. PENA (Eg.) means to reverse, turn back, return. The symbolists were figuratively commanding the waters to retire, and marking the boundary of the deluge.

The water of the Pool of the Two Truths was in one aspect the water of life and death. The Strathdown Highlanders have the Pool of Two Truths, although the ceremony of drawing from it has been changed from the vernal equinox to the winter solstice. On New Year's Eve they draw from the well called the "*dead and living ford*," from which a pitcherful is taken in profound silence, without the vessel touching the ground, lest its virtue should be lost. Early on New Year's morning the USQUE-CASHRICHD, or water from the "dead and living ford," is drunk as a potent charm that lasts until the next New Year's Day.[3]

In Cornwall there is a well in the parish of Madern called MADERN Well, and this name reproduces that of the Two Truths, called Mat. Mat-REN (Eg.) renders the name of Mat, and Mat was the more ancient name of AN, the place of the Pool of the Two Truths. REN (Eg.) also denotes renewal and making young; and this was the Well of Healing that flowed with the water of life.

In the Bay of Nigg (Co. Kincardine) there is a well, called DOWNY Well, and near it a hill, called DOWNY Hill, a small green islet in the sea. The passage to the isle is by a bridge, named the Bridge of a Single Hair, which young people cross over to cut their lovers' names on the green sward. DOWNY at the crossing suggests TENNUI (Eg.), the place of the crossing, hence the bridge; whilst NIGG is the variant of ANKH (NKH), the word for life and living.[4]

Two Egyptian names of the well, or abyss of source, are TES and TEPH. Both recur in TISSINGTON and DOVEDALE, where the wells on Holy Thursday are all decorated with flowers, and they have a particular variety of the double daisy, known as the Tissington daisy,

---

[1] Brand, *Wells and Fountains.*   [2] Dyer, p. 209.
[3] Stewart, *Superstitions of the Highlanders.*
[4] Brand, *Wells and Fountains.*

peculiar to the place.[1] The Two Truths are still found there in one flower in addition to the Pool.

Near Newcastle-upon-Tyne there are two sacred wells not far from each other. One is named RAG-Well, and the other is at a place called JESMOND. The RAG-Well is the REKH-Well, the well of purifying; and in Egyptian HESMEN is the natron, and the pool of purging, as one of the two waters. HESMEN is also a name of the menstrual purification. These two wells will be again referred to in further elucidation of the subject.

KARTI (Eg.) is a name of holes underground, therefore of wells. TERA (Eg.) means to invoke, rub, drive away, obliterate, and TERA-KARTI answers to the name of the Well of DRACHALDY in Scotland, much sought in Pennant's time for its waters of healing.

Near Tideswell, in Derbyshire, there is an intermittent spring called the ebbing and flowing well. The place is named BAR-MOOR. In Egyptian BAR is to be ebullient and boil up; MER is water; BAR-MER is the ebullient water. TIDESWELL is probably a form of TEPHTS-WELL; TET (Eg.) is an abraded name of the TEPHT, the well of source, whence the word TIDE.

Near Great Berkhamstead (Herts) there is a DUDSWELL. TUT, TEFT, TEB, and KEB, names of the old Typhonian Goddess, the Suckler of Source, are all names of holy wells.

GUBB'S WELL, near Cleave, in Devon, is the name of a chalybeate spring. KHEB was a name of the Sacred Nile; KAB is the water and the place of libation; KABH is pure, or purifying water; KHERP, or KHERF (Cleave), signifies the consecrated, holy place.

The well at Oundle in DOB'S Yard was reputed to drum against any important events. This is stated in the *Travels of T. Thumb*[2] who says, "No one in the place could give a rational account of their having heard it, though almost every one believes the truth of the tradition." Baxter, in the *World of Spirits*,[3] says he heard the well in Dob's Yard drum like any drum beating a march. It lasted several days and nights. This was at the time of the Scots coming into England. It drummed also at several other changes of times. Such a natural phenomenon would arrest attention. TEB is the Egyptian name for a drum. TUPAR is the tabor or tambourine. DOB is the Egyptian TEB, Goddess of the North and of the Tepht, the abyss or well of source here found in the yard at Oundle. She passed into the later HATHOR, to whom the drum or tambourine was given. If the well "drummed" periodically, and was supplied by an intermittent spring, that probably furnished the name of Oundle. UN (Eg.) is being periodical, and TUR means to wash, dip, purify. UN-TUR is the periodic purification.

There is a ROUTING-well at Inveresk, Midlothian, which is said to predict storms by the noise it makes. Rut or Ter (Eg.), with the

---

[1] Dyer, p. 211. [2] P. 174. [3] P. 157.

same sign, means to indicate time, and repeat. The Well of St. Ennys, in the parish of SANCRED, according to Borlase,[1] manifested its most salutary influence upon the last day of the year. SAN (Eg.) means to heal and save. SEN or SHEN also denotes the last day of the year, as the completion of the circuit. KHRIT (Eg.) signifies the victims, the fallen victims.

In the Isle of Lewis there is a well called St. Andrew's: it is in the village of SHADAR. It is used for healing and divination; the natives made a test of it to know whether sick persons would die of their ailments. They send one with a wooden dish to fetch some water; the dish is laid gently on the surface of the water, and if it turns round sunways, the patient will live; if whiddershins, he will die.[2]

In Egyptian, SHA is the pool, and TER signifies to question, interrogate, invoke, as well as to rub and drive away. SHA-TER is the divining pool or well.

We cannot here include the rivers of the world, but water is a thing so initial and vital that its naming ought to afford a crucial test of Egyptian nomenclature and of an unity of origin once for all. In the mythologies, Water is the first principle, factor, and type of existence. All naming of water originates in the feminine water of life, poured out by the genitrix; the water on which souls are nourished, and over which the spirit broods in creation. And Egypt supplies the words for water to all the chief groups of languages in the world.

(Egyptian.)

A, dew; **AA**, bedew; **A**, water; **HEH**, inundation; **IA**, wash, water, whiten, purify.

A, Akkadian.	HEIH, or HEUE, Chinese.	IA, Otomaku.
A, Norse.	AHU, Agaw.	YEH, Bali.
Aw, Kurd.	YUI, Khoibu.	Jo, Namsang.
EA, A.-S.	YEHO, Jakun.	YI, Yurak.
AWE, Zaza.	JUCU, Moxos.	I, Kamskatkan.
AYA, Tshampa.	I, Guarani.	I, Tupi.
AWA, Bramhu.	WAI, Maruwi.	IAH, Dizzela.
OEE, Banjak Batta.	AI, Sasak.	HAHA, Yagua.
OWAI, Savu.	EYAU, Catawba.	WAHI, Shina.
WAIJ, Cocos Island.	OI, Bima.	HOH, Jura Samang.
AIYAH, Korinchi.	WAI, Ende.	WAHE, Deer.
WAI, Bugis.	OWAI, Rotti.	HE, Warow.
WAI, Annatom.	WAYA, Wokan.	WA, Zyrianian.
WAI, Fiji.	WAI, Mandhar.	JA, Vogul.
EEIA, Masaya.	WAI, Bauro.	JA, Magyar.
OH, Santa Barbara.	WI, Mare.	YA, Burmese.
AHA, Mohave.	WAI, Malay or Polynesian.	YI, Mano African.
EAU, French.	AYA, Chankali.	YA, Gbese African.
JE, Susu African.	HOU, Manchu Tartar.	YAE, Myammaw.
JAH, Pessa African.	WOIO, Mandingo.	YOWA, Lohorong.
HIH or HWIH, Chinese.	AHA, Cuchan.	

AB (Egyptian), water, pure water.

AB, Persian.	ABH, Irish.	OBU, Western Pushta.
AB, Cashmir.	AB, Bokhara.	

---

[1] *History of Cornwall*, p. 31.　　[2] Martin, *Western Islands*, p. 7.

AK (Egyptian), the liquid mass of the celestial height. AKH, water.

ACH, Welsh.
WACA, Mayoruna.
AQUA, Iquito.
CAQUA, Salivi.
ACHYE, Miri.
KHA, Dieguno.
AUGR, Arnya.
OCK, Toba Batta.
AKI, Ternati.
OCO, Correguage.
ACUM, Chapacura.
OICHE, Gaelic.
ICK, Cornish.

EKU, Ebe.
EUAK, Guachi.
AGHO, Faslaha.
ACHO, Kaffa.
EK, Takeli.
AQUA, Latin.
HONACA, Mayoruna.
YGE, S. Pedro.
AK, Attakapa.
AICHU, North Tankhul.
OKA, Chocktaw.
AIGUE, French Romn.
YACU, Quiché, Peru.

YAKUP, Puelche.
UGH, Kashkari.
AKE, Ceram.
OCUDU, Betoi.
AECO, Ge.
EGA, Gafat.
ACHO, Gonga.
AKKA, Yangaro.
OKO, Isoama.
OUGE, Albanian.
YIGE, Soso.

ANKH (Egyptian), liquid of life.

ANINGO, Mpongwe.
NKE, Bamon.
NKI, Kanuri.
NGOOKKO, Aiawong.
NGI, Kanem.
NKE, Balu.
NKI, Ngola.
ENGI, Munio.
NGONGI, Maori.
INKO, Tocantins.

TANAK, Unalashka.
ENGI, Mumo.
AING, Madura.
NINHANGA, Jupuroco.
TANAK, Kadiak.
TANG, Kusunda.
NLANGU, Nyombe.
PANKHU, Sunwar.
NAK, Tonkin.
YING, Ostiak.

ONGOU, Fertit.
AZANAK, Eslen.
NIOGODI, Mbaya.
AING, Sumenap.
AMANGO, Nubian.
ONG, Lepcha.
MINQI, Okuloma.
RIANG, Tablung.
UHUNG, Nizhni Uda.
KIANG, Chinese.

AP (Egyptian), liquid, first essence. HEFN, to crawl along.

APH, Biluch.

AFON or AVON, Keltic.

ASH (Egyptian), wet, emission. KASH, to water, inundate.

UASH, Opatoro.
GUASS, Guajiquiro.

WASSER, German
GUASH, Intibuca.

ESK, Keltic.
UISGE, Gaelic.

BAH (Egyptian), Inundation.

PAH, Chemuhuevi.
PEH, Aino of Kamkatka.
VEHI, Manatoto.

BALA, Welsh ; effluence of a
river.
POH, Tesuque.

VEHI, Timur.
BECK, English.

BAI (Egyptian), water, visit, limit. BA, water, drink.

VAI, Marquesas.
VAI, Ticopia.
BE, Batta.
BI, Chanta and Baikha.

BU, Koi.
PA, Wihinasht.
VAI, Kanaka of the Sand-
wich Islands.

BAY, Mayorga.
BE, Tawgi.
BU, Motarian.
PA, Shoshoni.

HES (Egyptian), a liquid.

ESI, Dofla.
WESI, Estonian.
WESI, Fin.

WESI, Olonets.
EZI, Adampi.
ASI, Abor.

WESI, Karelian.
WESI, Vod.
EZI, Mahi.

IUMA or IMA (Egyptian), sea.

AHMAH, Cherokee.
AME, Sobo.
AME, Egbele.
HAMA, Purus.
EM, Palaong.

IMMEK, Labrador.
HUMA, Aymara.
AME, Bini.
AMU, Bode.
AMU, Karehare.

AMPO, Lutuami.
EMAK, mouth of the Ana-
dyr.

KARUA (Egyptian), lake, pond.

KHARR (river), Bengali.
CORR, Irish.
GOOR (salt water), Erroob.
KALING, Wiradurei.
GARRA (river), Ho.
GUR (running water), Akkadian.

KHAR (river), Uraon.
GHURR (rivulet), Arabic.
GOILA (fresh water), Redscar Bay.
GORO (river), Chepang.

GOUER (brook), Cornish.
COORA, Juri.
KELL, English, a well.
WERE, English, pond or pool.

KEP (Egyptian), inundation.

GAIPPE, Head of Bight.   KEPE, King George's Sound.   KAPI, Parnhalla.

KHEKH (Egyptian), fluid.

KIK, Talatui.
KIHK, Tshokoyem.

WOKH, Baraki.
KÜK, San Raphael.

HAACHE, Cocomaricopa.

MENA (Egyptian), to suckle; MENAT, wet-nurse.

MEENEE, Yankton.
MEENEE, Minetari.
MINI, Dor.

MIN, Mbofia.
MINEE, Dahcota.
MANT, Timmani.

MENYA, Kisama.
MANE, Begharmi.

MER (Egyptian), sea, basin, lake, water.

MARE, Latin.
MER, French.

MEL, Bolar.
MALUM, New Ireland.

MALAR, Lobo.
MOLUM, Port Praslin.

MES (Egyptian), product or source of river.

MAZI, Msambara.
MIZI, Lúchú.

MOSS, English, bog.
MOAZ, Kanyika.

MIZZU, Japanese.
MAZA, Kongo.

MU, MA, MUA, MEH, MERI (Egyptian), water.

MI, Hebrew.
MAY, Thounglhu.
MU, Korean.
MI, Tigre.
MI, Hurur.

MAYO, Syriac.
MOTE, Maori.
MU, the Tungús languages.
MEYA, Kasange.
MI, Arkiko.

MA, Arabic.
MAU, Coptic.
MA, Vilela.
MERE, English.

NAI, the; ARU, river: NAIRU or NILE (Egyptian).

NIRA, Tanema.
NIRU, Canarese.
NIR, Tulu.
NIR, Budugur.

NILATU, Rodiya.
NILLU, Telugu.
NUAL, Nalu.
NERO, Greek.

NIR, Kodugu.
NIR, Tuda.
NIRE, Kohatar.

NEM (Egyptian) water.

NAM, Siamese.
NAM, Laos.
NYIMA, Gurma.
NAMA, Greek.

NIMBU, Subtiabo.
NYAM, Dzelana.
NAMA (flood, torrent), Persian.

NYIAM, Guresa.
NAMUN, Bago.

NU, NNU, NA, NNUI, EN (Egyptian), water.

NU'OE, Cochin-china.
UNA, Quiché.
NÁN, Western Shân.
UNE, Saraveca.
NYO, Ankaras.
NIA, Antes.
NYU, Adampi.
NI, Dewoi.

NI, Kru.
NI, Geb.
NA, Yula.
NAINO, Timbora.
UNA, Naenambeu.
ONI, Guinau.
UNI, Omagua.
NYO, Wum.

NU, Erromango.
NOO, Accrah.
NI, Bassa.
NI, Grebo.
NA, Kasm.
NEE, Omaha.
NIHAH, Winebago.
NEAH, Osage.

NUPE (Egyptian), goddess of the celestial drink.

NEEBI, Ojibwa.
NEPEE, Knistinaux.
NEPEE, Skoffi.
NIP, Narragansetts.

NIPI, Illinois.
NEPPEE, Shawni.
NIPISH, Ottawa.
NEPEEE, Sheshatapoosh.

NIPPE, Massachusetts
NEPEH, Miami.
NEPPI, Sauki.

PAN or PANT (Egyptian), pool, mystical Red Sea.

BANUI, ordinary Javanese.
BENI, Udso.
PANI, Hindi.
PANE, Gadi.
PANNI, Mahratta.
PANI, Asam.
PANNI, Banga S.
PANI, Siraiki.

PANI, Bhatúi.
PANI, Gohuri.
PUNAL, Tamul.
PANI, Chentsu.
PANI, Tharu.
PANI, Pakhya.
PANE, Punjabi.
PĀNI, Gujeráti.

PANI, Bengali.
PANNAE, Ruinga.
PĀNI, Uriya.
PANI, Taremuki.
PANI, Bowri.
PANI, Kburbat.
PANI, Kuswar.
PANI, Urya.

REM (Egyptian), surge up, rise up as tears, to weep. URM, the Inundation of Nile.

LAM, Legba.
LEM, Keamba.

LEM, Kaure.

RIME (hoar-frost), English.

RU (Egyptian), drop of water, pool, gate, door, outlet, mouth.

LE, Shiho.
KE, Burmese.
LAU, St. Matheo.

LE, Koama.
LEH, Danakil.
RI, Rukheng.

LAU, Newar.
LUA, Hawsa.

SEKH (Egyptian), liquid.

SUX, Karagas.
SUCK, an Irish river.

SUX, Soiony.
SUCK, English, juice, drink.

SACK and SOAK, English.

SHE or SHA (Egyptian), a pool.

SHOI, Canton.
SHIU, Gyami.
SU, Kumuk.
SHIN, Akush.

SIN, Camacan.
SHUI, Mandarin.
SU, Turkish of Siberia.
SU, Osmanli.

SA, Mongoyos.
SIN, Minieng.

TEKH (Egyptian), liquid, drink, wine.

DECO, Coretu.
TI-CHI, Gyarang.

TKHO, Umkwa.

DAK, Ka.

TUA (Egyptian), kind of liquid.

Tu, Kioway.	Toue, Chepewyan.	Tii, Tengsa.
Tui, Kapwi.	Toah, Apatsh.	Ta-dui, Koreng.
Ti, Chepang.	Toya, Basá Krama.	To, Kutshin.
Ti, Mijhu.	To, San Luis Obispo.	Tu, Dog-rib.
Tu'i, Mrú.	To, Lule.	Too, Takulli.
Dui, Songpu.	Di, Magar.	To, Pinalero.
Tú, Slave.	Doi, Bodo.	Tei, Baladea.

TUR (Egyptian), libation, wash, distil. TERU, river-branch.

Taru (fresh water), Tobi.	Teloho, Gunungtellu.	Taru, Luhuppa.
Tarnar, Boraiper.	Taari, Tarawan.	Dwr, Kymric.

T (article), the; NU, water (Egyptian).

Tuna, Wapisiana.	Tuna, Waiyamera.	Tanni, Keikadi.
Tuno, Caribisi.	Tuni, Maiongkong.	Tannu, Modern Tamil.
Tuna, Accaway.	Tuna, Macusi.	Dannu, Iloco.
Tuna, Arécuna.	Tuna, Pianoghotto.	Tanni, Yerukala.
Tuna, Tiverighotto.	Dane, Irular.	Tannir, Malabar.
Tan, Rotuma.	Tona, Indians of Guiana.	Teign, Coropo.
Tain, Gaelic.	Don, Armorican.	

UAT (Egyptian), water, wet. UAT-UR (Egyptian), ocean, water.

Uatu, Botocudo.	Wit, Kondin.	Vat and Fad, Scotch (lake).
Ouata, Hottentot.	Wit, Vogul.	Wai, Maori.
Wut, Tsheremis.	Wata, Mairassis.	Ua (rain), Maori.
Üt or Öt, Narym.	Woda, Sclavonic.	Udor, Greek.
Wät, Mordvin.	Udra, Sanskrit.	
Watura, Singalese.	Vatu, Norse.	

UR (Egyptian), water, oil. AR, AUR, ARU (Egyptian), river.

Ari, Sanskrit.	Ar, Keikadi.	Yaru, Malabar.
Ayer, Malayan.	Er, Rutluk and Madi.	Ayer, Buton.
Arus, Malayan, current of water.	Ul, Yeniseian Group.	Ouar, Arago.
	Enarap, Abiponian.	Ero, Taneamu.
Uru, Otuké.	Jala, Sanskrit.	Aeromissi, Guebé.
Er, Gundi.	Ur or Errio, Basque.	Jal, Kooch.
Iera, Baba.	Jerie, Sumbawa.	Aru, Tamil.
Yeyer, Atshin.	Jojar, Poggi.	Ara, Malayalam.
Oira, Kissa.	Ira, Mangarei.	Arroio, Portuguese.
War, Papuan.	Ir, Naikude.	
Wire, Vanikoro.	Ar, Hindustani.	

It is not only that Egypt supplies the words for water; there might be no particular meaning in her having several names for water, but *each* of all these is a type, and most of them have a distinct ideograph, which shows the different relationships to water. The first sign of water itself is the phonetic N or single zigzag line of running water. Mu is water with three lines, that is the plural of water, the waterer, or the waters. Hi, water, is a canal of water; Mer, a limit of water, a reservoir, or an Inundation. A is water, as dew, with the sign of figuring forth. Ia (and these include forms of the word spelt with k) is water, and to wash. Rekh also is to wash and whiten. The Rekh-t is a laundress. Rukh-t, to wash, to full, has the water sign. This supplies the Australian Lucka (Carpentaria) water, and the Murray Ney-LUCKA for water is, in Egyptian, NUI-RUKHA, water for purifying. KAB is liquid poured out as a libation. KEP is a name of the Inundation. KHEN is water, a lake, an interior water, water chiefly as a means of transit, the water-way. RU is a single drop of water. ANKH only appears as a liquid life; Ankh permutes with NAKH. TEKH is a supply of liquid; NAM is a jug of water; NU, the water vase and receptacle; NU, the celestial water that descends. UAT is water, written with the papyrus sceptre, an especial sign of greenness, freshness, growth of plants; UAT is wet. UAT-UR (Eng., water) means the greatest principal wet. BA is water, as drink; AB, as the sign of purity; UBT is boiling water. HES is a mystical water of life, the feminine ANKH. HAN is tributary water; also the water of youth. TET denotes the water of the abyss of death. TA is the water of a tear, a type. TUR is to distil with water. SUR is drink. IUMA, for the sea, is the water (Ma) that comes and brings (Iu); it is the tidal form of water. IUMA is the earlier HUMA, whence HUMBER, the water that comes, and HUMID, the water becoming.

The common type-word for water, as AK, was almost worn out in Egyptian. It does exist, for AKHAB is pure water, and AB is pure. Also AK is the liquid mass of the celestial height. But it was worn down to an A for water, as in Akkadian. The ideograph of the Inundation has been read FENT, to stop, the nose. So read, as type of the waters, the feminine period, it means that water which is the antithesis of breath, and to stop the nose is the antithesis of breathing. FENTI, for the Inundation, is supported by the Lithuanic name for water, VANDU. KAMA is a name for water not applied directly in Egyptian. KA-MA is male water, and MA or MAI is the substance of the male. KAMAI was a gum and a precious oil in Egypt; the oil of Khem. This is one of the two waters of life when the life principles are both imaged by waters, and given to the male and female. ASH is wet; ASHR, a river. But ASH primally is a water of life. The ASH, as tree, is the tree of life. ASH, wet, is blood, and the variant SHAA is the substance born of, as KA-MAI is the substance begotten of. MENA, another type, appears as the name for the wet-nurse of the gods. So NUPE is the water-source personified, as the Lady of the Celestial Water. In the same way UAT (water) is a goddess of wet, or the water in the north, and the Uat, sceptre, the sign of the mystical water, is the special emblem of the goddesses. There is no goddess of wet primally except in relation to the mystical water, the source of life, which is essentially feminine, and most of the type-words may be traced to that origin. The NA (water) is red; ASH and TESH are the red source, with the sign of bleeding. KHEKH (fluid) has the same determinative. NAKHEKH is fluid, blood, essence of life, with the same sign. MU (water) is likewise the mother. KHEN is the water that carries or bears, as does the mother source. REKH, to purify, describes the water of purification. BA, water, is a means of being a soul. NDSAB for water in the Gabun dialects reads in Egyptian NET, invisible being, with the sign of blood, and SHAB is flesh and to form. NET-SHAB would denote the mother of flesh.

RU is a measure, a quantity, so much; MA and NA are water; hence REM or REN, an inundation or a deluge; REM, to rise and surge up as a tide of tears. These supply the type-names of REM and RANU. Water, as NU, NA, N, with the masculine article prefixed, forms the type-word for water, as PENA; with the feminine article, the type-word TUNA; and with the plural article, NAI, prefixed to AUR or ARU, the river, we obtain the NILE, NIRU, NORE, NIR, as names of water.

Sekh (Eg.), liquid, the root of all our USKS, ESKS, SEKS, ISCAS, OXES, UISGES, is identical with SUCK, or drink, which is derived by the suckling from the mother. Also the type-name AR, for water, is found in ART (Eg.), milk, meaning the liquid that is made, generated, for the child or Ar (Eg.). The Ar (Art), as milk, furnishes the ALL in Gaelic, meaning the white or wan water. The AL-AVONS may in some cases be the white or milk-like waters.

The present contention is that Blood was named as the primordial water of Nen, the Bringer, in relation to the source of life. The Egyptian NUNTER or NUTER, our Nature, the word for a divinity and type of periodic time, reappears as the name of blood in the oldest languages of India, as NETTAR in Tulu; NETTURU in Telegu; NETTURU, Canarese; NETRA, Kohater; NETRU, Budugur. The relation of Nuter with blood and periodicity is visible enough in the hieroglyphics.[1] It is typified in the Nuter, axe.

The water of Ouranos is the Egyptian URNAS, the mystical or celestial water of life, that is, blood (from which sprang Aphrodite). UR is the water or oil for anointing, really the blood.

The Assyrians called rain "ZUNNU," that is the Egyptian SHENNU, which means periodic water, as was the Nile Inundation, and the mystic water of feminine source. This latter is SEN, blood; NU, water. SINI in Kandin; ZAINI, Haussa; and SONA, Sanskrit, are names of blood. SEN has a variant in ZEM, Mose; SOMA, Gurma; ZEAM, Dselana; SEM-SEM (Eg.), the mystery, and the Well of ZEM-ZEM. The chief of all type-names for water are also the names of blood. This is most observable in the African languages. A few names are, AI, Khari-Naga; YEI, Yala; YE, Chinese; HI, Dumi; AJI, Mithan Naga; EIJE, Ako; HA, Sanskrit; ARU, Boko; ARA, Kupa; ERAH, Javanese; HARI, Nepaulese; KURI, Timbuktu; CROU, Cornish; KREW, Polish; CHORA, Malayalma; CHORE, Kurgi; GOR, Welsh; GORE, English; WERI, Fin; ULI, Kono; YELO, Kabunga; YELLO, Mandingo; WUL, Soso; KIL, Dsarawa;

---

[1] See p. 453, Birch, *Dictionary.*

KEAL, Koama ; MOSU, Undaza ; MAS, Ranyika ; MAHE, Shienne ; MAHASI, Songo ; MUAZI, Marawi ; NOO, Netela ; NAIH, Savu ; NAMA, Gbese ; NAMAI, Gbandi ; NYIEM, Barba ; NYIMO, Basa ; NYIYEM, Fulah ; NINYE, Bidsago ; NENYE, Wun ; US, Akkadian ; USI, Sunwar ; USU, Chourasya ; HUSI, Bahingya ; AZU, Nowgong ; ASRA, Sanskrit ; ASU, N. Tankhul ; ISAGE, Dsekiri ; SI, Ham ; SA, Gura.

The total list of names might have been lengthened indefinitely, especially by adding all the rivers entituled from these types, but that is unnecessary. The converse reading of the facts would imply that Egypt had gathered all these names of water from all the groups of languages in the world, including the river-names found in the British Isles.

SECTION VI.

EGYPTIAN NAMES OF PERSONAGES.

IN Egyptian a double form may be traced for the title of KING, German KÖNIG, Sanskrit GANAKA, Chinese WANG or ANG, Greek ANAX. ANK, ANUK, or NAK (Eg.) mean the I, the King, the living one. The types of this living one can be seen in the hieroglyphics. The word Ank may be, as in Chinese, abraded from Kank or Ka-nak, the Sanskrit Ganaka, or we may take Ganaka, Kanak, King, as a compound of Ka and Ank. Ank has a variant in Nakh, the powerful, or Power personified. One type of this power is the Bull. KA means the male image, and NAKH is power; the KA-NAKH, GANAKA, KA-ANK, KÖNIG, KING, is simply the type of Power, the first form being that of male potency, still extant by name in the English "Kingo" (Mentula) used by nurses;[1] then the male person, then the father, and lastly the monarch. The theory that the king never dies is obviously founded on the ANKH-title of the living, and the KA-ANKH is the type of life, the living one. ANKH, the living or the life, is, in the Maori, NGA the breath, and breathing one or the breather.

In this character the King represented the Bull, the Solar Divinity as Lord of Life, the erector as Khepr-Ra or Khem-Horus on the horizon, or Mentu, standing like the rock, in his type of the King or NGEI-NGEI (Maori), that is stretching forth, with the primitive sceptre in his hand, as the unhu or unkhu.

From the nature of the Allegory imported into the drama of "Punch and Judy," and its series of triumphs over all the ills of life, among which is one of Death being beaten to death, and another of the devil himself being outwitted, it looks as if Punch were a form of the ANK or NUK (Eg.) the Egyptian "I am," the Living One, the King. This, with the masculine article P is Punk, our Punch. In "Phrase and Fable"[2] it is stated that a statuette of Punch was discovered in 1727, with the long nose and goggle eyes, hunchback and paunch.

This is said to be the portrait of a Roman Mime named MACCUS, who was the original of Punch. Now Maccus agrees with Hor-Machus, i.e. Har-Makhu, the Sun of the double horizon or the Equinox.

---

[1] Urquhart's *Rabelais.*   [2] Brewer, p. 720.

This was the God Tum, whose especial title in the Temple of Pithom was that of the Ankh, the Living; he being the Sun of the Resurrection; written in Egyptian, his title is P-ANKH, PUNK, or Punch. We have him reduced to the status of Tom Thumb, and here there may be a link with the Italian Polichinello from Pollice, a thumb, the Tom-Thumb figure. The original type of the Nak or Ank will explain the humour of the Punch.

Punch and Nuk have their correlatives in Hunch, Bunch, and Junk. Punch means the short, fat, pudgy, thick-set fellow, whence the puncheon. So in the Xosa and Zulu Kaffir dialects a short thick-set pudge of a person is called isi-Tupana from Tupa, the Thumb. The "hunch" of bread is a thick lump; the Junk is also a short thick lump. Punch is typified by his HUNCH, and therefore he personates what the hunch signifies. Buncus is a Donkey in Lincolnshire, and the mystical Bull of Hu, called the BANGU, " Edewid Bangu,"[1] will enable us to recover the Solar Bull, the Neka or P-neka, as a form of Punch, of Makhu, of Hu, of Tum called P-ankh.

This might not be worth following but for the fact that the name of the Ank or Punch as the I, the A, 1, is the commonest form of the personal pronoun in the world.

It is ANK in Egyptian; ANOCH, Coptic; ANOKHI, Hebrew; ANAKU, Assyrian; ANAK, Kizh (Father); אנך Phœnician; NGS, Æthiopic; NGA, Kassia; NGI, Tumali, and NGA in a large number of African and especially the Negro dialects;[2] NGA-NGA, West Australian; NGAI, Port Lincoln; NGA-toa, N. S. Wales; NGAITYO, near Adelaide; NGATOA, Hunter's River; NGATOA, Wiradurei; NGA-po, Murray; NGA-pe, Encounter Bay; NGAPE, Lower Murray; NGAI, Parnkalla; NAIKA, Watlala; NGWANG, Kawi; NGO, Chinese; INK, Palouse; INGA, Limbu; ANG, Rung-chenbung; ANKA, Kiranti; ANKA, Waling; UNG-gu, Chourasya; UNG, Khaling; NGA, Bhramu; UNG, Dumi; ANG, Bodo; ANG, Garo; ANGKA, Dungmali; HANG, Thara; HANGA, the man of might, and IN-KOSI, Zulu-Kaffir; NGO, Abor Miri; NGA, Burman; NGAI, Tonga Naga; NGAI, Singpho; ONG, Laos; AING, Kol; ING, Ho-kol; ING, Bhumij; ING, Kuri; ING, Santali; ING, Mundala; INING, Cayus; NGAPPO, Aiawong; NGAI, Tarawan; AYUNG, Cherokee; AHAN, Pima; NYAH, Dieguno; NAH, Teruque; INAU, Guadal canar; INAU, Mallicollo; UNNO, Choctaw; UNNEH, Creek; NE, Chepewyan; NI, Shoshone; NO, Netela; EN, Tamul; EN, Tulu; EN, Rajmahli. To these may be added the Peruvian INCA; Maori HEINGA, the typical ancestor; Eskimo UINGA, the husband; Irish AONACH, the prince; Arabic AUNK; Malayan INCHI, the master; Gaelic INICH, the strong; American HUNKEY, the lusty, and Ako ONNUKU, the lively, active, equivalent to the Egyptian Ank. The root is expressed by the sound of NK or NG. This is extant

---

[1] Gwynvardd Brecheiniog, 12th century.   [2] Lottner, *Phil. Trans.* 1859.

in some of the African dialects as a nasal sound followed by a k-click, "UN-KA;" a nasal click still living in the Maori "NGA," which means to breathe, whilst NGETENGETE is to click with the tongue. This "NG" apparently formulates the earliest endeavour to utter by means of the nose and throat with breath and click the compound sound which was afterwards distinguished as N and K. And it was this with the "Ka" prefixed that furnished the name of the Ganaka, König, Kank, or King, and with the masculine article P, the name of Punch.

Ankh (Eg.), the living, appears to supply the ING terminal, as in Ætheling. ING denotes the son of. In the Saxon Chronicle (A.D. 547), "Ida wæs·Eopping" means Ida was the son of Eoppa, and the Ætheling was an old Saxon title for the Crown Prince, the Heir-Apparent. "The good true men of the land would have made king the natural heir, the young Chyld, Edgar Atheling. Whoso were next king by birthright, men call him Atheling: therefore men called him so, for by birth he was next king."[1]

It is a theory that the king never dies; he being the Ank or living. So was it in Egypt. And the son was the representative of the Ankh, as the Repa or heir-apparent, the visible link of continuity. When there was no natural heir, one was adopted, as the king could not die. Thus the Son was also the ANK, type of the ever-living, hence the "Ing," denoting the son and the sonship. The first ANK, king or Male "I," is the Son of the Mother; the branch of the Tree of Life. The royal son and prince in Egypt was the REPA, the shoot or branch, from rep, to grow, shoot, take leaf, sprout. Prince and branch are one at root, because the Repa was the branch. The child, the nursling, is the REN, and this with the suffix, which is both pronoun and article, the Ren (child) is the RENPA, the young shoot or branch. In Welsh this suffix is the prefix, and we have PREN for the branch, and PREN becomes the English BRANCH and PRINCE. The Branch, Prince, Child, or Ren, is the nursling of Rennut, the Gestator, child of the mother, named after her, and this is the Repa, Prince and heir-apparent. The Welsh have the Egyptian Repa in their PERYF, for the one who commands, especially used for the Pharoah. PREN is THE ren; Pref is THE ref; and PERYF is the Repa. Now we hear of the Welsh Princes before we hear of kings, because their beginning was with the mother and child, and the PREN as branch became the later BRENIN or king in the Welsh language, which title is extant, without the article prefixed, in the Breton ROEN for the king.

Our word Young or Jung, German Jung; Basque Jaung, the youthful god, our Jingo; Lithuanic Jaunas and Welsh Jeuancg, all include the Egyptian Ank (the king, the living one), as the IU.

---

[1] Robert of Gloucester, rendered by Earle in the *Philology of the English Tongue*, p. 268.

Iu means the coming one, and the Iu-ank, the coming life, is the Young. This is another form of the Repa, Branch, Prince, or Heir-Apparent to the throne. The Young God or the God Young (an English proper name) is the oldest in the world.

The diminutive of dear in dear-LING, the little dear, is probably derived from RENN (lenn), the Egyptian nursling. The Irish pronunciation is DARLIN or DARLINT, which adds the feminine terminal to the LIN. As Nursling the child is the RENN (Lenn), and not the diminutive of nurse, but the nursling (*i.e.* the renn) of the nurse. As Ren (Eg.), with the article prefixed, yields the Welsh PREN, the branch, or typical child, it is probable that this becomes the word BAIRN for the child, the Beryne in MORTE ARTHURE, and yields the name of BRENNIUS, the Prince, who was brother to Belin.

It is also possible that the acorn is not named as either the corn or the horn of the oak, but as the RENN (Eg.), the child, the nursling, the young, the type of renewal. RENPU (Eg.) is to grow, renew, be young, with the shoot for determinative. So read, the acorn is the young, the child, the renewer of the oak, or aak.

There is a plural in the Egyptian renn or reni for cattle, and if this does not supply the terminal syllable, in CHILDREN it may serve for the plural in EN, as in Housen; but apparently the REN accounts for both.

According to Borlase,[1] the Cornish people invoke the spirit Browny, when the bees swarm, to prevent them from returning to the old hive, and make them form a new colony. This connects the Browny with young bees and a new hive. Again the Browny, in faeryology, always disappears when old clothes are offered to him as a repayment. Now the Brownie or Brunie is also represented as originating in the young child that died unbaptized or un-NAMED. From this it may be inferred that the name of Brownie is derived from Pren, as in the branch, the young one, and from Rennu (Eg.), the nursling, with the article prefixed and modified into B, and as REN (Eg.) means to name, and RENNU is the nursling, and NU is No, it follows that the rennu was the young one not yet named, and if he died in this nameless condition as P-rennu, the un-named, he became the Brownie; hence the guardian spirit and guide of the Renpu, the young, in the shape of the young bees, and hence the Brownie's aversion to old clothes.

Egyptian also supplies the terminal in "RED," as in kindred, gossipred, or Ethelred. Red or Ret is the race. Ethelred is of the race of the Ethel; race is relationship, and one "RED" is used for relationship or kindred, although not limited to the blood-tie in Gossipred, a form of fosterage. RET also furnishes the variants of RED. Thus Ethel-red is the race of Ethel, and ret (Eg.) means to repeat, several: he is the repeater of the race of Ethel. So in Hundred the red (from ret, to repeat, several), denotes the repetition or enumeration of the hand or cent in the Hund-red.

[1] *Antiq. of Cornwall*, p. 168.

TEHANI was a title of the one who was nominated the Repa, the Egyptian heir-apparent to the throne, or of the Divine Father; found also in the Kaffir DUNA. This too the Kelts had; their heir-apparent was designated the TANIST. Also the coronation stone, a monolith erected for the crowning of a king, was called a TANIST, and the throne or elevated seat of the hieroglyphics is the Tan. Ast means great, noble, a statue, sign of rule, an image of the ruler seated. TEN-AST supplies the Latin DYNASTA, the prince, the ruler. A form of the TANIST stone, the coronation seat, is extant in the Lia-Fail, under the coronation chair, in Westminster Abbey.

The Egyptian Ank or Nakh, signifying the I as chief one; I in the highest form, the royal personal pronoun, wears down into the English "NICK-name." The Nakh-name is the individual name, and there are thousands of English working men with whom the Nickname is the only one they are known by, the sole title to individuality among their mates, the family name being sunk altogether in the *sobriquet* or nickname. The nickname, the royal name in Egypt, points back to the first attempts to individualise from the group (the Ing or Ankh) by means of a distinguishing epithet personally applied, just as some special characteristic or feature is still the source of the nickname.

From being a necessary cognomen, the nickname became degraded when applied in derision. This sense too is found in the word NEKHI (Eg.), meaning to deride, which expresses the existing status of the nickname.

It has been doubted whether the Sanskrit Rāj, to reign and rule, be a sovereign, exercise rule and sovereignty, could be derived from the Egyptian Râ. Nor would it if Râ were the primary form of the word. But it is not. The accented vowel represents a consonant found in Ka. Prior to the Râ, a title of the sun, the sun-god, and the Pharaoh, is REK, to rule, a name of time and cycle. The sun, as a ruler or Rex, was a law-giver of time, but not the first. The stars were the earliest time-rulers. The root Rek enters into the name of the Seven Rikshas (Rishis), and these were the first of all the rulers of time.

The celestial ruler, as Regulus or lawgiver, was also represented by the Constellation Kepheus in the north, and the Star Regulus (COR LEONIS) in the Lion. These were types of reign and rule ages before the Rek became the solar Râ. This Rek (Eg.) for Rule, however, is the original of the Sanskrit Rāj, to rule, reign, exercise sovereignty; the Vedic Rag, a king; Javanese RACHA, a divine image or type; Gothic Reiks; German Reich; Darahi Rak-uk, to rule or put down; Latin Rex and Rego; Gaelic Righ; Irish Rigan, a Queen. The word was worn down in Maori and Mangaian, as in Egyptian, to Râ, a name of the sun and day. The Rex or Regulus takes various forms. In English Gipsy he is the Rye, the Lord, or

Swell. In Akkadian, the Rak is the lady, and Racham in Hebrew is the womb.

The term LIEGE, French LIGE, Latin ligius, is but another form of Rek (Eg.) for rule. The liege lord in feudal law was the ruler over the tenant, entitled to claim his faithful service. The nature of the duty enforced was as absolute as the suit and service of the subject to any other form of the Rex or the Râ.

The Rukai (Eg.) are rebels, the culprits, rulers in the wrong. These are the Ruga of Ugogo, Central Africa, who are robbers and rascals, our English ROGUE in the singular.

A far earlier Rex than the Solarite King founded on Râ is the Rekh (Eg.), the reckoner, time-keeper, Mage, wise man, the knower, the architect, the washer, purifier, and WHITENER of men, i.e. the priest or Rook. The primus is the Egyptian name for rule in relation to time and period.

The same root rekh, to reckon or rule, is the only origin for the Logos as for the seaman's LOG. Instead of the Logos that was in the beginning, the Caribs have their LOGUO, the first man, who created the earth and then returned to heaven. Loguo and Rek, Rekh, Righ, Rex, Râj, Rajah, and Râ are all founded on reckoning and on time and rule. That which first completed a circle of time was the first lawgiver, and she, as the genitrix, was LEGIFERA, a title of Keres.

DRIGHT is an early name for the Lord or chief. This, like the rex and regulus, is primally the Egyptian Rek, to reign and rule; rekt, the pure, wise, the Magus, knower, intelligent one, with the Egyptian article T prefixed. TREKIIT is the REKHT, with the later sound DRIGHT; he is simply the right one for leader because pure and wise. The DRIGHT is the director. Ta ('Eg.) signifies direction, to go; hence the DRIGHT is the right-goer, leader, director. The people as DRIGHTEN are the directed, the followers of the DRIGHT. Breeching-time used to be a festival in a boy's life; and in the north a boy's breeches are termed DRIGHT-ups.

REK for time and rule is applied with the article in the word DRAG, and celebrated on the evening of a fair day, when the lads pull the lasses about; this is called DRAGGING (T-rek-ing) time, the time of their rule.

As the king is an image of the male power personified, the QUEEN is the feminine abode; the KEN or house in English, the KHEN or inner place (Eg.); KONA, Maori; CON in French; KUNA, old Norse; CHÁAN, Favorlang; KNAI, Dayak; Kunlen, Votiak; CHINA, Quiché; KUNS, Mandan; CUNHA, Lingoa Geral; GONI, Sanskrit; also YONI; GUNË, Greek; GWEN, Welsh; CUNHA, S. Pedro; CONIAH, Cayowa; CUNHA, Tupi; KENTO, Musentando; KENTO, Basunde. The Swedish QVENNA retains the oldest form, as in the Hebrew CHIVAN, the Queen of Heaven, and this as KEF-NU, the typical or divine KEF, identifies the queen with the abode of birth. The Kef (Eg.)

is a cave, a sanctuary, a place of concealment, and this we shall find in the Cefen Caves of Britain, first named by the Troglodytes, who dwelt in the Kef or Cefan of the Mother-earth.

Hathor is a queen of heaven, and she is the habitation of the child. Kheft or Aft is the feminine abode. The Khef is yet represented by the Cornish Coff (womb) and Chy, a house. There was but one human image for the heaven or the queen of heaven, who was mother of the gods. The King is the masculine potent, the Queen the female habitation, the house of life.

The English Empress of India was proclaimed to the natives as the " KAISER-I-HIND," and in the discussions we were told that KAISER was primally Persian. But the original of the Persian KAISER, German KAISER, Russian TZAR, and Roman CÆSAR is Egyptian. The hieroglyphics will show us what the title means.

" SER " is a most ancient and universal root, the Hebrew TZER is a divine name, the " Rock " of Israel. SER is the Egyptian word for the rock. Whether a divine or human title, " SER " or Sire, it is the chief, arranger, placer, disposer at pleasure. The title of the king of the Kheta was SIR; Assyrian SAR ; SIRE, was an old French title used by itself for the king only. The Quiché SCYRI was Lord over all. The Seren and Serene (Highness) are diminutives of Sir or Ser.

KHI is to rule, protect, govern, wield a whip. It was an Egyptian title. One form of the KHISER then is the wielder of the whip ; another Ser is the arm of the Lord ; another is the overseer. This is shown hieroglyphically by the camelopard sign of SER, the chief or high one. KHI signifies to rise up, elongate. The camelopard is the extended long-necked overseer, and a type of the SER or KHISER. SER is an epicine or neuter root, and the KHISER may therefore be either male, female, or neither, as SER is the ram, sheep, or eunuch. The KHI or whip emblem itself is, however, distinctly male in the monuments, as the sign of KHEM, PTHA, and OSIRIS. Still there are female camelopards ; and women have wielded the whip.

The latest living exemplar of the primeval KAISER is the TZAR, who still wields the knout (or did so lately) as the sign (Khi) of the SER. But the Tsar, the ruler with the whip, is a divine personage in the realm of the dead. In the Ritual (Ch. 126) we read of

" All the created just spirits who serve the TSAR."

The camelopard is the more appropriate sign of rule for the " KAISER-I-HIND," the tall overseer, and it is perfect ; KHI meaning to extend, elongate, rise up, be high, rule, govern, protect; and " SER " is the name of the animal thus described by " KHI." The camelopard would fitly symbolise the wide outlook of the British " KAISER-I-HIND."

The Ser is no doubt a worn-down form of the Kheser or Seser. The

Egyptian type of the Kaiser or Cæsar is the ideographic User, earlier Seser, and therefore still earlier Kheser, which furnished the Assyrian KASAR, the Arabic WAZIR, and Malayan WEZER. The sign consists of a head and vertebra, a backbone and brain, with the meaning of ruling and sustaining and maintaining power. Also the Ka sign is the determinative of Ses, the back and shoulders, the image of the Seser or Kaiser. With Kheser modified into Seser we have the Cæsar, in the Egyptian Sesertoses, the Seser himself, or typical Cæsar, who became the Sesostris of the Greek writers on Egypt.

Herodotus describes a pillar of Sesostris, which he says was engraved across the breast with words in the sacred Egyptian characters, signifying, " I acquired this region by my own shoulders." In those same sacred characters " Ses " means the shoulders and back, and has for determinative the shoulders with uplifted arms ; " tesas " signifies the very self. Consequently some one must have very nearly read the hieroglyphics of the Sesertosis.[1]

The Egyptian Cæsar or Seser appears on the monuments as early as Seser-en-Ra, in the Third Dynasty, and there is a female Seser by name—one only—in Ta-Seser, the wife of Siptah, at the end of the Nineteenth Dynasty.

The LORD is usually taken to be the Hlafford, as Saxon for loaf-author or bread-giver. And in Egyptian the REPA or Erp is the Lord. The ERP, the hereditary highness, has the earlier name of KHERP or KHERF, his Majesty or Lordship. The Saxon HLAF is the equivalent of the KHERF, the chief, first, consecrated, sceptred. HLAF for loaf is paralleled by KHERF, a supply, a sufficiency, to suffice, our CROP. Kherp abrades into Erp or Repa. The feminine Repa is the wife of Nile and Goddess of Harvest. Harvest is Kherp or Kherf, the supply or crop. This was furnished by the REPIT, the lady who as breeder was the primal bread or loaf-maker. The Hlafford, then, points to the Kherp for origin, not to the loaf, which became a later sign. But is it so certain that our word "LORD" is only the A.S. Hlafford worn down ? The Etruscan LARTH is the Lord, and that can scarcely be derived from the HLAFFORD.

The equivalent RURT (Eg.) identifies the meaning with going round, and the circle-maker. The circle-maker is a RURT as Ursa Major. RURT, the round, is a pill. The husband encircles the bride's finger with a ring in token of lordship. LORT-Monday, a name of Plough-Monday, is a day of going round. The LORD-size is the Judge of Assize who goes on circuit. The LARTH or Lord was the chief of a circle, not merely the loaf-giver. This signification is pointed to in the mummeries of the LORD of Misrule in the Inner Temple on St. Stephen's Day, when the Constable-Marshal went round and round and shouted " A LORD ! a LORD !"[2] Is the title of Lord, then, limited to the Hlafford ?

[1] B. 2, 106.　　　　　[2] Brand, "The LORD of Misrule."

The Egyptian REPA as Lord and Governor was our REVE by name, who as REVE of the shire became the Sheriff, and the REPA as a division became our RAPE. Sussex was divided into six Rapes with six Rivers, six Castles, six Forests, six REPA-ships or lordships. The REPA may perhaps lead us to the origin of the law of primogeniture and entail of property. In Egypt the Repa was hereditary lord and heir-apparent. He inherited the land with the throne, as representative of the Divine Ra. Also the Repa might be male or female; it was a name of Virgo, and with us the REEVE is the female of the RUFF (bird).

The flower is before the fruit, and the Lady REPA, who became the bearer of fruit and goddess of harvest or the leaf, has an earlier character in relation to the flower or flowers. The female REPA represents the Nile in one form and the harvest in another, as the Lady of the mystical Two Truths, the flowing and fixed, or flower and fruit. REP (Eg.) means to grow, bud, flower, and REPT or REPTI is the other of two forms of the word; Rep denotes the one that buds and flowers; REPIT, the one that fructifies and bears the fruit. This accounts for the number of ladies among the flowers, as the lady's finger, lady's slipper, lady's smock, lady's glove, and others.

The King's THEGNS, as the Thanes were called, were companions of the king. The term THEGN answers to the word Gesith, a companion. And in Egyptian TEKN is to accompany; tekh, to join together; teka, attach; whence TEKN, to be joined, and THEGN, one of a companionship or order of men who performed some personal service to the king. The THEGN, as one of the companions (Comites) of the king or chief in battle is illustrated by TEKEN, to accompany, be near, and stand fast. As Tekh further means to supply wine or serve with drink, and TEKHEN is to play on the harp, the office of royal cup-bearer and minstrel were no doubt early forms of THEGN'S service. To play on the harp is to accompany, and Tekh, to serve with wine, shows the TEKHEN or Thane as the server implied by the title of THEGN.

In the Saxon period there was a royal official who from handing the dish bore the title of Disc-Thegn.

The HERTOGA was an army-leader in war. HAR (Eg.) is the superior, the one set over, the over-lord. TEK (Eg.) has the meaning of attack and overthrow. In this sense the Hertoga is the leader and director of the attack, and lord of the army. In the Old English Chronicles (449) the first conquerors, Hengist and Horsa, are termed HERE-TOGAN. This was whilst their rule was limited to certain districts and lesser boundaries of land, and the word TEKA also means a fixed frontier. Thus rendered, the HER-TEKA is a border lord, or king of a county, the keeper of the boundary, whence the Hertoga and Hertogan.

In the same Chronicles (519) Cerdic and Cynric are called

EALDORMEN. But this title is by no means derived from the Jarl or Earl, both of which are modified forms of the Karl, Ceorl, or Churl.

In Lazamon's Brut we read: "Belin in Euerewic, huld corlene husting."[1] That is, Belin in York held a husting of Earls or Ceorls. The EAL in Ealderman is one with the WEAL in Wealhcyn. Wales is the earlier Gales, still earlier Kars, which became the shires. The Kar is the circle, orbit, inclosure; TER (or der) is all people who dwell in the Kar. The Welkin is the circle round. A Weal is a wicker basket for inclosing fish. The EALDER is just the equivalent to garter, and Kilder in Kilderkin, and the Ealderman is the man of the whole Kar or inclosure. As ter is the frontier and extreme limit, the Ealderman may have been the boundary-keeper of the Kar, Gale, Weal, as representative of the ter (Eg.), that is, the whole, as land or community. Kar modifies into Har, as in Hertoga, and Al, as in Ealdorman. The Alderman still represents the ward, and Ward is philologically one with guard and with weald. Ward also means good keeping. The Ward was the little world that needed the warden. The Weald as forest land had its warden, and the Ealdorman is the warden or warderman who represents the ward and weal, and is chosen for that purpose. Weal means to choose, and the Weal-der-man, Ealdorman, is the man still chosen to represent the whole, entire, the Ter or commonweal.

The Earl, Ceorl, Churl, or Jarl, is derived from the Kar inclosure, and, primarily, the name denotes one who belongs to that circle or inclosure before it was given to him to whom the circle belonged.

Our word RUNE denotes a mystery, something secret, and this is connected with ROUNE, to whisper, and the runes which were mystical in character. To roun or whisper, face to face and "mouth to ear," and "the word at low breath," is a Masonic commandment. To ROWNE in English is to name in a whisper; in Egyptian to REN is to call by name. The ROUND were officers appointed to inspect the watches, and called "Gentlemen of the Round." In relieving guard the soldiers still ROUNE or whisper the password, and the Gentlemen of the Round are related to this secret REN-NING. The REN inclosure of the royal name became the English Round. The Egyptian NA-RUNA and NA-AA-RUNA who are officers of name and of great note, but of an unknown office, are possibly the same as our Gentlemen of the Round.

HAN (Eg.) is honour, sanctity, royalty, majesty, and rule. Har or Her is the lord, chief, superior. HAN (Eg.) also denotes territory, a field; the HANUTI are labourers. HAN (Eg.), moreover, means tribute, and to conduct as tribute. These include all our "Honours;" the "HONOURS" of cards as the majesties, royalties, highnesses; the honour or sanctity of character, considered in points, affairs, and debts of honour belonging to the unwritten law, and the Honour as a

[1] MS. Cott. Calig. A. 9, lines 4765–6.

superior seignory under which other manors or divisions of land were held by the performance of service.

The MARMOR was a style of high nobility among the Gael; he was a great officer of justice. Mor, great; MAOR (Gael), officer of justice. The MERE is the Mayor, and there is an English MER, who is a bailiff or superintendent. This is the Egyptian MER, a superintendent, prefect, overseer, or governor. An official called the MER governed the people of the quarries at Turuau, the mountain-quarry in Egypt.

The MER was not only an overseer and superintendent, but an architect. The architects of the Egyptian Pharaohs, who were the royal sons and grandsons, were called the MER-ket. And we are told by the Barddas that "MORION lifted the stone of the Kettai." MORION is said to have been the architect of Stonehenge, Gwath Emrys, or the MUR Ior. MUR (Eg.) means the circle, as does the KETUI. KET (Eg.) denotes the builder, and the Kettai are the builders. MORIEN was chief of the KETTAI. Now, as a negro is still known as a MORIEN in English, may not this indicate that MORIEN belonged to the black race, the Kushite builders? The name of the MARMOR himself appears in the *Travels of an Egyptian*, where the question is asked, "Didst thou not meet the MARMAR?"[1]

MERU (Eg.) is a cow, a goddess, and a form of Hathor, the cow-headed genitrix. With the feminine terminal the Mer is MER-T, and the MART is still a Gaelic name for the cow. The MERT (Eg.) is a female attached; MAR-T denotes feminine relationship and office. MER (Eg.) is love, and the MER-T are persons attached, the lovers. This, in English, is married; but in Egyptian the female MER (mer-t) was the person attached as consort in the divine (or the pre-monogamous) marriage. The great goddess Pasht is designated MER-Ptah, Ptah's beloved; she was literally his MORT. And the poor vagabond's "MORT" that trails after him, dog-like, on tramp as a female attached, has the same name and is the living representative of the goddess, the divine MER-T. She is also called his Doxy. In Egyptian TEK means to attach, and Tekai is a name for the adherer or person attached. Tekhi (the Doxy) was a goddess of the months. This is sadly typical of the old divinities, when we find them in modern dress, and the divine names in current usage. The MER has become the MORT and MERETRIX, and with a change of terminal the MARQUE of the French Argot; the FILLE, the FEMME PUBLIQUE, the MARCA, MARQUIDA, MARCONA of slang phraseology.

The MERT has got be-mired, and the MUT, the divine mother, has had a like descent into the mud. Her living representative is the English "MOT," the harlot. These names of the creative motherhood and consort of the creative power surely ought to plead for a little more pity and charity for those who bear them now!

[1] Section 4, line 9, *Records*, v. ii. p. 107.

MER also means to die, hence the MERTI are the dead. And in the British mythology, the dog of Pluto is called Dor-MARTH, the gate of sorrow. It was the door of Hades, the entrance for the dead or the MERTI. Maarau (Eg.), to grieve, corresponds to the word MARTH, for sorrow. The dog's mouth was the door of Hades. So Kerberus in Egyptian reads Kherb-ru, a first or model form of the mouth, gate or door. This dog of Gwyn ap Nudd and the Greek Kerberus is a figure in the Ritual. Just where the MERTI entered the underworld at the western corner, the angle of the pool of fire, sat the dog having "the eyebrows of men," "Eater of Millions is his name," as the door-keeper and watcher who devours the fallen at the angle of the west.

The mouth of the dog was not the only form of hell-door; the Bull was another type; the Crocodile of the west was depicted as swallowing the Akhemu Ureta, or setting stars, hence the expression in the Ritual, "Back, Crocodile of the West, who livest upon the Akhemu who are at rest; what thou abhorrest is upon me."[1] Here the passing souls are identified with the setting stars, and the crocodile of the planisphere does duty in spirit-world.

The root in Kan, our Can, sometimes assigned for the name of king, is extant in Egyptian as Kan, to be able, courageous, valiant, a victor; Khun-su is thus the brave child, the victor-son. This title is the British Cun, a chief. There .is also another sense of the canning or cunning man, only found in Egyptian. The cunning man, as wizard, is not only the knower and diviner, he is likewise the averter of evil and of bad luck. KHENA (Eg.) signifies to blow away, breathe, puff away, repel and avert evil. KHENN also means to rest and believe. The Khennu, in the modified form of Shennu, is a diviner. The Khennu then is the one who is able to divine, the averter of evil and bad luck, hence the KENNING-stone and the cunning man or woman.

Tum was the judge of the dead in the hall of dual justice. He was the god of the darkness; his was the all-seeing eye. Tema (Eg.) means to make justice and truth visible, to announce. Tum, the name of the god, also signifies to spy out. Now in Ireland the wise man, the diviner, the seer in the dark, and discoverer of lost property, is a TAMAN. "I know," says Vallancy,[2] "a farmer's wife in the county of Waterford that lost a parcel of linen. She travelled three days' journey to a TAMAN, in the county of Tipperary: he consulted his black book, and assured her that she would recover the goods. The robbery was proclaimed at the chapel, a reward offered, and the linen was recovered. It was not the money, but the TAMAN that recovered it."

TUMAU (Eg.) also means to restore. The TAMAN is founded on

---

[1] Ch. 32.   [2] *Col. de Reb. Hib.* No. 13, 10.

the type of Tum as seer in the dark, maker of truth and justice visible, and the restorer of that which was lost or concealed.

The English "DUMB-wife," a name of the feminine fortune-teller, is a kind of Taman, and the Irish spelling will recover the original form of Dumb in Tum (Eg.), to announce, reveal, or make the hidden truth visible. In Egyptian Tum means dumb, mouthless; hence dumb means tum in relation to the dumb-wife who announces and makes known. Conjurors are proverbially born dumb. Tum, as representative of the lower sun, as Hak (Kak), is a form of Harpocrates, the dumb child, or mystic word, who points to his mouth, and is the antithesis of Makheru, the true voice. The "Dumb" cake used for purposes of divination is the Tum cake that reveals, announces, and makes known the secrets of futurity.

The thumb, the lower member of the hand is named after the god of the lower world. This too was a type of Tum, the diviner in the dark. Hence the allusion of the witches :—

> " By the pricking of my thumbs
> Something wicked this way comes."

The thumb foretold.

In another bit of gesture language the thumb was bitten at a person to convey a meaning without words. The type of Tum made the truth visible.

In Egyptian the Mage, the wise man, the illustrious and revered man of learning, is the AKH. The AKHEN is a recluse. AKHENNU denotes those who praise, salute, glorify. The variants KHEN and KHENNU signify belief, intelligence, news, and seclusion in the sanctuary. The Akhen lived on in the Irish AIGHNE, for prophet. In English AKENN means to reconnoitre, make observations, to discover, and one of the Irish names of the Round Tower is the Turagan. The tower we may take to mean the "Tur," which in Egyptian denotes the building on a frontier, a limit, to hinder, a wall, fortress, or prison. AGAN may signify the recluse, the men who dwelt apart in seclusion, the Magi, who watched the heavens and reflected the light of their knowledge round the land from the sanctuary of Druidic lore, the Turagan or Round Tower. The Persian Magi, said Pliny, might have been taught by the British Druids. Tir in Welsh and Irish is territory, and the Tiragan may have denoted the land of the Akennu or the Khen, the sacred soil of the dead that was protected by the Tower. We have the Doole or dule, a kind of Tel (Arabic, mound), as a conical heap of earth which marked the limits of parishes or farms upon the downs, probably the Cairn of the dead before towers were erected. So Tillie-Beltane is the little hill or eminence of the Baal-fire.

The Lich-gate at the entrance of the churchyard, as the place where the corpse is set down first, serves to connect the Lich, Gothic LEIK;

English LIC, with the sense of LIGGING, or lying at rest, as in LIGAN, Gothic, to lie ; Frisian LIGA, to lie; old Norse LIGGIA, to lie ; Latin LEGERE, to lay ; Russian LOJIT, to lay; Servian LOJATI, to lay ; Greek λέγειν, to lay, lay to sleep. The Lich, as the dead (corpses), are those that lig or lie at rest. But in the hieroglyphics the Rekh (Lech) are the dead in another sense ; they are the pure wise spirits with the Phœnix emblem of renewed life rather than the bodies that moulder in the ground. But the name was applied to these that ligged or lay there. The dead are those who know ; they are the supreme knowers. The wise spirits, as the Magi, are also the Rekh. And on this ground it is claimed that the Lichtun for the churchyard is not merely the Tun (place) of the dead. Just outside the city walls of Chichester, on the east, are what the common people call the " LITTEN " schools. The name is abraded from LICHTEN, but not necessarily limited in meaning to their standing on the LICH-field. The Rekh, the knowers, as priests, were the dwellers with the dead, and they taught in the sanctuary, but the Rekhten or Lichten school originates with the living Rekh. The same word has various applications.

LECHTEN is identical with LEIGHTON (Buzzard), the TEN (Eg.) land, region, seat of the REKH, the learned, the Magi. Sart (Eg.) signifies to sow seed, in this case, the seed of knowledge ; the Sert is a keeper. The Buau (Eg.) are the chiefs, heads, archons. The Buau-Sart of Leighton are thus the chief men of learning, who dwelt in the Tun of the Rekh, not of the dead only, but of the Magi.

The ROOK, a name of the parson, identifies him with the Egyptian REKH, the wise man, the knower, the Mage. In the Breton language a priest is called BELECH, agreeing with Prekh (Eg.) as the Rekh unless we read it Bel-akh, the Mage of Baal. The Rekh supplies the Leech. According to Strabo, the British Druidesses or wise women were also styled " BRIG of the Judgments," i.e. with the article prefixed, the REKH, and as they dressed in black, they too were ROOKS. In Akkadian the woman is the Rak.

The ancient professors of witchcraft and raisers of the dead were called WICCA and SCIN-LÆCA. The SCIN-LÆCA was also a species of phantom or apparition, and the name was used for the person who had the power of producing the Manes. SCIN-LÆCA is said to mean the shining dead. But SCIN-LÆCA, as the evoker of the dead, does not mean shining. It is the Egyptian SKHEN, to cause to alight, be manifested, give breath or beingto ; modified SAN, to charm by magic. The original of WICCHE, as in Wiccraft (Witchcraft) is HEK or HUKA (Eg.) magic, to charm, evoke the spirits of the dead.

The " Ur-HEKA " is found on the monuments as the Great Charmer to the King, and Master of Magic. UKH is the name of a spirit, as is the AKH, the Spirit, or Manes, and their invokers the Mages, are also named Akh, earlier Hek. So the dead called LÆCA, Egyptian

REKHI, are one with the spirits and the Magi, from REKHI, pure spirits, Mages or Magicians. Our Wiccacraft was their HUKA-craft.

In the Saxon period the Mage, Magician, or Sorcerer was known as the DRY, and his craft of magical evocation was called Dry-craft. Yet the name is unknown to any of the cognate dialects. It was therefore a word, like so many more, adopted from the British, and out of this the Saxons formed a verb be-drian, to bewitch or en- chant; but they found the Dry already extant. It is yet to be found in the Gaelic Draoi as the Magician.

In Irish, DRAOI-Acht is the name for Druidic law, or the law of the Druids. Draoi represents the Egyptian " TRI," which signifies the invocation, evocation, and questioning of spirits. Tri is to invoke, adore, question, with the determinative of a person making the Invocation. As the full form of the word is TRIU, and "It" is Heaven, a passable Druid (Triu-it) might be derived as the Tri or Dry, who was the Invoker of Heaven. But the name "Druid," whilst retaining the Draoi or Tri meaning, as well as that of teruu (Eg.), stems, roots, which associated the Druids, in the mind of Pliny, with the Greek name of the Tree, is more probably compounded from Tru or Triu, the Two Times related to the Two Truths of Egypt, the root of all knowledge. Teriu (Eg.) denotes the twin-total, the whole, the All. An Irish name for Druidism is MAITHIS, and that includes the Egyptian dual Truth called Mati, which, applied to time, is the Teriu or Two Times at the base of all reckoning.

Another meaning of " It" (Eg.) is to figure forth, and portray, as the artist. In this sense a Teru-it might be the hieroglyphist who drew in colours. The Tru-It would also be the teacher of time and period, and the fulfilment of the cycles. This is the explanation of the Roman report of the Druidic proficiency in the science of astro-. nomy, which was the science of celestial chronology, or circle-craft; a "TRU-IT "-IC (Druidic) science.

The Egyptian priest called a "KARHEB," the Reciter of the Festivals, answers somewhat to the meaning of the Druid. The Kar-Heb as explained by de Rougé is the chief of the "UTISTS" or conductor of the sacred rites. The full form of the name Ut is "Hut," with the star determinative of time and fate. Ut (Hut) also denotes magic. TRU is time, and the Tru-Ut or TRU- HUT, as an order of time and destiny, corresponds to the astrologers, magicians, or Druids of Egypt. It is therefore likely that the Druidic name is a modified form of Tru-Hut, especially as the Welsh Hud associated with them signified their mystery.

This HUT or HAT supplied an important formative in compounding words such as Priest-HOOD, Man-HOOD, Mother-HOOD, Boy-HOOD, Widow-HOOD, Maiden-HOOD. In Brotherhood, Sisterhood, and Priesthood, applied to religious bodies, it is the original HUT, an order. HUT-ER (Eg.) is to join together. In the form of HAD the word denoted faculty, degree, office, quality. In the Saxon Chronicle[1] the function bestowed on the candidate when consecrated was termed Biscop-HAD. The bishop's power of consecration was called HADIAN, and in ordaining for Holy Orders he was represented as conferring "HAD," which made the initiate one of the holy order, or HUT, *i.e.* HOOD. This quality of BISCOPHAD is Egyptian as well as the order or Hood. For "HUT" also signifies to consecrate by touch, as the bishop bestows the HAD; and HTI-an is purely Egyptian for being consecrated. It has no relation to an Anglo-Saxon speech independently of that which the Angles and Saxons found in Britain.

The Bishop too is Egyptian. There was an order of the SHEP (Shept)[2] found also as SEP, persons belonging to religious houses. The BUA or BUI is the head, the Archon. The BUI-SEP, BUI-SHEP or BUI-SKHEP is the archon of the SKHEP, SHEP, or SEP, the learned, teachers, enlighteners. The SKHEP are images of the other life, and these denote the religious nature of the order. SKHEP signifies to transfer and transform. The Bishop, in ordaining, claims this power of consecrating by touch when he bestows the "HAD."

KHEPR is a name for the Creator, as the Transformer, imaged by the beetle, so that the bishop is an archon of the HUT order, of the SHEP, SKEP, or SKHEPR (Khepr), the Scarab-God of Egypt.

Ambrose, Archbishop of Milan, proclaimed Christ to be the Good Scarabœus, and it is in keeping that the Bishop should be the priest of Khepr.

"SHAP" is our word shape. It means to figure, image forth; and this the bishop does in antique fashion, with his hand in blessing. He still holds up two fingers without knowing what they stand for. The symbolic value of those two fingers will be better understood when we have treated of the Biune Being, and the Two Truths.

The two fingers have the same meaning as the SHEBTI image, the SHAPTI plumes of the solar disk, or any other symbol of the two-fold truth. The duality came to be applied to the life here and hereafter. But the apron is a relic of a more primitive significance.[3]

The Welsh bards call the Druids NADREDD, rendered Adders. And the word is the name of the serpent in Welsh, but the Nadredd were not merely adders; the adder or serpent was a hieroglyphic symbol of the divine, and the word for divine is "NUTER," of which the serpent is a determinative, therefore the serpent in Egyptian was Nuter. NET signifies serpent. NEDDER is an English name of the adder, and NA is an Egyptian article, the. But NUTERUT is a temple, the Divine House. AT or ATA is a father, a priest; and another, NUTERAT, is a kind of priest, a holy father, a prophet or a prophetess.

---

[1] E. 1048.   [2] Denk, ii. 143, B.
[3] APRON.—In the paragraph on the "Basu" (p. 118) it should have been pointed out that one form of the BES-skin is still worn by the soldier as the bear-skin BUSBY.

And it is here claimed that the Druidic NADREDD were identical with the NUTERAT of Egypt, the priests and priestesses, prophets and holy fathers who served in the Nuterat or Divine Houses.

There is in Scotland both a castle and a parish of KIN-NEDDAR. Khen in Egyptian is a sanctuary of rest and faith, and Nuter is divine. KIN-NEDDAR is thus a divine sanctuary, a NUTER-AT.

The ASC is a Gaelic name of the Druid or adder. ASC permutes with Sekh in Egyptian. SEKH (Eg.) is the scribe, to write, and writing. SAKH is to adore, pray, understand, and the name of the shrine; SAAKH, is the intelligence or intelligent spirit; SAKH the illuminator, inspirer, and informant in person. This is the ASC as Druid; he is the informant, hence the verb to ask or seek to be informed.

This SAKH is possibly the original of the name of the Saccæ and Saxons. We shall find the later typical names were religious before they became ethnical. SEKH-SEN (Eg.) would read the fraternity of the learned men or priests. The name of hate used by the Irish for the Saxon invaders is SASENACH, and it may include the Egyptian Sesen, to fight, distract, torment, whence the Sesen-akh would be the great distractors and disturbers, as fighters.

The Egyptians had an order of priests, the name of which is written with the jackal, SHU. Lepsius[1] has read it "SA"; "SA" is to recognise or perceive, English see, and these would thus be the seers. Still the jackal, the seer in the dark, is "SHU," the Hebrew "SHUAL" (the fox); and the Welsh bards call their diviners, an order of priests, the "SYW,"[2] a word signifying that which is circling. They were also designated SOWS. In Egyptian Sow is SHAAU or SHU, and "SHAUNU" is a diviner; these were the SHAAUS or SYWS, the diviners. Some divining faculty was ascribed to the sow; it is yet said that pigs see the wind.

The SUES or SUAS was a well-known name of the Kabiri. The Egyptian SUA is a priestess and singer. The SUAT (plural, like the Welsh SYWED) also appear in the male form, as choristers or glorifiers. The British Keres, Keridwen, assumed the character of the HWCH, a sow, the multimammalian mother. HWCH is Hog, independently of sex, hence the boar. In the hieroglyphics the HEKU is also the hog, and HEK is the Ruleress, as was Keridwen the HWCH. HEK also means charm, magic. The HWCH was a magician, like the Greek Hecate.

Here is a good test, as it seems, of Egyptian origin. The name of the British Merlin permutes with Merddin. So, in Egyptian, MER and MERT are identical. Both signify the circle. REN (Eg.), is to name. MER-REN means the circle-namer. Merlin is reputed to have made the Round Table for Uther Pendragon, which descended to

---

[1] D. 117.　　　[2] Cyvoesi i. ; Davies *Mythol.* 467.

Arthur when he married Guinever. The Round Table was the name-circle of the twelve signs, and later of the twelve Knights.

Ten (Eg.), is to complete, fill up, terminate and determine. The Tennu are lunar eclipses; Tennu, is to go round. Mert-ten, the equivalent of Merddin, is the reckoner and determiner of circles and cycles. A type-name of the same value as Tru-it (Druid), for one who is learned in circle-craft, the old English name of Astronomy. This will serve to show how Merlin and Merddin may be interchangeable names.

The scribe is already named in Egyptian by Skha or Saak, writing, to write, and the writer. But the first writing was cut or scratched, hence Sekha means to cut, incise. Crafa (Welsh) is to scratch; Breton, Krava; and this agrees with Kherf (Eg.) a first form or mode of figuring and modelling. With the Welsh prefix Ys we obtain Yscrafa to scrape, also the words scrape, sculp and scribe. Thus our first scribe is the scraper of bone and the sculptor of stone.

Kherp (Eg.) to figure with the causative prefix becomes Skherp or the Scarab type of Khepr, the Skherper, scraper or figurer, the former who preceded the writer, as figuring was earliest. Khepr as Skherp, the Kherp, was the Scribe before writing was invented, and his type is the Skarab, English Scarbot. Hens are said to Scrab a garden. Scrab to "claw hold," is identical with the Scarab or figurer. In Devon Scrapt signifies slightly frozen, that is, beginning to form or be Kherpt, figured. Scrap, English, is a plan, a design, from Skher (Eg.) to plan, design, picture; the Egyptian terminal p turns Skher into Scrap. A curious relic of Skherping, figuring, inscribing, is extant in the Devon word Scurrifung. Khepr the figurer is the Creator, the former as Generator. Khepr and Kherp are variants of the figurer. Skurrfung, to couple, lash tightly together, signifies Futuere. Ankh (Eg.), means to pair, couple, clasp together, and corresponds to "ung." Scurrif is a form of Skherp, Kherping, crafa-ing, carving, and Scurki-fung is creating in the sense of figuring the Child. Wilkinson found a picture of Khepr (Ptah) engaged in making a drawing of the Child as Horus; this was the Creator as the figurer forth, in the character of the Kherp or Scribe, who was the earlier skherper or sculptor.

The word Skeptic is derived by etymologists from the Greek, Skeptikos, an examiner, and inquirer, from Skeptomai, I look, I examine. Then we are assured that in Greek the root Spek was changed into Skep and accounts for it. Spek is to be found in the Sanskrit Spa'sa a spy; in Spashta, clear, manifest, and in the Vedic Spa's, a guardian.[1] In the hieroglyphics Skeb (Eg.), means to reflect. Skhep is to clear up, enlighten, illumine, render brilliant. Sap also means to examine and to verify; Saph, to examine and reckon up. Thus we have Sap, Saph, and Skhep all meeting in the same meaning. Further, the Seps is an ancient form of the As, the

[1] Max Müller, *Lectures*, first series, p. 258.

great, noble, the ruler, protector, or overseer, the original SPAS, as guardian, or overlooker. SEP is to judge, and the SHEPS was a sort of judge. There is no need, therefore, of converting SPEK into SKEP in Greek to account for their being there, or to derive SPEK from the Sanskrit SPAS. The Sceptic or Skeptikos is derived from SKEP, to examine, verify, and elucidate, and TEK (Eg.) that which was hidden and had escaped previous notice.

The name of the Bride is as old as the ceremony of capture in marriage. PRIOD in Welsh means appropriated, and owned. PRIOD takes the form of Bride in English; Braut in German. The BUARTH (Kym) is a cattle fold ; BUARTHO is to fold cattle ; they are thus, as it were, BYRE-D. The Bride is also BYRED, folded, owned, captured with the noose or tie of marriage round her. Here alone is the origin, PRI (Eg.) is a girdle or tie, to slip, wrap round with the tie sign of binding, and the terminal t makes the past participle in Prit, Priod, or Bride.

The designation of "HUSSY" is assumed to be an abbreviation of housewife. Yet in some counties it means simply a girl, and is never applied to the housewife or a married woman, but to the girl in anti-thesis to the house-wife. HUS is the HES (Eg.), House or feminine abode personified in Isis (Hes). SI (Eg.) is a child, either male or female. Hes being feminine identifies the child, and HESSI (Hussy), is a girl. SI is an abraded Sif (a Child), so HUSSY has the form of HUSSIF, which is not the House-wife, but the house-child (Sif) the Girl, as we find it. Wife is derived from Khef (Eg.) the Cornish KUF, a Wife. House-Kuf is House-Wife, and House-Sif, the House-Child.

What is the meaning of the title in STEP-Mother? No satisfactory explanation has ever been offered. She certainly steps in and stops a gap. But the step-son or step-daughter do not. In each case, however, there is an adoption. And the Egyptian STEP (or setp) signifies to choose, select, try, be chosen, adopted, be active, attentive, assiduous. The despised title of STEP was royal and divine in Egypt ; the Pharaoh was crowned as Step-en-Ra, the chosen, adopted, approved of the God Ra, and he was the STEP-God. The STEP-Mother is the adopted one, and so in each relationship the STEP means adopted. The Egyptian Genesis of the word is SET with the p suffixed. Set means to transmit, to extend, and the SETP is the transmitter. The STEP or adopted relation was to ensure transmission. This was the Setp of Egypt, the adopted for transmission and continuity. We have an application of the word in the " STAB," a hole adopted or selected by the rabbit for securing the transmission of her litter. There is also an occult significance in the STEP-Mother ; she is one of the feminine Triad, and one of the two divine sisters, the wet-nurse. STEF (Eg.) means to menstruate ; SUTB to nurse and feed with the sign of divine.[1] The mystical origin must have been known when

---

[1] Lep. Denk. iv. 63, c.

the flower of the violet was first called the STEP-Mother, as it still is in England. The primitive STEP-Mother nursed the child before birth. And to this occult origin may be attributed a considerable share of the odium attached to the name of the STEP-Mother, who has suffered for her symbolical character.

The word "WIDOW" is one that has caused much speculation as to its origin, but all the light which is thrown on the early family life of the Aryans by deriving Widow from the Sanskrit VI-DHAVA, man-less, or without man, which would have applied equally to all un-married women, whilst the Widower would likewise have to be a form of the man-less, vanishes in presence of the Egyptian "UTA," to be solitary, separated, divorced, as a woman. This reaches from the centre to the circumference of the meaning. In this sense the Widow is far older than Marriage, and a first form of her is UATI, mother of source, the wet-nurse. The second Widow was the woman put out and set apart, divorced from the herd or camp for seven days. The third Widow is a woman who has lost her husband. Names like these originate in primaries, not in the tertiary stage of application.

UTA, to be separate, divorced, set apart, is synonymous with WITE to go out, and with the word out. The WITE-law is the outlaw.

UÂ (Eg.) signifies the one, alone, solitary, isolated. The t is the feminine terminal; also ta is typical. UTA is the heron or crane, as the widow, the solitary, isolated one, distinct from the gregarious birds. The goddess UAT or UATI is the divine widow, in the form of the genitrix, who was the one alone in the beginning, the one who brings forth the gods; she who was mateless, the Virgin Mother of Mythology.

UTA (Eg.) has an earlier form in FUTA, to be separate, divided, set apart for certain reasons, as the word shows. FUT and AFT are variants, and AFT is the mother of flesh and blood, the Widow of Mythology, whom we shall find at the head of all the divine dynasties, as sole genitrix of gods and men. Fut is found in the Irish FEDB, Bavarian Fud, for the widow; Gothic Viduvo; Latin Vidua. In the Welsh Gweddw, the single, solitary, widow, or widower we have the gutteral prefix to the Uta (Eg.). Thus the widow is the Uta, Fut, and Khut, each of which has its still earlier point of departure in the name Kheft (Eg.), the ancient mother which deposits the Gweddw on one line, and the Fedb on the other, and shows how the Egyptian precedes both. This old genitrix, the Typhonian Great Mother, as Kef, survives in our English Wife.

In Gipsy language, the female, as wife, girl, daughter, is named CHAVI. KUF in Cornish English is the wife. The letter W comes to represent the K G or Q in many ancient words, hence KEF (older Kuf) is the Wife. KEF is the wife who was the earlier widow, before her son had become her consort. One of her titles is the Great

Mother of him who is married to his mother; the great one who bore the gods; this was when she was separate, or a widow, *the one alone*.

The mother of Romulus was the Virgin Mother, Rhea Silvia the Vestal; but he was also said to have been reared by Acca, who was designated the Harlot of Laurentum. The Virgin and the Harlot are two names of the same character in Mythology; the mother of the gods, who bore without the male, and was the prototype of the Widow.

All that belonged to the first formation of thought was afterwards decried, denounced, and derided. Its types were condemned to serve as images of evil, evil being chiefly discovered in the superseded conditions, out of which the advance had been made. The Ass was one of these living types that have suffered ever since. Woman has been degraded in various characters for her early supremacy in typology, one of these being that of the stepmother, another that of the widow. Her type in mythology is pre-monogamous. Her other name, as in the Book of Revelation, is the Great Harlot, because she had been the Great Mother, who produced without the proper, that is, later fatherhood. In her sacred aspect she was the Virgin Mother, in her degraded one the Whore. The synonym of KHARAT (Eg.), the widow, in the Gaelic CALAT, means the prostitute. For this the widow suffers, and the opprobrium descends to her children. The Russian proverb, " Do not marry a widow's daughter," the meaning of which the Russians do not profess to understand, remains as a relic of this bad character inherited by the widow from the most ancient type of the genitrix. In this way we shall gradually learn that mythology is a mirror which still reflects the primitive sociology.

All profane words were considered sacred at first. Things now held to be vulgar and unclean were the divine verities of an earlier time. They were the gods and goddesses of mythology, and the mysteries, who are now but the cast-off rags and refuse, the dross of refined humanity. Who that hears the profane term of " BEAK " applied to a dignitary of the Bench by any vagrant offender would imagine that the title is a divine sign of rule? Yet "BAK," in Egyptian, denoted a god. The BAK is the Divine Hawk of Har (cf. beak of a bird), and sign of the sun-god; with the whip attached it was a symbol of the highest authority. BECC, a constable, is the earliest form in English.

The oldest known Egyptian statue, one that was found by Mariette in the newly discovered temple of the Sphinx,[1] wears a wig which may have been the type of one worn by a PUISNE judge. The wearer sits for the portrait of one. Is the PUISNE judge named in Egyptian?

The PUISNE judges are the four inferior judges of the Court of Queen's Bench, who are compelled to go on circuit. SHEN or SHENI

---

[1] Portrait, *Journal of the Anthropological Society*, June 1874.

(Eg.) is the circuit. PUI is the article, the; PUI-SHENI is the circuit. SHENI likewise means the common crowd, the multitude, to avert, turn away, abuse. PUI further signifies to be; PUI, to go. If the Egyptians had PUISNE judges who went on circuit to redress wrongs, PUI-SHENI (later SENI) would express the character of our Puisne judges, when itinerating in a circle, or on circuit. The dropping of the S in pronunciation is no proof that the title is from the French *puis né*, subsequently used.

We have a PUISNE Court under the name of SENE, an ecclesiastical, therefore most ancient, foundation, in which the abuses of the church Reeves were corrected. It was a court of appeal.

SENAGE is the names of fines and payments levied in the SENE Court. SENHAI (Eg.) signifies to bind, conscribe, review, levy. HAI-T is a court. SEN-HAIT would be a court of appeal with power to review, to loose or bind, an early form of the Senate. Sen being second, the SEN-HAIT is the second court of two. Sen (Eg.) means second, and Puisne is also a second brother, and a secondary form of a judgeship. So that we have Puisne (Pui-Sen) (Eg.) for the circuit, also Pui-Sen, the second; the Sen-hait is the second court of two, as it still is in the second house of legislature, our Commons.

HA (Eg.) is the chief, ruler, governor. HAT denotes various forms and symbols of the ruler, as the mace, the upper crown, or throne. The Egyptian HAT or HUT is the highest of the two crowns. The HAT sign of the ruler is extant along with the mace in the English House of Commons, and is in the last resort the same emblem of authority. The Speaker puts on his hat as the extremest sign of his ruling and governing power, in the lower of the two houses, answering to the neter-kar (or hell) of the two heavens; the double house of the sun. He is typified by his Hat, and is thereby a Hat in person. Great Hat is likewise a sacred title among the Jews. The Hat is put on to compel, and HAT (Eg.) means to terrify and compel. Hat signifies to be called or ordered in English, and in Egyptian it means order. The cry of "Order, order," is thus "Hat, hat," and in extreme instances the Hat is forthcoming. Hatt (Eg.) is a salute, and in saluting we take off or touch the hat.

The college "GYP," we are informed, derives his name from the Greek vulture because he preys like a vulture. The GYP waits on gentlemen as a porter. GYP, as in Gipsy, is KHEB in Egyptian, and Kheb is a name for one who is in a lowly position, the title of an inferior. KHEB has an earlier form in Nakhab, Akkadian NEKAB, or the Khab, as we say, the Gyp. This title is not identified in the hieroglyphics, but in the cuneiform NEGAB is the porter, as is the college gyp. There must be many Egyptian words connected with these old college foundations of Cam and Isis. The name of the CANTAB is said to be abbreviated from Cantabrigiæ. But the term CANTAB may also be derived from some form of a religious service.

as in the hieroglyphics KAN-TEB is a servitor or dependent. The word implies religious service. The determinative denotes a hall, a foundation. KAN is service; TEB, to pray, prayer. KANTEB may also mean a member of a family, house, hall, or temple.

The fagging in our public schools is an extant form of slavery or compulsory service explicated by the Egyptian word FEKH, capture, captives, to be a captive. The same word means reward, and this points to the service of the FAG being the captor's reward. The captive FEKH is now represented by the under-schoolboy, who drudges for the upper one's reward. Five heads of "FEKH" or captives were given to Aahmes as his share of the spoil.[1]

Nor do we derive the name of the PAGE as a servant from the Greek παιδίον.[2] The Page in East Anglia is a boy servant, especially an underling to a shepherd. This is the Egyptian BAK, a servant who is a labourer. The PAGE likewise is the bricklayer's boy, and the BAK also works in stone. There is no form extant in PAK, but there is in Fekh, the captive, the bound one, as the PAGE was bound, and word for word the FEKH and BAK are our FAG and PAGE. The glorified form of the page in livery is found in the BAK, a type of Horus, the prince, the BOY, the youthful Sun God, and finally BOY is the modified form of both page and bak. We might say at first sight that page or boy and the page of a book had no relationship, but the fact is there is nothing unrelated at root. The page of the book we may derive from Pakha (Eg.), to divide, a division, one of two, and this may also be the meaning of the name of the boy as one of the two sexes. Khe (Eg.) is the child, and P is the masculine article, the. P-KHE yields the male child—Bekh also denotes the male—and with the feminine terminal T we have the name of the goddess, the lion and cat, as Pekht, a form of the biune being.

One meaning of our word HIND is to be in an abject, evil, enslaved, or accursed condition. The earlier forms HINE, HEYNE, HEAN, French HAINE and HONI, relate to the condition of the person; this may be abject, poor, evil, or other shapes of humiliation. All the meanings of HINE are extant in Egyptian. HAN is evil, envy, malice, hate; AN, to be afflicted, sad, oppressed; HANNU, to rule and flog; HANRU, to stint and starve; UN is to be bad, to want, be defective. These sufficiently denote the evil condition which may vary indefinitely. The UNT (Eg.) is a person whose condition as the HIND is a washerwoman. The "UNNU" appear as persons of an unknown condition, but apparently dark.

UN, AN, HAN, and HANNU, then, are Egyptian for conditions of misery, poverty, want, wretchedness, and serfdom. Hence the word in English includes the evil condition of HAINE or HONI, and the condition of serfdom in the Hind. But the word HIND also signifies

---

[1] *Inscription of Aahmes*, lines 20, 21.
[2] Sayce, *Science of Language*, vol. i. 342.

periodicity. HAN, AN, or UN, mean a cycle of time. UN or UNT is the goddess of periodicity. The periodic type is also found in the HIND-calf or one-year old. Thus the labourer bound annually is the HIND.

The primal illustration of the bad and evil condition of the hind as revealed by the hieroglyphics is that of the feminine period named "UN" or "AN," the period, defect, deficient, open, bald, afflicted, murmuring, waiting; the period of purifying, from which comes the name of the washerwoman. This was a first form of periodic evil, and the lady who represents this is extant as the lady of HANE, rendered hate, the lady of darkness, the MATER DOLOROSA, the negative of two characters assigned to the Great Mother. An infertile feckless female is still a HEN-wife.

In the hieroglyphics "KAT" means to go round, circulate, travel round and be round, and to work. So the CADIES, a body of messengers and porters extant in the last century at Edinburgh, were men who went round in doing their work. The Scotch market CADIE and milk CADIE still go their rounds. To CADDLE or CUDDLE is to clasp round. A CADAR placed over a scythe in mowing surrounds the swathe of corn. A CADE is a cask, also round. To CADGE is to bind round; the CADGE is a circular piece of wood. A KID, faggot, is bound round. The CADGER plies his round. The KID or COD incloses round.

In Scotland both gipsies and tinkers are called CAIRDS. In the hieroglyphics KARRT is a furnace and an orbit, so that the CAIRD may be named as the Brazier from carrying his furnace, and the gipsy from his going and coming round, the one being the tinker by trade, the other the nomad. KARTI is the plural form of KART, and it means "HOLES"; these are stopped by the Caird as tinker. TEN in Egyptian is to fill up and stop; KAR is a circle or hole, so that the tinker is the hole-stopper. The commoner form of the tinker in Scotch is TINKLER. This is likewise found in the name of the furnace or brazier as "KRER." "Tin-KRER" with the jet of flame determinative obviously denotes the mender by means of fire, our Tinkler.

Kar (Eg.) is a course, the sun's course or daily round. This is our word CHAR applied to the CHAR-woman who works by the day, the course or char. Her orbit, like the sun's, is completed in a day. CHARRED is completed, as in the saying "that Char is charred." Daily CHARES or CHORES are duties done as the day comes round. The same root gives us our quarters of wheat; there is no statute measure in which four of these make a whole; five quarters are one load. The QUAR-ter (ter Eg., all) is the total that is KARRED as in the Quart.

The Scoundrel may be derived from the Gaelic Sgonn, the vile, bad, worthless, and Sgonn from Skhennu (Eg.), a plague, a torment, to treat with violence. This word has Typhon, the devil, for determinative.

But the droil or drel is the knave, the worthless one, and equivalent to Sgonn, as if it only reduplicated the Sgonn in Scoundrel, which suggests another derivation from Skhennu, to treat with violence, torment, and torture, apparently connected with the cucking-stool. This was used at one time as a sort of choking stool called the Goging Stool, the Gog being a bog or quagmire. Criminals were choked in quagmires. In Germany, cowards, sluggards, prostitutes, and droils in general, were suffocated or nearly so in bogs, and the cucking stool is a remnant of this kind of punishment employed in Britain. In the PROMPTORUM PARVULORUM,[1] ESGN or Cukkyn is rendered by STERCORISO. ESGN is a form of Skhen, to treat with violence ; hence it seems probable that the Scoundroil was the droil, the knave, and rascal, who was placed on the goging-stool to be choked, whence the term would be applied to one who deserved that treatment.

The Fiend, we are told by the Indo-Germanic philologers, is a participle from a root, FIAN, to hate ; in Gothic FIJAN ; and it comes from the Sanskrit root PIY to hate, to destroy. That is, F is derived from P, and FIJAN from FI. Nothing of the sort. The sound of the F or FU was possibly uttered thousands of years before the human lips were sufficiently close together to pronounce the P.

The true root of the word meaning to destroy is FEKH (Eg.), to capture, ravish, burst open, denude, destroy, whence the Sanscrit Piy. But the Fiend is not derived from Fekh. The Fiend existed in Egypt as FENT, a worm, as BENT, the ape, as the Pennut or "abominable Rat of the Sun," and as Typhon, the Devil. Typhon is equivalent to FEN-T with the article reversed. FENT is the nose, and the Fiend is proverbially of a bad smell. FENNU (Eg.) is dirt, and this is connected with the bad smell and the FENT, as nose-symbol. FAINICH in Gaelic is to smell; PENCHIUMAN, Malayan, is sense of smell. FENKA (Eg.) is to evacuate. This is the English FUNK, to cause a bad smell, a stinking vapour ; Irish FANC for dung ; Maori PIHONGA, putrid, stinking ; Sanskrit PANKA, mud, impurity, slough, with which we may parallel VANCH (Sans.), to move slyly, secretly, stealthily, and go crookedly, to deceive, delude, cheat.

It is the Chinese FUNG-YUE, considered by Morrison to be too indelicate to translate except by calling it Breath and Moon. PENKA (Eg.), to bleed, disjoin, make separate ; FENKA, evacuate, show the menstrual nature of the Fiend. The FIEND and FONT are here identical. Thus FEN denotes dirt, filth, and the FIEND is a personification. PEN (Eg.) is to reverse. FANE (Eng.) are foes and enemies, the Fiend is the adversary. BAN (Eg.) is evil, and the Fiend the evil one. Lastly, we have the Egyptian FENTI in our "Old Bendy," an English name of the Devil. Fenti is a god of the nose ; but the real fiend was female, T being the feminine terminal of Fen or Ban, the evil. Nothing can be more misleading than words when divorced

[1] MSS. Harl. 221, British Museum.

from things, and the nature of the eschatological or modernised FIEND will not determine the origin of the name.

The Greek form of Typhon has never been found on the monuments. But we can see why. The FENT of the calf's head was worn down to FET. Fent, the nose, was also worn down to Fet. That is, an ideographic Fen became a phonetic F. Thus Tef would read Tefen or Typhon, and Fen-t as the calf's head is T-fen. With the snake F, read Fen, Fet, the worm, is the Fent as in Coptic, and Fet, to menstruate, is Fent. The calf's head, Fent or Fet, is the sign of periodicity, the first of the two feminine periods or truths. There is a goddess Ahti, with a calf's head and the body of a hippopotamus. That is a form of Fenti or Typhon. Ahti means the double house, as the place of birth.

Typhon or Fent became our Fiend, partly in relation to a certain physiological fact, whence the ill-odour of the Fiend and the red complexion of Typhon. Hence the Irish fin or fion means *that* colour, as red ; and fana in Arabic is the name of a doctrine of annihilation identical with the dissolution into primordial matter (blood) that takes place in the Egyptian pool of Pant (Fent), where Typhon was at last located as the Devil or Fiend in the Fens of the Ritual.

The name of Old NICK has never been satisfactorily accounted for. It is said to have been borrowed from the Danes, who had an evil genius in the shape of a sea-monster. The Swedish NEKAN is an evil spirit of the waters that plays deluding strains of music. No matter how it got into Europe, the NICK or NECKEN is Egyptian. NEKA signifies to delude, provoke, be false, criminal, evil, and NAKEN has the same meaning. The Neka personified is the monster of the deep, the dragon of the waters, the Apophis serpent, the eternal enemy of the sun and capturer of souls. Typhon (a form of the Apophis) was red. He dwelt in the mythical Red Sea or Pant. His companions are described as being red in the face. The Osirian asserts (Ch. 42) that the redness of their faces is unknown to him ; and Wormius says the redness in the faces of drowned persons was ascribed to the Neka (or Old Nick) having sucked the blood out of their nostrils.[1] A modified form of the false deluder of the waters exists as the English NICKER, a syren. The Yula (African) NEKIRU is the devil. The same meaning as NEKA (Eg.), to delude, play false, provoke, deride, be impious, is found in the Cornish NICKA-NAN NIGHT, the night before Shrove Tuesday, when boys were permitted to play all sorts of impish tricks upon the unwary. Nun or Nunu (Eg.) is the little boy, the ninny, whose night was that of Nicka-nan.

The name of MAN is said to be derived from a root meaning to think, so that man was originally distinguished from the animals as the thinker. If so, the child is indeed the father to the man in naming, for thinking, in the modern sense, is altogether a late application of the word man or men which means, in Egyptian, to fix,

[1] *Mon. Dan.* p. 17.

memorise, or memorialise, *i.e.* to MIND, in the Scottish sense of remembering. Man was first named from his physical attributes rather than his mind. The hieroglyphic type of the man *par excellence* is the bull, the potent male. This is the MAN or MEN, for the earlier dative MEN, now confined to the plural, had the Egyptian form. MEN means the fecundator, the male, the bull. It is the title of AMEN-generator or Khem. The name may be derived from Ma, the true, the firm-standing. One phonetic M is the male emblem; NU means the type or likeness. Also MA or MAI is the seminal substance, that which has and gives standing and stability. Men also means to erect a stone monument. And here we have the connection between the name of Man and that of Stone. The man and stone are often synonymous. The stone is Men (Eg.), a true type of stability. The stone is the ideograph of erecting, and the man is named from his virile power. Men permutes with Khem as a male prefix, the bolt sign being read both ways. Khem, the erector, the bull, the physical male, signifies the potent, powerful prevailer, as the man. Khem, rectum (Chem in Chinese) and erector, is the MA-NU, or true type. Now as Khem is Men, it seems likely that the Latin homo is a modified form of KHEMU, the master. Khemu, to prevail, be master, have potency and authority, answers to the Latin HOMO, to be stout and brave. The worn-down HEMU (Eg.) means the woman, the typical seat, abode, place, which has, however, an earlier form in KHEMU, the shrine, the habitation of the child, the womb. The name of the Amenti (Eg.), the earlier Menti, is based on Men, the Man as the physical founder ; and the Menti, the region of the dead, is literally the " re-foundry," where the pictures show the regenerator in the likeness of the generator, Men-Amen.

To be Khemt (Eg.) is to become the HOMME FAIT, and that identifies the HOMME with Khem. In the monuments, Khem, Men-Amen, and Mentu are three deified forms of " Man," as the generator and their names are found in the following specimen lists of languages, for that of Man.

KHEM.	MEN and MENTU.
KOM, Vogul.	MAN, English.
COMAI, Oregones.	MANU, Banga, S.
NGOME, Mare.	MANA, Kirata.
KAMI, Burmese.	MANUS, Kambojia.
GOM, English.	MANAS, Darahi.
GUMA, Gothic.	MUNS, Bhatúi.
CHAMAI, Koreng.	UMAN, Kasia.
KAIMEER, Erroob.	MANHAI, Tharu.
HOM, French Romance.	AMUNU, Mangarei.
HOMO, Latin.	MANUSHA, Sanskrit.
AMHA, Irish.	MINYAN, Namsang.
AMME, Sibsagar-Miri.	MANUT, Pali.
AMI, Khari-Naga.	MUNTU, Wakamba.

It appears to me that the name of COUSIN may have been derived by a shorter way than from consororbrinus to consobrinus, and thence to cossobrinus, cos'rinus, cosinus, and finally to cousin. In Egyptian KHU is a title, apparently of relationship; SEN is blood; Khu-sen would be a blood title, a title of blood-relationship; SEN, as second, would indicate the status of second in blood or the cousin. A certain "country cousin" will sustain this derivation. Khu found as a title has the meaning of birth, to be born, as in Khab. Khab-sen or Khu-sen, then, signifies to be sister-born, as Sen is the sister in an important mythological relationship. The Great Mother takes the form of the two divine sisters or the dual goddess as Isis and Nephthys, or Keridwen and Ogyrven, and the dual type is found in Urti, Pehti, Merti, and Senti, as the twin-sisters. We shall find in these two sister-goddesses the two women whom the Kamilaroi tribes claim to descend from. When this type was applied sociologically the children born of the Sen (sisters) were the first cousins. Sen (Eg.) is the name of the Two Truths as Blood and Breath, whence the two sisters personified from Pubescence and Gestation as the Senti.

Here language tells its tales. The Sen is the sister, but the earlier Khen is a concubine. Also the Khen denoted the uterus itself impersonated by the naked and outcast goddess Kên, Kefn, or Katesh. The English cousin has its earlier form in the Italian CUGINO, which goes back to the Khen and the concubine in the premonogamous stage, but still showing the uterine relationship. This gino or Khen is the Con in consobrinus.

The duality of Sen is preserved in the Latin Con, as shown in Consort, Congress, and other forms of being TOGETHER; and the Egyptian Khen proves the feminine type; also Khen, the Concubine, is equal to Sen, the Sisters, and KHENEM is to unite, join together. The Cousin in Spanish is SOBRINO, without the Con. The equivalent SIF (Eg.) is the Child, and REN (Eg.) means the name, or to name. Thus Sif-ren or Sobrino is the child-name. KHEN-SIF-REN is the uterine child-name, as it denotes the name given to the child on account of uterine and blood-relationship on the sister's side, which is perfectly recoverable in SEN-SIF-REN, the child named from the two sisters, or from the earlier Khen-sif-ren, from the two concubines Egyptian here shows the roots from which different words have been compounded to express the same meaning, and these have been in this, as in so many more instances, compounded independently of one another, instead of having been derived from each other, as philologists have hitherto assumed.

Parent is said to come from the Latin PARENS, a father or mother, breeder or nourisher. Both, doubtless, derive from one root, the Egyptian REN or RENN, which will tell us more about the meaning. RENN is the nurse, to nurse, and the nursling. If we take PA for the male, as begetter, then PA-RENN is the father. Ren, however, signifies to name, to call by name, and to rear. T in

Egyptian is the participial or feminine terminal, and REN-T, although not found on the monuments, except as the name of Rennut, the Gestator, is the named, or the namer, wherever found, and shows the formation of PARENT from PA-REN-T. PA, then, is the producer, REN means to name, nurse, rear up. The PARENT is not the mere begetter, but the namer and bringer up of the child. In Welsh ERN is a pledge, and the REN (name) was conferred as a pledge of fatherhood.

It is difficult to realise the ancient mystery of the name and naming. The natives of the Aru Islands only asked to have the *real name* of the traveller's country, and then they would have the means of talking about him when he was gone.[1] Such was once the paucity of speech and economy of words! This reminds us of the story told by Dr. Lieber, who was looking at a negro feeding some young birds by hand, and asked if they would eat worms. "Surely not," replied the negro; "they are too young; they would not know what to call them."

The name was a representative likeness. The name (SEM) in Hebrew is in Egyptian a representative sign. There is a point at which the child and the name are one, as they are in the Egyptian RENN, the child and the name. It is the same with the Word and the child. KHAR (Eg.) is both child and Word. Both are uttered; both are issue. The name with the Hebrews is one with the god, and the SEM or divine name of the Chaldeans was a person.

With the Egyptians the personal name was sometimes synonymous with the KA, a spiritual image or double of the self. The child of the parent was likewise an image of her or him. The name (REN) is identical with the son as the representative sign of the parent. Oaths were sworn by the name and by the son, and the name and son are equivalent as types of continuity.

The Hebrew Metatron is the Angel of the Name. He is called by that title, say the Rabbis, because he is a messenger. This is best explained by the Egyptian METAT, to unfold, hence to reveal; and REN, the name. Metat-Ren is the revealer or manifester of or as the Name.

"The Angel Metatron," says one Rabbi, "is the King of Angels."[2] As the angel he represents the NAME, "for my name (is) in him."[3] He is called Metatron, according to the Rabbis, because that name has two significations. These meet in Egyptian as the nursling and the name. "Blessed be the name, the honour of his kingdom for ever and ever."[4] "He shook the urn and brought up two lots—one was written for the NAME, and the other was written for Azazel, the goat that departs, the devil."[5] The mystery of the Metatron was caused by the sonship, which belonged at first to the motherhood.

---

[1] Wallace, *Malay Archipelago*, ch. 31.
[3] Ex. xxiii. 21.
[5] Mishna, treat. 5, ch. iv.

[2] Zohar, f. 137, c. 4.
[4] Mishna, treat. 5, chap. iv.

In the Egyptian Ritual "knowing the name" and uttering its "open sesame" are the means of passing through the door of the dead on the way to the land of life. "I will not let you go," says every part of the door and entrance, "unless you tell me my name;" and a knowledge of all the names is equivalent to salvation. The souls in the dark valley claim the aid of the god on account of their knowing the name whereby they invoke his help.

The same superstitious regard for the name survived as a Christian doctrine. The name is still as much the word of magic power as it was in the parental religion of Egypt, and precisely on the same ground. Here is the original doctrine. The seventy-five forms and manifestations of the Sun-god RA are synonymous with his NAMES. In certain glosses on the Ritual RA is described as "creating his NAME as lord of all the gods," or as producing his limbs, which become the gods who are in his company. The son Renn is the typical name, and the "Name as lord of all" is identical with the son. So the formula of invocation, "for thy name's sake," signifies for thy son's sake, the son and the name being equivalent and synonymous, as in the Egyptian RENN.

In John's Gospel the son and name are identical, and the Son of God is the name of God.

"Father, glorify Thy NAME. Then came a voice from Heaven (saying) I have both glorified (it) and will glorify (it) again," in the person of the Christ.[1] "O Father, glorify Thou me . . . . I have manifested Thy NAME."[2] So in Revelation the Son is one with the Name.

These things are alluded to in passing because it has to be shown that the primitive typology is yet extant in the eschatological stage, and we have to go back to the first phase of religious doctrines and superstitions before we can possibly understand them. But here was the later difficulty. The issue, whether as Word, Name, or Child, proceeded from the mouth. The mouth was feminine; it was the mother; it was RENNUT, the emaner of the RENN. The child was her representative sign. And when the male became the parent, human or divine, here was a mystery connected with naming, ready made. Hence the Jews would not mention the names of either Jehovah or Baal, for reasons, as will be shown, connected with the motherhood and the early sonship.

In the ancient British system ten or a dozen Britons had their wives in common, particularly among brothers and fathers and sons, but the children were held to belong to him who had taken the VIRGIN to wife. The Virgin is RENEN (Eg.), a form of Rennut, and through her and the first husband we find the parentage passing from the motherhood to the individualised fatherhood not yet sufficiently ascertained.[3]

---

[1] Chap. xii. 28.　　[2] Chap. xvii. 6.　　[3] *Cæsar*, book v. chap. 14.

In Sumatra the father is distinguished by the name of his first child, and is proud to sink his own personal name in that of his son. Even the women, *who do not change their first names*, are sometimes honoured by being called the Mother of the Eldest Child, as a matter of courtesy.[1] They retain the status of Rennut, the Virgin Mother, with their first names.

With the Kutshin Indians the father's name is formed by the addition of the word TEE to the end of his son's name ; thus if Que-ech-et has a son whom he names Sah-nen, the father calls himself Sah-nen-tee, and his previous name, as that of a son, is then forgotten.[2] The son was first, the fatherhood was secondary, because the first naming was from the mother, and here the sonship still dominates the naming. The Welsh Tad for the father's name implies this secondary or TWOED status of the male parent.

The Kaffir custom of HLONIPA is related to this mystery of naming and to the descent of the child from the mother. The Kaffir women and their children avoid mentioning their own father's name. They also refuse to pronounce or make use of words which have for their chief syllable any part of the name of the father, father-in-law, or para- mount chief, and HLONIPA is the designation given to the custom of avoiding the name of the male.[3] HLONI means maiden modesty, bashfulness, sense of shame. The present writer sees in HLONI the Egyptian RENN, the Virgin, which answers to the sense of modesty, bashfulness, and shame ; to the name of the Nursling and to REN, to call by name, or rather to RENNU, which contains the elements of REN, name, and NU, not, no, without. PA, Kaffir, means to give, and HLONIPA is not to give the name. It looks as if the male was still treated as the RENNU or yet unnamed child of the Virgin, the son of the mother who, in Mythology, became her consort.

The first parent then was female ; she is personified in the goddess Rennut, the primordial producer, as goddess of the eighth month, and of harvest. PA-RENNU reads the abode of the nursling, and Pa-RENNUT is the parent, as the abode of the child. She was the first namer (PA-RENT), and was portrayed as the Serpent-Woman, and the serpent that encircles round as the type of gestation. This serpent-ring was the first shape of the REN, a cartouche that inclosed the royal name, as Rannut had enfolded the Renn, her nursling. Then, with PA for a masculine prefix, we have the male REN-T (Rennut) as the namer of the child, and, with the s terminal, PARENS.

When we call to mind the solemnity of the ancient rite of naming amongst what are considered the uncivilized peoples, this original significance of the word Pa-ren-t is of great sociological interest. The Pater might beget, but the male parent assumed the position of the

---

[1] Marsden's *History of Sumatra*, p. 286.
[2] *Smithsonian Report*, p. 326, 1866.
[3] Davis, *Kaffir Dictionary*, Ulonipa.

family-man and husband. The child that was RENNED by him took
RAN-k, as in old English a knight was a " RENK." He was RAN-KED
by a name. To be renowned is to be recognised by name. A Runt,
Ronyon, randy, rannel, ranter, each imply conspicuous naming.

The word Name itself is probably derived from " Ken-am " as
Num is from Khnum. Ken (Eg.) signifies a title, and Am means
belonging to; the name being a title belonging to. Am (Eg.) also
denotes letters-patent. Thus the Ken-am would be not only the title
belonging to, but the patented title, the name by lawful right.

Still earlier ma or am is the Mother, and Ken-am or Ken-ma would
be the mother-title. In hieroglyphics the word may be Khnem or
Nem, according as the sign be read Khen or N, syllabic or phonetic.

The Khnem, like the Renn, is the nurse, the educator or bringer-up.
Khenems denotes a title, name, function, relationship, tutor. And as
Khenem also means to smell, to perceive, to select and choose, it is
apparent that the animal's knowing its young by smell was a recognised form of Khnem-ing or neming.

This word Khnem for a title, and parental function will answer
for the Greek ὄνομα, Sanskrit NÂMÂ, Gothic NAMO, Finnic NIMI,
Lap NAMM, Ostiak NIM, Votiak NAM, Permian NAMID, Vogul NEMA,
Samoyed NIM, Latin NOMEN and AGNOMEN, Persian NAM, Switz
NAM, Maltese NOM, Jukao NIM, Gaelic NIM, Malay NAMA, Avanish
NAMA, Guzerati NAMA, Malabar NAMAN, Birmese NAMADO, Tamul
NAMATTIN, Telugu NAMADHYAN, Wohaks NIMUD, Hoch Indian
NAMADHEIAN, and many more ; and as nam and nef interchange it
will include the Magyar NEV, old Norse NEFN, Welsh ENW, Ostiak
NIPTA, and Mahratta NAWE. Also as the syllabic ken deposits both
K and N as phonetics, KNAM will include the Kaffir GAMA for the
name. And with the modification of KA into SHA and SA, which
occurs in Egyptian, we have the SHEM or SEM as the Semitic
type-name.

KENAM or Nam, to repeat and renew, may indicate something
of the mystical identity between the name and the child. NIM in
Toda is the plural pronoun of the second person. NEMA in Sanskrit
means " the other." NAM is the personal pronoun in Akkadian ; it
also denotes state and status, and NAM-ad signifies paternity, AD
being the father, an equivalent of the English NAMED.

If " name " be identical with "Nam " (Eg.) it has a bearing on the
universal practice of not mentioning the dead by name. Nam (Eg.)
means to repeat, a second time, a second condition, seconding, and to
renew. This was the status of the child named by the parent. In
Shetland no dead person must be mentioned by name, because the
ghost was supposed to come in response ; and this seems to have
been the general reason for the custom. To repeat the name was
synonymous with a renewal of the person. The Dyaks will not
repeat the name of small-pox lest repeating and reproducing should

be synonymous. We have a survival in the popular "talk of the devil he's sure to appear" employed on meeting with any one who has just been named.

It is commonly assumed and asserted that the names of PATER, VATER, FATHER, are derived from a root PÂ, to protect, support, nourish. In Egyptian PÂ came to mean the male, the men. But the accent has to be accounted for. It represents a missing f or p. When restored we have the word PAF, which means breath, the English PUFF. PAIF (Eg.) is breath, wind, or gust, the earliest form of ghost or soul.

It is the universal testimony of language that Being and Breathing are synonymous ; to breathe is to be, and to be is to breathe. BHU (Sansk.), BA, Zulu ; BO, Vei ; BU, Zend ; BA, Egyptian ; BASU, Assyrian, signify to be. And to be, in Egyptian and Sanskrit to BA, is to be a soul, the earliest soul being the breath. Ba, be, bo, bhu, however, do not reach the root of being, they had a prior form ending in f, p, or b. In the hieroglyphics FU is ardour, dilating cause, dilatation, to become large, vast, expanded ; FÂ is to bear, to carry the corn measure, a symbol of pregnancy. FUF, PUF, and BAF are not developed forms of FU, PU, and BA ; on the contrary the latter are the reduced forms, and what have been taken for original roots are not primaries at all.

This root of Being and Breathing is found in Bhava (Sans.) for being, state of being, existence, origin, production, and it is related to breath, wind, winnowing, purifying and making bright (light, fire, soul and breath are synonymous)— in PAVANA ; in the Malayan PUPU for generation ; in the Swahili BEBA to carry the child. Also, the series of senses in which BHAVA is used relating to being are to be found in the Hebrew BAVA (בוא) to go, come, coming, arriving, appearing, coming to pass, fulfilling, to live. Here the fundamental sense is based on the swelling and puffing in pregnancy, hence BAVAM (בום) to be bellied ; BAVEL (בבל) to bub, to sprout and bring forth. PEVA (פוא) means to breathe and blow. Galla, BUBE, wind or breath. Maori PUPU, to rise, as breath or mist. This is the Egyptian PAF, breath, a gust of wind ; PAF or VAPour. PAF is determined by the sail puffed out with wind, and before sails were woven the sign of PAF was the pregnant woman. Hence the PEPlum, or sacred chemise of Athena, was suspended as a symbol on her vessel, in the manner of a sail ; hence also the word sail is the same as soul, and the sail is a symbol of the soul. The true root of PURUSHA, according to the primitive thought, is extant in Sanskrit as PERU, swelling, causing to ferment ; that mystery of life called SHETH and KEP by the Egyptians. The root of being then is puffing, and this is the original significance of PAPA (Eg.), PEVA, Hebrew, and Sanskrit BHAVA ; Gaelie FOF, to swell, and English FUF, to puff. Bab (Eg.), to exhale, describes the process as one in which the water passes into breath, whence came the doctrine of the Spirit brooding and breathing over the waters, the BAEV which in Sanchoniathon is one of the two principles of life; Môt being the other: the BEBA, Swahili, spirit; BWDACH, Welsh, a spirit; PUPA, Spanish, a spirit.

In the hieroglyphics the root BABA, PAPA, FAVA was generally worn down to BA, PA, and FA, with an accented vowel, and sometimes the accented vowel is all that is left. Thus PÂTI for two handfuls shows there was a hand called PAF and this is found as FA and A for the hand. PAPAKA (Maori), the crab, is word for word the same as ÂPSH (Eg.), the tortoise, the hard-shelled breather in the waters and preserves the earliest forms in PAP for ÂP and KA for SHA. This necessitates a good deal of restoration.

Fortunately the younger languages often preserve the older forms. Take the word people, or as we have it in English pepul. In Egyptian mankind, the human race or species are the PÂ, that is the PAF, PAP, BABES, or people. The PAF are the breathers and offspring of breath. To PEPE (Eg.), is to engender, and the word came to be applied to the male parent, the PAPA, but can be proved to have had no primary application to him.

The Bigenitrix was the Mother of Breath and of blood. As Mother of Breath (paf), she inspired her soul into the child and PUFFED and swelled in bearing it. This can be seen by the word AHTI, which is the name of the Genitrix, of the womb, and of a pair of bellows. The female was the first PAF-ER, or breather; PAF is breath, gust, ghost, road, way, we still say the family-way; PABA, in the Ritual,[1] is the soul, and PAPA or PAFA to breed and bring forth is the primordial word for being extant in the Sanskrit BHAVA, Hebrew BAVA, Mangaian PAPA, the Great Mother; Chinese FUPA (Genitrix), Ashanti BABESIA the woman as bearer; Swahili BEBA, Russian BABA, for the Grand-Mother; Japanese BABA San, "O BABA San," the old woman; Greek BAUBO; Meang BABI, the Mother; Chilian PAPA, Mother; English Gipsy, BEBU, the Aunt; Hindustani BUBU, the favourite concubine; the BAPHOMET of the Templars; Amoy PO, the old Mother; the Zula BEBA, the mystery called BOBO in Irish Keltic; the Welsh Pobo, producer of life; the Egyptian PABA for the soul, also the "baba" (Eg.), a collar with the nine "bubu," or beads denoting gestation. The dual character of the Genitrix personified as BHAVANI, or PAPA, FUPA or BABA is expressed in the hieroglyphics by PÂ-PÂ, i.e. PAF-PAF, and it means to produce or engender, and to bring forth; in other equivalents to PUFF and PUP, with the determinative of the female bringing forth. She was the dual PAP or PÂ, the primary PAPA.

The present writer holds that f is an earlier sound than p and b, and that the primitive pa-pa is extant in the Tobi VAIVI, for woman; the FAKAOFO FAFine, woman; and the Ticopia FEFinetapu, woman: This FAF is modified in the bushman T'AIFI, for woman. In

[1] Ch. 165.

Saparua woman is called PIPinawa ; BIBini in Sumenap and BABini in Ceram, all forms of the feminine PAPA the producer. AFFA in Danakil is the mouth and AFFAN in the Galla dialect. Avi in Tamil is spirit, literally as breath and then as life. Af (Eg.) means born of ; and in Af or Eve was impersonated the maternal breath of life. This is a deposit of Kaf on the one side or Faf on the other ; the meeting point of both. Faf as the Egyptian Fâ (faf) denotes the seed-bearer, the swelling gestator. And Faf-faf is the oldest form of paf-paf, pâ-pâ, papa, pa, and ba.[1]

The Scottish Gaels have the birthplace by name in the "Isle of PABaidh."

"There came a woman of peace (a fairy) the way of the house of a man in the island of Pabaidh, and she had the hunger of motherhood on her. He gave her food, and that went well with her. She staid that night. When she went away, she said to him, 'I am making a desire that none of the people of this island may go (die) in childbed after this.' None of these people, and none others that would make their dwelling in the island ever departed in childbed from that time."[2]

PAF-PAF had to be reduced to PAPA before the name could be applied to the father. When the male was discovered to be or was imaged as the breather of soul and his type was set up as the PAF-er or breather and author of being one-half of the name of the bigenitrix the PAF-PAF, PA-PA, is assigned to him as the PA. Thus the root of the name father, vater or pater is not PA but PAF and the PAFTER is the engenderer of breath (paf) as primary being. Baf (Eg.) to inspire, give breath, and Bat, for the father as the inspirer are later forms of paft. Ar is the child, and to image or make the likeness. Pafter modifies through pâter into pater. But the mother was the first breath or bread-maker. For bread is synonymous with breath. The cake and bread are written PÂT and PPAT, the latter being the same as our word puffed ; the cake or bread is dependent on breathing and puffing, and the name like that of the human species, PAF or BAB, is derived from the nature of breath and breathing.

The Greek FUO (φύω) to be, renders the Egyptian FUI or FÂ (faf) to bear or carry, with the ideograph of the bearing mother, who FU'S, or fufs, that is, dilates with new life, just like bread. In the texts[3] the Infinite God is characterized as FU-NUN-TERI, or dilation without limit, as an image of the Divine Being, and in others the divine substance is described as bread, in one[4] the circle of the gods are a

---

[1] FÂ-FÂ. Certain words, such as "KHA-KHA," "NU-NU," "RER-RER," "NKA," and "FÂ-FÂ," have been thus dwelt upon for a purpose beyond the present, as in them, the present writer considers, we are approaching the origines of speech to be illustrated hereafter in the "TYPOLOGY OF SOUNDS."

[2] Campbell's *Popular Tales of the West Highlands*, vol. ii. p. 25.

[3] Harris, *Magic Papyrus*, 3.　　　　[4] Stele naoph of Turin.

vast loaf of bread, a sort of infinite puff, or Bap, as the food of souls. The Kosa and Zulu Kaffir PEFUMLO is the breath and the soul of man, and PAPA, the English puff, is a kind of fungus, a type of lightness like the puff tart, or cake.

The symbolical bread-maker was the inspirer of soul ; the Bâ (Eg.) (earlier Baf) is both bread and soul, and PAF is breath. Thus bread was named from breath or soul, and PAF as the PUFF, supplied the name of light pastry. BÂ or BAF found in P-PAU (Eg.) for bread and a cake, is the Scottish BAP. FÂ (faf) with the genitrix carrying the corn-measure is yet earlier. Earliest of all was the great mother fuffing, puffing, pup-ing as the producer of being, the breather of the waters imaged by the fish, the frog, duck, or pregnant hippopotamus long long ages before bread was baked. We shall find that she was the first baker who produced thirteen to the dozen, and the only baker who ever had the owl (hieroglyphic Ma) for a daughter.

The Hebrew Bra to create like the word CReate implies the circle as the type of all beginning. Pra and per (Eg.) mean to wrap round, surround, go round, and be round. BER the eye is an image of the circle and circle-making. But before the circle was drawn by the human hand, came the watching of the figure made by nature. First there was the RU, the female emblem, the uterus. Next the swelling in pregnancy, the bellying and rounding forth in creation.

This being the result of paf and puffing it is natural to find that the word Bra, Pra, or Per has an earlier form in BVR (Heb. בר), the same as Bar for the hollow, pit, mystically the womb, the bubbing, puffing, or breathing place, hence BVL (בול), to bubble, to swell, to bring forth, that is, when modified, to bell, belle, belly. Bealing like bearing is to be big with child. This was the earliest creation. In a later stage the male was considered the Breather. Bat (Eg.), to inspire breath thus denotes the Begetter. In the Mexican picture writing, the breath passing from the mouth, is typified by the male image of the breather, the Bahu (Eg.), as symbol of the inspirer of breath. This sense is extant in English, " Thinkst thou to BREATH me upon trust," says a female character in Heywood's *Royal King* (1637), breathing and begetting being synonymous. When the male was the Breather the female became the Breathed. The name of the Genitrix who carries the seed-basket as Neft, means the Breathed, the Seeded.

PAUTI (Eg.) is the name for the divinity who recovers the lost character in the primary spelling PAFTI. PAF is breath and TI reduplicates ; PAFTI is the reduplicator of breath. This proves the appropriateness of the first Pafti being figured as the female, the inspirer of breath into the embryo. She alone is represented by PAF-PAF (pâ-pâ) the one who puffs and pups, or who produces and brings forth. When the male was recognized as the spirit-giver he is the

BÂ-T, that is BAFT, the Breather or soul-inspirer, and the member, the bahu is BAFHU, the breather, literally, the puffer of spirit or aliment of life.

From this it follows that PUTHU or Ptah is primarily Pafthu who, as father, is personified in the breather or inspirer of spirit, and named as the masculine Aft, of the four corners, who took the place of Aft, the Genitrix, at the head of Egypt's deities. The name is extant in Hebrew as Pevth (פוח), to be puffed, extended, said of the female, and of that which is made manifest and revealed by swelling. Puth (Eg.), *i.e.* PUFTHU is to open the mouth wide, to pant as the lioness. PEVEL (פול) is to be swelling, big, bellied, gestating. And PUT (Eg.) to stretch the bow out is equivalent to PUFT. The PAUT or PUT, a company of nine gods, also the name of No. 9, the nine months of gestation, is PAFT or PUFT, and this in accordance with the Gnosis expresses the full period of gestation or the total number of the nine gods as the extent. PUT, *i.e.* puft, is to figure and form, create the type; this too is based on swelling and puffing as did the Genitrix, in shaping the child, the primordial image of breathing being.

The male breather of soul shares the type-name of Papa with the female producer as in the Songo, Bola, Pepel, and Sar'ar name of PAPA; Soso Fafe; Tene Fafa; Baga Bapa; Ife, Ota, Kareharc, Ngodsin, Kamuka, Kiriman, Nalu, Kano, Turkish, Sonthal, and Carib BABBA. In a later stage the produce was called the Pup, as the young of the dog; Pube, Cimbrisch, for the boy; German, Bube; Swiss, Bub, and Swiss Romance, Boubo, the boy; Latin, Pupa, a girl, Pupus, a boy, and the North American Indian PAPPOOS.

Because the Sanskrit JIVE means to LIVE, be alive, revive, it is forthwith classed with various other words that signify life, living, to be alive. Such include the Latin, VIVO; Greek, Bίος; Old German, QVEH; Gothic, QVIVS; Sclavonic, SCHIVA; Lithuanic, GYWAS; and English, QUICK. Here the Egyptian shows us two distinct ideas of life at the root of these words. The Latin VIVO is not of the same parentage as JIVE or QVEH, ôr the Russian Givoy. The Sanskrit J represents an earlier K just as JINA, the "overcomer of all things" a title of Buddha; JINA, the victorious, triumphant, represents the Egyptian KAN or KANU, the brave, able, victor. The root on this line is found in KEF (Eg.). KEF or KEP was the most ancient mother of life, the Egyptian Eve. Now KEF means the mystery of life related to the mystical water; it is called the mystery of the Nile, of fertilization, and fermentation; one of Two Truths in the primitive Physiology, that of the Water or rather the Blood which formed the flesh (Af). This was one aspect of the mystery, one phase of the life.

KEF passed into Hebrew as GHIV, later GHI, the life, living; Russian, Givoy, living. The nature of this origin will explain why life and the beast are synonymous in the word GHIV or GHIVA.

The life in this sense was the blood, the life of the flesh. KEFA was the mother of flesh and the beast personified. From KEF come the O. G. QVEH and Lithuanic GYWAS. But the Slavonic SCHIVA corresponds to SKHEP (Eg.), to make live, and SHEB, flesh, bodily shape. This relates to the first of the Two Truths.

VIVO is a form of PÅ-PÅ or PAF, the wind, gust, ghost or spirit; the life of breath and primordial form of soul. This second of the Two Truths of life is shown by FAF (fâ) to bear, with the corn-measure on the head of the Genitrix, signifying gestation. It is the Irish-Keltic BOBO, the mystery; the Welsh PABO, the producer of life; Galla B'UBE, wind, breath, or BUFA, to puff; Egyptian PABA, the soul; English PUF, to blow; Tamul AFA, breath, as vapour, and later spirit of life; Polynesian PU; Greek FUO, to be; Vei FE, to blow, breathe, kindle the fire; ba (Zulu), to be; Bû, Zend, to be; Egyptian Ba to be a soul. With the article T prefixed to pu, to be, tepu (Eg.) means the first and also to breathe or blow. Buffaloes are called TEPU (Eg.) from their blowing. This suggests the origin of the word BUF-falo from the same meaning. In Maori the bud, shoot, growth or blowing of flowers is Tupu, and Tupu in Mangaian signifies from the very beginning, when things first began to TUPU or blow.

Languages without the verb to be in one sense have it in the more concrete form of to live. The negro says the thing no "lib" there, for "it is not there." To live in the Latin Vivo is to BE in the sense of to puff or breathe. The word LIFE is synonymous with REP (Eg.), to grow, bud, blossom, and bear like the tree, or as Repit the Genitrix, the LLAF-dig. To be QUICK is to be living, and this means also to be pregnant, and is traceable to the mother's quickening. To quicken, in English, is to ferment with yeast. The word QUICK is KHI-KHI in the hieroglyphics, to beat with a whip or to fan, to make go; and this whip, early fan, is the symbol of spirit and breathing, whilst its fellow of two, the AUT (Crook) signifies matter. These are types of the Two Truths carried in the hands of the gods.

Our word Mother is not derived from the Sanskrit MA, to fashion, but from the Egyptian name of the mother as MUT. Mut means the mother, the Emaner, the mouth (she was the mystical mouth of the breath of being); MUT the chamber, place, abode, the womb; MUHT, the fulfiller from Mut to fill full, be full, complete, No. 9. The form MUT also means to fix and establish. AR (Eg.) is the child, or the likeness, the type of a fulfilled period, the thing made. Thus MUT-AR is the place, the gestator, founder, and emaner of the child.

The name MAMA is also Egyptian. The word signifies to bear and has the determinative of the female carrying the Modius or corn-measure on her head, the hieroglyphic of gestation.[1] In various other African dialects MAMA is the mother. This name has been

---

[1] Lep. Denk, iv. 63, C.

supposed to be the spontaneous and universal utterance of infancy. Ideographically it is written with the type of the mother bearing seed; phonetically, with a double boat-stand. The Mamuti or Mam-Kuti is the cabin of the boat. This doubly identifies the Mama as the bearer.

Max Müller has suggested that the Sanskrit BHRÂTAR, Zend BRÂTAR, Greek φρατήρ, Gothic BROTHAR, Irish BRATHAIR, Slavonic BRAT, Cornish BRAUD, English BROTHER mean the one who carries or assists. The earliest namers and myth-makers, however, made the female the carrier or bearer. If our principle of naming the male be true, it follows that the word BROTHER is the same word as BREATHER. The Egyptian "PRUT" is our English breath, and seed.

PRUT signifies to void, emane, manifest, pour out, shed seed, with the male sign for determinative. This identifies the PRUT-AR as the male FRUIT-er, the BREATHER, seed-shedder, our Brother. The soul was the heat, the fire that vivifies as well as the breath. Thus the Breton BROUD, the Welsh BRUD and Irish Bruth take the form of heat in place of breath, both being synonymous. with spirit. The Irish "broth of a boy" is a man of spirit and eruptive vigour. The Brother includes both meanings, as the male child. We have to do with words in their primary sense. The relation of the brother is secondary. The meaning of sister is just as primitive, and as closely connected with the BREATH. These words come to us from a time when to breathe was to be, and breathing dominates the imagery used as symbols of the male and female.

The word SISTER is derived etymologically from the Egyptian "SIST." SIS is breath and to breathe. SES-MUT is the breathing Mother. The ideograph of breathing is the brood, i.e. breeding Mare. SEST is the same, the She and Her represented by a brood-mare going, galloping. SES-T means breathed; it is the participial form of SES, breath, and to be breathed is equivalent to breeding, and thus names the breeder. SEST likewise is the preparing house, the House of Breath, over which Nephthys, called the Saving SISTER presided.

THE phonetic S represents an ideographic SF, and by interchange of U and V the Egyptian Ses, to breathe, represents the "Svas" of the Sanskrit SVASAR for sister. ŚVI means to swell, increase; ŚVASA is breath, breathing, inspiring the breath, or soul of life. ŚVAS is the root of many words denoting breath and breathing, puffing, as in pregnancy. Sanskrit preserves the earlier form of Ses, as Sfs or Svas in Svasar. Sesar (Eg.) is the breather, and Ses-ar is to breathe the child, whence the sister is the one who inspires life, is the breather or breeder like the Sest-Mut. Only in Egyptian do Ses and Sest meet, and only Egyptian could furnish both Svasar and Sister as the two distinct names, derived on two different lines of development.

SEST (Eg.) is the Mare, as Mother, and AR (Eg.) is the likeness of, *ergo* SEST-AR is. the image of the mother, a repetition of the same sex, whence the sister, just as the brother is the likeness or type of the father, as male breather, the inspirer of life. The Egyptian Ar for the child, likeness, type, image, to make, serves in each case to illustrate the Paft-ar (Bât-ar, Pat-ar), Mut-ar, Broth-ar, and Sist-ar.

In Egyptian SF, later SU, is the Child; NU is a male type. SIFNU or SUNU means the Male Child. This was worn-down as a word to SUN, to be made, to become, SEN the second, the other, the *alter ego*, or second self. The form SFNU is implied in the Sanskrit SÛNU, the accent representing the missing consonant, and in the English SON the o represents the f of Sif (Eg.) the Child. In Egyptian, however, the two components of the word (also the word SUN) are anterior to the distinction of male and female.

The origin and meaning of the word Daughter have exercised me more than most words. The derivation from Duhi, Sanskrit, to milk, which makes the daughter the milkmaid, fails to fathom it.

The female as woman, wife, and mother is designated from the womb. WAIMO, in Finnic, is the woman, and in Lap and Scotch the WAME is the womb. In Egyptian the female is HEM or KHEM with the determinative of the womb. Mut, the Mother, is identical with Mut the mouth or uterus. Uterus and udder are equivalents and in Sanskrit VÂMA is the udder. The wife is also named from the same origin, as is shown by the Cornish KUF, COFF, and KEBER; English Gipsy, CHAVI; Chippewa, KIVA; Maori, KOPU; Egyptian, KHEP; Malagasy, KIBO; Old Bohemian, KEPP. The female is identified on this ground in Anglo-Saxon as the Wif-Child, that is, the Womb-Child. The Wife-Child is the " Kuf"-Child, that is, the Womb-Child in Cornish and other languages. The Daughter is certain to be simply distinguished from the male on the same principle. The name is found as the Sanskrit DUHITAR; Greek, θυγάτηρ; Zend, DUGHDHAR; Lithuanian, DUKTERE; Bohemian, DCERA; Lap, DAKTAR; German, TOCHTER: English, DAUGHTER; Gothic, DAUHTER. Following the clue of the Wife-Child leads us to the name of the oldest Mother, Khept, in whom the womb and goddess are one. Khept modifies into KAT, and this supplies the likelier root of the word Daughter. KAT (Eg.), is the womb; KOHT, in Esthonian; QATU, in Fijian; QUIDA, Old Norse; QUITI, Alemannic; QUITHEI, Gothic; UCHT, Gaelic; Cut, English; CWYTHER, Welsh. Ar (Eg.) is the Child, and Tu the feminine article. Thus TU-KAT-AR is the Womb-Child or child with the womb. KAT is also the Goddess, the Seat, the hindward one, the Cow, the Bearer, the female in various forms, therefore it is inferred that daughter is compounded from Tu-kat-ar as the female child. In English, daughter is also represented by DAFTER, and Af (Eg.) means born of; Aft is the feminine abode, the womb, the ancient Mother from the first. The Mother and Daughter are one in Mythology. The

Mother of the God is likewise called his Daughter.   Hathor is the Hat (earlier Kat, earliest Khept), the abode of Ar, the Child, and yet she is also the Daughter.   If we take the form of Kat, as in Kat-Mut, Hathor is Katar, and with the feminine prefix Tu, Tu-kat-ar, she is the θυγάτηρ, Dughdhar, Daughter, Dauhter, or Duhitar, the Divine Milkmaid, or rather nurse, as the Cow-Goddess.   This feminine type of KHEPT, KAT, or HAT, may be followed in language generally thus—

KAT	. Karagas . . .	. Woman.
KOTA	. Kwaliokwa . . .	. Woman.
KITHIA . . .	. Chetemacha . . .	. Woman.
KHATUN . . .	. Mongol . . .	. Woman.
KHOTON . . .	. Pelu . . .	. Woman.
KODAR . . .	. Wokan . . .	. Woman.
KITEIS . . .	. Malaguaya (Cf. Greek, CTEIS)	Woman.
WATA . . .	. Baba . . .	. Woman.
WATOA. . .	. Peba . . .	. Woman.
WAT-WAAT .	. Keh Doulan . . .	. Woman.
WADON . .	. Javanese . . .	. Woman.
OAT . . .	. Kwaliokwa . . .	. Woman.
OUTIE . . .	. Guachi . . .	. Woman.
ITTHI . . .	. Pali . . .	. Woman.
ETI . . .	. Sekumne . . .	. Daughter.
AIAT . . .	. Sahaptin and Kliketat .	. Woman.
MACATH . .	. Minetari . . .	. Daughter.
MEYAKATTE. .	. Crow Indian . .	. Woman.
TAKATA . .	. Annatom . . .	. Woman?
MU-HATA . .	. Kisama . . .	. Woman.
MU-HETU . .	. Songo and Lubalo . .	. Woman.
SAPAT . .	. Riccari . . .	. Woman.
TSAPAT . .	. Pawni . . .	. Woman.
SET . . .	. Amharic . . .	. Woman.

These are derived from one original Khept, as were the names of the hand, because the uterus and hand are permutable types and both represent the parent power as female.

SECTION VII.

TO "clear oneself by an oath" was a recognised form of speech with the Egyptians, and a mode of covenanting.[1] The word "ARK," for oath, means to bind. To clear oneself by an oath is a common form of speech with English boys; one of these being "By Goll." This is the holy Cornish oath. The hand is held aloft whilst the oath is taken. Goll means the hand, or rather the fist, for the hand is clenched in token of covenant. The equivalent Ker (Eg.) is the claw, to claw hold,[2] to embrace ; and in Suffolk Golls are large, clumsy, claw-like hands. The Irish swear this oath in the form of be-GORRE. The custom denotes the making of a covenant and swearing by the hand, in the primitive condition of claw, when laying hold was literally seizing with the claw. By hook or by crook was then by the fingers in the TALON stage. Ark (Eg.), to bind, is symboled by a tie, which is later than the claw. Goll is a worn-down form of Gafael (Welsh), the grasp, grip, hold and Gaffle a hook ; Gaelic Gabhail, for seizing and holding as a tenure, whence to GALE a mine is to hold it on lease, to rent it ; and one form of holding land is by the law of Gavelkind. The Languedoc Gafa, to seize or take, is the Egyptian Kefa, to seize with the hand, or claw, as the first instrument for laying hold. In a following stage the claw, ker, or goll, as hand, is the Tat, or Tut, and the word means to image, typify, unite, and establish. We still covenant in shaking hands or daddles, as the fists are called, and the vulgar "Give us your FIST" or DADDLE is equivalent to "Give us your TAT" (Eg.), that is, the hand, as a type of establishing a covenant.

[1] Brugsch, *Zeitschrift*, 1868, p. 73.

[2] This name of the claw is also that of the hand in various languages, especially the African and Indian, as the KERE in Mano ; KORA, Gio ; KOARO, Basa ; EKAROWO, Eafen ; KARA, Mobba ; KHUR, Dhimal ; KAR, Sokpa ; GALA, Mantshu ; KHAL, Tobolsk ; KALIOCK, Lopcha ; KARAM, Malabar ; KAR, Hindustani ; KHUR, Dami ; GHAR, Mongol ; CHEIR, Greek ; CIOR, Irish ; and KIERS, the finger, in Gura ; AKARU, the finger, Biafada ; AGRA, the finger, in Sanskrit. In the earlier form of Ker, the claw, we recover KHEPR, the beetle-type of laying hold.

"By Gigs," is a common oath or exclamation, as in *Gammer Gurton's Needle*, "Chad a foule turne now of late, chill tell it you, by GIGS." As an oath is a form of covenant, this is probably the Egyptian KEKS, which means a binding, to entreat, to bind, with the noose determinative of a bond or covenant.

There is an English custom still extant of touching the seal of a deed with the finger on signing the document. Sidney Smith said the ancient family of his name did not bear arms, but sealed with their thumbs. The Statutes of Akkad decreed that, if a Son had said to his father, "Thou art not my father," and had executed a form of deed and "MADE A MARK WITH HIS NAIL to confirm it," he was to pay a fine. The clay tablets or DUPPI, still show the print of a finger-nail on them in place of a seal. In the Egyptian word Teb we have the name of the seal-ring, the brick, and the finger, and Teb means to answer, be responsible for. Taf (Eg.) means attention. To tap is a sign of calling attention. The tabor is played by tapping. The Teb (Eg.) is a drum. To tap is the same as to dab or to dub. To dub a man a knight is done by giving him a tap. The ancient method was to tap him on the side of the head or give him a box on the ear. Thus a box is synonymous with a tap. So, in Egyptian, Teb is a box. The Teb are the temples of the head. To dub is to clothe, ornament, equip, as the knight is dubbed. Teb (Eg.) is to clothe and equip. The box on the ear, or blow given in dubbing, is explained by Teb (Eg.), to answer and be responsible for. Teb, to place instead, be the substitute, shows that the person dubbed was to be henceforth representative of the king or queen, and be responsible for any blow aimed at them. Teb (Eg.) is to seal, and by the process of dubbing the knight was sealed to the royal service. Tebn (Eg.) is to rise up, and the one who is dubbed is told to rise up a knight. The oneness of a box and a tap leads up to the meaning of the Christmas Box and Boxing Day. The Teb, as box, is a sarcophagus, the sign of an ending. Teb, or Tep, is a point of commencement of the Teb or movement in the circle of the year. Boxing Day is the first day of the solstitial new year. The meaning of Teb, the box, is to close, shut, seal. Teb is a recompense ; and on the first day of the new cycle gratuities were given for past service, and to secure and seal it for the future. The box comes in as the type of inclosing, closing, sealing, whether delivered on the ear, or in a gift of money.

The cow's tail is an emblem worn behind by the male divinities of Egypt. This was their QUEUE, in Egyptian Khef, for the hinder part. The tail was a type of the goddess of the hinder part, and when she was put back by the male, the tail or Queue was worn behind the male, who represented the front. Hereby hangs a tale. For the tail, at last, deposits the Q as the letter with the tail, which is still an image of Khep or Kefa, the Goddess of the Hind Quarter.

The tail was formerly extant, as in France, in the thongs of hide called QUEUES to which seals were attached in legal documents; also the end of a document where the seal was attached was called the Queue, and when the deed was witnessed and sworn to, the finger was laid on the Queue—a mode of sealing modified from that adopted in the initiatory rites of the "Sabbath," and the worship of the Goddess of the North, which is still retained, however, in "kissing the book." The Queue is also extant as the QOPO of the Zulu Kaffirs, a tailed girdle worn round the loins, after the manner of the gods of Egypt. When the Zulu messenger of a court of justice is sent on official business as an UM-SILA, he carries the white tail of an ox as his sign and Um-sila of authority. The Zulus still make a notch or tally, in scoring, called i QUOPO, the same in name as the Peruvian QUIPU or knot, a figure of ten in reckoning. The present cue, however, was only to point out the survival of a type of the old genitrix (called the "Living Word") in our tail-letter Q.

A game is played in Gloucestershire with a ball, which, the two parties ranged on opposite sides endeavour to strike to the two ends of the course to secure the goal. It is called NOT; supposed to be from the knot of wood of which the ball is made. But NUT (Eg.) means the limit, the goal, the end, all. This is the likelier derivation of the name of the game.

HANDY-DANDY is an ancient game played with the two hands. Cornelius Scriblerus says handy-dandy is mentioned by Aristotle, Plato, and Aristophanes. It depends on a thing being changed from hand to hand for the guess to be made as to which hand the thing is in. HANTI (Eg.) is the returner, from han, to turn back, return, pass to and fro; tenti is to reckon, how, where, which one; TENTI also means separate, in two.

When money is given by a newly wedded pair for the poor to drink their healths, it is called "HEN." This may be interpreted by HAN (Eg.), tribute, to bring; and HAN, young.

In shelling peas at a Peascod Wooing it was a great object to be the finder of a pod containing nine peas. That is a hieroglyphic of the Pet Circle of Nine Gods and the nine months of gestation. The kitchen-maid who finds this pod of nine will place it on the lintel of the kitchen-door, and the first man who enters is to be her husband. When a youth had been jilted, it was a Cumbrian fashion for the lasses to rub him down with peas-straw, the lads doing the same to a girl deserted by her sweetheart.[1]

In the Marriage Service of the English Church, printed in the York Manual, the bride pledges herself to be "BUXOM" to her husband. In the Sarum Manual she engages to be "bonere and buxom in bedde and at borde." These are explained in the margin

---

[1] Brand, " Peascod Wooing."

by "meek and obedient." Buxom and bowsome came to mean obedient and pliant, but that is not the primary sense. The earlier spelling is bucksome, to be blithe and conjugal, and this reaches the root. Bukh (Eg.) means to conceive, engender, *enfanter*, fecundate, be fecund; Sam means similitude, likeness of; bucksome is Bukhlike, and the promise is related to fertility and ensuring of offspring. Of this we can adduce a remarkable ideographic custom. One form of " PET " in Egyptian is a foot and to stretch. Pet and Pesh also mean the same thing, to stretch out. So in English the game of put-pin is likewise called push-pin. The root meaning of "Pet" is to stretch, to reach, to attain, no matter what the mode may be. Hence " Pet," the foot, is stretched forth in walking. " Pet," the bow, is stretched in shooting. Pet, No. 9, is the full stretch of the measure by months in gestation. The pot-belly is at full stretch of roundness. Anything putrid has gone to the utmost verge. The putting-stone used in curling is that which stretches the player's capacity to the utmost in putting. Hutchinson, in his *History of Durham*,[1] speaks of a stone cross near the ruins of a church in Holy Island, which was called the "PETTING Stone." When a marriage took place, it was customary after the ceremony for the bride to step upon this stone cross, and if she could stride to the end of it, the marriage was supposed to be fortunate. To be lucky and fortunate in marriage always meant child-bearing. And this petting or stretching stone was the means of divining or pre-figuring the future with regard to the woman's fertility and willingness to be fertile. The promise to be bucksome was represented by the act of stretching or petting. Every one of these primitive meanings has been turned to male account. This has been taken for a promise to bend and be yielding in the servile sense. It was not so. The woman simply pledged herself to motherhood as well as wifehood.

The sign of the Sut-Heb festivals, held every thirty years, is the double-seated boat. It was, as the name implies, the Heb of Sut, although given to Tum in a later system of myth. Heb means a festival. The double-seated boat is a form of the Neb (basket) sign, which indicates the whole, all, composed of two halves. Our symbolic custom of HOB-NOBBING repeats this festival of conjunction. Hob-nobbing is celebrating. Moreover, the ancient hob of the chimney corner actually reproduces the image of the Sut-Heb, the double-seated Heb of Sut. It was called the hob, and on either side of the fire there was placed a settle, the double seat or Set of the Heb, where people could hobnob together. Grose says, " Will you hob or nob with me ? " was a question formerly in fashion at polite tables, signifying will you take wine with me, and if the party challenged answered Nob, they were to choose whether white wine or red. These were the two colours of the two crowns, upper and

---

[1] Brand, on Wedding Customs.

lower, of the two halves of the circle which was symboled by the double-seated boat, and Neb, as before said, is a twin-total.

The Essex labourers divide a jug of beer into three "pulls" at it. These three draughts are called "NECKUM" (1), "SINKUM" (2), and "SWANKUM" (3). Khem (Eg.) means to have power over, potency, prevail over, possess by force. Nek (Eg.) denotes the first, as the I or "A one." Nek-khem is equivalent to "My first pull." Neked (Eng.) is a small quantity. Nakhn (Eg.) is a little one. Sen (Eg.) means second ; Sen-khem, the second strong pull. Skhen (Swan) is to imbibe, multiply, and render victorious, therefore the finishing draught.

A tea-drinking among Oxford students is called a "BITCH-party," but the designation does not imply any slur on the sex ; it is rather an apology for the beverage. It is at root a sign of modesty and bashfulness. It is true that "BITCH" is a term of reproach or worthlessness, and is generally limited to one sex. But a Bitch-party is simply a name for the POOR-THING-NESS, the humble status of the celebration. In like manner the German swipes are "BOSCH." A small beer, or mead, made in the North is BOTCHET. BOTCH, a failure or shortcoming, is a form of the word. BUDGE is a Suffolk term for dull and poor. BATCH-flour is coarse flour. BATCHworth is the lowlying place, or hinder part. To BUDGE is to give way, succumb, accept the inferior station. The BADGE is a token of this inferiority ; it was the sign of slavery, the mark of the serf, the brand of possession, as much so as the brand on the sheep. Hence the term, the badge of slavery. Its true survival is yet on the livery button. Later, it was elevated to a place of dignity in heraldry, and worn as a trophy by the conquering superior instead of by the conquered inferior. But even there it keeps its humbler station ; badges being a subsidiary kind of arms.[1] The Batcheler (A.-N.) was a young man who had not attained the honours of knighthood ; the Bachelor is one who has not reached the dignity of marriage or mastership. He too is a BITCH-party, and of lowly estate. A Bitch-party, then, is so called because it is a poor thing of humble status, where the drink is weak and the proceeding slow. The total meaning may be found in the Egyptian word "BETSH," weak, slow, lazy, lowly, and humble. The Betsh or Bitch party has a divine origin, and a primordial form of it is found in Egyptian mythology as the "Children of Inertness," who dwelt in Am-Smen, before the firmament was lifted, as the first eight gods, and did not keep correct solar time, as there are 366 days in the sidereal year ; whence the 366 bells attached to the robe of the Hebrew High-Priest. They consisted of the seven stars of Typhon (Ursa Major) and the dog-star Sut. These same Children of Inertness are described by Taliesin as the "sluggish animals of Sut" or Satan ; the primal Bitch-party being Typhonian, and belonging to the lower region of the hinder part in

---

[1] Montague, *Guide*, p. 48.

the north. By aid of the Swabian PETZ, we can recover the Bear, the type of the original Betsh or Bitch, the Lap Pittjo, the genitrix of the Betsh-party, or " Children of Inertness."

Lightfoot says that in the Scottish Highlands, when an infant is born, the nurse takes a green stick of ash and thrusts one end of it into the fire, and as it burns, she receives the sap oozing from the other in a spoon, and administers the liquid to the child as its first sustenance.[1]

One writer suggests that the reason for giving ash-sap to newborn children is, first, because it acts as a powerful astringent, and, secondly, because the ash was potent against witches.[2]

Another affirms that it was because some thousands of years ago the ancestors of Highland nurses knew the *Fraxinus ornus* in Arya, and had given its honey-like juice as divine food to their children.[3]

We need not go so far, however, to derive the sacred character and living virtue of our symbolic ash. The ash was the Egyptian tree of life, the Persea fig-tree named the ash. " Ash " signifies emanation, emission, the creative substance of life. The rowan tree is the typical ash, with its sap within and red berries without. Ruhan in Egyptian is a shrine, therefore a synonym of sacredness. The rowan is named the quicken tree, from quick, alive, pregnant. It is therefore our tree of life called the ash, precisely the same as the tree of life in Egypt.

Sap in Egyptian is to spit, evacuate, as does the ash wood in the fire; to make, create. Saba is food and aliment. The ash-sap was a form of the essence of life in Egypt as in England.

In Germany it was a custom to tap the ash tree in spring, and drink the sap as an antidote to the venom of snakes. The serpent having become a type of the inimical power, the ash was in every way fatal to it. The common belief was that snakes could not rest even in the shadow of an ash tree, and that a single blow struck with an ashen wand would prove fatal to the adder.

"BAAKABAKA" in Egyptian signifies upside down, topsy-turvy, with a man standing on his head. This sense of reversal is extant in the Hebrew "Bakbuk," a bottle or pouring out of a bottle. Buge and beck in.English mean to stoop; Baka (Eg.), is to squat down, also to pray.

Still more primitive is the survival in the boy's game of "Buck buck, how many horns do I hold up?" in which one boy stoops with face down in a reversed position to guess the numbers. He bucks and sets a back at the same time. This is probably the game which we see played in the monuments with one prostrate figure face downwards, the others holding up their hands as if pounding his back, which is permitted when the guess goes wrong, and may have been called

---

[1] *Sylvan Sketches*, London, 1825, p. 24.   [2] *Choice Notes*, p. 24.
[3] Kelly, *Curiosities of Folklore*, p. 145.

BAAKABAKA, now rendered by the English "Buck-buck." Puka (variant of huka) means magic, conjuring, divining, *i. e.* the object of the game. Another mode of this divining is played by children in Devonshire, and is called Buggy bane, or Bucka bene; the following rhymes are repeated by one of the players:

> "Buggy, buggy, bidde bane,
> Is the way now fair and clean?
> Is the goose y-gone to nest,
> Is the fox y-come to rest,
>   Shall I come away?"

Bug or Puck is a supposed hobgoblin. Bugan is a title of the devil. In the hieroglyphics "Pukha" is a name of the infernal locality.

In Jersey the cromlechs or tumuli are called Puck-lays, or places of Puck. This name identifies them with the Egyptian Pukha, of the underworld, the pit-hole or Sheol, as the place of the dead.

The name Pukha is probably from P-uk-ha, the dead-house. Uk or Akh (Eg.) is the manes, or the spirit, whence the Uk or Puck, bug or bogey. The rhyme, like the questioning of Buck-buck, denotes conjuring or divination; the game being played in the dark. The goose of Michaelmas and the fox (jackal, Apheru) of the vernal equinox, symbols of Seb and Anup, witness to the astronomical allegory. Between these two lies the locality of Pukha, the infernal region. Pukh or Pekht as cat-headed goddess was stationed there to look after the Apophic monster. The appeal is apparently made to the Puck, as spirit of the dark extant as Shakspeare's sprite of the night.

BIDDE formerly meant to require, and BANE to proclaim or make known publicly. "Buggy, buggy, bidde bane," is thus a demand made to the spirit or hobgoblin of the dark, the underworld, to answer the questions propounded, and the chant is a magical invocation.

The present writer, when a boy, was taught that a Typhonian monster lurked in the dark places of deep waters, called "Raw head and bloody bones." The name should assimilate the monster with the Apophis or red dragon in the Pool of Pant, the Red Sea. Raw-head is a name of the devil. Also, when bathing, it was considered the correct thing to urinate on the leg before wading in the water; a supposed antidote to cramp or the lyer-in-wait in the water. Now in the eighty-sixth chapter of the Ritual in which the deceased is crossing the valley of the shadow of death and the waters, and the pool of fire, we read: "What do I say I have seen? It is Horus steering the bark. It is Sut (Typhon), the Son of Nu, undoing all he has done." The allusion is to the evil enemy of the sun and of souls, who lurks under the waters. Then follows the statement, "I am washed on my leg. Oh, Great One! I have dissipated my sins; I have destroyed my failings, for I have got rid of the sins which detained me on earth." He has performed an act of lustration and purification by the washing on the leg before entering the water, and this is symbolical of getting rid of his failings or in the eschatological sense of dissipating his sins which detained him on earth. It does look as if the bather did the same thing, although unconscious of the symbolism.

In the parish of Altarnun, Cornwall, the people had a singular method of curing madness by placing the patient on the brink of a square pool filled with water from the Nun's Well; he was plunged suddenly and unexpectedly into the water, where he was roughly baptized, and repeatedly dipped until the strength of the frenzy had forsaken him; he was then carried to church, and masses were sung over him. The Cornish people called this immersion BOOSSENNING, from Beuzi, in the Cornish-British and Armoric, signifying to dip or drown.[1]

In Egyptian we find BES (mau), inundater;[2] Besa, an amulet, protection, and Besi to transfer. SAN means to heal, save, charm, immerse. BES-SAN therefore agrees with Boossenning as a process of immersion for healing, charming, protecting, and preserving.

The Egyptians were accustomed to enrich their tombs with valuable writings: Mariette, who recovered the Serapeum from its burial-place, an ocean of fluid sand, says, " On certain days of the year, or on the occasion of the death and funeral rites of an Apis, the people of Memphis came to pay a visit to the god in his burial-place. In memory of this act of piety they left a STELE, that is, a square-shaped stone, rounded at the top, which was let into one of the walls of the tomb, having been previously inscribed with an homage to the god in the name of the visitor and his family." [3] These documents were found to the number of 500. The custom was not confined to the tomb of Apis.

With this we may parallel the practice of the ancient Britons of depositing in their burial-places a wooden rod with Ogham letters on it. This also was shaped four-square, and called a FÉ. A VE, in Icelandic, is a sanctuary; the FAI, a painted figure; the PEI, in Chinese, is a stone tablet erected to the dead in a tomb or temple; PEI is also divine, or inexplicable; PE denotes eternal life; BAI, in Irish, means death; FAY, English, doomed to die; FEI, Chinese, is to be grieved, to mourn, bewail; PEI, Mantshu Tartar, is to cry " alas!" BOIYE, Galla, is to cry, howl, weep. The Carib BOYE is an invoker of the gods. In Egyptian FUA means life, full, large, dilating life; BA is to be, to be a soul. FE also has the sense of to bear, or be borne by the genitrix who was typified by the tomb, and ÂII or FAII signifies to raise up. FÉ is no doubt a form of fay and faith, and some of the sarcophagi of the Eleventh Dynasty contained the writings which especially embody the Egyptian faith as found in Ch. 17 of the Ritual.

Possibly the name of the Ogham represents the Egyptian Aukhem.

[1] Borlase, who cites Carew, *Nat. Hist of Cornwall*, p. 302.
[2] *Champ, N. D.* 209.   [3] *Monuments of Upper Egypt*, p. 93.

The Ogham is a monument with the letters cut in stone, and figured round a circle. Aukhem (Eg.) means indestructible. Akh (Eg.) denotes a circle and to turn round. AM means belonging to. The Akh are the dead, manes. Khem also means the dead, and the Oghams are found as the monuments of the dead.

In Egypt it was the ceremonial custom to cast sand three times on the remains of the deceased, and with us this survives in the "Ashes to ashes, dust to dust," and the earth cast thrice upon the coffin lid. It is likewise popularly supposed that, wherever a corpse has been carried, the way becomes thenceforth a public thoroughfare. And although no extant statute may now warrant such a belief, the writer does not doubt that such was once the custom from the stress laid by the Egyptians on everything concerning the road of the dead, the eternal path.

According to Chief Justice Hale, the sources of the common law of England are as hidden as were those of the Nile. It was for so long an unwritten tradition, whose sole record was the proverbial memory of mankind, when priests and lawgivers were instead of books, and through them tradition spoke in the living tongue. This has bequeathed to us an inheritance of use and wont, the larger liberties of which crop up continually as an unwritten tradition, not verifiable by the Roman or Norman code of laws.

The Curfew Bell is a suggestive example of ancient things retained under the mask of later customs and names which often conceals the face of the past altogether. There is no doubt that a COUVRE-FEU law was enforced by William I., having the meaning of Cover-fire. But the custom was neither of Norman origin nor enforced as a form of servitude, nor had it the only meaning of cover-fire. The ordinance directed all people to put out their fires and go to bed ; noble and simple alike. Nor was the Curfew solely an evening bell. A bell was formerly rung at Byfield Church at four in the morning and eight in the evening ;[1] also a bell was rung at Newcastle at four in the morning. In *Romeo and Juliet*[2] we read :—

> " The second cock hath crow'd,
> The curfew-bell has rung, 'tis three o'clock.''

A cover-fire bell tolled at three in the morning !
Again, in *King Lear*[3] :—

" This is the foul fiend Flibbertigibbet ; he begins at curfew, and walks to the first cock."

In Peshall's *History of Oxford*[4] it is said the custom of ringing the Bell at Carfax every night at eight o'clock was by order of King

---

[1] Bridges' *History of Northamptonshire*, i. 110.   [2] Act iv. scene 4.
[3] Act iii. scene 4.   [4] P. 177.

Alfred, who commanded that all the inhabitants should, at the ringing of that bell, cover up their fires and go to bed.

The Curfew, then, was pre-Norman, and it was rung in the morning as well as at night. As pre-Norman, Curfew is earlier than Couvre-feu, and if the word signified Cover-fire, it would have to apply in a double sense, as we find in Cure, to cover, and Kere, to recover, the fire covered at night being recovered in the morning. But cover is not the earliest sense. The bell was rung night and morning, the beginning and end of the course; that is, the Char or Karh in Egyptian, the course of the night. Karh is night. Kar, the lower, under of the two courses of time; HRU being the upper, the day. In this sense the Curfew applied to fire is the bell that announced the beginning and end of the Kar of darkness, or fireless time. But the Cur-few is not limited to the meaning of fire. "Few" also denotes quantity and measure, therefore Cur-few, or Char-few, may have primarily indicated the length of the Char, or Kar, or course of the night; and the bell may have announced the sunrise and sunset, like the morning and evening gun, before ever it was the Couvre-feu Bell.

The Welsh had an ancient coöperative custom called the Cymhortha, in which the farmers of a district met together on a certain day to help the small farmer plough his land or render other service in their power. Each one contributed his leek to the common repast, and the leeks for the occasion were typical, for they were the sole things so contributed.[1] In Egypt the Mert were persons attached or joined together for a common purpose, such as a community of monks. Ka is labour and land. Now in the old quarto *Hamlet*[2] the word COMART occurs, and is usually understood to mean a joint bargain. Co-mart (Mert) is co-attached, co-bound, hence the covenant and the co-agreement. If this be the origin, Comart is at root the same word as the Welsh Cymhorta, explained by the Co-merti. It also follows that the name of Comrade is a deposit of both, the Comrade being the co-attached person, fellow, one of the Merti. The form of this primitive commune is yet extant as the Ka-merti (Eg.), the Merti on the land, in the Russian Mir. The leek answers to the onion, the Egyptian Hut, and it was the express token of the Co-mart or Cymhorta. Huter (Eg.) means to join together, as did these men of the Hut (Leek). "A regular Huter" is a vulgar phrase applied to a common woman. The word Hut (Eg.) also means to bundle, or a bundle, with the "ter," determinative of time, and to indicate; and this same word for bundling means to touch, to consecrate. Whether the Leek (Hut), a form of the "had," is emblematic of the Welsh bundling, the present writer knows not, but this same word "Hutu" signifies one-half; and in the Guernsey custom of "Flouncing," and other forms

---

[1] Hampson's *Kalend.* i. 107 and 170.   [2] Act i. scene 1.

of betrothal, the practice denoted that the lovers were half-bound, in token whereof, if the man changed his mind, the woman could claim one-half (Hutu) of his property, and *vice versâ*.

Many popular and hitherto inexplicable customs relate to the keeping of time and period. A complete year-book of the heavens might be made from these celebrations ; the larger number of them belong, however, to the time of the vernal equinox. It is now intended to take a " run round " the year, in the order of the seasons, the illustrations being limited to what is here called the Egyptian naming, following the philological and ideographical clue.

An old custom was extant in the Isle of Man when Train wrote,[1] called the Quaaltagh. Young men went from house to house on New Year's Eve singing rhymes and wishing the inmates a Merry New Year. On these occasions a dark-complexioned person always entered the house first, as a fair person, male or female, was considered unlucky for bringing in the New Year. The Quaaltagh signifies the first foot that crosses the threshold on New Year's morning. In Egyptian the Kar is the lower half of the solar circle out of which the sun begins to ascend with the New Year : also Karh means night. Takh means a frontier, and to cross. The Kar-takh (Quaaltagh) is therefore emblematic of the sun's crossing from Hades. This sun of the night and winter was the Black God, Kak or Hak, who must be represented by a person of dark complexion to complete the symbolical significance of the custom. Persons with dark hair are in the habit of going from house to house in Lancashire to bring in the New Year auspiciously. Light persons are as good as prohibited. And so in keeping with the dark is the feeling that the most kindly and charitable will refuse to give any one a light on New Year's morning lest it should bring ill-luck on the giver. The nature of the Black God here represented by the dark-complexioned man has to be expounded in this work, the present volumes of which are limited to the comparative matter.

A grotesque manorial custom was extant in the time of Charles II. At Hilton, Staffordshire, there existed a hollow brass image representing a man kneeling in an indecorous position known as the Jack of Hilton. The image had two apertures, one very small at the mouth, another larger at the back. When filled with water and put to a strong fire, the water evaporated as in an æolipile, and vented a constant blast from the mouth, blowing audibly, and making a sensible impression on the fire. There was an obligation upon the lord of Essington, the manor adjoining, to bring a goose to Hilton every New Year's Day and drive it three times round the hall-fire which the Jack of Hilton was blowing all the while with his steam. The goose was then handed over to the lord paramount of Hilton. One wonders if this was related to the Chase of the Goose, a great mystery with the Egyptians.   The Jack, we may be sure, was a representation of Kak, the God of Darkness, and Sun of the Underworld.   Hil-ton, read as Egyptian, is the upper seat, and Hes (Eg.) means the captive subject, ordered to obey.

Among the Scottish peasantry the first Monday of the year—Handsel Monday—is a great day for " tips."   The young people visit the old for tips.   The tips of Handsel Monday are the equivalent of Christmas boxes in England.   Tep is the Egyptian word for the first, and for commencement of the circle.   The tip, however, is unknown by name in the North.[1]   Handsel Monday was the day on which labourers and servants changed their places, and were engaged by new masters.   Those who stayed on were treated by the farmers to a liberal breakfast.   It is in short the labourers' day, and in Egyptian the Hanuti are the labourers.   This plural is the equivalent of " hands."   Sel (Eng.) is self.   The equivalent Ser (Eg.) is private, reserved, and sacred.   Thus Handsel is sacred to the Hanuti's self, or the labourer's own day, as recognized.

To han'sel money is to spit on the first that is received—a custom still common.   The spitting is a mode of consecrating or anointing. Hant (Eg.) is the name for a rite of consecration, which in han'selling is the spitting.   The word Sel (Ser), in addition to sacred, private, reserved, sole, also denotes some kind of liquid, cream, or butter, and is evidently connected with anointing.   If we take the word to mean "han" rather than hant, then Han (Eg.) signifies to bring, contribute.   Han is the divinity of bringing called the Bringer, and as the money thus han'selled is consecrated to the bringing of more, the han'sel may be devoted expressly to Han the Bringer.   The deity is all the more probable as we have old " OUNSEL " as a name for the devil, the final status of the earlier god.

New Year's gifts in England were formerly called " XENIA."[2] Khen (Eg.) is the act of offering, and some kind of festival.   Kennu means plenty, abundance, wealth, and Khent is to supply; it also denotes the circumstances of a festival.

The Boosy or Boosig is a trough out of which cattle feed ; commonly it is the manger in front of them.   In some counties it is called the Booson, which in the earliest form would be Boosigen. The Sekanu (Eg.) is a trough.   Buh signifies in front or before.   Buhsekhanu is the trough in front.   Sekhan abrades into Sen.   At the wassailing on the vigil of Twelfth Night a large cake with a hole in the middle used to be made by the farmers' wives in Herefordshire, and with much ceremony placed on the horn of the finest ox.   The ox was then tickled to make him toss his head.   If he threw the cake behind, it became the mistress's perquisite, and if before, in what was termed the Boosy, the bailiff claimed the prize.[3]   This adds the

---

[1] Chambers, *Book of Days*, i. 52.   [2] Brand, New Year's Day.
[3] Brand, on Twelfth Day.

Apis Bull to the Boosig, and an illustration of the Two Truths of Egypt, male and female, before and behind.

In Cumberland and other northern parts of England, Twelfth Night, which ends the Christmas holidays, is devoted to dancing and sport. The supper concludes with the eating of a large flat oaten cake, which is baked on a griddle, and sometimes has plums in it. This is called a THARVE-cake.[1] In Egyptian Terf is the word for sport, dancing, being lively; and terp is not only the name of a cake, but of certain ceremonial rites of Taht, the reckoner and recorder of the gods. The English festival ends with the burning of a tar-barrel, a common mode of terminating popular rites. Ter, in Egyptian, means to indicate, the end, extremity, *finis*.

The image of winter was burned on the 12th of January, 1878, at Burghhead, near Forres, where there is an ancient altar locally known as the "DOURO." A tar-barrel set on fire was borne round the town, blazing, and then carried to the top of the hill and placed on the Douro. When the barrel crumbles down, the fishwomen try to snatch a lighted brand from the remains, with which the cottage fire is kindled, and it is lucky if this fire can be kept alive the whole year through. The ceremony is called a CLAVIE. The Douro answers by name to Teriu (Eg.), the two times, and the complete circumference of the round of the year. Clavie is a form of Kherf (Eg.), first, chief, consecrated, to pay homage, the primary form and model figure of a thing, the typical ceremony.

The Plough is a name of the Great Bear constellation. In the Fool-plough performances the characters are seven in number, six males and Bessy, or six males and Cicely, the Fool, Cicely, and the Fool's five sons, the number of stars in the constellation. Cicely is also a form of the Irish Sheelah-na-Gigh. Lort-Monday is a name of Plough-Monday, on which day the mummers and morris-dancers used to go round and entertain the people with shows and plays. Lort in Egyptian is Rert, to go round, make a circuit. Amongst other characters exhibited, as we gather from an old song of the mummers,[2] was the hobby-horse, a dragon, and a worm or snake. In the hieroglyphics Rert is the name of a snake, a sow, and the Typhonian water-horse, the hippopotamus. Thus the old Typhonian genitrix and Goddess of the Great Bear or Plough is identified by five of her types with Lort-Monday, and by the going round. The bear and unicorn were forms in which the mummers were sometimes disguised, and both were types of Sut-Typhon. The fox's skin was worn in the shape of a hood, the fox being a symbol of Sut. The Fox (jackal) and Bear are Sut-Typhon. The great character in the "fool plough" is the Bessy, who used to wear the skin of a beast. The Bes or Basu was a skin worn by the priests in Egypt. The besau was also a sash with ends

---

[1] Dyer, p. 29.   [2] Brand, "Fool Plough."

behind, as Bessy wore the calf's tail and the fox's tail. The Basu is some kind of beast, as the leopard. Bes in Arabic is the cat or lynx. The cat-headed Pasht was a feminine form of the Bessy or beast.

In a Yorkshire representation of the Fool Plough[1] the character of the Bessy is taken by a commander-in-chief called Captain Cauf's Tail, who is the orator and dancer, and one of the titles of Taurt of the Great Bear is Bosh-Kauf (Ape), in which we find the Bessy and the Cauf are identical, as they were in the mumming. As we read the matter, the Bessy is primarily the Goddess of the Great Bear. Bessy's tail denotes the hinder part; the Pes (Pest) in Egyptian and English is the back, and the Goddess of the Great Bear represented the hinder quarter. The word Bes signifies to bear and transfer, pass from one place or shape to another, be proclaimed and exhibited. In the Ritual the sun is said to transform into a Cat, that would be, into the Bessy. The meaning is that the sun was re-born of the genitrix represented by the beast, whether as Rert or Pasht, Hathor, the Beast, or Bessy. This may account for the death of Bessy, who is killed by the six youths in white for interfering while they make a figure of 6 with their swords. The hexagon was a figure of the four corners and the upper and lower heaven, possibly connected with the Pleiads as the typical six. Bessy represented the No. 7. At Hollstadt, near Neustadt, a plough-festival is still held in the month of February once in seven years, and the plough is drawn by six maidens corresponding to our six youths in white.

The cat was one of the Druidic types. The "Paluc Cat" is spoken of, and was thought by Owen to be a tiger. So the Basu kind of beast may be cat, tiger, or leopard. Again it is called "Cath Vraith," the speckled cat. Taliesin says the spotted cat shall be disturbed, together with her *men of a foreign language, i.e.* her priests.[2]

The cat was both male and female, Cath Vraith, and Cath Ben Vrith, and the Sun-God became female in making his transformation into the cat. The Druidic cat was likewise a symbol of the sun, and Taliesin, who is assimilated to the solar divinity, recognises the sun's transformation into the cat type, just as we find it in the obscurest, most remote, and rarest matter of the Ritual (Ch. 17). Speaking of one of his transformations in the solar character, he says, "I have been a cat with a speckled head on a tripod"—"Bum Cath Benfrith ar driphren"[3] (or on something with three branches). The spotted cat denoted the double nature. The Welsh were in possession then of the Two Truths, with all that the fact implies, which has yet to be explicated.

"Ploughing the fields" was one of the things to be done in making the "working figures of Hades" (Ch. 6). In the Ritual (Ch. 1) we

---

[1] *Costume of Yorkshire*, pl. 11 (1814).   [2] *Welsh Arch.* p. 73.
[3] *Ibid.* p. 44.

also find the "Festival of ploughing the earth (Khebsta) in the land of Suten-Khen," which answers to Bubastes, the abode of Pasht, the cat-headed, a form of the Basu, Bessy, Beast, or Bosh Kauf. The men who follow the plough on Plough-Monday are called Plough Jags. Jag is Khakh (Eg.), meaning to follow. In Norfolk the ploughman is a plough-jogger.

In many churches a light was set up before an image, and termed the "Plough Light," maintained by the husbandmen, old and young, who went about and collected the money on Plough-Monday. The image, no doubt, represented the lady of lights, whose first type was the constellation of the seven stars, from which was derived the typical seven-branched candlestick. That was as Typhon, the old beast, who gave birth to the son as Sut, and who was the Sabean type of the genitrix. She was followed by the cow-headed type of the beast in Isis-Taurt, the lunar genitrix, and lastly by Pasht, the lioness type of the beast, as the solar genitrix. The death of Bessy, while interfering with the hexagon, probably represented her supercession in a later chart of the heavens and the bringing in of the six Pleiads.

A custom was formerly observed at Ludlow on February 3rd. The corporation provided a rope thirty-six yards in length, which was given out at a window of the market-house as the clock struck four. A large number of the inhabitants then divided into two parties—one contending on behalf of Castle Street and Broad Street wards, the other for Old Street and Corve Street wards; both strove to pull the rope beyond the prescribed limits; when this had been done, the contest ceased.

This is a mystery.[1] The measure of thirty-six yards relates the rope to the thirty-six decans of the zodiacal circle, and four o'clock to the four quarters. The game of pulling the rope was the drama of the two lion-gods of the horizon. The equinox was imaged by the scales or balance, and two powers were described as contending for the victory up or down at the level place. We read in the Ritual (Chap. 17) that the day of contending of the two lion-gods was the "day of the battle between Horus and Sut, when Sut puts forth the ropes against Horus, and Horus seizes the gemelli of Sut." It is the battle of north and south, darkness and light, evil and good. The Osirian, using the same imagery in the Ritual (Chap. 39) exclaims, "I make the haul of thy rope, O sun. The Apophis is overthrown! Their cords bind the south, north, east, and west. Their cords are on him." The cords of the four quarters. The same conflict occurs between the lion and the unicorn (the type of Sut), "a-fighting for a farthing," or for the circle imaged by that coin. Kherf is a title of the majesty of the sun-god, and one of the streets is named Corve Street, that is, in Egyptian, the street of his majesty the Horus. In support of this, the lion-gods who contend are called the Ruti, the

Twins of the Ru, the horizon. Ludlow also has its rock of the horizon, the place of the Ruti, castle-crowned, and, in Welsh, LLEWOD is a name of the lions. The ceremony has the look of being belated from the day of the winter solstice, and of belonging to the division by north and south which preceded that of the east and west.

On Shrove-Tuesday the Highlanders make bannochs called the bannich bruader or dreaming bannochs. These are eaten for the purpose of divination, the eater being supposed to see the beloved one in his sleep. At (Eg.) signifies sleep, image, type. Pru (Eg.) means to see, appear. Pru-at is to see, appear in sleep, and Bruader is dreaming.

On Shrove-Tuesday, at Westminster School, a verger of the Abbey in his gown, bearing a silver baton, issues from the college kitchen followed by the cook of the school, in his white apron, jacket, and cap, carrying a pancake. On arriving at the schoolroom door, he announces himself as "the cook," and, having entered the schoolroom, he approaches the bar which separates the lower from the upper schoolroom, twirls the pancake in the pan, and tosses it across the bar into the upper schoolroom among the boys, who scramble for the catch, a reward being dependent on securing the cake whole.[1] The hieroglyphic cake is the sign of the horizon, the place of the equinoctial colure, the line of the crossing. The pancake is tossed across the line. The line separates the lower from the upper of the two halves of the solar course in the two heavens. The sun in crossing the colure completes the circle of the year and the symbolical cake must be secured whole. To toss a thing up is to cook it or chuck it up. This is done by the Cook. Moreover, the balance or equinox is the Khekh (Eg.). Tossing the pancake across the line is also an Irish custom. A fine is imposed if the cake be broken in the process.

The Jack-o'-Lent was a puppet set up to be thrown at for sport.

In a ballad called "Lenten Stuff," Jack-o'-Lent wears the "headpiece of a herring." On Easter Day at Oxford the first dish brought to table was a red-herring depicted as riding away on horseback, set in a corn salad.[2] There used formerly to be held on Shrove-Tuesday, at Norwich, a festivity in which the seasons were represented, and Lent was clothed in white and in herring-skins, and the trappings of his horse were oyster-shells. The fish had been adopted into the Christian iconography, but the symbol is not to be understood there. The Messiah, Son, was born in the Fishes; born of the goddess with a fish on her head (Athor), or a fish's tail (Derketos and Ichton), when the sun was in Pisces at the time of the spring equinox, 27,000 years ago; at least the imagery belongs to that time, not to the sun's entrance into Pisces, 255 B.C. When the sun passed forward into the sign

---

[1] *Book of Days*, vol. i. p. 237.
[2] Brand, Easter-Day. Aubrey, MS. Ashmolean Museum.

of the Ram, the fish was done with, as it ceases to be eaten at the end of Lent. It was rejected and made a mockery of, a puppet to throw sticks and stones at, or set on horseback to ride away. It was a fish in April, a fish out of water, a Geck, the Khekh which in Egyptian had modified into Kha, the fish.

In France the April fool is called the April fish. This can be read astronomically. It would still hold good if the custom only dated from 255 B.C.; but it more probably belongs to the fixed year of the zodiac, in which the spring equinox occurred with the sun in the sign of Pisces. When the sun had left that sign, the fish was the type of the past, the passed-time, synonymous with pastime, and the fish of April was out of date.

In Lancashire, May-eve was at one time celebrated by all kinds of mischief and practical jokes. One of these consisted in exchanging the sign-boards of different tradesmen. They were representing the sun in his exchange of signs. Formerly there existed the following custom at Frodsham and Helsby. The bourne of the two parishes of Frodsham and Durham was a brook, and in walking or beating the parish boundary the " Men of Frodsham" handed their banner across the brook which divided them from Helsby, in the parish of Durham, to the " Men of Helsby," who in their turn passed over the Helsby banner.[1] This also enacted the change of signs by an exchange of banners. Helsby shows the place (By) of the Kar or circle completed at the crossing.

The phenomena of the seasons were followed and reflected in this way seriously, religiously, at first, and at last in fun and frolic. This was so in all lands; in none was it more faithfully followed than in ours, and although the Christian re-adapters of the past have obscured much of its imagery as with a coating of lampblack or a whitewash of new names, it could not be obliterated. The Jack-o'-Lent is another symbol connected with the fish.

In the Egyptian mythology the region of the Eight Gods is named Sesennu. The lunar deity Tahuti is lord of this region, which is also called Smen, the Hebrew Shmen, N°· Eight, and Hermopolis. It was the place of return or facing round for both sun and moon. Sesennu also reads the eight Nu or gods. In the zodiac Smen was the locality where the solar Son was established in place of the Father, hence the solar and lunar birthplace. The region of the eight great gods was in the sign of the Fishes in An. The fish in Egyptian is the REM.

Now Halsted, in his *History of Kent*,[2] describes an ancient custom of the fishermen of Folkstone, who used to select eight of the largest and best WHITINGS out of every boat when they came home from that fishery; these eights were sold apart from the rest of each " take," and the money was devoted to make a feast on every Christmas

---

[1] Dyer, p. 210. [2] Vol. iii. p. 380.

Eve, called a "RUMBALD"; the fish were likewise named Rumbald Whitings. The custom is extinct, but Christmas Eve is still called Rumbald Night. It has been suggested that this usage was in honour of St. Rumbald. Saints in general are modern sign-posts put up in place of the ancient symbols on purpose to mislead. The Rumbald connects itself with the fish and the number Eight. In Egyptian Rem-part, the equivalent, means proceeding, emanating from Rem, the fish-region of the Ritual, where it is plural as "Rem-rem," our fishes. The number Eight connects the fishes with Sesennu, the eight gods, and their region of the same name. The institution belongs to the passage of the sun in the Fishes, or out of that sign, at the time of the spring equinox, shifted to Christmas when christened anew. A Rum-duke, the name of a grotesque figure, some faded symbol or other, is older than the saint, whilst Rum-fustian is a drink made with the yolks of twelve eggs, and therefore zodiacal, but with no rum in it. Bale is a pair; the sun in the Fishes would be Rem-baled. "Rumbalow" belongs to an old refrain, sung no doubt at the Rumbald. There is a broth called Balow or Ballok made from two fish, the eel and pike, *i. e.* the jack, obviously connected with the sign of Pisces.[1] Ballow is a goal, and the Fishes were the goal of the sun, and Rem-ballow would signify the goal attained in the Fishes, the twelfth sign, hence the number of eggs in the Rum-fustian. Ballow, the goal, is likewise called "Barley" in children's games. BER (Eg.) denotes the goal as the summit. The eight whiting correlate the Rumbald with the region of the eight great gods.

Ash Wednesday begins the penitential season of Lent, when the devout mourned their sins in dust and ashes. How ancient is the name may be judged from the word Ash (Eg.), a cry, plaint, answer, turn, invert, with the sign of a man praying, or invoking heaven. Asha is also applied to a festival.

The rectorial tithes of the parish of Great Witchingham were held (in 1835, by P. Le Neve Foster) under a lease from the Warden and Fellows of New College, Oxford, and a bond of covenant to provide and distribute to the poor inhabitants two SEAMS of peas, containing in all sixteen bushels.[2] These were distributed on Ash Wednesday amongst all the people, rich and poor, who happened to be in the parish on that day.

In the ancient reckoning one kind of measure ran into another, and each was a part in the total combination. In the present instance, the two Seams are equivalent to the half-year or circle, if we consider the measure in time belongs to the equinox as it does, Peas and Pasch being identical; the Peas are typical of the division, and were divided because they divide in opening, and Pasch (Eg.) or Pekh is the division which divided the year into two halves. The Seam is a quarter, and there was an Egyptian goddess of the western quarter named SEM.

<hr />

[1] *Forme of Curry*, p. 12.   [2] Dyer, p. 96.

The seam is a quarter of corn, also the quarter of an acre. Two seams are a half of some unknown total, like the quarter of corn, which has no whole in our corn measure. Sem (Eg.) means a total, and here the quarter which is a total in itself is a Seam.

The churchwardens of Felstead in Essex were accustomed to distribute, as the gift of Lord Rich, seven barrels of white herrings and three barrels and a half of red, on Ash Wednesday and the six following Sundays, amongst ninety-two poor householders of the parish, in shares of eight white herrings and four red a-piece.[1] The number twelve correlates the twelve herrings with the twelve signs, and the ninety-two with the number of days in the three months, as in March, April, and May. Also the number 91—92 would be the divisor of the year into four quarters. Such a custom belongs to the times when reckonings were enacted like other forms of symbolism, and facts were recorded by means of acts which were a mode of perpetuating remembrance to supply the want of letters.

The Bomonese in Central Africa are tattooed with twenty cuts or lines on each side of the face; these are drawn from the corners of the mouth towards the angles of the lower jaw and cheek-bone; six on each arm, six on each leg, four on each breast, and nine on each side above the hips, with one cut in the middle of the forehead. The total number is ninety-one. These groupings also correspond to the one year with its twelve months and four quarters of ninety-one days to the quarter.[2]

A Colorado woman described by Spix and Martius wore a circle on the cheek, and over this were two strokes. Down her arms the figure of a snake was depicted.[3] The serpent signifies renewal, and the two strokes obviously denote reduplication of the circle or the Egyptian two times, and these were true hieroglyphics. The name of Tattoo in Egyptian (Tattu) means the Eternal. Tattu in An was the place of eternizing or establishing for ever.

The wife of a Bectuan chief was seen by Lichtenstein wearing seventy-two brass rings, the number of demi-decans in the zodiac. What did they symbolize? Why, that she impersonated the whole circle of the heavens, as did the ancient mother who embraced and gave birth to her solar Lord, her RA, or HAR. Such customs did not originate in the mere ornamentation of human bodies, but were the means of reckoning, and of registering facts for use. Picture-writing was precious because of its purpose. No amount of suffering was considered too great. The significance had to be branded into the memory. When the boundaries of certain parishes are beaten, and the boys are bumped against the stone, they are told it is to make them remember. These customs contain the earliest acting drama on the world's stage. The players were bringing on to us what they

---

[1] Dyer, p. 94.   [2] Denham, vol. iii. p. 175.
[3] *Travels in Brazil*, vol. ii. p. 224.

knew with no other means of preserving and communicating their knowledge.

The herrings connect the reckoning with the sign of Pisces. When the sun emerged from this sign in the fixed year, it was the place of the spring equinox, the point of issue from the three water-signs. One wonders whether the Lord Rich was one of the Rekhi, the knowers, the Mages, whose name of Rekh means to reckon, keep account of, and to know?

Three Egyptian words will tell us more about the customs of Valentine's Day than all the falsehoods concerning the saint.

It is, says Bourne, a ceremony never omitted among the vulgar to draw lots which they term Valentines on the eve before St. Valentine's Day. The names of a select number of one sex are, with an equal number of the other, put into some vessel, and after that every one draws a name, which, for the present, is called their Valentine. Va (Eg.) or Fa means to bear; Ren is the name and to name; Ten means to determine. Thus the day of Va-len-tine is that of determining whose name shall be borne by each person in this mode of marriage by drawing lots. Valentine's Day is the day of coupling, and the custom points to the time when chance rather than choice was the law. Marriage is still said to be a lottery. The custom of sending caricatures on Valentine's Day is probably based on asserting the freedom of choice, and making a mock of chance.

St. David's Day is observed by the officers and men of the Royal Welsh (23rd) Fusileers, by the eating of the leek, every man in the regiment wearing a leek in his "BUSBY." The officers have a party, and the drum-major, accompanied by the goat, marches round the table carrying a plate of leeks. Each officer or guest who has never eaten one before is bound to mount a chair, and, standing with one foot on the table, eat a leek while the drummer beats a roll behind his chair.[1] Hut, the name of the onion, is not only applied to that type of the Sun-god, the goat is also Hut or Hutu in Egyptian. St. David has now taken the place of Hu, and all the toasts are coupled with his name. But the onion, Hut, is the sign of Hu, and the step from chair to table identifies the act with the worship of the ascending Sun, the winged disk or Tebhut. Tebhut is likewise a name of the table in Egyptian. The goat, designated a Hut or Hutu, like the onion, is a symbol of Hu. This name of the goat and onion, Hutu, signifies one half-circle, and in the solstitial year which commenced with the sun's entrance into the sign of the Crab the ascending half of the year began with the sign of the Goat. It looks as though the name of David were a modernised form of Tebhut or Tevhut, the lord of heaven and giver of life, the great solar type of commencement in a circle.

Simnel or Mothering Sunday is the mid-Sunday in Lent. On this

[1] Dyer, p. 110.

day a cake called the Simnel Cake was eaten. It was the custom for apprentices to visit their parents especially on this day, and the practice was termed going a-Mothering. The simnel cake is also known as the mothering-cake. In the *Dictionarius* of John de Garlande, compiled at Paris in the thirteenth century, it appears thus : " Simeneus-placentæ-simnels." In the fifteenth century the form of the word is symnylle. The placentæ were signed with the image of the Virgin Mother [1] or of the Child. Now Sem (Eg.) is a representative sign, and nel (ner) means the mother and the vulture. Sem-nel is the mother-sign. The mother-sign was the Fishes in the Hermean zodiac. In this sign the mother as the fish-goddess (Athor-Atergatis) brought forth the child. In this sign (in An) was the tree of life from which Athor poured out the waters, holding the cake in her hand, the mother-cake, the placenta-symbol of birth, the simnel cake of our Mothering-Sunday.

Another derivation, however, is the more probable. Smen (Eg.) was the place of establishing the child in the seat of the father. Smen means to prepare, set up, constitute, in the region of the eight gods, where the son was established. The cake was a symbol of this establishing, the type of the land attained by the youthful sun-god. The son or child is the El (Ar), and the simnel cakes with the Christ on them would be the sign of the newly-established son, whence their connection with Mothering-Sunday, and the festival of the young people who went to see their mothers. At Bury, in Lancashire, from time immemorial, thousands of persons from all parts assemble to eat the simnel cake on Simnel Sunday. And this practice of meeting *en masse* in one town is confined to Bury. The origin of the custom is entirely unknown. Bury is evidently a representative of Para, the sacred name of An (Heliopolis), the birth-place of the Solar God, where we find the cake, the mother, and the child.

Pa is written with the open house or the bird with open mouth. And in Bury nearly every shop was formerly kept open on this day in the most unaccountable defiance of the law respecting the closing of shops during religious service on Sunday.

Passion Sunday, the Sunday preceeding Palm Sunday, was formerly known as Care or Carle Sunday, as may be seen in some old almanacks. On this day Carlings were eaten, carlings being explained as peas boiled on Care Sunday. Careing Fair was held at Newark, 1785, on the Friday before Careing Sunday. It is also called Whirlin Sunday in the Isle of Ely, and cakes were eaten called Whirlin Cakes. Whir and Kar are interchangeable. Carling and Whirlin cakes were provided gratis at the public-houses, and rites apparently peculiar and sacred to " Good Friday " were celebrated on this day, which the Church of Rome called Passion Sunday. Yet it

[1] Dyer, p. 113.

was an ancient popular festival in England, having no relation to a mourning. In the old Roman calendar a "dole of soft beans" is set down for this day. This is the same as the dole of peas boiled soft called Carlings. "Our Popish ancestors," says Brand, "celebrated, as it were by anticipation, the funeral of Our Lord on this Care-Sunday, with many superstitious usages."

Lloyd, in his "Dial of Days," observes that "on the 12th of March, at Rome, they celebrate the Mysteries of Christ and His Passion with great ceremony and devotion." They celebrated many mysteries in Rome undreamt of in Protestantism; this of Careing Sunday being one.

What with the beans in Rome and the beans and peas in England, we may call it the Feast of the Lentils.

The festival of the lentils was Egyptian, and consecrated to the elder Horus, the child that died, not the Horus who rose again. Isis, according to Plutarch, either conceived or was delivered of Harpocrates about the winter tropic, he being in the first shootings and sprouts very imperfect and tender; which is the reason, they say, that when the lentils begin to spring up, they offer him the tops for first-fruits.[1]

Plutarch, however, has mixed up the two Horuses. Har-p-Khart was conceived in the month Mesore—in the African Galla language the lentil is named MESERA, and MASURA in Sanskrit—and a sort of pulse was presented to his image in that month (Plutarch); his death occurs about the time of the winter solstice, when Isis made search for him, and the sacred cow was led seven times round about her temple. This was in the seventh month of the sacred year, Phamenoth (Pa-Menat, the mouth of the wet-nurse). The reason for this, says Plutarch, is because the sun finishes his passage from the winter to the summer tropic in the seventh month. In the Alexandrian year the feast of Phamenoth had receded to February 25, and about this time was the feast of lentils, brought on by Rome, and celebrated on the 12th of March. The tender shoots of the lentils offered to Har-ur are imitated by the peas being steeped until they were soft and tender in making Carlings.

Gregory says there is a practice of the Greek Church to set BOILED corn before the singers of those holy hymns which were sung in commemoration of the dead, or those which are asleep in Christ, and that this rite denoted the resurrection of the body, and he quotes Paul, "Thou fool! that which thou sowest is not quickened except it die."[2] The parboiled wheat, the steeped beans and peas, and the tender shoots are all one as types.

The hymns also were identical in their character with that song of Linus heard by Herodotus in various lands sung in memory of the divine victim; the song of the RENNU or nursling, the child Horus, who in death was Maneros. The child and seed are identical as Sif

---

[1] Of Isis and Osiris.   [2] *Opuscula*, 128, ed. 1650.

(Eg.), and the first Horus as the seed was buried in the earth, during the typical three or the forty days, when the seed quickened and the transformation took place by which he became the Horus-Khuti of the resurrection.   In India it is yet held to be the most propitiatory of all good works to personate the buried Seed by entering alive into a vault and remaining there whilst a crop of barley, sown in the soil overhead, springs up, ripens, and is harvested, which takes about the length of the forty days of the Mysteries and of Lent.   Rich Hindoos usually perform the forty days' rite by proxy.

In going "a-souling" on "All Souls' Day" in Herefordshire, the oat-cake, called SOUL-MASS CAKE, used to be received with the acknowledgement,

> "God have your saul,
> BEANS and all." [1]

The oats had superseded the beans, but this type of the seed survived.

The lentil takes its name from that of the Renn (Eg.), the nursling child and tender shoot.

The feast of the lentils, then, belonged to the elder Horus, he who was born of matter, and was always the elder, the sufferer, and the child, because the type of the dying sun.   Har-p-Khart is Har the child, the Crut, the dwarf or puny weakling, in short, our Carling. Carline is a name of a woman that does not bear.   The Carling is the foundation beam of a ship or the beam on the keel.   Har-p-Khart corresponds to both.   He was the basis, but also typified the infertile sun.   The truth is the adapters of the ancient festivals and celebrations to the new theology were hard put to it in adjusting the times of the two Horuses to the one Christ.   For the Egyptian Messiah was double, as will be demonstrated.   And the feast of the lentils was dedicated to the first-born Horus, whereas the Easter festival was consecrated to the younger, the god who rose again.

This is the one of whom Plutarch observes in continuation of the account of the lentils offered to Harpocrates :—" They also observe the festival of her (Isis) after-birth, following the vernal equinox." The after-birth was the younger Horus, the god of the Easter resurrection. The suffering Messiah was represented as passing through a feminine phase, and as weeping tears of blood.   This was signified by the wound of Tammus, and the Kenah image used by the women of Israel in their lewd and idolatrous mourning for Adonis.   Apis was passing through this period during the forty days of Lent when he was visited by the women alone, who stood before his face and raised their clothes to show him their secret parts ; they who were forbidden to enter his presence at any other time. [2]   The action was

---

[1] Brand, *All-hallow Even.*   [2] Diod. Sic. b. i.

symbolical of the feminine nature of the mystery of the biune being of whom so much has to be explained.[1]

The lentils are identified by name with the season of Lent, just as the carlings are with the French name of Lent, Carême. Lent itself is named from the Egyptian Renn (Len), the nursling child. Lent is the time of the great mystery of the transformation of the child Horus into the young hero of the resurrection. Hence the Mothering Sunday of mid-Lent. The Renn, so to say, becomes the Renpu. Pu adds the masculine article to Renn, and Renpu means the young shoot, plant, or branch. The first Horus (the Renn) was of a feminine, dwarfish kind of nature, the type of the winter sun. This, in a feminine form, would be the Ren-t, our Runt, a dwarf. He was a deformed dwarf, hence the child. In his transformation he is the Renpu, the renewed and renewing Youth.

The branch or shoot of the palm is the Renpu, and this, too, is an extant type in our palm branches of Palm Sunday.

Palm Sunday follows Car Sunday, and the palm shoot follows the carlings, the tender shoots of the lentils in Egypt, the ideograph of a new cycle of time.

Care Sunday was the ancient Passion Sunday. The passion, which lasts seven days, was the transformation of the god or the soul; "he is transformed into his soul from his two halves, who are Horus the sustainer of his father and Horus who dwells in the shrine," as it is written in the Egyptian Gospel.[2] The seven days correspond to the cow being led round the temple seven times.

TAHIN is the Turkish name of an oily paste still made use of for food by Eastern Christians during Lent: that is, during the time when the eye of Horus was being formed which was called the Tahn, and was made of tahn, a substance typical of preservation or salvation. This was in the place of preparation and of re-uniting the Osiris from his two halves, the two Horuses. The process of preservation by the TAHN is described in the seventeenth chapter of the *Ritual* as that of being steeped in Resin or TAHN.

A very ancient form of the Genitrix who gave re-birth to the Child-Horus, the Runt as the Renpu, the fresh shoot of eternal life, was Rennut, and her name is that of the season during which the mothering, the passion, and the transfiguration take place, the Romish Lent.

The day before Good Friday is called Shere Thursday and Maundy Thursday. Shere Thursday is the last day, the day selected for the Last Supper of the Lord, and Sher (Eg.) signifies to close, to shut. Sheri also means a rejoicing, and to breathe with joy. Maundy

---

[1] In the worship of the biune being called Venus-Barbatus, "*Videre est in ipsis templis cum publico gemitu, miseranda ludibria et viros muliebria pati, et hanc impuri et impudici corporis labem gloriosa ostentatione detegere.*" Also see Arnobius, *Adv. Gentes*, 5, 7.; and *Dictionary of the Bible* (Smith), v. iii., p. 1434.
[2] *Ritual*, ch. 17.

interpreted by Menti (Eg.) tells the same tale of the ending. Menat means the end, repose, death, or having arrived.

In Northumberland and Yorkshire Shere Thursday was known as Bloody Thursday, and in Egyptian Tsher is blood, gore, red blood, bloody, also the name of the red calf or heifer of sacrifice, the lower of the two crowns, and the desert land. The "red calf in the paintings" is alluded to in the chapter of transforming into a Phœnix.[1] In ancient times, we are told, the people clipped their beards and polled their heads, and the priests shaved their crowns, on this day. The hair-symbol is important. The Child-Horus, the Carling, wore a single lock of hair, the type of childhood. This was put away on arriving at maturity, when he transformed into the fully pubescent God. As the child, he was the non-pubescent Horus, as the second Horus he became the Sher. Share in English denotes the pubes of a man ; in Egyptian Sher denotes pubescence. Sheru (Eg.) is barley, because it is bearded, and the word signifies the adult, the youth of thirty.[2]

The Child-Horus was the beardless youth, the mere Carling with the curl of childhood, either boy or girl. The sun of Easter is the virile, pubescent, full-bearded, no longer the wearer of the Horus lock, but the adult Sher, represented as a youth of thirty.[3] Sher has an earlier form in Kher, to be due ; Kher, the word or logos. Passion Week is called Char or Care Week ; the Char, as in Egyptian, is a completed course, and on Char-Thursday the circle clasped on Good Friday was completing, and being "CHARRED." Khar (Eg.) also signifies the animal destined for the sacrifice ; and in England, on Shere-Thursday, the altars were washed (for the new sacrifice).[4] Khar modifies into Har, the lord, who was the hairy or full-bearded solar god, represented as being buried for three days in the underworld, and mourned with the same ceremonies as those of our Shere-Thursday, or rather as those of the three days. Thus we have the two types in the Kar (Shere) and the Carling, the one being the diminutive of the other ; and as that modifies into Har, we have the two Horuses in their right relationship by nature and by name. This will explain why the Christian ritual traverses the same ground twice over. The Church of Rome continued both Horuses and all their symbols faithfully enough. For example, the time was, as late as the year 1818, when Bloody or Holy Thursday was celebrated by the typical burial of the Christ on that day ; and in the Sistine Chapel and other churches the Host in a box, i.e. the real flesh and blood of Christ, was laid in the sepulchre the day before the rite of the Crucifixion was performed. " I never could learn," says the eye-witness, "why Christ was to be

---

[1] *Ritual*, ch. 84. 　　　　　　　　　[2] *Denkmäler*, iv. 51, B.
[3] " Youth of thirty." Cf. Gen. xli. 46 ; and Luke iii. 23.
[4] Collier's *Eccles. Hist.* vol. ii. p. 157.

buried before He was dead."[1] They were worshipping the double Horus of Egypt, as will be proved in a later part of this work; and this necessitated the beginning on Thursday, for the fulfilment in three days, as it was in mythology, and as it was in Rome, where the resurrection took place on Saturday.[2]

The mystery of the Child-Horus, who always remained a child, is also the mystery of St. Nicholas and of the boy-bishop. Nicholas is the chosen patron of children, and is himself the child. In the *English Festival*[3] it is said " he was christened Nicholas, a man's name, but he keepeth the name of the child. Thus he lived all his life in virtues with his child's name, and therefore children do him worship before all other saints."[4] His child's name! the name of the child! and yet a man's name! In Egyptian Neka is the typical male, virile power, the bull. Ras (las) is suspended. The suspended virility marks the child, the unvirile, infertile sun, the Child-Horus of Egypt. Nicholas was a survival of the Child-Horus, who was the Neka-las in person. In cathedral churches in Spain, when the boy-bishop was elected, there descended from the vaulted roof a cloud that stopped midway and opened, whereupon two angels issued from it with a mitre and placed it on the boy's head. This is a replica of the crowning of the Child-Horus by the two divine sisters Isis and Nephthys.[5] The Child-Horus is Har-Skhem, lord of the shut-place, the secret shrine. The mouse was one of his emblems. And this character of secrecy and of working in secret is extant in the child's Saint Nicholas.

The writer is forced to confess that every great day of festival and fast and every popular ceremony and rite pressed into the service of the Christian theology were pre-identified in these islands. No true account of many of these has ever been given; of others we have nothing but downright lying, as needs must be in a thorough course of systematized fraudulence and imposture such as was practised by the Romish Church.

The return of Palm Sunday has, from time immemorial, been celebrated in a peculiar manner at Hentland Church, Herefordshire. The churchwardens presented the minister and congregation with a bun or cake, and formerly a cup of beer. This is partaken of within the church, and the act is understood to be one of good-fellowship, implying a desire to forgive and forget all animosities in preparation for the Easter festival.[6] Hent-land suggests an Egyptian name. Hent (Eg.) signifies rites, consecration. Hen is one's neighbour or familiar friend; an equivalent of our " forgive and forget." Hen also means to bring tribute, and Hent is the priest; here the church is called Hent-land.

---

[1] *Rome in the Nineteenth Century*, vol. iii. pp. 144, 145.
[2] *Ibid.*   [3] P. 55.
[4] *Liber Festivalis in die S. Nicholai.*   [5] Sharpe, *Egyptian Myth.* fig. 23.
[6] Dyer, p. 128.

A singular custom existed for ages at Caistor Church, Lincolnshire, and Sir Cullen Eardley, in 1836, petitioned the House of Lords for its abolition.   The estate of Hundon appears to have been held by the lord of the manor subject to the performance, on Palm Sunday in every year, of the ceremony of cracking a whip in Caistor Church. The whip was taken every Palm Sunday by a man from the manor of Broughton to the parish of Caistor, and while the minister was reading the first lesson, the whip was cracked three times in the church porch.   At the commencement of the Second Lesson the man approached the minister whip in hand, with a purse at the end of it, and kneeling opposite to him, he waved the whip and purse three times, and continued in a determined attitude until the end of the chapter.   After the ceremony, the whip was deposited in the pew of the lord of Hundon in Caistor Church.   There is no reference to the subject in the title-deeds.   The estate was held under the ancient tenure of demesne.[1]   These *dateless* customs have all been Christianized and dated ; the present one has been supposed to refer in some way to Peter's repentance and the cock crowing thrice.   With this we parallel certain FACTS derived from Egyptian which may possibly throw some light on the mystery.   The whip is a most important hieroglyphic.   Hun (Eg.) means to rule and to flog, also territory.   Hun then is rule-of-whip.   Ten (Eg.) is place, seat, or land.   Hunten is the seat or land of whip-rule.   Khi is the whip.   It is the sign of rule, and means to rule, govern, screen, protect, and cover. Ster is a name of the dead laid out and lying together.   Khi-ster then signifies "Protect, screen, cover the dead laid out together." From this we may suppose the land of Hundon (the whip-land) was held on condition that the owner protected and gave shelter to the buried dead.   Hence Caistor Church was built and named as the latest place of protection for the dead.   Ster, the couch of the laid-out dead in the monuments, becomes our Min-ster.   Mena (Eg.) is the dead.   Mena-ster, the couch of the laid-out dead, is our Minster, Cockneys persist in calling Westminster "Westminister," and that represents the Mena-ster, the Egyptian couch of the dead.   The whip is as good a hieroglyphic in Caistor Church as the ideographic Khi, to rule over, screen, cover, and protect.

So interpreted, the tenure of demesne is obviously typical. Temesu (Eg.) is the name for the division, or a division of land, and Nu is a divine or sacred type ; Temes-nu is literally demesne, the Egyptian e having been an earlier u.   The oldest tenure of land was typical of service to be rendered to the dead.

At a place named STOOLE, near Downpatrick, a ceremony is performed at midnight.   Crowds of worshippers assemble to do penance, kneeling and crawling on their knees.   The men, without coats or hats, ascend St. Patrick's Mount by steep and rugged paths, on their bare

---

[1] Dyer, p. 128.

knees, many holding their hands at the back of their heads. This they do seven times over. At the top is St. Patrick's Chair, formed of two large flat stones set upright on the hill. There sits an aged man who, while they repeat their prayers, turns them round three times. The penance is concluded by the devotees going to a pile of stones called the altar.[1] The name of the place, "Stoole," is identical with the Ster (Eg.), a couch or seat, and the other meanings of the word, coupled with the nature of the ceremonies, suggest that this must have been a most ancient form of the Ster or burial-place of the dead. The seven times also appear to connect the Mount with the goddess of the seven stars, the Great Bear, who was the first form of the seat, and abode of the living and the dead. According to Polwhele, there used to be on Start Point, in South Devon, the visible remains of a temple that belonged to the goddess Astoreth, and he connects the Start with her name. In Egyptian, Ster-t is the participial form of Ster, to be laid or stretched out. The Start might be named from the way in which it is stretched out. It is the Start Point, and the Ster was the couch of the dead. Start or Stert may include the As (chamber, resting-place) of Taur or Taurt. These high places were burial-places, and the dead used to be carried long distances to be interred on the headlands, where the stone sanctuaries once stood. Caistor Church had taken the place of the Stoole and the Start.

Although out of date here, it may be mentioned that in Northumberland it was customary on the 24th of June, to dress up stools (the seat) with a cushion of flowers. A layer of earth was placed on a stool, and various flowers were planted in it, tastefully arranged, and so close together as to form a cushion. These were exhibited at the doors of houses and at the crossings of the streets and corners of lanes, where money was solicited from the onlookers for a festival in the evening.[2] The stool was a form of the Ster, the seat which represented the genitrix.

In the *Witches' Sabbath* the eye-witnesses tell us how they joined hands and formed a circle standing face outwards, and how, at certain parts of the dance, the buttocks were clashed together in concert,[3] in the worship of the goddess of the hinder quarter; and at one time a ceremony was observed at Birmingham on Easter Monday, called "clipping the church," when the first comers placed themselves hand in hand with their backs to the church, and thus gradually formed a chain of sufficient length to embrace the building.[4] In our Easter and Pasch we have the same season doubly derived from Hest and Pasht, two Egyptian goddesses. The term Easter denotes the division (er) of Hest, the British Eseye and Egyptian Isis, who was the earlier

[1] *Hibernian Magazine*, July, 1817.   [2] Dyer, p. 327.
[3] De Lancre, *Tableau de l'Inconstance des mauvais Anges et Démons*, p. 209.
[4] *Every Day Book*, vol. i. p. 431.

Taurt, whence Hes-ta-urt, Astarte, Ishtar, and Eostre. She was the Sabean-lunar genitrix. Pasht is the later solar goddess, whose types were the cat and lioness. Her name denotes the division of Easter. Both Hest and Pasht, as well as the earlier Typhon, were typified by the seat, the hind quarter, which became the seat of worship, as the Church, just as Stonehenge had been the seat of Eseye.

The gammon of bacon and leg of pork, which are still eaten at Easter, are typical of the goddess of the hinder thigh, who brought forth the son, whether as the Typhonian Khepsh or the lioness Kheft. The pig however identifies Rerit, the sow, the Goddess of the North Pole and Great Bear, the oldest form of the genitrix in heaven, whose son was Baal, and whose bringing forth was solstitial, whereas the solar time of birth was equinoctial.

About the end of the sixth century it was discovered that the difference in point of time between the British Pasag, as celebrated by the natives, whether we look on them as Christians or pre-Christians and the Easter ceremonies as observed in Rome, was an entire month. This means that the festival had been kept in the British Isles for 2,155 years previous to the sixth century, and the people were behind solar time to that extent, on account of their not having re-adjusted the times of the feasts, fairs, and fasts, by which the reckoning was kept.[1]

It was a popular superstition in England that the sun danced on Easter Day. In the middle districts of Ireland, says Brand, the people rise on Easter morning about four o'clock to see the sun dance in honour of the Resurrection. He also mentions a mode of making an artificial sun dance on Easter Day in a vessel full of water set out in the open air, in which a reflected sun was seen to dance. This custom was practised by the present writer's mother, who little knew what a good heathen she was! We read in the Ritual, "I do not DANCE like thy form, oh sun! borne along in the river of millions and billions of moments."[2] "Thou hast LODGED DANCING"[3] is said of the sun of the horizon, that is of the level, the sun of the equinox, who was called Har-Makhu. The dancing may be interpreted by the scales or balance (Makha), and the nodes of ascent and descent. The sun dancing on Easter Day is at the poise of the equinox. Maka (Eg.) means the dance, and the Makha-level was the place of dancing, and Khekhing up and down.

"Apheru dandles me," says the Osirian.[4] Ap-heru is the equal road, that is, the equinoctial level, and dandling is the same as the dancing of the sun on Easter Day; the image being founded on the scales or balance figured as going up and down and dandling the child new-born as an immortal at this the place of rebirth into the higher life.

---

[1] Gieseler, vol. I. p. 54; Cummianus, cited by Ussher, *Sylloge*, p. 34.
[2] Ch. 15.　　　[3] Ch. 15.　　　[4] Ch. 44.

Two farms in the township of Swinton, belonging to Earl Fitzwilliam, change their parish every year. For one year from Easter Day at twelve noon till the next Easter Day they are in the parish of Mexborough, and then till the Easter Day following, at the same hour, they are in the parish of Wath-upon-Dearne, and so on alternately.[1]

This is the same dancing at the time of the vernal equinox. Mak (Eg.) is mixed; Maka, to dance; Makha, the balance or level. Mexborough may be named from this, and the alternation marks a boundary line answering to the boundary, in time, of Easter. Buru (Eg.) means the cap, tip, roof, the eye, which was made at this place in the planisphere. Swinton equates with Shen-tun (Eg.), the seat, throne, high-place of the circuit (Shen) which was clasped at the equinox.

Cole, in his *History of Filey*, says, on Easter Day the young men seize the shoes of the females, collecting as many as ever they can. On the next day the girls retaliate by getting the men's hats. Both are redeemed afterwards at a meeting held for the purpose.[2] These shoes and hats correspond to the Two Truths of the lower and upper heaven.

Changing of clothes or signboards, and mixing of the sexes is a form of Mak-ing from Mak (Eg.), to mix, and has the same meaning as the two farms changing their parishes. A form of this Mak-ing is made use of by Hamlet in his " Miching Malecho," an evil Miching, or double-faced performance. The Mak-ing or mixing had strange illustrations in the ancient religious festivals, as may be gathered from the Hebrew practice of תבל or בל-בל in בת-דבלים.

One form of the Makha and Mak-ing is to be found in the sport called hocking.

The meaning of the word hock or hoke in the ceremony of hocking is, according to Chambers's *Book of Days*,[3] totally unknown, and none of the derivations hitherto proposed deserve consideration. It is an Easter festivity in which the men hock the women on Monday, and the women hock the men on Tuesday; hocking consisting in binding or stopping people with ropes. Tuesday appears to have been the principal day, and on this the women bore the rule. The Egyptian hok, hek, or hak, denotes a time of festival. The Hakr[4] is shown by the twin lions to be the equinoctial festival. Hek signifies rule, dominion, and is a form of hooking and holding. In hocking the men rule one day, the women the other, by binding them. But it was an essential part of the ceremony that the men should lift the women up in their arms, and the women in their turn should lift the men. This alternate heaving was the analogue of the dancing sun, and the balance of the equinox, and the change from the lower to the upper hemisphere. It is represented on the monuments by the " Kabat," a

---

[1] Dyer, p. 168. [2] Dyer, p. 163.
[3] Vol. i. p. 499. [4] *Egyptian Saloon, B. M.* 573.

legend of two dancers doing the mill by raising each other up and down.[1] The GAVOT dance is an extant form of the KABAT.

The imagery and place of the equinox can be identified as Egyptian. The cake is an ideograph of the horizon and the cross figured on it of the crossing. The cake then is a symbol of the equinox.

Honey-fairs are celebrated in Cumberland and other parts of the North, with no relation to honey. They are a kind of wake, with dancing and other sports, held a week before Christmas. The honey, or hinny, called a "singing hinny," is a cake. The fair marks a re-peating period. Han (Eg.) is the cycle. Hani means to turn and return. Hani is the solar bark; Hannu, the scales. If the Honey fair had got belated from the time of the autumn equinox to the week before Christmas, that would calendar the lapse of over five thousand years. The cake, however, as the pancake, belongs to the horizon of Easter. Khekh (Eg.) is the balance, level, equinox. In English the cock is the tongue of the balance, as is the Khekh in Egyptian. Making cockledy bread is related to the equinox. Aubrey and Kennett describe the game. "Young wenches have a wanton sport which they call moulding of cockle bread, by getting upon a table-board and gathering up their knees as high as they can, and then, wabbling to and fro as if they were kneading dough, they say these words:

> " My dame is sick and gone to bed,
> And I'll go mould my cockle bread,
> Up with my heels and down with my head,
> And that is the way to mould cockle bread."

A Westmoreland version reads:

> " My grandy's sick, and like to be dead,
> And I'll make her some cocklety bread."[2]

This, however, was not the only way. The present writer, when a child, was received by a group of country girls as one of their own sex, and initiated into the mysteries of their games, which retained relics of the most primitive symbolical customs. Making cockledy bread was one of these. "Up with my heels and down with my head" shows the reversal or transformation to be found in what we term khekhing. It also denotes the bringing of head and heels to the level or Khekh. And that this was the significance is shown by the other practice of lying down flat on the floor and rolling to and fro. Each one of the party did this in turns whilst the rest sat round in a ring. The ring was zodiacal, and the wabbling to and fro was the ascending and descending motion of the balance. They were doing their scales. It was the same thing as the Kabiric custom of doing the mill by two persons raising each other up and down as in a pair of scales, called the Kabat, the same as the Kapat of the Abipones,

---

[1] Champollion. See Pierret, " Kab-t."   [2] Brand, " Cockle Bread."

who danced all night first on one foot, then on the other, swinging round a half-circle on each. The Kabat survived in the Easter custom of lifting. The Dame or Granny who is sick represents the Great Mother as the bringer-forth.

There is an endowment in the parish of Biddenden, Kent, of ancient but unknown date, for making a distribution of cakes to the poor every Easter Day in the afternoon. The source of the benefaction was twenty acres of land in five parcels. The cakes made for the purpose were impressed with two female figures, side by side and close together. An engraving of one of these may be seen in the *Every Day Book*.[1] It was believed among the country people that the figures were those of two maidens named Preston who had left the endowment, and it was said they were twins born in bodily union and joined together.[2] The gift being on Easter Day tends to identify the cakes with that of Easter, and it may be with the two characters of the motherhood, the two divine sisters, who, as Isis and Nephthys, bring forth the Easter child. At Easter the two houses of the sun were twinned, forming the Beth or Both. It may also be that Biddenden derives its name from this origin. Pet-ten-ten (Eg.) would denote the region or place of the division of the circle of heaven. Also the Egyptians made a kind of cake called the Baat or Boths.

At Bury St. Edmund's on Shrove Tuesday, Easter Monday, and the Whitsuntide festivals, twelve old women form two sides for a game at trap and ball, which is kept up with great spirit till sunset.[3] This is the same contention in another form, and still more interesting because doubly feminine. Bury (Eg. Buri) means the top, cap, roof, supreme height.

The Egyptian name of the balance would seem to have given the title to MAGONIA, a mythical region once believed to exist in cloudland. Agobard, Bishop of Lyons in the ninth century, says there were people in his time insensate enough to believe that there was a region called Magonia, whence ships of cloudland came to take on board the fruits which had been beaten down by tempests as the wrecks of earth. The sailors of that upper deep were fabled to be in league with wizards who had power to raise the wrecking storms, the fruit of which was shipped off to Magonia.[4] This ascension and declension of the scales between the two solstices is evidently at the bottom of such a tale of upper and lower as is told by Gervase of Tilbury,[5] who relates that a native of Bristol sailed from that port for Ireland, and his ship was driven out of its course to the remotest parts of the ocean. It chanced one day that he dropped his knife overboard, and it fell through the skylight of his own house at Bristol and stuck in the table in the midst of the family dinner, so directly did it descend

[1] Vol. ii. p. 443, 1827.
[3] *Every Day Book*, vol. i. p. 430.
[5] *Otia Imperialia*.

[2] Dyer, p. 165.
[4] Grimm, *Deutsche Myth.* p. 604.

from where his ship was sailing overhead. This would originate in some astronomical teaching, just as we might say if the knife fell straight through the earth, it would come out at a given point in Australia. It was a mode of describing the antipodal positions of the solstices and the sailing of the sun's bark through the upper signs, in relation to Makha or the equinoctial plane and the region of Magonia. Magonia as the place of the scales would, at the time of the autumn equinox, be the landing level during the season of the equinoctial gales and Typhonian tempests.

The Egyptian MAKHA, and the Irish MAGHERA (Co. Down), where the maypole was formerly erected at the crossing; the Moslem MECCA, and the Greek MAKARIA, an abode of the Blessed, and the MAKARON NESOI, or Islands of the Blessed, were each and all based on and named from the Makha of Magonia, as the landing-stage of the sun and the souls from the passage of the underworld. MEIGH is an Irish name of the Balance or Scales.

We find the Cake also under this name. When Dulaure wrote his work it was the custom at Clermont and Brives in France to make Easter Cakes in the image of the female, and these were popularly known as MICHES.[1] The MKATE in Swahili is a cake, and in English a Micher is a cake or peculiar kind of loaf.

The Guising Feast, or Gyst Ale, was commonly held in the spring about the time of Lady Day, when rents were paid and servants were engaged for the year. The Gyst is really the hiring or covenanting, and the Ale was the periodic festival. Kes (Eg.) is to bind and be bound, to envelop with slight bands, and Khes is a sacred rite. The Gyst was the binding or covenanting at the most hallowed time of the year still known as Lady Day. The Marlocking, or frolic, and rough horse-play of the same season, supposed to illustrate the manuring of the fields with marl, is more probably derived, like Gyst from Kes, from Mer (Eg.), to bind, attach (marry), will, and Lekh (rekh), to reckon, know, relationship. The Marlock is the periodic merriment and celebration of the newly made covenant or binding; a form of the statute fair.

Amongst other Hocktide customs kept at Hungerford, in Berkshire, is one connected with the Charter of the Commons for holding the rights of fishing, shooting, and pasturage of cattle on the lands and property bequeathed by John O'Gaunt, Duke of Lancaster. The day is known as Tuth Day. The tything or tuth men proceed to the high constable's house to receive their "tuth," poles, which are commonly bedecked with ribbons and flowers. These tuth men visit all the schools and ask a holiday for the children. They call at various houses and demand a toll of the gentlemen. The tithe levied on the ladies is a kiss, and in the streets they distribute oranges all day to the children. The high constable

---

[1] Dulaure, vol. ii. pp. 25-57.

is elected at the annual court held on this day, and one of the customs is for the constable's wife to send out a plentiful supply of cheese cakes to the ladies of the place.[1]   The Tut, or Tat, was an Egyptian magistrate ; the Tut is also a symbolical image, a type, and a ceremony.   Tat means to establish and to signify.

The palm with us is the sallow or willow, and this serves for the same symbol as the palm-shoot of Taht or Tekh, on which he marked another year (Renpa).   It was the custom on Ascension Day for the inhabitants of parishes to perambulate and beat the bounds.   At the commencement of the procession willow wands were distributed, especially among the boys ; at the end of each wand there was a handful of " Tags," as they were termed, and these were given away in remembrance of the event, and as honorary rewards for the boys to remember the boundaries.[2]   It was a practical mode of tecch-ing or teaching.   Tek (Eg.) means to fix and attach.   Tekh was the name of the divine teacher who registered the years and cycles on the branch of palm, which was thus represented in England by the slip of sallow.   The peeled willow wands were called Gads, and the gad is an English measuring rod ; thus the wand with the tags was another emblem of Tekh, the measurer of earth and heaven and preserver of boundaries.

At Leighton Buzzard, Beds, the children of the township, bringing green boughs in their hands, assemble each year at the market cross on Rogation Monday.   There a procession is formed, headed by the town crier, and usually accompanied by the guardians of the charity lands.   They proceed to a number of different stations situated on the boundary of the land belonging to the poor, and at each of these a boy is made to stand on his head with legs extended. A book is held over this figure of the cross or crossing, and a reader recites, in a loud voice, a description of the benefaction, its purpose and extent.   The children receive one cake each, and the boy who is inverted and bifurcated receives two.   Here, again, is the cake and the crossing with the beating of the ancient boundaries, the double cake corresponding to the Dual Truth.

A festival called BEZANT, so ancient that no authentic record of its origin or meaning exists, was formerly held at Shaftesbury or Shaston on the Monday in Rogation week.   The borough stands on the brow of a high hill, and, owing to its situation, was, until lately, so deficient in water that the inhabitants were indebted for a supply of this necessary of life to the people of the hamlet of Enmore Green, lying in the valley below.   The water was taken from two or three tanks or reservoirs in the village and carried up the steep ascent on the backs of horses and donkeys, and sold from door to door.   The Bezant was an acknowledgment of the privilege made on the part of the mayor, aldermen, and burgesses to the lord of the manor of Mitcombe, of which

---

[1] Dyer, p. 191.   [2] Hawkins, *History of Music*, ii. 112.

Enmore Green forms a part.   The Bezant was represented by a kind of trophy consisting of a framework about four feet high.   On this were fastened ribbons, flowers, peacocks' feathers, and it was also hung with coins, medals, and jewels and plate.   On the morning of Rogation Monday a lord and lady of the festival were appointed, and these, accompanied by the mayor and aldermen and the mace-bearers carrying the Bezant, went in a procession to Enmore Green. The lord and lady performed, at intervals, a traditional dance to the sound of violins, as they passed along the way.   When the steward of the manor met them at the green, the mayor offered for his acceptance, as the representative of the Lord, the Bezant, a raw calf's head, a gallon of ale, and a pair of gloves edged with gold lace.   The steward accepted the gifts, but returned the Bezant, and permission was accorded to use the wells for another year.   No charter or deed exists among the archives to explain the ceremony.[1]   The calf's head is presented as an offering to the steward on account of the water privilege.   The calf, in Egyptian, is named Behs, and ent (or nt) signifies to be indebted and bounden, to present tribute, or make an offering.   Behs-ent is the offering of a calf.   The hieroglyphic of the calf's head is the sign of breath.   Water is one of two life-principles ; breath the other.   The calf's head is the typical acknowledgment that they were indebted to the folk of Enmore Green for very life.   Only an Egyptologist can know how aptly the two are juxtaposed according to Egyptian symbolism.   An (En) means a valley, and Mer is a pool, trough, cistern, or reservoir of water.   An-mer answers to Enmore.   Shau (Eg.) is the high dry place, and this is an abraded Shaf ; tes is the dense, hard rock. Thus Shautes, or Shaftes, is the waterless rock.   Shaston is appa-rently Shafteston, corresponding to Shaftesbury.   Ton equates with bury, and throws light on it.   Ten (Eg.) means the elevated seat, and Burui is the cap, height, summit.

At the Bel-tein celebration of the 1st of May in the Highlands of Scotland it was customary to make oatmeal cakes, upon each of which nine nipple-like nobs were raised, each one being dedicated to a different being supposed to be the preserver of their herds and flocks.[2]

The Egyptian Put was the festival of the 9th or Nine.   Put is the company of nine gods.   The Baal-fire belongs properly to the summer solstice, coincident with the beginning of the Inundation, when three months overflow and nine dry months made up the year.   A cake is the ideograph of land, and the nine nobs, like the nine Bubu of Isis the Gestator were equivalent to the nine months of dry land.

What is the origin of the belief that there is a peculiar virtue in the dew of the first day of May ?   It was at one time religiously regarded like the fabled fount of living waters, that made the bather young

[1] Dyer, pp. 205. 206.   [2] Pennant's *Tour in Scotland*, p. 90.

or renewed the beauty for ever. One of the commonest English customs was for people of both sexes to rise early and wash their faces in May-dew to make them beautiful. The present writer was one of a faithful few in his boyhood who performed the ceremony without attaining the supposed result.

Our English dew is probably the Egyptian "Tuau," some kind of liquid. But if the U be a modified V, "Dev" still represents the Egyptian "Tep," a drop, the dewdrop, "Tep" seed, "Tef" the divine source; TEF to drip. One of the earliest observations enshrined in mythology was that of the condensation of breath into dew. Dew is both breath and spirit. In Toda Div is breath, in Zend it is spirit, both meet in the Egyptian Tef, seed, source, and this was the first dew, the dew of life, dew of heaven, dew from above. One Egyptian name of this dew of source is Mai, the semen; our English May, the seminal month of the year. The may or hawthorn is one of the first trees to blossom as the first fruits of Spring. Our word Haw is the Egyptian "Hau," signifying first-fruits and rustic or countryfied. Thus interpreted the dew of May is an external emblem of the Mai of masculine source considered as the fount of life and water of immortality.

This, however, was later, the first water of life was assigned to the female nature, and poured out of the tree by Nupe or shed by the wet-nurse Mena, Maka, or Mâ, our May, and by TEFnut.

On May-day in the Isle of Man, there was a Queen of the May elected, likewise a Queen of the Winter. Each was supported by their respective followers who marched and met on a common where they fought a mock battle. It was a celebration of the turning back of Winter in presence of Summer. There was a procession of Summer, sometimes composed of little girls, locally called the Maceboard —an assumed corruption of May-sport![1] The "Maceboard" went from door to door with a small piece of green ribbon, asking if the inmates would buy the Queen's favour, the token of triumph over the Winter. Now Mesh (Eg.) is to turn back or the turning back, Pert is the name of Winter, and Mesh-pert, the equivalent of MACEBOARD means the turning back of Winter. Green was the symbol of rebirth. Our May customs, games, rites, and ceremonies belong mainly to the equinox, and this contention of Summer and Winter equates with the battle of Horus and Sut at the crossing; the proper date would be the 25th of March.

In Hasted's *History of Kent*, it is related that a singular and most ancient May custom was extant at Twyford in that county, although nothing was known of its origin or meaning. Every year the people elected a "Deputy to the Dumb Borsholder of Chart," as it was called. This Dumb Borsholder was always first called at the Court-Leet holden for the hundred of Twyford, when the keeper of the

[1] Dyer, p. 246.

image for the year held it up to the call, with a neckcloth or hand-kerchief run through a ring fixed at the top.[1]

The dumb Borsholder was made of wood, about three feet and half an inch long with iron ring atop, four more at the sides, and a square iron spike at bottom, four and a half inches long to fix it in the ground. It was made use of to break open, without the warrant of any justice, either of a certain fifteen houses in the precinct of Pizein-well, on suspicion of anything being unlawfully concealed there. The Dumb Borsholder claimed liberty over these fifteen houses, every householder of which was formerly obliged to pay the deputy one penny yearly. This Borsholder of Chart and the Court-Leet was discontinued and the Borsholder put in by the Quarter Sessions, for Wateringbury, afterwards claimed over the whole parish.

Chart represents the Egyptian Karti, the dual Kar or circle which was divided equinoctially at the Pool of the Two Truths. The plural chart exists in Kent where we find the Two Charts called the Great and Little Chart.

The place of the Leet was Twyford, the double crossing, an equi-noctial name. At Twyford the river Medway receives two of its affluents, one rising in Kent, the other in Sussex ; and here the Pool of the Two Truths (in An) is represented by the Pizein Well.

Pi-shin is in Egyptian the circuit, the twin-total of the Two Truths typified by the two waters or by the Pshent Crown and Apron ; it is the equivalent of Twy-ford. The mapping out is astronomical and identifiably Egyptian.

At the place of the well of the two waters, was the hall of double justice. And at Twyford was held the Leet. A Leet is a meeting of cross-roads, a type of the equinox, and the Leet in the legal sense is the hall of Justice.

Lambarde says that which in the West Country was at that time, and yet is, called a Tithing, is in Kent, termed a Borow. A borowe (A.-S) is a surety, to be a pledge for another ; the (A.-S) borgh, a pledge. Borwehood is Suretyship, and the Ealdor of this Tithing, who is also known as the Borsholder in a Tithing of ten families, was the Borow-Ealdor, Borgh-Ealder, the Surety and Pledge-Ealdor, who was responsible for the security of his borh, borge, or borough. The Borowe-Ealdor became the Borsholder and finally the Bosholder.

He was the one who gave pledge and surety as a substitute for the rest. A doctrine of the Messiahship is bound up with this suretyship. Horus, in one character, was the pledge and substitute for others. In the chapter of coming forth justified [2] the Osirian says, " I come forth. . . I have crossed the earth at the feet of spirits, a SUBSTITUTE, because I am prepared with millions of charms." The sun-god, who descended into the Hades or crossed the earth, was represented as the suffering substitute, the one who pledged himself

---

[1] Vol. ii. p. 284.　　　　[2] *Rit.* ch. 48.

or his word for the safety of all. When he went down he promised to rise again, and when he re-ascended he was as good as his word, the word made truth, the justified Makheru.

The Kart or orbit of the sun was divided into upper and lower heaven, and in the nether Kar, the "bend of the great void"[1] are the fifteen gates of the House of Osiris, through which Horus, as "Tema" the justicier, has to pass, and issue from the fifteenth gate on the "day of the festival of the adjustment of the year," that is, of the spring equinox.[2] These fifteen gates were probably luni-solar, fourteen belonging to the half-circle of the moon, the fifteenth being added in the luni-solar half month of fifteen days. These correspond to the fifteen houses over which the Borsholder claimed lordship and liberty in his half of the Kart in the precinct of Pizein Well. Horus Tema was but the deputy of his father, and he breaks his way through the fifteen gates, "correcting the fugitives," "chasing the evil," and "slashing the enemies of Osiris," as the deputy of the Borsholder had the right to break into the fifteen houses without warrant of any justice. The Bors was lifted up in court by a hand-kerchief or neckcloth passed through a ring fixed at the top of it. Amongst other IDENTIFICATIONS of himself with things, Horus, in the Fifteenth Gate, says, "I am the strap of the hole (or ring) which comes out of the crown," evidently to lift it by.

The image called the "Dumb Borsholder" was the Deputy's sign of rule, and probably represents the Tum Sceptre, the sign of strength. Every year in the Hundred of Twyford they elected the Deputy to the Dumb Borsholder. The Deputy impersonated the solar son. Every year in the Myth the father, as Atum or Osiris, was represented by deputy in the suretyship which became the Messiahship of Eschatology. This deputy was the son, the Nefer-hept, the Prince of Peace, called in Egyptian the Repa, or heir-apparent, the Governor for and in the place of the father. Also, in accordance with this are the other facts that one of the titles óf Horus is "Lord of Khent-khatti," that Khent-katti is a designation of the Har Sun, as "Lord of Kem-Ur, dweller in Katti;" that the "Stone of Ketti," one of the three vast labours of the Kymry was erected in Kent, and the Kymry were the first known inhabitants of that county. The custom being equinoctial had, like so many more, got behind with the lapse of time.

So inseparable are the cross and circle that, at Northampton, the ceremony of beating the bounds is termed "beating the cross." The crossing and the four quarters are synonymous. The four quarters, in Egyptian, are named Fetu or Fatu. The "Furry Festival," cele-brated from time immemorial, at Helston, in Cornwall, on the 8th of May, was an equinoctial festival, as shown by the illustrations of crossing. It was held as a general jubilee. People who were found

---

[1] *Rit.* ch. 147.   [2] *Ibid.*

working on that day were compelled to leap across the river Cober, or fall into it.   The Cober answers to the Egyptian Khepr, the transformer and god of the crossing where the transformation occurred. We have the mount of transformation of the one water into two rivers in the Irish Kippure; the image of transformation in the Cyfriu, and here the Cober, the river of the crossing, supplies another type of the passage and change of Khepr.   At Helston the people danced what was called the Fadé dance, claiming the right of crossing and passing wherever they chose, up and down the streets, and through and through the houses.   This answers to Fetu (Eg.), the four quarters, and it is suggested that that is the meaning of the Fadé dance. FUDU, in Zulu Kaffir, denotes a peculiar kind of dancing: a VITHI in Sanskrit, is a sort of drama.   The festival is called the Furry, supposed to have the same meaning as the fair.   The word and its true significance are probably represented by Peru (Eg.), to go out, go round, show, appear, see, sight, manifest, explain, with the ideograph of the year.   Michael is the patron saint of Helston, and he is the British form of Har-Makhu or Khepr-Tum, the sun of the double horizon, and equinox.

Helston has a tradition which shows the place is named as the stone of Hel, that is Har (Eg.) the Solar Lord.   The stone placed at the mouth of hell is contended for, as was the body of Moses by Michael and Satan,[1] or the advantage in the scales by Har-Makhu and Sut.   This marks the annual conflict of the Mythos localized at Helston, the Furry festival and Fadé dance being held in commemoration of Michael's or Har-Makhu's victory.   The 8th of May is 3,000 years behind the correct date.

An old distich says :—

" Shig-shag's gone and past,
   You're the biggest fool at last,
   When Shig·shag comes again
   You'll be the biggest fool then."

Shig-shagging belongs to the time and motion of the equinox, the Khekhing already expounded.   It is here coupled with the fool, the Gouk, or Khekh, but is now applied to Oak-apple day.   At Tiverton on " Shig-shag Day," the Black God or Black Jack, has been transformed into Cromwell dressed in black with a blackened face, and called " Master Oliver," who is made sport of.   After him follows a young child, borne on a kind of throne made of green oak-boughs.[2] These now represent, but did not originate with Cromwell and the Second Charles.   They are but a survival of imagery re-adapted to a later purpose, just as the whole masquerade of so-called heathenism has been re-christened and continued.   We shall find the young

---

[1] Murray, *Handbook for Cornwall*, 1865, p. 301.
[2] *Notes and Queries*, 1st. ser. vol. xii. p. 100; *Every Day Book*, 1826, vol. i. p. 718; Dyer, pp. 302-305.

child in the tree, as the messianic branch, is one of the oldest types in the world, and in the solar allegory it belongs to the time of the vernal equinox.

Croker chronicles a custom observed in the south of Ireland on the eve of St. John's Day, and some other festivals, of dressing up a broomstick as a figure, which is carried about from cabin to cabin in the twilight, and suddenly thrust in at the door or window to startle the people of the house. The fright caused by this apparition was productive of merriment. The dressed-up figure was called a BRE-DOGUE. Prut (Eg.) denotes manifestation, appearance, and Ukhu or Akhu is a spirit, the Manes. Prut-ukhu is a spirit-manifestation or apparition of the dead.

The festival of the Solstitial division, or Ten, is celebrated in the Isle of Man on the 24th of June, which is termed Tynwald day. The ceremony of the Tynwald Hill is described in the *Lex Scripta* of the Isle, as given for law to Sir John Stanley in 1417. "This is the constitution of old time, how ye should be governed on the Tynwald day. First you shall come hither in your royal array, as a king ought to do by the prerogatives and royalties of the Isle of Mann, and upon the Hill of Tynwald sitt in a chair covered with a royal cloath and quishions, and your visage to the east, and your sword before you holden with the point upward."[1] The Barons beneficed men, deemsters, coroners, and commons were to be ranged around the royal seat, according to their degrees. This was on the one side to hear the causes of crime and of complaints, and on the other to hear the government of the land and the royal will annually proclaimed. Wald is an English word, which signifies government. The Tyn we interpret by Egyptian. Ten is the royal seat, cabinet, or throne-room. To ten (ten-t) is to take account, reckon, each and every. The Hill was the Ten, elevated seat, or throne, said to have been built of earth brought from each of the seventeen parishes of the island, just as to ten (Eg.) means to fill up and complete the total. The Ten, with the article suffixed, is the ten-t, the throne, royal chamber, or other room of the king; and to the present time a tent is erected on the top of the Tynwald Hill on the Tynwald Day, and arrangements for the rites are still made according to the ancient custom.[1] The ceremony belongs to a time when the year began with the summer solstice, and the king turning to the east shows his assimilation to the solar god.

The "Blue Peter" is a flag with a blue ground and white figure in the centre; it is hoisted as the signal when a ship is about to sail. It notifies to the town that any person having a money claim may make it before the vessel starts, and that it is time for all who are about to sail to come on board. Peter is supposed to be a corruption of the French PARTIR. The Blue Peter is a time-signal.

[1] Cumming's *History of the Isle of Man*, pp. 185, 186.   [2] Dyer, p. 325.

The present suggestion is that Peter is the Petar (Eg.) meaning time, to explain, show, regard, look at.   Pru (Eg.) the equivalent of blue, means go forth, come forth, proceed.   The Petaru a slip of papyrus is extant with us as the slip or broadside of the street patterer.   Petar-er is Egyptian for announcing by word of mouth. Petar (Eg.) is some form of measuring and reckoning, and has the palm-shoot for determinative, a symbol of time and period.   We have a third form of the petar.

At Nun-Monkton, Yorkshire, on the Saturday preceding the 29th of June, called Peter's Day, the villagers at one time mustered together and, headed by musical instruments, went in procession to Maypole Hill, where an old sycamore stands, for the purpose of "rising Peter," who lies buried under the tree.   The effigy is a rude one carved in wood—no one knows when—and clothed in a ridiculous fashion. This was removed in its box-coffin to the public-house near, and there it lay on view.   Then it was thrust into some outhouse and no more thought of till the first Saturday after the feast, when it was taken back to the tree and re-interred with all honour.   The ceremony was designated the burying of Peter.   In this way the risen Peter presided at his own feast.[1]   This also is now claimed as the Petar, an image of time, the time being the Summer solstice when the sun begins descending to his burial.

A bundle of reeds tied up is an ancient ideograph of Tur, a time, and to indicate, and of Rut to repeat.   Bundles of reeds, rushes, and flags, were tied up and carried in procession at our old country English rush-bearings and wakes.   On the Sunday next after the feast of St. Peter the parish church of Farnborough, Kent, is annually strewn with reeds.   The day is called by the inhabitants of the village Reed day, and the local tradition affirms that a Mr. Dalton was once saved from drowning by reeds.   A mural tablet in the church sets forth that this gentleman left a perpetual annuity of 13s. 4d. chargeable upon his lands at Tuppendence; 10s. to be given to the preacher of a sermon on that day, and 3s. 4d. to the poor.[2]   But the tablet does not corroborate the tradition.   Now to go no farther back than the canonized Peter ,the hieroglyphics enable us to see that Reed Day may have been connected with his escape from drowning. if not with Mr. Dalton's.   The bundle of reeds tied up is the symbol of a time and a repetition.   With the article p prefixed it is Peter by name, meaning to show, explain, the time.   The time is that of the Midsummer solstice, the crown and climax of another year, when the new year in Egypt was announced by the inundation.   Rut (the reed) is our word reed, and the reeds are the sign of tur, Petur, or Putar, to regard, look at, explain, show, the time of repetition, i.e. the solstice.   The whole matter may have been connected with the Dalton name.   Dal represents tar (Eg.), and ptar the interpreter of time and period, whilst

[1] Dyer, p. 333.   [2] *Maidstone Gazette*, 1859.

ton (tun) means to fill up, terminate, determine, as was done by the reeds. Tr-tun also reads the high time or tide of midsummer.

We also have the Petar or Betar candle. At the festival of St. Giles, whose day is September first, betar candles were burned in his church at Oxford. These are mentioned in the Proctor's accounts so late as the early half of the sixteenth century. They are called Judas Betars and Betars for Judas' light. This apparently associates them with the lights at the betrayal ascribed to Judas, which would be in keeping with the meaning of Petar, a time, to discover, show, explain, reveal. In Egyptian symbolism the candle, Ar, was a type of the Eye-of-Horus and is called the Ar-en-Har. The betar was made so as to give forth a strong smell in burning. A form of this candle used in the Coventry Mysteries was made of resin and pitch.[1] Also the eye or candle of Horus was made of tahn or resin. The first Horus, the Khart or Khar, was the cripple deity, and Giles or Gele, as his name is also spelt, was the patron saint of cripples. He was a cripple himself who refused to be cured of his lameness. His church in Cripplegate, London, still represents the idea. Now it seems to me that the betar or petar candle tends to show, reveal, explain, that it was an extant form of the Ar-en-Har candle, and eye of resin, and that the Cripple Gele or Giles was the Cripple Khar (or Khart), who was also the Kherp as the first form of the child Horus.

On July 25th, St. James's Day, it is the custom for the rector of Cliff, in Kent, to distribute at his parsonage, annually, a mutton-pie and a loaf, to as many persons as choose to demand them. The amount expended in costs is about £15 a year. Nothing is known of the origin of the gift. But it happens that the name of the living, Cliff, corresponds to the Egyptian Kherf or Kherp, which means to supply a sufficiency, an offering of first fruits, the first or model form of a thing. Kherp also means to steer. And this would apply to the first day of the oyster-fishing, as on St. James's Day it is customary in London to begin eating oysters. The Egyptian sacred year opened about this time, the date given being July 20th. With this custom of Cliff we may compare that of the " Clavie," in Morayshire, where we find the procession used formerly to visit all the fishing-boats in making the circuit of the boundaries.

The upper crown of Egypt is white, the lower red. When the sun entered the abyss, the white crown, put on at the time of the vernal equinox, no longer applied. One name of the abyss, the deep, is the Tes, and at Diss, in Norfolk, it is the custom for the juveniles to keep " Chalk-back day on the Thursday before the fair day, held on the third Friday in September, by marking each other's dresses behind with white chalk." [2] At this time the sun, the enlightener or whitener, entered the region called the hinder part of the circle. The

---

[1] Sharp's *Coventry Mysteries*, p. 187.
[2] *Notes and Queries*, 1st ser. vol. iv. p. 501.

upper crown is the Hut, of Hu, the white god, and the red is the crown of the lower sun. In the Universities of Oxford and Cambridge, WHIT-sunday was especially observed as a SCARLET day.[1] Further, we shall find that Thomas is a representative of Tum, the wearer of the red and white crown, and to him has been assigned the onion of Hu, and one of the old cries of London was "Buy my rope of onions ; WHITE St. Thomas's onions," the white (Hut) Hu being a form of Tum or Tomos. The chalking on the back denoted that the rule of the White God of the White Crown had ended.

The peasantry in the parish of Bishop's Thornton, Yorkshire, object to gathering blackberries after Michaelmas Day, because, as they say, the devil has set his hoof on them. The triumph of Typhon, the Egyptian Satan, began with the autumn equinox, when the sun entered the lower signs, and Osiris was shut up in the ark of the underworld. The black fruit then passed into the possession of the dark power, the Apophis-serpent of evil, who, as Satan, made his trail over the berries. The berries thus rejected are often the ripest and finest, yet the superstition holds its ground against the temptation of selfish gain.

In Suffolk the harvest men have a custom called ten-pounding, the origin of which term is unknown, but it has nothing to do with the number ten.[2] The reapers who work together agree to a set of rules by which they are governed during the time of harvest. When any one breaks them, a mode of punishment is practised called ten-pound-ing. The culprit is seized and thrown down flat, and stretched out at full length on his back and held down. Then his legs are lifted, and he is pounded on his posteriors. Ten (Eg.) means to extend, spread, and stretch out. This describes the signification of both ten-pounding and tunding.

The Welsh had a symbolical play on Allhallow Eve, in which the youth of both sexes sought for a sprig of ash that was perfectly even-leaved, and the first of either sex that found one cried out "Cyniver," and was answered by the first of the other sex that succeeded, and this was an omen that the two were destined to become man and wife.[3] The meaning of the word is unknown. KAB (Eg.) signifies two or double. Nefer means good, beautiful, perfect. Kab-nefer (Cyniver) would express the meaning of the cry when the TWO perfect leaves are found. Nefer (Eg.) also signifies a crown, the youth, puberty, and to bless. Nefer is a divine title, expressive of the highest good and absolutely perfect one. The "Un-nefer" is the good or perfect being. It may be the word Cyniver is an abraded form of Ken-nefer. Ken (Eg.) is to accompany, go together. Ken-nefer means that perfect match sought for in EVEN leaves. This Cyniver suggests that the name of the beautiful Gweniver is derived from Khen-nefer, the

---

[1] *Kalendar of the English Church*, 1865, p. 73.   [2] Forby's *Vocabulary*.
[3] Owen's *Welsh Dictionary*, voce *Cyniver*.

beautiful Khen, queen, accompanier or mate. Khen (Eg.) is the boat, the ark, the feminine abode of the waters, and Gweniver is the lady of the summit of the water in the triad of Arthur's wives.[1]

The festival of Hallow Eve is observed in the Isle of Man by kindling of fires with all the accompanying ceremonies to prevent the baneful influence of witches. The islanders call the festival Sauin.[2] Sahu (Eg.) means to assemble and perambulate, to set up, charm, drive away the evil, and the island was perambulated at night by young men who stuck up at each door a rhyme in Manx, as the charm against the evil influence.

We learn from Martin[3] that the inhabitants of Lewis worshipped a deity known by the name of Shony. On "All Souls' Eve" of each year, the people round the island gathered at the church of St. Mulvay, and brought their provisions with them. Each family furnished a peck of malt, which was brewed into ale. One of their number was chosen to wade into the sea up to his waist, to carry a libation to the god. He then cried, with a loud voice: "Shony, I give you this cup of ale, hoping that you'll be so kind as to send us plenty of sea-ware for enriching our ground the ensuing year," and then threw the cup of ale into the sea. This act was performed in the night-time, and at his return to shore the company assembled in the church, where a candle was kept burning on the altar. They stood silent for a little while, then the light was put out, and they all of them went into the fields and spent the rest of the night in merriment.

Shony presided over the tides that deposited the sea-weed and drift on the land, and Shennu (Eg.) is to fish and gather from the waters. Num (Nef), the God of the Inundation, was likewise the Lord of Shennu. Shony, interpreted by Egyptian, was a deity of the tempest and the tide answering to the Lord of Shennu and the Inundation. The sacrifice was offered to SHONY on the eve of "All Souls'," and SHENI (Eg.) signifies the crowd, myriads, the million or millions, and also the region beyond the tomb.

A modern writer has advanced the theory that religion began in the worship of dead ancestors. Unquestionably the image of the dead did take its place sooner or later as the object of sacrificial offerings, and in the case of the "TENF" (Eg.), the ancestor, which is determined by the mummy figure, we cannot dissociate the human ancestor from the wave-offering or TENUPH of the Hebrews. Yet, according to the system of thought and theory of things unfolded in mythology and symbolism, and enforced by the imagery of the Egyptian Ritual, the sun as father, he who descended into the grave or the lower heaven every year, and was renewed in the person

---

[1] Davies, *Mythology*, p. 187.
[2] Train, *History of the Isle of Man*, vol. ii. p. 123, 1846.
[3] *Western Isles*, p. 287.

of the son, was the first ancestor whose death had any sacred significance. He is the old man, the ancient of days, the past of the two Janus-faces in the images of Time, whose place was on the inner side of the closing door whence issued the radiant youth for universal welcome. And while the "Tenuph" or typical corn was waved to and fro in token of the waving wheat, and in welcome to the sun ascending from the lower signs, it was the wave of welcome and farewell; welcome to Horus and farewell to Osiris, the father, the ancestor, who had passed away in giving birth to the offspring; whether the transformation was imaged by the sun or the grain, it was the dead ancestor who had reproduced himself in the offspring now waved and offered to the manes. This belongs to the genesis of ancestor-worship, according to the data now collected and correlated.

At least it is certain that the solar, lunar, and stellar imagery furnished the types by which the primitive men expressed their feelings and intimated their hopes. The first ancestor of the fifty claimed by Tahtmes III. in his ancestral chamber is Ra, the sun-god. Solely on this foundation was the throne of the monarch built and the name of Ra as monarch conferred. Ra was the first ancestor worshipped because the earliest type of the fatherhood. Ancestor-worship applied to the fatherhood could not have existed when men did not know who their fathers were. Long before that time the bones covered with red ochre, and the embalmed body, the mummy, represented the ancestor of the soul in the primitive cult. The mystery of "Semsem" (Eg.) applied to the re-genesis of souls is based on the solar myth, and this is related to the ceremonial celebration of "All Souls' Day."

In the course of our explorations we shall find that "All Souls'" Day is common to the various mythologies, as the one day of the year on which the ghosts of those who have died during the year assemble together, and prepare to follow the sun through the underworld as their leader into light. In the Mangaian version of the myth, if some solitary laggard fails to join the crowd of "All Souls" at the time appointed for the annual gathering and exodus, the unhappy ghost must still wait on and wander until the next troop is formed for the following winter, dancing the dance of the starved in a desert place where desolation seems to be enthroned.[1]

On "All Souls' Day" a solemn service is held for the repose of the dead by the Church of Rome. The "Passing-Bell" used to be rung on this day, or, as it is sometimes called, the NOANING Bell. NUN (Eg.) means negation, not, is not, without, and in English to NOAN is to toll the bell. In some counties they say "the bell NOANS," when the knell is rung; it proclaims that the person is not (Nun), and the living are bereft (Nun), and the bell NOANS. "Old Hob" was carried round from All Souls' Day to Christmas; the head of a horse (the grey mare) enveloped in a sheet. The Irish

[1] Gill, *Myths and Songs of the Pacific*, pp. 157-9.

kept the festival of SAMHAN, called OIDCHE SAMHAN, when it was believed that all the souls which had passed away during the year were assembled together and called before the god Samhan to be judged, and then passed on to their reward in the abodes of the blessed, or, according to the modern report of the Druidic cult, to be sent back into re-existence on earth to expiate their sins in the flesh.[1] Samhan and Samana, in Sanskrit, mean to bring, unite, and join together. SAMANA is coming, meeting together, collection, and union. Samyana denotes the carrying out of a dead body. SOMEN in English, SAMYN in Scotch, and SAMEN in Low Dutch have the meaning of assembling and joining together. Sem (Eg.) is to combine, join together, unite, go in, a total, the "All" Souls; Sem, to conduct a festival, to traverse and PASS. SMEN (Eg.) signifies to determine, constitute, make durable, fix, and establish. The SMEN were the primodial eight gods, the Ogdoad of mythology, founded on the seven stars of the Great Bear (our Old Hob) and the Dog-star, the "Children of Inertness," the "Betch-party" in Am-Smen who ruled before the firmament of Ra was uplifted. In the solar mythos the son was annually established in place of the father in SMEN, the place of preparation of "All Souls," and their regeneration in the mystery of Sem-sem followed, and was founded on that of the sun and earlier stars.

"The Osiris lives after he dies like the sun daily, for as the sun died and was born yesterday, so the Osiris is born" or the soul is reborn.[2] In the same way the annual sun was the type of the soul in the gathering of "All Souls," that assembled on the day appointed to pass from earth to heaven along the shining track. To recur for a moment to the mummy type of transformation, the Shebti or double shape, it can be shown that this figure also represented the risen Christ of mythology.

The Christ is said to mean the anointed, but it cannot be that grease is the root-meaning of so mystical a name. It is so, however, for all that has hitherto been expounded. Chriso, Chrisei, Christes, Chriesthai denote an anointing with oil or unguents, and the Christ in this sense is literally the greased. Various languages show the same result. But the root which yields grease supplies Kr worn down in Egyptian to Ur for oil, and to anoint. IR, in Welsh, is oily, unctuous matter; IRA, Cornish, to anoint; URO, Fijian, fat, grease;

[1] It is doubtful whether the doctrine of transmigration and reincarnation of souls is not a Hindu and Greek misinterpretation of the Egyptian doctrine of trans-formation, and a mis-rendering of the typology. In the Egyptian eschatology of the Two Truths of flesh and soul, blood and breath, the "Sen-Sen" still dominated in their expression, and if the first soul, the mummy, that represented the flesh body of earth, did not transform into the second or pure spirit, was not re-gene-rated, it was resolved in the place of dissolution just as if the flesh were resolved again into blood, and the blood formed the Red Sea, the Pool of Pant or Primordial Matter. Now, in their "Abred," the Druids possessed this same region of source and dissolution. This subject will be pursued in a chapter on the *Ka Image and the Mummy Type.*
[2] *Rit.* ch. 3.

EWIRI, Oloma, palm-oil; HORU, Maori, red ochre ; KORAE, Maori, to anoint with red ochre and oil; GUHR, English, a kind of ochre ; OCHRA, Greek, coloured earth; GERU, Hindi, a kind of red ochre ; ICHIRA, Manyak, oil ; GIRA, Kra, palm-oil ; KIRA, Basa, palm-oil ; EKURO, Ako, palm-oil ; UKARA, Bola, palm-oil; GEIR, Gaelic, to anoint with grease; CHRIO or CHRISO, Greek, to rub over with colour ; CHRIESTHAI (Greek), to rub over with colouring matter ; CHRISTES or CHRISTOU, one who colours, smears, or bedaubs.

Anointing the living with oil was a mode of consecrating, but the dead were consecrated first, and red ochre was one of the earliest substances employed, as in the Maori custom of preserving the bones of the dead, which were exhumed periodically, scraped, and re-anointed or rather re-clothed with HORU, or red ochre. So the Hurons celebrated their feast of "All Souls" once every ten years, when the dead were taken out of their graves, no matter in what condition of corruption, cleansed from worms,[1] and carried once more to their homes. They were collected from near and far for the ceremony, and then were all laid in the earth together.

The human bones in the British mounds of Caithness were found to have been coated over with red earth. This, which was practised by other races, was the earliest mode of embalming and anointing the dead, who were KARAST in their covering of ochre ; the red earth, being an image of the flesh, preserved a kind of likeness to life. These were the men of the later palæolithic age, who had rudely begun the art of embalming the dead, which culminated in the production of the Egyptian mummy, as the KARAST, in the Karas. Karas denotes the burial and embalmment, and the corse embalmed (anointed) becomes the KARAST.   Karas is equivalent to the English corse for the dead body ; Gaelic and Irish Cras, the body ; Greek Chros, the body.

The Karas as the place of the mummy embalmed is extant in the Irish CREAS, a shrine ; CROISÉE, French, transept of a church ; the CROUSTE (French Rom.), Arabic KURSIY ; Turkish KYURSI for the pulpit ; Irish CREAS and French (Rom.) CRES, for a grave.

The KAR-AS is the sepulchre as resting-place below.   And because the circle and cross, as in the Ankh, were typical of life, the Karas and ancient graves were often cruciform, and the dead were laid there with their arms crossed, hence the identity of the Karas and the cross ; also of the KARAST and the crossed with the Christ, in the sense of the crucified.

The mummy image is the reduplicated shape, as the Shebti, the *alter ego*, other self, or literally the double of the dead.   It was a type of tranformation, and as such stood up in the Karas as the re-arisen image of the corse that lay below.   It was the risen KARAST.

---

[1] Rabbi Isaac declares that a worm in a dead body is as painful as a needle in a living one.

Chrestos is a Greek term applied to the sacrificial victims, denoting them to be auspicious, and signifying good luck. This was the Chrestos or Karast, the Maneros of the Egyptians, the divine victim who, " in the likeness of a dead man," was carried round at the festival, not, says Plutarch, to commemorate the disaster of Osiris, but by way of wishing that things might prove fortunate and auspicious.[1]

In the African Pepel language KRISTO means an idol or divine image, and in this the worshippers had their Christ independently of the Greek or of Christianity. It represented the primitive type of the mummy or Mamit, as did the figure of the deceased in Egypt or Assyria, the one that was embalmed and anointed as the KRAST, the Egyptian original for the name of Christ. But, to return.

On St. Leonard's Day each tenant of the manor of Writtel paid to the lord for every pig under a year old a halfpenny, for every yearling pig one penny, and for every hog above a year old twopence, for the privilege of pawnage in the lord of Writtel's woods. The payment was called Avage,[2] or Avisage. Aph, in Egyptian, is the hog or boar, and Aph-age would be boarage. Also, Sekh is to remember, remind, memorize, and Sak, to bind, direct, order, execute. So read, Aphisak is the tenure of pawnage in the woods. As many of these payments show that the tenure was religious, the name of Writtel may denote an ancient religious house or lands.

A belated equinoctial custom is apparent in the hundred of Knightlow, where a certain rent is due to the lord called Wroth (or Warth) money, or the Swarff Penny, payable on Martinmas Day in the morning, at Knightlow Cross before sunrise. The person paying it has to "go thrice about the Cross and say ' The warth money,' and then lay it in the hole of the said Cross before good witness "; the forfeiture for non-payment in the prescribed manner being 30s. and a white bull.[3] The cross is a certain sign of the equinox.

WRATH is the name of a pillar, a prop, *ergo* the cross, as the Hindustani URUT is a cross-beam, the Irish UIRED, a pillar, column, or stone cross. The Egyptian RUTI are cross-shaped, as gates,[4] and the horizon, or crossing, is also the RUTI. The same word " RUT " means to engrave in stone, figure, retain the form, the earliest writing, and it passes into the name of writing. But this custom of payment at the Wrath or Cross must have been a survival from the Stone age, when there were no written documents. The cross is a sign of the Kart (Eg.), the orbit, or circuit of the two heavens, and Wrath is equivalent to Kart. The payment was made at the cross because the course was completed, and cross and course are synonymous. Also the stone cross served the same purpose as the making of the cross for a signature of covenant. The " SWARFF

[1] *Of Isis and Osiris.* Herod. b. ii. 78.
[2] Blount's *Law Dictionary*, 1717.
[3] Dugdale, *Antiq. of Warwickshire*, 1730, vol. i. p. 4.
[4] *Eg. Sal. Brit. Mus.* 254.

penny " probably denotes the KHERF penny (Eg.), an offering of first-fruits by which homage was paid, now represented by the GLEBE. The Scottish WRATH for food and provender tends to identify the offering as the provision penny.

The four cross-quarter days of Whitsuntide, Lammas, Martinmas, and Candlemas are doubtless the most ancient quarter-days, or gules, as witnessed by the rents still paid on them, especially in Scotland ; and as these were markings of the solstices and equinoxes, they are now some 3,000 years behind time. Lammas, for example, preserves the Egyptian Rem, or measure of extent. The determinative of Rem is the arm as type of the extent ; and the charter for Exeter Lammas Fair is perpetuated by the sign of an enormous glove, which is stuffed and carried through the city on a long pole decorated with flowers and ribbons. It is then placed on the top of the Guildhall as a token that the fair has begun, and when the glove is taken down the fair terminates.[1] The glove takes the place of the hand or arm, the sign of Rem (LIM-it), the measure, and the hieroglyphic is the same whether on the top of the Exeter Guildhall or the Tower of Anu, or in the caves of Australia. One form of the Rem (Eg.), measure, is a span, that is, a hand, used as we measure by the foot. The human body supplied the first hieroglyphics, and these were after-wards supplemented by the productions of man. So the glove follows the arm and hand. It was customary, at one time, to give glove-silver to servants on Lammas Day, but this was not the only limit in time thus marked. Gloves were likewise given on New Year's Day, as well as glove money. The word glove still retains the value of Kherf (Eg.), a first form, a model figure, a primal offering.

Tander and Tandrew are Northamptonshire names given to St. Andrew, supposed to be corruptions of the Christian name. St. Andrew's or Tander's Day used to be kept with ancient rites and ceremonies, amongst which was the exchange of clothes, the men being attired as women, the women habited as men.[2] The day has receded to November 30, but the change of raiment identifies the custom as belonging to the equinoctial crossing. The type of the " Saint " Andrew is the cross. An (Eg.) means to repeat, to renew the cycle, and is the name of the crossing where the cycle was renewed. Teriu (Eg.) is the two times, the circumference. Andrew, like so many more saints, is an imposter, a personification of the cross, which has been assigned to him as his symbol. Hence it comes that the Maltese Cross, called by the name of " Saint " Andrew, is found to be the ideograph of the old god Anu of Assyria ; and neither he nor his emblem, nor the Egyptian two times (Teriu), represented by the cross, could be derived from the Christian Andrew. The singed and blackened sheep's head that used to be borne in procession before

---

[1] *Every Day Book*, vol. ii. p. 1059.
[2] Sternberg, *Dialect and Folklore of Northamptonshire*, p. 183.

the Scots in London on St. Andrew's Day was probably the antithesis to the ram of the spring equinox,[1] just as the black bird of autumn is opposed to the bird of light. Tander's Day regulates the commencement of the ecclesiastical year. The nearest Sunday to it, whether before or after, constitutes the first Sunday in Advent, and Tander's Day is sometimes the first, sometimes the last festival in the Christian year.[2] This, again, relates the day to the equinox, and keeps up the dance of the crossing, but at the beginning of the lunar year, still kept and correctly adjusted by the Jews about the time of the autumn equinox.

Shau, in Egyptian, is the English sow. The word Sha also denotes all forms and kinds of commencement, beginnings, and becomings. Now the people in the parish of Sandwick, in Orkney, kept what they termed Sow-day on the 17th of December, upon which day every family that had a herd of swine killed a sow.[3] The Egyptians, according to Herodotus, held the swine to be impure, but they had their Sow-day. One day in the year (at the full moon) they sacrificed swine to the MOON and Osiris. He knew why they did it, but thought it becoming not to disclose the reason.[4] The sow was a type of Typhon, and the time of Typhon began at the autumn equinox. Anent this time we learn that from Michaelmas to Yule was the time of the slaughter of Nairts.[5]

It appears to me that the Nairt here slain in the Typhonian time was an infertile animal named from its not breeding. Narutf (Eg.), the variant of Anrutf, is the barren, sterile, infertile region in the *Ritual.* Neart also is an English name of night. Nai-rut (Eg.) denotes the negation of the race, or non-fertility. Sow-day was so ancient that there was no tradition concerning its origin, and if the 17th of December represented by the natural lapse of time that 17th of Athyr (September in the sacred year) on which Typhon shut up Osiris in the ark,[6] the custom was, indeed, most ancient.

The tinners of the district of Blackmore, Cornwall, celebrate "Picrous Day," the second Thursday before Christmas Day. It is said to be the feast of the discovery of tin by a man named Picrous. There is a merry-making, and the owner of the tin stream contributes a shilling a man towards it.

Tin in Egyptian is Tahn, which is also the eye of Horus, and the halfway of heaven, that is, the equinoctial division where the eye constellation is found. The division is Peka (Eg.), our Pasch or Easter. Res (Eg.), to raise up, is also determined by the same sign, the half-raised heaven. Pekh-res (Eg.) is the half-way heaven

---

[1] Brand, St. Andrew's Day.   [2] *Book of Days,* vol. ii. p. 636.
[3] Brand, Martinmas.   [4] B. ii. 47.
[5] *Laws and Constitutions of Burghs, made by King David the First at the New Castell upon the Water of Tyne, in the Regiam Majestatem,* 1690, ch. 70; *Of Buchers and Sellers of Flesh.*
[6] Plutarch, *Of Isis and Osiris.*

of the equinoctial division. As before said, many of the equinoctial festivals were transferred to the time of the solstice, as the initial point of the uprising.

St. Thomas's Day is observed in some places by a custom called " Going a Gooding." The poor people GO ROUND the parish and collect money from the chief people for the keeping of Christmas. Formerly a sprig of holly or mistletoe was presented to those who bestowed alms. Going round, peregrinating, is the essential meaning of Gooding. The good-time is the periodic festival. Khut (Eg.) signifies to go round, travel circularly, make the orbit, circuit, circle, cycle.

Har-Khuti, god of both horizons, is the deity of going round, the good or Khut god. The devil has the character of the goer round, and he is called the good man. The fairies go in circles, and they are the "good folk." Going gooding is the same as going gadding round about. Khut is to shut and seal, to catch and keep hold. And in the customs of Valentine's Day, catching and clasping of the person is a salute equivalent to the salutation " Good Morrow."[1] The going round from house to house to sing the " Good Morrow, Valentine," is identical with the going a-gooding. One form of Har-Khuti, the god of going round, is Tum, whom the Greeks called Tomos. Gooding is based on going round, making the circle as a symbol of a completed cycle of time. In this sense the last Sunday in Lent is designated " Good-pas Day ; " the six Sundays being called Tid, Mid, and Misera, Carling, Palm, and Good-pas day. Khut-pesh (Eg.) is the extent of the circle-making. The Khut as place was the horizon of the Resurrection. And the " Good " Friday is the Khut Friday. Nor is the hieroglyphic missing.

The Khut-ring is a seal and sign of reproduction, restoration, and resurrection, and the kings of England, according to Hospinian, had a custom of hallowing rings with great ceremony on Good Friday to be worn as an antidote to sickness.

The greater number of popular customs and festivals belong to the vernal equinox, although some of these have been shifted to the Winter solstice to celebrate the later new year, and others have got belated through not being readjusted in the course of time.

Train, in his *History of the Isle of Man*,[2] relates that the Christmas waits go round from house to house at midnight for two or three weeks before Christmas. On their way they stop at particular houses to wish the inmates " Good morning." The fiddlers play a piece called ANDISOP. Anti (Eg.) is to go to and fro. Sop (or Sep) is a time, a turn, as is midnight and the turn of the year, the solstice. But the true time of to and fro was equinoctial. The dancing, mocking, and mixing were all connected with the vernal equinox and the sun's ascent from the underworld.

Formerly it was a custom in Somersetshire for the youth of both

[1] Brand. [2] 1845, vol. ii. p. 127.

sexes to assemble beneath the thorn-tree at midnight on Christmas Eve or on old Christmas Day, and listen for the bursting of the buds into flower. It was said by one village girl that " as they came out, you could hear 'um haffer."

The word haffer has been given by Halliwell and others as meaning to crackle, patter, make repeated loud noises. But it is more likely a derivative from HFA (Eg.), to crawl like the snake or caterpillar. Thus hearing them haffer would be to hear their stealthy movement in opening, their heaving. A form of the word Hâu (Hfau) means first-fruits, and yields our name of the haw or thorn-tree. The ceremony had doubtless been put back to Christmas.

In the Scilly Isles the young people had a pastime on Christmas Day called Goose-dancing, in which the sexes changed clothes with and wooed one another; vieing with each other in politeness and gallantry.[1]

In Egyptian Kes is to dance, also to bend and sue, entreat pronely, abjectly. Khes is a religious rite, and means to reverse, turn back, and is connected with the turning back of the sun from the lower solstice.

Not long ago the festivities of Christmas commenced at Ramsgate, Kent, with a strange procession, in front of which was carried the head of a dead horse, affixed to a pole four feet long; a string was attached to the lower jaw and pulled frequently, so that the head kept snapping with a loud noise. The people who accompanied the horse's head were grotesquely habited, and carried hand-bells; the procession went from house to house with the bells ringing and the jaws snapping, and this was called going a-HODENING.[2]

Our word head is the Egyptian Hut, head and height. Hutr (Eg.) is the going horse. The winged Hut was a symbol of the sun, and the horse was also adopted as a type of the swift goer. Hutu (Eg.) means one-half or halfway round the circle. One Huti image is the demoniacal head on a staff, the ideograph of throat and swallowing. The action of the horse-jaws suggests that of swallowing. Huter is a ring, and they made the ring in going round a-hodening. The Hut (Eg.) is the good demon. And the horse-head was typical of the Hut and of the horse constellation, Pegasus, which the sun at one time entered at the turn of the Winter solstice some five thousand years ago. Uttara-Bhadrapada is the twenty-seventh lunar mansion in the Hindu asterisms, partly in Pegasus. This was the point at which the sun began to mount, hence the winged horse.

The horse-head was the Hut (Hutr), the good demon threatening and terrifying and overcoming the powers of darkness. The horse was a substitute for the ass of Sut-Typhon, which was condemned at a very early period in Egypt, so early as to be almost absent from the monuments except as the symbol of Typhon. If for a moment we

[1] Troutbeck, 1796, p. 172. [2] Brand, Christmas.

restore the ass, then this "hodening," with the horse's head and snapping jaws is the exact replica of the jaw-bone of the ass with which the Jewish solar hero slew the Philistines.   The Hebrew mythology made use of the ass instead of the horse ; the ass on which the Shiloh rode, the Shiloh being the young hero, the avenger of his father ; in the Hebrew myth Shem-son.   The singing of carols at Christmas is still called hodening.

It is not known why our ancestors chose the 26th of December, called St. Stephen's Day, for bleeding their horses, but people of all ranks did so.   Aubrey[1] says, " On St. Stephen's Day, the farrier came constantly and blooded our cart-horses."   Tusser refers to the same custom.   The Pope's stud were also bled and physicked on this day. Now in Egyptian STEFU means to sacrifice, to purge, purify, and refine ; this includes bleeding and physicking.   Stefu is also a name of the Inundation, which in the mystical aspect was the periodic flow of blood.   The blood-letting was probably a comminated form of sacrifice, hence we find it is called " sacrificing."

The game of " Snap-dragon," played by children at Christmas, belongs to the solar allegory.   Raisins are snatched out of the blue flame of burning spirits or from the keeping of the dragon.   The word snap is the same as Snhap (Eg.), to take hastily, but Snab (Eg.), fire, sparks, to burn, is the more appropriate, and it renders snap-dragon as the fiery dragon.   Snab (Eg.) also signifies configuration, and Snab, to retreat and flee, expresses one part of the performance.

The yule log in Cornwall is called a Mock ; in English the Mock is a stump or root of a tree.   This is the old stock of the symbolical tree of the old year, which was renewed from the branch annually. Log or Rek (Eg.) is time, reckoning, rule.   Mak (Eg.) is to regulate, and the Christmas Eve was regulated and reckoned by the log, in this instance by burning it.   But the Mock is more than the stump or root of a tree.   It is the name of the wake or watch ; the children being allowed to sit up to watch the log a-burning and drink to the Mock, and keep up " Mag's diversion."   Makh (Eg.) means to watch, think, consider, and this was the watching.   Also the name shows that the festival was removed from the equinox to the time of the solstice, as the Makha (Eg.) is the balance, scales, the emblem of the equinox. The Christmas tree will be especially treated in the " Typology of the Tree," but it may be necessary to say a word here in season.

A writer in the *Revue Celtique*, Mr. David Fitzgerald, has lately argued that the TREE BAAL, and not the divinity, is the origin of the name of " Beltene " for May.   He says :—" The theory that the first element is the name of an old solar or fire god has many adherents yet, not by any means confined to the class of the superficial and half educated. The following, however, would seem to be the true explanation. First, the Northern antiquaries seem to have been quite accurate in

---

[1] *Remains of Gentilism ;* MS. Lansd. 226.

seeing a representative of the world-tree in the may-tree, or may-pole, and the Christmas tree. The usage yet survives in Galway, Donegal, Westmeath, and elsewhere of planting a may-tree or may-bush (*Crann-Bealtain, Dos-Beltain*) on the dunghill or before the farmhouse door, and eventually throwing it into the bonfire. The name of the festival, *Lá Beltene*, was the same as *Lá Bile-tenidh* (or *Bele-tenidh*), Day of the Fire-Tree, and came from the bonfire and may-tree usage."

Philology by itself can settle nothing from lack of the ideographic determinatives, hence the eternal wrangle over words when divorced from things. Baal may denote the god or the tree, the star (Sothis) or the pyramid, or several other variants.

Lá Beltene, to begin with, is the day of the Baal-fire, and LÁ (Rá) is the Egyptian name for DAY. The tree is earlier than the sun-god, who was born anew at the time of the vernal equinox, and Beltene applied to May is but a belated equinox. The log of the old year is now burned at Christmas, when the birth of the branch, shoot, or divine Child is celebrated. This festival belongs to the end and re-beginning of the equinoctial year, the 25th of March. The god then reborn was the solar son, the new branch of the old tree. But there was a still earlier solstitial beginning and ending of the year determined in Egypt by the heliacal rising of Sothis the Sabean Bar, or Baal, who was born as the Child of the Mother, one of whose types, we shall find, was the tree. This time corresponded to our Midsummer.

The boundary of each Cornish tin mine used to be marked by a tall pole with a bush at the top of it, and on Midsummer's Eve these were crowned with flowers.[1] The tree of the year and the boundary had typically blossomed anew at the time of the summer solstice. And at Whiteborough, a large tumulus with a fosse round it, near Launceston there was formerly a bonfire on Midsummer Eve, with a large pole in the centre surmounted by a bush, round which the fuel was piled up for burning.

The tree as Bar, Baal, Bole, or Fur, is a symbol of the god Baal which can be bottomed in Egypt only, where the imagery is yet extant.

The tree was a type of Baal before pyramids were built, and there the pyramid had·superseded the tree, as the symbol of Baal or Bar, that is, Sut, Sebt, or Sothis. Ber, the supreme height, the roof, determined by the pyramid, and star, is identical with Bel or bole for the tree, and the tree as Baal is a type of the god Baal whose other type is the pyramid. In proof of this the tree-type still interchanges with the pyramid for the Christmas symbol. In Germany the pyramid is a form of the Christmas tree, and in England small pyramids made of gilt evergreens used to be carried about in Hertfordshire at Christmas time.

[1] Brand, Midsummer Eve.

In the neighbourhood of Ross, Herefordshire, it is customary for the peasantry to carry about a small pyramid on New Year's Day built up of fruit and leaves, which takes the place of the tree.[1]

The pyramid is an ideograph of Ta, to give, and the pyramidal tree is loaded with gifts for the children. The FIR-tree is pyramidal, named from the same root, and chosen for its shape as a Christmas tree, or a FIRE-tree.

This permutation of tree and pyramid shows we have both types of Baal. The fir-tree adds another application of the name, and it agrees with Afr (Eg.), the name of fire. The fire-tree adds another type of Baal, the fire-god, who has at least three names signifying fire, the fire of the Dog-star. The Baal-fire then, it is repeated, belongs to Midsummer and the rising of Sothis, visible as the Dog in the tree, and the emblems imply the cult to which they belong in whatsoever land they may be found. The log now burned at Christmas was represented by the tree, or fire-wheel, or besom once burned at Midsummer, not because the sun was then about to descend in the circle of the year, but because the star had risen that opened the new year; the fire in heaven was once more rekindled, the time and tide of plenty had come again, another branch had sprouted on the eternal tree, and the merry-makers wore the young green leaves, and burned old brooms, and relighted the sacred annual fire, the Need-fire, as it was called, which can be interpreted by the Egyptian Nat, to compel or force, as in the Old German NOT-FEUR; Nat, to salute, address, exhort, bow, incline, hail, help, and save. Nat is also a name of the heifer-goddess Isis, and in the year 1769 a heifer, the type of Nat (Neith), was sacrificed in the Need-fire kindled at that time in the Island of Mull. This was the offering and tribute likewise called Nat (Eg.).

For years it was a subject of wonder to me why Egyptian offered no explanation of the name of fire found as TAN, in Welsh; TEINE, Irish; TEINE, Gaelic; TEEN, or THUN, Chinese; DANU, in Hindustani; TENA, Soso; TEENE, Salum, firewood; TEINE, Irish, a firebrand; TINE, Cornish, to light a fire; TINE, English, to kindle a fire; and TINDLING, for firewood. Each of these is a worn-down form of a word represented by the Welsh TEWYN, in which the w stands for a K, and the full word is found in the Persian TIGIN for fire. In Egyptian the root AKH means fire, and in the African languages, AKAN, Bode; IKAN, Anan; AGUN, Udom; OGON, Akurakura; UGONI, Rungo; EKANG, Haraba, denote fire; and UKUNI, in Swahili; IKUNI, Matalan; EKUNI, Meto; OGUNO, Egba; TEGENA, Soso; IGINIO, Aku; EKUAN, Afridu, are the names of firewood. Akh is an abraded Kakh, as in CHECHI (Swahili), a spark; KOKA (Ib.), to set on fire; CHIK (Uraon), fire; KAGH (Persian), fire; and T'JIH or T'KIH (Bushman), fire, the T being a click. With this click, or the

---

[1] Fosbroke, *Sketches of Ross*, p. 58.

Egyptian article prefixed to the Akh, fire, we obtain TEK, a spark, to spark, and sparkle. KAR-TEK is a title of the Goddess of the Great Bear and Mother of Baal, meaning the spark-holder. Now, the Baal-fire, the Need-fire, was always sacredly reproduced FROM THE SPARK in the annual ceremony, and "TEK-EN" means the fire at the spark. The word TEKHEN is extant in Egyptian for winking with the eyes—that also sparkle. Teken accounts for the Persian TIGIN, and Welsh TEWYN, on the way to TEINE, TINE, TIN, or TAN, for the fire of Baal which was kindled from and was representative of fire as the divine spark. The DAWN of the Druids and Barddas was the divine spark of inspiration, the fire from heaven, and TANE in Japanese is the creative fire, ferment, cause, origin.

It was a custom formerly and not many years since in Leeds and the neighbourhood for children to go on Christmas Day from door to door, singing and carrying a "WESLEY-BOB." This was made of holly and evergreens, formed like a bower, with a couple of dolls placed inside, adorned with ribbons. The Wesley-Bob was kept veiled or covered until they came to a house-door, when the two dolls in their leafy niche were exhibited during the singing of a ditty.

At Huddersfield the children carry what is there termed a " WESSEL-BOB," consisting of a large bunch of evergreens, hung with fruit, and decked with coloured ribbons. They sing a carol of "Wassailing."

> " Here we come a wassailing
> Among the leaves so green."

Wassail is said to mean Wish-health or wholeness. The "Bob" in the Wesley-Bob answers to Beb (Eg.), a niche, to go round, circulate. Beb-t (Eg.) is a branch, and a place. Ba means wood, leaves, and B, a place. The Bab or Bub, is a place or niche made of green leaves, carried round.

But the form Wes-ley may denote another origin than wassailing. Uash (Eg.) is to invoke, call; English, wish. Lui (Rui) is the door. The Wesley-bob is carried from door to door, and the good wishes are there expressed. At Aberford, near Leeds, the two dolls were borne about in the same way, but the bob was called the Wesley-box. Box is the Bekh (Eg.), the birthplace. It was a name of the solar birthplace in which the genitrix brought forth the child. The two dolls, no doubt, represented the mother and the child brought forth at Christmas instead of at the equinox.[1] Elsewhere the Wesley-bob appears as the "vessel" called the "Vessel-cup," which merely reduplicates the name. When the vessel or box is uncovered, the carol of the "Seven Joys of the Virgin" is sung.[2] The cup, vessel, bekh, and beb, each typified the womb of the Genitrix.

The Christmas-box and Boxing-day are supposed to be named

---

[1] Dyer, pp. 483, 484.   [2] Ibid. p. 464.

from the begging boxes in which gifts were deposited on Stephen's Day. Boxing-day is begging-day. In Bedfordshire there formerly existed a custom of the poor begging the broken victuals of the rich the day after Christmas.[1] It is still the day on which the annual begging is done on a national scale.

In Egyptian Beka means to pray, solicit—that is, beg. Bak is a name of the servant, the labourer, the menial. Boxing-day is the servant's day. The X represents an earlier K, and BEK-ing is identical with begging. Back, to bow, is a cognate of the same group answering to Beka (Eg.), to pray. The boxes used by the Romans for receiving the contributions at rural festivals were called Paganalia, the box being a type of praying or begging. Bak is the variant of Pag, and the Pagan is not only the peasant in the country, but the servant, the labourer, the Bak. The Latin Pagus is a division the same as Pek-ha (Eg.), and the Bak-ing, or beg-ging-day, commemorates the solstitial division of the year, the Pekha of Christmas, which had been removed from the time of Pekh, or Pace, our Easter. The name of Bakshish is in use from Egypt to India, and is doubly connected with the name of the gifts sacred with us to the Boxing-day. Egyptian will tell us why. SHUS is the name of the servant and follower. The Shus-en-Har were the servants and followers of Horus in pre-monumental Egypt. Baka (Eg.), to beg, pray, and Shus (shish) the servant, yield Bakshish as the present solicited by the servant and follower. Shus (Eg.) also means food, and Bak-shus will read the servant's food. With us the gift is given at Christmas under the Christian dispensation. But in the hieroglyphics the BAK hawk represents the Har of the resurrection, who was brought forth from the Bekh at the time of the spring equinox, hence the Egyptian Boxing-day was equinoctial.

Hogmena, or Hogmanay, is the Scottish name for the last day of the year. The Hog, or Hock with us, as in Egyptian, denotes a time and a festival. Hak (Eg.) is a time, a festival, and the double Lion shows it was at the end of a year, the equinoctial year, whereas our Hogmena is solstitial. The Hog-Colt, or Sheep, is the one-year old. The Hock-Cart is the last, the harvest-home load. A shilling is a hog, twelve pence, as the year consists of twelve months. Mena (Eg.) denotes the end and death. Hogmena is the end or death of the year, the time, the completion of the cycle. Hence the festival; and because it is the death of the old year, the festival is pre-eminently a celebration by the young.

In Scotland the children go round begging food, oatcake being the principal offering. Each child used to be presented with one quadrant section of oatcake, which identifies the corner of the circle. The

---

[1] Dyer, p. 493.

cakes were expressly prepared beforehand. The Egyptian Hak-ing, so to say, is begging. Heku, is to supplicate ; Hekur, to hunger ; Hekau is food. The children cry—

> " Hogmenay,
> Trollolay,
> Give us your white bread and none of your gray."

Hekau equates with Kamhu as the name of some kind of bread,[1] and Kam-hu reads black-white, the equivalent of gray. TRU-RERU-RÂ (or Tru-lelu-lay) means time, children (or companions going round), to give, and the whole may be rendered—" Hogmenay, the end of the year, is the time for gifts to the children who go round." The giving of gifts to the young is emphasised with an appropriate moral in these words :—

> " Get up, guid wife, and binna sweir,
> And deal your cakes and cheese while *you are here ;*
> For the time will come when *you'll* be dead,
> And neither need your cheese nor bread." [2]

The demand is compulsory, and the bread and cheese are termed Nog money. Nog is the Egyptian Nek, to force compliance.

In a Derbyshire masque at Christmas, the mummers perform a play of St. George, in which he fights with and slays a character named " Slasher." The doctor is called in and applies his bottle to the fallen Slasher's mouth, which brings him to life again. Then the Slasher is addressed : " Rise, Jack, and fight again ; the play is ended." [3]

They had a custom at Ashton-under-Line of shooting the Black Lad on horseback. He was supposed to represent a black knight who formerly held the people in bondage, and treated them with great severity.

The Scotch " Quhite Boys of Yule," perform a drama of St. George, in which Black Sambo is the opponent of the good divinity. The black knight and Sambo are reliquary representatives of the Akhekh of darkness, the oldest personification of the Typhonian monster. Horus the George of Egypt, as the opponent of darkness, was the white god. These contests are forms of the battle between Horus and Typhon in the Eschatological phase, and of light and darkness in the earlier time.

The " Quhite Boys " represent spirits, and in Egyptian Akhat, the equivalent of Quhite, means white and a spirit, the white sun-god, Horus, or Hu, into whom the black Kak transforms.

Mummers disguised as bears and unicorns were particularly prominent in the grand scene of Christmas mumming,[4] and the

---

[1] Lep. Denk, ii. 28.
[3] Dyer, *Popular Customs*, p. 469.
[2] Brand, New Year's Day.
[4] Dyer, p. 461.

unicorn and bear are the types of Sut and Typhon, the oldest form of
the Mother and Son in Mythology.

Christmas mummers in Hampshire are called Tip-Teerers. Tip-
ter (Eg.) means the commencement of the season.[1] Tep is the first
and ter a time Loaf-stealing was one of the practices of the Tip-
Teerers, and Teb (Eg.) is a loaf. They were dressed up fantastically
and danced. Teb is to dress up, clothe, crown, and tep means to
dance. It was necessary that the mummers should be transformed
as the winter solstice (or spring equinox) was the time of transformation. This was effected by the two sexes exchanging their clothes.

This scene of transformation is as sacredly preserved in the Christmas pantomime. And the exchange of sex illustrates the same transformation as is illustrated in the book of the dead when Osiris goes
into Tep, and is transformed into his soul, from the two halves, who
are Horus the sustainer of his father and Horus who dwells in the
shrine, or " the soul of Shu (male) and the soul of Tefnu " (female) ;[2]
these two constitute the one, and are symboled by the mummers'
change of dress and blending of sex.

The going from house to house to partake of Christmas cheer
indicated the going forth of the sun or Osiris from the lowest sign.
The blackened faces were symbolic of the dark depths in which the sun
had been buried. Masking, disguising, blind-man's buff, blackening,
bowing, and bobbing, all forms of suppression and effacing of self,
were characteristic of the Christmas mummeries, in keeping with the
lowly and benighted state of the sun. It was a common superstition,
that at twelve o'clock on Christmas Eve the oxen in their stalls would
be found on their knees, all things preserving the lowliest attitude.
In antithesis to these, the summer solstice was the sign for carrying
about the giants in the midsummer pageants. The giants were
represented on stilts. In Marston's *Dutch Courtezan*, one of the
characters says, " Yet all will scarce make me so high as one of the
giant's stilts that stalks before my Lord Mayor's pageants." The
morris dancers are raised upon stilts. Their chief time is May-day.
The celebrations of the equinox are for the sun that rises up from the
water boundary. Some of these have got belated so far as May.
Mur (Eg.) is the water limit of the land, and RES is to raise or be
raised up ; also it is the name of the south toward which the sun is
ascending.

Our transformation scene at the Christmas-tide is merrily made to
call up the light of laughter in the young-eyed. The Aleuts in their
cold north region took theirs more solemnly in terrible earnest. Their
traditions tell of certain mysterious dances held by night in the month
of December. They divided the sexes, the men being placed far

[1] *Great Har. Pap. An. of Ram*, iii. pl. 53-12.
[2] *Egyptian Faith*, ch. 17, Ritual.

apart on one side, the women on the other, this being the solstice. In the midst of each party a wooden figure was set up. Then they all stripped naked, except for the wearing of a huge mask which limited their sight to a small circle about their feet. It was death to lift the mask, or for the one sex to look on the other, and they danced on the snow naked to the arctic night, before the image.[1] This was their mode of mumming, this was their celebration of the transformation of the sun in the passage through the Meska. It was the incarnation of the Child, for even while they danced it was held that a KUGAN descended and entered the symbolic figure. This was the spirit of the renewal often figured as the Messiah and Saviour Child of other mythologies.

In Egyptian Khu is a spirit, and Khen, to alight, rest, reveal. This was the significance of the KUGAN. When the pantomime was over the image was destroyed, the masks were broken and thrown away. The Kugan is the transforming spirit, who, in Egypt, was the beetle-headed Khepr, and it is noticeable that in the Xosa-Kaffir dialect a peculiar kind of sacred beetle is called a QUGANE. The mask was also used by the Thlinkeets to place over the face of the dead.

Our English pantomine still preserves the imagery of the Egyptian Ritual, and scenery of the Meska, the place of re-birth. The Meska is our Mask, and the mask plays a great part in the panto-mime. The Meska was the place in which the Mum (the dead) transformed. The Mime in the mask represents the mum or mummy. In the Bask the Mamu is the ghost or hobgoblin, and to mam or mum is to mask in a hideous manner, in fact, to personate the dead, as was done by the African Mumbo-Jumbo. In German the ghost or bugbear is the mummel. The "masks" of the pantomime are the mummels, or mummers of the underworld, who undergo their change or transformation. The two worlds, lower and upper, are represented, and the change from the one to the other is portrayed in the transformation scene with its emergence from the domain of gnomes, fairies, giants, sprites, into the upper world of common day or daylight, the fun and frolic, dancing and feast-ing of which are symbolical of heaven. And the gods are still there in person. The great Mother in her ancient type of the Dove (Columbine) and the Ancient of Days, the old father or pantaloon ; the clown and harlequin are the two brothers Horus, the clown, Kar-nu (Eg. inferior type) is the elder or child Horus, and harlequin is Har, the younger, the spiritual type ; Har of the resurrection with the power of becoming invisible, or a spirit among mortals.

A few things in common at starting may be sufficient for the

[1] Bancroft, vol. iii. p. 145.

foundations of languages, colonies, and civilisations, which grow up unlike in their surface features.   What they had originally in common may be either outgrown or mossed over.   If they grow at all there will be divergence amongst the branches, although they spring from the same rootage. Here, however, the supreme surprise is the amount of Egyptian material still extant as English ; but more than enough will have been produced in illustration of the " hieroglyphics in Britain," the Egyptian " origines in words," and the Egyptian naming of our " personages " and symbolical customs.

Aye keeping their eternal track,
    The Deities of old
Went to and fro, and there and back,
    In boats of starry gold.

For ever true, they cycled round
    The Heavens, sink or climb;
To boundless dark a radiant bound,
    And, to the timeless, Time:

Till, mortals, looking forth in death
    Across the deluge dark,
Besought the Gods to save their breath
    In Light's celestial Ark.

To the revolving Stars they prayed,
    While sinking back to Earth;
"*In passing through the world of Shade,*
    Oh, give us thy re-birth!"

And, ever a Sun, beyond the Sun,
    Quickened the human root
With longings after life, that run
    And spring with heavenward shoot.

Their yearnings kindled such a light
    Within them, so divine,
That Death encompassed them with night,
    To show the starrier shine.

SECTION VIII.

EGYPTIAN DEITIES IN THE BRITISH ISLES.

ACCORDING to Cæsar, the Druids worshipped Mercury in particular, and possessed many images of the god. They regarded him as the inventor of all arts, looked to him as their guide of ways on their journeys, and considered him to have great influence over mercantile transactions. Next to Mercury, they adored Apollo, Mars, Jupiter, and Minerva. "Concerning them, they have almost the same opinion as other nations, namely, that Apollo wards off diseases, that Minerva instructs them in the principles of works and arts, that Jupiter holds the empire of heaven, and that Mars rules in war. To him, when they have determined to engage in battle, they generally vow those things which they shall have captured in war. When they are victorious, they sacrifice the captured animals, and pile up the other spoils in one place. The Gauls declare that they have all sprung from their father Pluto; and this, they say, was delivered to them by the Druids."[1]

"Cæsar's statement, that the Druids worshipped Mercury, Apollo, Mars, Jupiter, and Minerva, is of the same base metal as the statement of more modern writers—that the Buddhists worship the Trinity, and that they take Buddha for the Son of God." So says Max Müller.[2]

Cæsar, however, as we shall find, knew what he was talking about. Such divinities as were familiar to him he identified by their nature, and transliterated their names according to Roman equivalents. So far as it went, Cæsar's statement is entirely true and trustworthy. It is supplemented by Richard of Cirencester, who writes of the Britons thus: "Among their gods, the principal object of their worship was Mercury. Next to him they adored Justice (under the name of Astarte), then Apollo and Mars (who was called Vitucadrus), Jupiter, Minerva, Hercules, Victory (who was called Andate), Diana, Kubele, and Pluto."[3]

[1] Cæsar's *Commentaries*, b. vi. ch. xvii. and xviii.
[2] *Chips*, vol. iii. p. 251.   [3] B. i. ch. iv. 4.

Dionysius Periegetes sings :—

> " Upon the ocean's northern coasts are found
> Two British islands fronting to the Rhine,
> Where in the sea he disembogues his stream ;
> Of these th' extent is vast, no other isles
> To the Britannic justly can compare :
> Islets adjacent lie, wherein the wives
> From the Amnites' distant shore perform
> Due rites to Bacchus thro' the livelong night,
> Deck'd in the dark-leav'd ivy's clustering buds,
> While the shrill echo of their chaunt resounds :
> Not so, upon Absinthus' Thracian banks
> Bistonians hail the harsh Iraphiote ;
> Nor thus, around the dark-gulf'd Ganges stream,
> The Indians with their sons on Bacchus call,
> Noisy and loud, amid the festive scene,
> As shout these women, ' Evœ ' to their god."

Geoffrey, in his *History*, tells us that Walter, Archdeacon of Oxford, a man of great eloquence, and learned in foreign histories, offered him a very ancient book in the British tongue, in which he found the story of Brut, the first king of the Britons, written in an elegant style, and continued 'down to the time of Cadwallader. This he rendered into Latin.

The present writer can have no difficulty in accepting the tale of the book brought from Brittany, and translated into Latin by Walter, Archdeacon of Oxford. Walter, as a Latin scholar, did as Cæsar did, he read the myth, common to various races, because each had derived it from the same original, as if it were Roman. He knew the name of Brutus, the legend of Troy, and the tradition of Æneas, but was unacquainted with the mythos as British, or with the character of Prydhain, the youthful solar god, whose reign was established when the celestial Troy was overthrown, that was, when the Sabean cult and reckoning were superseded by the Solar, the dragon-tyranny overturned, the Son of the Virgin Mother elevated to the supreme place as the Father in Heaven, and the idea of a divine fatherhood exalted over that of the earlier motherhood. Geoffrey Latinized the British mythos. The Troy city that was overthrown is still figured in the children's games in Cornwall and Wales, and consists of seven circles round a centre cut in the grassy sod. These represent the seven regions of Dyved, the seven encirclers of the Great Bear. The seven belonged to the great mother, and with the son added, as Eshmen (8th), the earlier Sabean Sut, this was the Egyptian Troy or Teruui, the circumference, a form of Sesennu, and No. 8. It can be identified by the shield or circle of Pridwen, in which the Ogdoad of Arthur and his seven companions escape from the Deluge. The naming of London as the New Troy (Trinovantum), when collated with the gate and tower of Belin, can be restored to its proper place in the mythos. Belin is the son, considered as the child of the virgin mother. In one version of the " Hanes Taliesin " the speaker says: "I have endured hunger for the Son of the Virgin. I have been in the White Hill in the Court of Cynvelyn, in bonds for a year and a day. I have had my abode in the kingdom of the Trinity,"—this in enumerating the manifestations of the Word or Announcer of the various cycles of time.

Belin is said to have made a road from Totness to Caithness, and another from Southampton to St. David's.[1] And when he had made the burgh of Kaer-Usk, he went to London, the burgh he greatly loved. He "there began a tower, the strongest of all the town; and with much art a gate thereunder made; then men called it Billingsgate."[2]

The New Troy was established by the young sun-god, considered as the child of the mother. "I am come here," says Taliesin, "to the remains of Troy."[3] Also, New Troy, the White Hill, and the Gate of Belin will supply a possible rendering for the name of London. Renn (Eg.), the virgin, pure, unblemished, is the equivalent of white, as in linen (Renn); thus Renn-ten (London) is the White Hill.

Helvellyn is another Hill of Belin, and retains his name so long as mountains stand. "Now and evermore the name standeth there,"[4] says Lazamon's *Brut* of Billingsgate, the Gate of Belin, in the account given of the tower built by Belin, the good king who lived there as the prince of peace and plenty, the Nefer-Hept of Egypt. Belin, like Sutekh and Sat-ren, means the little or child Baal, who was Sut, the star-god, in the Sabean mythos, and Pryd (or Brute) in the solar. So interpreted, London is the royal seat (Tun), the throne-room of the Renn. The Renn is also the nursing mother, who was Rennut in Egypt, Luna and Selene in Greece, and the Keltic Luan, the moon-goddess. Tradition tells us that a temple of Luan once stood where St. Paul's now stands. Thus London is the Tun of the virgin (Renn) and child of a pre-Christian religion. Belin is the diminutive of Baal, and he is the mythical builder of London. Belin is the Nursling, *i.e.* the Renn, and London was his seat and throne, or tun, as witnessed by Billingsgate; therefore the most probable derivation of the name of the city is from Renn-tun (Lenn-tun), the throne of the child, who was Belin in the British Mythos. The child who came to the "remains of Troy" was Pryd, who is represented by Geoffrey as Brute who came *from* the remains of Troy.

The sun of the resurrection, *i.e.* of the vernal equinox, is the potent because pubescent son. Hence he is the bearded or the long-haired god in many mythologies; the elder Horus being the wearer of the lock of childhood. This is illustrated by Arthur in the story of the giant who had made himself furs formed of the beards of kings whom he had slain, and who commanded Arthur to cut off his beard and send

---

[1] *Brut*, MS. Cott. Calig. A. 9, lines 4829-35, Madden.
[2] *Ib.* lines 6050-60.
[3] *Hanes Taliesin.* [4] Lines 6060-4.

it to him as a tributary offering.[1] Belin and his brother BRENNES are a form of the "two-halved youth," the dual Horus or double Anubis of Egyptian mythology. Brennes or Brennius is but the Latinised form of the Pren (branch), that is, the Renn (Eg.) or nursling child of the mother, a type of which is extant in the Wren as the little king, the prince of all the birds, the Breton Roen.

The wren was hunted to death on Christmas Day, and on the day following it was carried about, suspended by the foot in the centre of two hoops crossing each other at right angles. A procession used to be formed in every village of men, women, and children, who sang an Irish catch importing the Wren to be the king of all birds.[2] Now the hieroglyphic noose for the feet of cattle is a Ren. In this ceremony the dead wren was typical of the first of the two brothers who died and rose again, or was transformed.

The process of modernizing the ancient fragments by re-casting in the classical or Biblical mould is sufficiently apparent. In the " Poem in praise of Lludd the Great " the name of the god Aeddon (Hu) has become Adonai. How easy is it to turn round and claim this to be the Hebrew Adonai, and then to infer that he was derived from the Hebrew writings! Aeddon is identical with Adonai and with Adonis, and Tammuz, and Duzi, and other forms of the same god in divers lands, who were independently derived from the Egyptian Aten, long before the Jews brought their version of the mythos out of Egypt. The same perversion of the original imagery is manifest in free-masonry through this re-casting of the ancient matter in the Biblical mould. It was this process of interpreting the fragments by the Hebrew rendering of the same original matter that put Davies irre-coverably on the wrong track. The process may be followed into the Christian stage in which the Christ is substituted for Prydhain.

All this and much more is admitted without determining the true nature of the ancient British relics, which can only be done by collecting their correlatives, and showing how they belong to an original system of thought, of mythology, of typology, of eschatology, which it is now proposed to identify as Egyptian at first, and pre-Hebraic in the isles. The Welsh text of these fragments cannot be adequately rendered without their mystic meaning being understood and allowed, instead of laughed at.

It has to be shown that the most ancient form of the great mother was the goddess whose image in heaven is the Great Bear. Her name in Egypt runs through the gamut of Taurt, Khepsh, Kheft, Aft, Apt, Khebt, Kheb, Kep, Ap. Her name in full is synonymous with that of Egypt. Af signifies born of ; Ap is the first ; Aft the abode and the four corners. She was represented as the hippopotamus, the cow, or horse of the waters. Her name is likewise that of the north. Lower Egypt was to the north, and her constellation revolved about the pole of the

---

[1] Geoffrey, x. 3.   [2] Vallancy, *Collectanea.*

north. The celestial north is the oldest place of birth in mythology. Kheb means to give birth to ; Khep is the womb ; and Khebt was the birthplace in the north personified. From Egypt (Lower) the name of Kheft extended to Phœnicia, Kheft or Keft being the Egyptian name of that country, which was called so from the north. The Hebrew name of Japhet, Japht, or Apht (יפת), corresponds to the Egyptian Aft, earlier Kheft, the lady or queen of heaven, and is represented by Aipht, the Welsh name of Egypt. In the account of the generations of Noah in the Book of Genesis, Japhet and his descendants are the people who migrate and take possession of the isles. "By these were the isles of the Gentiles divided in their lands ; every one after his tongue, after their families, in their nations."[1] The present writer believes that books have been written, although he has not read them, to identify Gomer with the Kymry of Wales. After long seeking and by very different roads he arrives at the same result, and anticipates having the pleasure of seeing the ripple of derision ironed smoothly out of the faces of the scoffers and scorners this time by the weighty pressure of hard facts.

Kheft is the north, the birthplace, the genitrix. She was the bearer, the great or pregnant mother, hence her type of the water-horse, the ark of life amid the waters before any artificial means of carrying had been invented. Kheft, by a well known law of language, becomes Khêt. So Kheft, the abode of birth, becomes Ket-Mut, the Mother Ket or Kat. Kat, like Kheft, means the womb. Kat softens into Hat, the habitation of the child, named Hathor. By this process we are enabled to claim our British goddess KÊD as the old Mother Kheft figured as the living ark of the waters. Kêd is the Arkite goddess of the British Druids.

The readers of Davies' writings will remember how like a ridiculous dance witnessed by one who may be deaf to the music seemed all his wearisome references to the goddess Kêd and her ark, which was also a mare—as bad as any nightmare. Yet there was original meaning in the whole matter, although its expounder knew not how to put or prove it.

Kheft or Aft was the goddess of the four quarters of the first circle made in heaven. Aft, the abode, is also the four corners. This circle of the four quarters was repeated in the Caer or sanctuary of Kêd, called the "quadrangular enclosure." The Druids, her teachers, are described as Druids of the Circle, of four dialects, coming from the four regions.[2]

Taliesin, in praise of Lludd the Great, recognises the Kymry of four languages, and says, "the Kymry of four languages shall change their speech." These four quarters of the Kaer and four languages of the Kymry affiliate them to Aft of the four corners, the earlier Kheft, or Kêd. Taliesin says that Necessity produced nothing

---

[1] Gen. x. 5. [2] Cynddelw, *In Praise of Owen Gwynedd.*

earlier than Kêd (Keridwen), and the primary sacred order in the world was that of her priests; the claim in relation to the four quarters is exactly the same as that made by the monarchs of Egypt and Assyria.

In the first period of the Round Table, Kêd is represented as living in the time of that Arthur whose symbol in the heavens was the Great Bear, and whose harp was the constellation Lyra.[1] Arth is the name of the Great Bear, and of Arthur it is said, "Aythur-ap-Arth-Hen against foeman's attack and injury made the blade (for use) in battle," which identifies him as the son of the old Arth, the genitrix, goddess of the Great Bear.[2] Khebt, the hippopotamus of the waters, became the fish-goddess as Derketo, the Syrian mermaid. And in an old Christian poem which was palmed off as one of Taliesin's, the fish that swallowed Jonah is called "Kyd." The writer asks, "Who brought Jonah out of the belly of Kyd?"[3]

The ark of Kêd is described as passing through the dale of grievous waters, having the fore part stored with corn, and mounted aloft with the connected serpents.[4] The Bard Cuhelyn sings (eighth century) of "the goddess of various seeds," and of the "enjoyment of the society of Kêd," and the poems sung in her praise by the "chanters of Caw." This is in allusion to the Druidic mysteries. She is the goddess of corn and of those who carry ears of corn, as did her priests. Ta-urt, a name of Khebt, may be translated the corn-bearer, the genitrix being represented at times with the *modius* or corn-measure on her head. This "TA," illustrated by the Akkadian Umme-DA, the bearing-mother, means the ENCEINTE. Ta-urt is the ENCEINTE, or great mother, and corn was one of the ideographs.

The Cornish Hav, English Gipsy Giv, English GOFE, for corn harvested, Sanskrit Yava, Lithuanian Javai, Egyptian Sef, Greek Zeiá, lead back to Khefi (Eg.), harvest, the Kaffir Kwebu, an ear of corn, and to Khepsh, Khept, or Kefa, the name of the genitrix as the primal corn-bearer. So is it with the name of Ta-urt. TA is corn, but, as in the Akkadian Umme-da and the Maori To, it also denotes pregnancy. Da-mater (Demeter) is the mother of corn, but the external is not the original sense, and she was the great mother, the gestator. The corn or seed was an image of life. One of Kêd's names is Llâdd, to cut, reap, and mow, which corresponds to Rept (Eg.), the goddess of harvest and lady of corn; and this name of the lady of corn was certainly not derived from the Saxon HLAFDIG. LLAD (Welsh) means to confer favours, gifts, blessings; and the favourer, the giver, was the lady, the good lady, Welsh LLADAI, Gaelic LEUDI, from REPTI (Eg.), the lady of corn and goddess of harvest.

Enough to show that Kêd WAS, as the fragments of the Barddas and

---

[1] *Hanes Taliesin.*
[2] Iolo MSS. 263; Rhys, *Welsh Phil.* p. 325.
[3] *Welsh Arch.* p. 43.
[4] Gwawd Lludd y Mawr, *Welsh Arch.* p. 74.

Druids claim; much more might be cited to prove that she has been. As we have seen, the Welsh call the Great Bear by the name of Arth, the Irish Art, its Egyptian form minus the Ta. The Bear was originally the water-horse image of Ta-urt, the Typhonian great mother. In this connection with the horse we are enabled to identify Taurt with the prefix to the name.

On one of Camden's coins, No. 32, there is a female head with the legend Direte. History, says Davies,[1] mentions no queen or city of this name, but in our old orthography Direit, and in the modern Dyrreith, is a title of the mystical goddess who is introduced by the name of Dyrreith in the ancient British poem called the "Talisman of Cunobeline." She is a goddess, and takes the form of the Mare to carry the hero to battle and victory. It is said "Cunobeline, the indignant, the lofty leader of wrath and that divine allurer Dyrreith, of equal rank with Morion, shall go under the thighs of the liberal warriors."[2] This was the bearer, the ark of the waters, and Ta-urt is the chariot and bearer of the waters. It is now claimed that our Druidic Direit, the goddess whose symbol was the mare, the crosser of the waters, is a British form of Taurt by name, which doubles the identity of Khebt and Kêd.

The word Tasc is frequently found on the British coins, sometimes Tascio, Tascia, or Tascie. This has never been satisfactorily accounted for, although Davies rightly connected the word with corn and the corn-bearing vessel of Kêd. The great mother of mythology is depicted as the corn-bearer whose solar son is Hu, or Corn, and whose earlier Sabean child was Sut, a name of seed or corn. The ear of corn is frequently depicted on the coins, along with the mare or mother. It is omitted, however, from one coin,[3] but in its stead we find the word Dias inclosed within a frame. Dias in Irish signifies an ear of corn; it here takes the place of the ear of corn, and in Egyptian "Tes" is corn and food made from corn, also a tie or bond. Moreover, "Tes" is a part of the style of the great mother, meaning the enveloped form and very self; Ta-su signifying the bearer of corn, or the child.

Tas in English is a mow of corn, the tasker, a reaper of corn. Ticcan in Cornish is a handful of corn tied up as a sheaf. On a coin published by Whitaker, and reproduced by Davies, the goddess appears in the dual shape of the mother and mare, in the act of going, like the hieroglyphic Ses-Mut. The coin bears the legend, "Tasc-ia-no-va-lin." Tasc is corn, IA or IUA (Eg.) is a boat; Nu means a divine type; Fa (Eg.) is to bear, carry the corn, with the corn-modius on the head—a symbol of pregnancy; and Lin is the water. The title thus interpreted is that of the divine boat which carried the corn across the waters as the Ark of Kêd. This figure was the water-horse in Egypt, and is the mare in Britain. Both are

---

[1] *Myth.* 609.   [2] Davies, *Myth.* p. 620.   [3] No. 13, Camden's Tables.

personifications of the womb, or the mother great with child, called the Great mother.

Whitaker also publishes a coin [1] on which there is a child borne upon a dog with the legend, " Tasc No Va," the same as Tascianova, with the boat, IUA (Eg.), omitted. This reads the bearer of divine corn, or corn typically carried. The dog connects the legend with Sothis or Sut, the Dog-star, the first child borne by the great mother, whom we shall try to identify with the mythical Arthur, whose star is the Great Bear.

The noose hieroglyphic held in the hands of Ta-urt is the ARK sign of reckoning, the end of a period, a cycle of time. So, on the back of the mare (coin 6, Camden), the vase of Kêd is portrayed, the Pair or cauldron that boiled for a year and a day, and is, therefore, the symbol of a cycle, the circle and period of the water-horse or Bear. The one equates with the other. Kheb (Eg.), the water-horse, gave us the hob, or hobby-horse, of the mysteries and mummeries.

Memorials of the goddess Kêd exist in many symbols and things that bear her name. She was the genitrix. Kat (Eg.) is the womb, and the cat, or cut, is a type-name of the female emblem. The chat, cat, or gat, is alluded to by the Wife of Bath, who says " Gat-toothed I was, and that became me well." Her teeth were cut, indented, suggestive of the gate, the opening. Cate signifies lecherous; hence the gat-tooth is the liquorous tooth. Kêd is the typical female in other applications of her name. The chid-lamb is the female lamb; the chideress, or chidester, is a female scold; also the cat in the same sense. She was the cat, and is still the cat with nine lives; the nine months of gestation during which she kills the rat nine times. She was the abode of birth, and the cot, cottage, quod, cathedral, and city, are forms of the abode named after Kêd. She was the first house, and the kit is an outhouse for cattle; the cat a shed for protecting soldiers. She was the lady of corn, Keres, and cod is an enclosure of seed and a seed-basket. The cod-fish is the prolific seed-fish; the milt of a single cod-fish has been calculated to contain one hundred and fifty thousand million animalcules : the eggs of the female are numerous in proportion, and this furnished a type of Kêd, and was named after the prolific great mother. We have her namesake in the kid or cod of peas, with which some ancient customs are connected. The " scadding of peas " was a Scottish custom of boiling the common grey peas in the kid and shelling them afterwards. Skhet (Eg.) means to be enclosed, shut up, as in the pod. It was also a custom to woo and to divine with the peascod. The old chap-book called *Mother Bunch's Closet Newly Broke Open* gives instructions for seeking a green peascod in which there are full nine peas, and writing on a slip of

---

[1] *History of Manchester*, 2nd ed. vol. i. p. 342.

paper, "Come in, my dear, and do not fear"; this is to be placed inside the kid and laid under the door; the first man who enters is the predestined husband. Kêd was the seat and sustainer of being, and the cat is still used as a support. The Cather (har or ar Eg. is the child) is a cradle. Kêd was the first cradle of the har, child. She carried the tie or noose in heaven as one of her symbols, a sign of surrounding, inclosing, catching, tying up, being pregnant. In the children's game of cat's cradle we have Kêd's cradle, and the string twisted on the fingers is analogous to her tie; the game is hieroglyphical. She was the typical ship that carried us over the waters when she let us out of the ark, at the end of nine months, or ten moons, and the cat is still the tackle of a ship. It was also carried on board as the Cat-o'-nine-tails. How profound the worship of Kêd in Britain must have been may be judged by the religious feeling with which we preserved the Cat-o'-nine-tails! so dominating is the symbol. Kêd was a goddess of fire and ferment in one phase; one of Ta-urt's titles is spark-holder or reproducer; and her emblem is the Chad-pot or feminine fireholder used by the lace-makers and straw-plaiters. Lastly, she became the devil in Egypt, and Quede is an English name for the devil; Quaid, Scotch, evil. She was degraded to this secondary stage of deity when the male was made supreme; *en revanche*, Kêd seems to have become our St. Catherine, or Catern, who had her festival on Catern Day. Kêd became our Kate. Ern, in Welsh, is a pledge. Ern (Eng.) means to flow, run; and on Catern Day, Kate (Kêd) is pledged in the flowing bowl called the "Catern Bowl" by the Chapter of Worcester.[1] Khat-renn (Eg.) would be the renewer of the race, the dandler of the child. Khat also means going round, wheeling round, as did the stars of Kêd, and the wheeling or circle-making of the Great Bear becomes the Catherine wheel of the spark-holder, still imaged by the firework. This was the wheeling round perpetuated in the ceremonies of Catern Eve, as in London, when, at six o'clock of the day, there used to be a procession round the battlements of St. Paul's, accompanied with fine singing and great lights.[2]

On her day, in the Isle of Thanet, the carters place a small figure across a wheel on the front of their cart-sheds. But this has no relation to the popish imposture of St. Catherine and her wheel. It represented the circle of the year divided at the autumn equinox, and the celebration once dated September 25 instead of November 25, or more than 4,000 years earlier. In the "Cadair Keridwen,"[3] the chair of "Gedwidedd" identifies the goddess with the celestial circle of revolution and the stability dependent on cycle-making. Hence her wheel, for she is called the "Goddess of the Silver Wheel." Catherine has been

---

[1] Dyer, p. 429.
[2] Strype, *Eccles. Memor.* vol. iii. pt. ii. p. 507 (1822).
[3] *Welsh Arch.* p. 66.

made the patron saint of weavers, knitters, and lace-makers, and Kat (or At), in Egyptian, denotes weaving and knitting. Khet is to net, weave, a woof. The Kat (Eg.) is the loom, the primal form of this being the womb (Kat or Kêd). The genitrix was the first weaver and knitter, the Ankh-tie being a primitive symbol of her work. Kêd (Kheft), in going round, made the noose and did the netting. Catherine is the typical virgin. To coiffure St. Catherine, or to braid her hair, is to remain a virgin. Renn (Fg.) means virgin, and Rennut was the virgin mother, that is, the mother who bore before the fatherhood was acknowledged, as will be sufficiently explained. Kat-renn (Eg.), in this sense, denotes the virgin womb. The lace-makers of Buckinghamshire hold merrymakings on Catern Day, and eat a kind of cake called " wigs "; [1] that is, they eat the symbol of hair instead of coiffuring it for the virgin condition. Hair was a sign of puberty, and it was tied, snooded, or wigged up, at marriage. The merrymaking celebrates the other of the two characters assigned to the mother. Also the name of Kheft was given to an Egyptian headdress. The vessel of Kêd called a Pair, English pail, Egyptian Par, a pail, survives in the milkmaid's pail called, after the goddess, a Kit. The " milkmaids' dance " is yet performed on the first of May, the kit, or pail, being dressed and decorated for the occasion. At Baslow, in Derbyshire, there is a festival of dressing the kit now and again observed. The kits are fancifully and tastefully ornamented with ribbons and festooned with flowers. They are carried on the heads of the young women of the village, who parade the streets attended by the young men, preceded by a band of music. The day is ended in dancing.[2] One name of a fiddle is the kit. Baslow reminds one of Bes (Eg.), to bear, to carry. The milkmaid with the kit on her head is an image of the bearer, one of whose types is the cow. The cat being a type of Kêd and a name also of the fiddle may have a serious bearing on the rhymes of

" Hey diddle diddle
The cat and the fiddle,"

and the cow that jumped over the moon may be the cow-goddess of Ursa Major, Kêd, who was anterior to and higher in heaven than Luna. Also, a " diddle " is a young duck or a young pig, and both are types of the old genitrix ; and the little dog, or Canis Minor, imaged Sut, the son of the cow-headed goddess, or Hes-Taurt, later Astarte and Eostre.

The Seat of the Goddess of the Seven Stars was represented by the seven hills. She is the mythical beast, whose seat is the seven hills of Rome ; the Mount Meru, with its seven steps or divisions, is a form of the sevenfold hill. And at Great Grimsby the divinity Kêd sat enthroned on the symbolical seven hills. Two of these

[1] Dyer, p. 426.   [2] *Journal of Arch. Assoc.* vol. ii. p. 208, 1852.

seven, Holm and Abbey Hills, are joined together by an artificial bank, known as KIT'S Bank, once a landing-place.[1]

Teb, Kef, and Kheft are names of the goddess of the hinder part, the back, the north, the place of the mount, the fundamental seat. And in English, teb is the fundament; also the extreme end and outlet of a cart. Keb is the Peak of Derbyshire, with the same meaning. The Cefn or Keven is a ridge also called the Back, the hinder part. The Chevin is a ridge in Wharfdale, and Chevening is on the great ridge in North Kent. To this naming from the North, the hindward quarter, we may assign the Back as in Saddleback.

Kêd is synonymous with the good. The good time or tide is named from Kêd; Gut-tide is a name of Shrovetide; the good-day or holi-day was the day of Kêd. The good wife is a form of the goddess. Kedy, a familiar name with the British Barddas, is our "Goody." A goody was an old woman that wore the red cloak, and the red crown was worn by the genitrix in the lower world. One "Good Woman" was a sign of a woman without a head. This too can be understood as a form of Kêd or Kheft, who represented the lower and hinder part; it was synonymous with the lower crown of Neith—minus the Hut or head—the crown of the north and hieroglyphic of her name.

Our word URE means use, custom, practice, and Ur (Eg.) is the first great, oldest, principal; this gives the primal sense of ure, as use and usage. Ured means to be fortunate, that is, fruitful. Urt (Eg.) is to be gentle, meek, peaceful, bearing or pregnant. The Urt is the crown with asps, a type of maternity. Urt was the great mother, who in mythology is the goddess of luck and fortune. In Egypt she was personified as Ta-Urt, the pregnant Urt. She was depicted as the hippopotamus, with big belly and long drooping dugs of breasts, more like udders. In English Ur is a name of the udder. The con-stellation of Urt was Ursa Major, and this most ancient form of the genitrix is identifiable in Ireland, where the name of the Great Bear is known as Art, and in Britain as Arth. And Urt is a name of Khebt, our goddess Kêd.

As-t or Hes-t, the great mother, who is personified as the heifer, the seat, the house, couch, or bed, reappears by name as Ast, a title of Kêd as the greyhound bitch, the female dog being a type of her, as were the dog and dog-star of Isis in Egypt. One of Kêd's stone monuments in Cardiganshire was named "Llech yr Ast," the flat-stone of the bitch.[2] A place near Tring, in Herts, called Astoe is probably the circle of Ast. Ast, the bitch, is a form of the As or Hes, the seat, bed, chamber, abode of birth belonging to the female, personified in Astarte, Ashtareth, Ishtar, Asterodia, Eseye, and Eostre.

[1] Cliver, *Monumental Antiquities of Great Grimsby*, p. 39.
[2] Gibson's *Camden*, cols. 772, 773.

Gwal y Vilast, in Glamorganshire, is the couch of the greyhound bitch. In the Mysteries Keridwen, the great mother, is represented as transforming herself into the swift greyhound bitch and pursuing Gwion the Little.[1] In the story of Saneha, an Egyptian tale of the Twelfth Dynasty, it is said of the swift hero, " His limbs are like (those of) the greyhound of the great goddess."[2]

So in the hieroglyphics the bitch-dog Khen is the image of the inner abode, the Khent or womb. And this brings us to the meaning of Kêd's name of Keridwen. She is the Khen. Wen has a prior form in Gwen, the equivalent of Khen. Gwen, the lady, is Khen, the hall, interior, or boat. Gwenhywyvar, the lady of the summit of the water, is the Khen (Khen, to image, navigate, carry on the water) personified as Keridwen. This Khen is written with the boat-oar assigned to the goddess ; Kher-it (Eg.) is the figured oar, and Khart is the child. Keridwen, as the vessel of the child and the oar, is the ark of life.

The sow was a primitive type of the great mother as the suckler, the Dea Multimammæ and goddess of the Great Bear. It was cast out of Egypt as unclean, but its name of Shaau shows it to have been an image of primal being, whilst the primordial name of Hathor as Shaat is the same as Shat, the sow. The sow was also an image of Taurt as Rerit. This was one of the shapes of the British genitrix Kêd, and is a proof of her being the goddess Khebt, the good Typhon. The ship or vessel of Kêd that carries the corn is typified by the sow called Hwch (hog) in one of the Triads. Hwch is also an epithet for the ship. In English hug means to carry.

The sow of the Welsh Druids was born in Dyved, and she went to the Black Stone in Arvon ; under this she laid a kitten which Coll threw from the top of the stone into the Menai. The sons of Paluc in Mona took it up and nursed it to their injury. This became the celebrated Paluc cat.

Even a sow that gives birth to a cat may be explained by Egyptian symbolism, for Shau, the name of the sow, is the same as Shau, the name of the cat, and the two are interchangeable types of the genitrix. The Druidic cat or tiger is spoken of as a large ferocious beast. In the *Ritual* we have the " cat in the house of Pet, whose mouth is twisted when he looks because his face is behind him."[3]

The cat and the ass are called the " Sayers of Great Words " in the Hall of Two Truths. The cat is primarily a feminine type, that of Pasht or Pekht, the mortal enemy of the rat. When the Solar mythos was adapted to the symbols and imagery previously extant, and the sun-god became supreme, the sun in one phase took on the female form, and in the passage of the Pool of Persea is made

---

[1] *Hanes Taliesin*, ch. 3.   [2] *Records of the Past*, vol. vi. p. 138.
[3] Ch. 125. Birch.

to transform into the great Cat or Leopard.[1] The Greeks had the same representation, although the symbols had become a dead letter to them.

Apollo is designated the rat-killer; but why, the Greeks cannot tell us. A story is current about a priest of his, one Crinis, who neglected his sacred duties, whereupon the god sent against him a devouring swarm of rats. The priest repented, prayed for protection, and Apollo slew the rats.

Apollo is the rat-slayer because the evil Apophis in the Egyptian mythology takes the rat for one of his types. The rat is a form of the destroyer, "the abominable rat of the sun," as it is called. This is the rat that ate the malt that lay in the house that Jack built. This was the rat that was put a stop to by coupling. and as rhyming is a kind of coupling, this may be the origin of the Irish practice of "rhyming rats to death." Remn (Eg.) denotes the limit and stopping-place, whence the rhyme.

The Cat into which the sun transformed (or was catted) "on the night of the battle made to bind the wicked," when the cat attacked the "abominable rat of the sun," seems to have been represented in the rites at the Witches' Sabbath, for we are told that after the supper or Eucharist there stepped out of a statue standing in the midst of the assembly a black cat, as large as a goodly-sized dog, which advanced backwards towards them, having the tail turned up. Then the company gave the cat the kiss in ano, the hindward salute, a common formula of the Witches' Sabbath, and sat in silence, with all heads bowed towards the cat. Then the lights were put out, and, like the Israelites, they rose up to play. After which there appeared a figure, half sun, half_cat.[2] This was probably the trans-formation scene in which the great cat was re-transformed into its solar splendour.

The kiss in ano is equally the kiss in Annu, for Annu (Eg.) also signifies behind, and in Annu occurred the scene of transformation into the cat, when the Egyptian mystery of Sem-Sem was enacted in the darkness. The Witches' Sabbath serves to enlighten the obscurity of the Ritual and its mythological allusions. In the Hellenic Cosmogony the Sun is said to create the Lion; the Moon creates the Cat. This likewise is an illustration of the Egyptian imagery.

In the two bulls issued by Pope Gregory the Ninth (1232 and 1233) against the Stedingers of North Germany, he charges them with their heathen practices, and amongst other secret ceremonies used on the initiation of a convert, he says that a shining personage appeared from the dark corner of the chamber, the upper part of his body being luminous as the sun, making radiant the whole room,

---

[1] *Rit.* ch. xvii. Birch.
[2] De Lancre, *Tableau de l'Inconstance des Mauvais Anges et Démons.* 4to. Paris, 1613; pp. 67-72.

while his lower parts were rough and hairy, and like a cat;[1] an image of the sun above and cat below that perfectly reproduces the solar symbolry of the Egyptian Book of the Dead,[2] where the sun in Annu makes his transformation into the cat.

This, then, it is claimed, is the speckled cat into which Taliesin, assimilated to the sun, says he had been transformed: "I have been a cat with a speckled head upon a tripod," or a tree. Nor is the cat an isolated symbol, but carries with it the total cult and all its doctrines. The cat-headed solar goddess Pasht followed the sow-goddess, and in the British Mythos the sow brought forth the cat. The cat-headed goddess Pasht is designated "Menhi," and she is also called Ur-Heku, the old ruleress, or the great magic power. With the article prefixed, P-ur-ukhu, the great magic power or the old ruleress, is the equivalent of the Paluc cat of Menai.

Not a refrain, burden, or rhyme of the old popular nursery lore but had a meaning once, and became a permanent possession on that account. Things thrown off without sense do not become matterful by repetition. They live after the sense is lost, because of the meaning they once conveyed.

> "Ding, dong, dell,
> Pussy's in the well,"

says the distich, and pussy as the goddess Pasht is found there in the Well or Pool of Persea, and at the bottom of the well we are to find Truth. That same well was the Pool of Maât, goddess of the two-fold total truth. Our Pussy even is a diminutive of Puss, because SI and SIF (Eg.) denote the child.

There is a goddess Uati on the Monuments, very ancient, but little is known of her. She is identified, however, with the Buto of the Greeks. Uati is the goddess of the north. Uat (Eg.) is the name of the North and of Northern Egypt. Khebt is also the North and Northern Egypt. Thus Uat and Khebt are synonymous. Khebt, Goddess of the Great Bear, is likewise the Goddess of the North. Buto and Leto are one with the Greeks. Leto is Urt (or Urta), a name of Khebt, the old genitrix, and it follows from this that Uati is a continuation of Khebt, as Goddess of the North and of the Great Bear, the British Kêd, mother of the sun-god Hu.

Now we are told by Pliny that the British ladies, married and unmarried, stained their bodies with woad, and danced naked in the open air.[3] This was obviously in the performance of certain religious ceremonies, but it has given rise to many false notions about the ancient Britons being painted savages. One name of woad is wad. Wad is a Cumberland name for blacklead. Woad is also written ode. Kettle is a name of purple, and the purple orchis is called the

---

[1] Baronius, *Annales Ecclesiastici*, tome xxi. p. 89.
[2] Ch. xvii.   [3] Bk. xxii. 2.

Kettle-case. Kit is also to smear or daub. Woad implies a form written with K. In Egyptian Khu is paint. Khu-t will be painted, and we have need of this for determining the nature of woad. Pliny says the woad was black, and wad is blacklead. Cæsar called it blue. Jornandes affirms that it was produced from iron ore.

In Egyptian Uat is the name of bluish-green, as stone for paint, and green plants and herbs. It is also the name of a blue cosmetic in the great Harris papyrus. And we can see that Uat in Uatmes, the name of Collyrium, merely means colour or paint, and Uatmes is black paint. It is also found as a brown colour. Uat as the name of water included both green and blue. My own conjecture is that Woad was green when wet, and that it dried on the flesh in a kind of blue tint. Green and blue were the two colours wanted for the types of lower and upper, earth and heaven, whilst the transformation of the wet-green into the dry-blue would be the most perfect realization of the symbol possible. Again, one of the great objects of worship was Hu, the solar god, whose sign is a tongue, and the tongue is found to be painted blue and red, two colours that interchange with blue and green, the image of the heavenly and the earthly, or spirit and flesh, soul and body.[1]

Uat in the dual form is Uati, goddess of wet and heat, the two truths of water and breath, the two factors of being also personified by Hu, as the Sphinx. Madder is a name of woad, also of paints. As Egyptian, Mat-ter would read the twin-total of truth. Uat is found to be the colour of vestments worn in certain religious ceremonies of the Egyptians. This is identical with the woad vestment of the British devotees; and the woad has the name of the goddess of the north, who was Uati and Khebt, our British goddess of the north, who was Kêd. We have our Wad's-den not far from Gad's-den; two forms of the name of Kêd.

Painting the rosy flesh with a blue tint was clothing the earthly with the heavenly, and the ladies, maiden and matron, who danced and showed their colours dedicated to Hu and Uati made the same sign that we still perpetuate on the Union Jack. In the Ritual [2] Uati is described seated with Pasht in the great quarter, the greatest of the heaven, i.e. in the north.[3] And in another chapter, with a vignette of the deceased adoring, we read: "Oh, great land, I have come from thee, I have prepared, I have irrigated the meadows. I am the bull painted (drawn) BLUE, the lord of the fields: the bull called (by) Sothis at her time. . . . Oh Uat (Blue-Green), I have come, putting on my clothes. I have put on me the WOOF OF THE SUN when within the heaven. Oh, Usert (Sustenance) at the head of the place where Hu was born! Oh

---

[1] As on the sarcophagus in the Amhurst collection. Copied in colours by Bonwick in *Egyptian Belief and Modern Thought.*
[2] Ch. xxiii.   [3] *Ibid.*

divine land of corn and barley, I have come from thee. I have stopped my arm from working at my service in thee, who art called ruler of purity—pure mistress."[1]

In this the worshipper of Uati and Hu says he has painted himself of a bluish-green hue, and put on the woof of the sun for his vestments. The bull of the sun wears the blue woof of heaven as his clothes. He has stained himself, as it were, with woad, like the worshippers of Hu and Kêd in Britain. Mistress of purity was a title of Kêd; she was said to be pure as the crescent moon, and fair "as the snow which the cold has polished upon the lofty peak."[2]

We have the name Uat in Watchet blue, now given to a palish kind of blue. But the original mixture of blue-green was worn by Sabrina (described by Drayton),[3] who sat as a queen in Neptune's throne wearing

> "A watchet weed, with many a curious wave,
> Which as a princely gift great Amphitrite gave."

In mythology the son of the mother becomes her husband and his own father. This is the relationship of the god Hu to Kêd. His name of Hu-Gadarn is rendered Hu the Mighty. But such titles as this and that of El-Shadai, the Almighty, are all too vague for the primitive thought. Gadarn is susceptible of a fine rendering in Egyptian. Renn is the child, the nursling of the great mother called the old dandler, who is Kêd, and the Welsh Ern, a pledge, agrees with Renn, the nursling, as the child of Kêd.

The Druids called Hu the overseer, and on the Mithraic sculptures this solar overseer is pictured in place of the disk, afloat overhead on wings, with the serpent attached. Pliny said the Druids of Britain might have taught the Magi of Persia. But both drew from the parent source.

The magical banner of the ancient British was emblazoned with the same device of sun and serpent, and the Two Truths were likewise identified by the presence of Hu and Kêd, the father and mother who supported the disk and serpent.[4] One emblem of Hu (Eg.) was the tongue, from which he has been called Taste personified. But the tongue means more than taste. Stockius observes that a tongue was the type of flame. The tongue denotes the Word, utterance, mystic manifestation.[5] The tongue-emblem of Hu is represented on the Tokens of Cuno along with the mother as the mare, Hu being the male deity.[6]

The 56th Triad asserts that the god Hu had already instructed the race of the Kymry in the art of husbandry and the cultivation of corn, previous to their removal and separation from the old land.[7]

---

[1] Ch. cx. Birch.   [2] Hywell.   [3] *Poly-Olbion*, song 5.
[4] Oliver, on *Initiation*, p. 229.   [5] *Hor-Apollo*, b. i. 27.
[6] Gibson's *Camden*, tab. i. fig. 3.   [7] Davies, p. 107.

Hu, whose name in Egyptian signifies corn, also means spirit, aliment, and sustenance, and he is the giver of wine and generous liquor, who presides over the festive carousals. "After the deluge he held the strong-beamed plough, active and excellent, this did our lord of stimulative genius."[1] Hu (Eg.) means both aliment and genius. The god Hu on the Monuments is the good Demon, the winged sun, the sun in the act of shedding. In the Ritual he is said to be one of the gods attached to the GENERATION of the sun.[2] He is seminal as well as solar, hence Hu represents the seed of life, the giver of corn.

The great emblem of Hu is the Tebhut, the sun on expanded wings called the great god, lord of life. A British priest invokes the god under the title of Hu with the wings.[3] He too is the Tebhut. Hu was depicted as the driver of his three oxen, and in the hieroglyphics Hu signifies "to drive."

Hu is the bull, the mighty bull, and the one bull takes the Triadic form in the three bulls that draw the Avanc out of the lake. The triad of bulls which is the three-in-one is the analogue of the Egyptian male triad.

In the *Ritual*[4] we read, "These gods who are attached to the generation of the sun are Hu and Ka; they are followers of their father Tum daily." KA or KAK was the god TOUCH. We still swear by touch in the sayings "true as touch" and "touch-true." Kak, the blind God, went by touch, and touched home as the one who reached the boundary. Kak, in Eskimo, and Kakoi, in Japanese, mean boundary. The boundary of Kak or TOUCH is extant in TICH-field and TICH-bourne, the bourne in the latter being a translation of TICH. Tum (Atum) is the solar bull, the powerful bull, and Hu and Ka complete the bull-triad. The death of the bull as a sacrifice of virility was represented in the Druidic mysteries.

"The assembled train were dancing after their manner and singing in cadence with garlands on their brow, loud was the clattering of shields round the ancient cauldron in frantic mirth, and lively was the aspect of him who, in his prowess, had snatched over the ford that involved ball which cast its rays to a distance, the splendid product of the adder shot forth by serpents.[5] But wounded art thou, surely wounded, thou delight of princesses, thou who lovest the living herd. It was my earnest wish that thou mightest live, O thou victorious energy! Alas! thou bull, wrongfully oppressed, thy death I deplore."[6]

Aeddon is a name of Hu, and Atum is said to be "the duplicate of Aten."[7] The priest of Hu was distinguished by the title of Aedd

---

[1] Iolo Goch. fourteenth century.   [2] Ch. xvii.
[3] Davies, p. 121.   [4] Ch. xvii.
[5] Cf. Pliny's account of the production of the serpent's egg or stone, to note another instance of the Roman and Greek ignorance of the ancient symbolism.
[6] Ancient Welsh poem. Davies, p. 576. See also p. 172, *Ib.*
[7] Stele of the *Excommunication*.

after the god Aeddon, and in the hieroglyphics the At is a divine father, a priest.

Having shown the identity of the British and Egyptian Hu and Aeddon with Atum (earlier Aten or Adon), it is now intended to suggest that the triad of Atum, Hu, and Kak is the British triad of Tom, Hu, and Jack.

Atum in one character is the setting sun; he sets from the land of life.[1] He is the sun of AUTUMN, to which season he has bequeathed his name. He is the god of the underworld, also named from him as the tomb. He was the lame and lessening sun of winter, and it is touching to think of the ancient deity who was the great god of heaven and earth, the great judge of the dead in the lower world, from whom we derive the primitive name of the judge as the Demster, and judging as dem-ing or condemning, and lastly damning, actually reduced to the condition of Tom Tiddler; but so it is. Tum was the winter sun : the slow-moving, long-in-coming, feeble sun, who as Tom Tiddler is Tom the Toddler, a sort of simpleton or Tim-doodle, who moves with slow tiny steps, and is twitted for being the lazy one, from whom his gold and silver may be filched with impunity. One game at his expense is played on the eminence up which he, the lower sun, has to mount and send off the mockers. Tum, the setting sun, is depicted as crossing the waters by means of the cow. In one instance, this sun, as lord of Hab, is making the passage, as it is called ; and it is said of the deity, " Thou hast rested in the cow, thou hast seized the horns, thou hast been immersed in the cow Mehur."[2] The sun was re-born at the thigh of the cow.[3] How has the myth been minified by our faeryology in which Tom Thumb is described as being swallowed by the cow and re-born from it ! Yet the matter is the same. Tum is the name of the lowest member of the hand, the thumb. Tum was the red as well as the lower sun. Both colour and lowliness meet in the plant called Thyme. The Damson also is the redder, lesser plum. Tom is a close-stool, botTOM, the lowest part. Tawm, a swoon, a sinking down. The Temples are the lower part of the head. The Taum is a fishing-line which goes under. The Tom-tit, a name of the wren, builds underground, the other Tom under the grass. The Tommy-loach stays in holes, the Tom-cull, Miller's Thumb, or Bull-head, lurks under stones. Tom-tiler is a hen-pecked husband who knuckles under ; the tame and timid derive their appellations from Tum, the lower, hinder. Timings are the dregs of beer. Tum was the negative, sterile under, hindward sun. Hence Tum signifies no, not, negative. He completed the circle of the day and of the year, hence Tum, to announce, or Time, which depended on termination ; thus " tumt " is

---

[1] *Ritual*, ch. xv.
[2] *Trans. Bib. Arch.* vol. v. 293. *Ins. of Darius at the Temple of El Khargeh.*
Birch.
[3] *Rit.* ch. xvii.

total or timed. Tum the lower, hinder, and secondary, are among the meanings of the word, and these have been curiously applied in the formation of English, and in words not found in Egyptian, though shaped in its mould. Tum, as the lower, is the name of our underworld, the tomb. Toom means empty, hollow, void. From Tum, the winter sun, comes the word and meaning Dim. This is echoed in many other languages, as DIM, in Akkadian, a phantom; TUMMA, Fin., dull, slow, dim; TUMME, Esth., dim, dark, slow; TUOM, Lap., dull in action, slow, and dim; DUM, Danish, obscure, dull, and dim; DIMBA, Swedish, haze, fog; TUMU, Shoshone, winter; TOMO, Wihinasht, winter; TAMN, Kanuri, to complete, finish, end. In the Xosa and Zulu Kaffir dialects, DAMBA means to grow less and less in bulk, and a person who totters with unsteady gait, whether from drink or weakness, is called DAMBU-DAMBU. To tumble is to go under, to dimple is to dip under. To be in the dumps is to be down. Trees are timber when cut down. A Timp is a place at the bottom of a furnace through which the metal runs. A Dump is a deep hole in water, supposed to be bottomless. The Ducking-stool was also called a Tumbril. The helmsman at the hinder end of the vessel is a Timoneer. A Tim-Sarah is a kind of sledge with wheels behind, and a Tim-whiskey is a chaise all bottom and no head. A Tom-noddy and Tim-doodle are foolish, deficient persons, and Tom, as the sign of the lower lesser, or little, attains the point of culmination in Tom Thumb.

A Tom-toddy is a tadpole. Here, too, is an image of Tum. One type of the sun crossing the waters was the frog-headed Ptah, the father of Tum, and our Tom-tadpole reproduces the son. Tom-toddy, or TUTTI, is literally the secondary type found in Tum, the son of Ptah. Tom-tut (in Egyptian the image of Tum) is also a kind of bogy. Children in Lincolnshire are frightened by being told of Tom-tut, a supernatural being that still haunts the nursery ; and persons in a state of panic are called Tut-gotten. In Norfolk the same bugbear of naughty children, and the especial demon of dark places, is known as Tom-poker. Possibly this title actually enshrines the motherhood of the God. Tum was the son of Pekh, the cat-headed goddess,[1] and Pekh-ar is the son of Pekh ; Tum was Pekh-ar, as Osiris, the son of Hes, is Hes-Ar.

At Bromyard, in Herefordshire, among the ceremonies performed in the first hours of the new year, is a funeral service said over " Old Tom," as the departed year is called.[2] Here the transformation of Old Tum is applied to the year and made solstitial. In the Egyptian cult of Tum it was equinoctial, the old Tum changed into the young Iu-em-hept. When the devil appeared to the Witch of Edmonton, he called his name Dom.[3] That is Tum, the solar deity of darkness, who

---

[1] *Rit.* ch. xvii. [2] Fosbroke, *Sketches of Ross*, p. 58, 1822.
[3] *Sphinx and Œdipus, or a Helpe to Discourse*, p. 271, 1632.

becomes the devil of eschatology.   Though dead and buried and
transformed into the devil, the spirit yet lives, and still bears the name
of " Old Tom," in a kind of gin.

Tum, as the lower, was the sun of the left hand.   And there was a
custom of drinking over the left thumb connected with passing the
bottle round the table according to the course of the sun, and the
left hand or lower sun was the solar god Tum.   Also instructions were
given that in a fit of convulsions or shortness of breath you should
hold your left thumb with your right hand, and the god Tum was the
breather of life, the breath of the mouth.   Children were taught to
fold the thumb inside the hand as a charm against approaching danger;
it was an English custom to double the thumbs of the dead within
the hand.   This was a typical mode of laying hold of Tum, the
great judge of the dead.

It is reported that during the battle of SOLFERINO Victor
Emmanuel, King of Piedmont, kept his right thumb doubled in
his hand for good luck.[1]   He was typically holding on to the god
of justice.

The good old Irish names of O'Tom and O'Jack, common in an
Irish poem by O'Duvegan, of the early fourteenth century, are of
divine origin after all.

Tum's two manifestations are in the persons of Hu and Ka or Hak,
earlier Kak.   In the pictures of the lower sun (Tum), crossing the
Ament in the boat of the solar disc, Hu is at the prow and Hak
(Kak) is at the helm.[2]

Tom is our impersonation of all that belongs to the lower, minimized,
dull-looking, lagging winter sun, and Jack is his natural antithesis.
If Tom goes down, Jack leaps up, or springs, as illustrated by the
Jack-in-the-box ; or ascends, as by the bean-stalk.   Jack is the lively
lad, the spirited, full of spirit, a spirit or sprite.   Jack-bandy is a
name of the spritely minnow.   The Jack-a-dandy dances on the
ceiling, Jack-a-lanthorn is the dancing will-o'-the-wisp, Jack-in-the
green dances in true Egyptian colour on May Day.   Green with them
was the hue of reproduction from the invisible or spirit-world, and
Jack as the sprite of the May dances invisibly in green.   We still
perpetuate the symbol in the colour called "invisible green."   The
flesh of both Ptah and Num was painted green, the hue in which the
spirit of life emerges from the underworld.   The Aztec divinity
Huitziton was represented with his head decorated with feathers, his
arms in the shape of tree-trunks with branches, while from his girdle
green leaves fell or flowed downwards.[3]

The god Ka is the more ancient Kak, whose name modifies into Ka
and Hak.   Akh (Eg.) is a spirit, lively, gay, and the word also means to

---

[1] Thomas Wright, on the Worship of the Generative Powers, appended to
Payne Knight's *Essay*, p. 151; privately printed, 1865.
[2] *Book of Hades*, first division, plates.   [3] Bancroft, iii. 400.

rise up and illumine. Akh is to elevate, suspend, adjust. An image of this is extant in our Jack-in-the-box, who is suspended on springs and who "jacks up" with a broad smile to illumine with merriment, being now reduced to a solar symbol for the nursery. Jack dances on May Day in green leaves, and Akha is to be verdant, green. Jack is the quick, clever, sharp, hence the knave. Akar in Egyptian is to be quick, clever, sharp, always ready, just as we say "Jack's the lad," the character aimed at by "cheap Jack." Akh is a spirit, the creative or virile spirit in Egyptian, Assyrian, and Hebrew. It is the evil spirit in Japanese. In English the Jack used to hold the spirit. Jackey is a name of gin and strong ale. It is said or sung of our ancestors that they "took a smack of the old black Jack, till the fire burned in their brain." The Jig is the lively dance, full of spirit. The Egyptian Akh, to lift up, suspend, is embodied in the Jack instrument for lifting a weight, and Akh, to turn round, is imaged by the Jack suspended and adjusted to revolve in the chimney. The Jacks used to jump up in the spinnette of old, Jack Ketch hangs up, the Union Jack is run up, the Jack struck the bell when the hour was up. Jack in his box leaping up with a laugh or a broad grin is a type of the sun or of the soul ascending from the nether world. Here is the fellow picture from the Egyptian Ritual :[1] "I rise up as a god from men. I prevail as ye do with that god taller than his box. I have sat in my place on the horizon." Just as the Jack leaps up. "Oh, taller than his box ; lord of the crown Atf!"[2] That is the head-dress of Jack! The Akh in his box is literally the Jack, and this spirit taller than his box is the Akh. Jack-in-the-box is an Egyptian hieroglyphic of resurrection, and this gives the significance to "Jack's Barrow," a large tumulus in the parish of Duntesbourn, Gloucestershire. The name was like a warrant for rising again. A place near "Jack's Barrow" is named "Jack's Green." Jack represents the spirit of life in spring, in the act of springing up or in jactation. One image of this was the ascending sun that rises up, aspires, illumines, puts a spirit of youth in everything, or "Jacks up," rises, revolves, and reigns higher and higher, and ranges from little Jack who climbed up the bean-stalk, to the place of the giant, or to the top of the great circle, over which reigns Hu with wide-expanded wings as the god in his disk, or, to reverse the process, from him who rides on the heavens by the name of Jach, down to our Jack of the bean-stalk, and the box. Possibly Jack, the Akh who jumps up out of his box as the young god, the sun of the spring equinox, is extant as the veritable "little Jack Horner." The divinity was represented as the young one, the Ar or Har. His place was the corner, and he is described as being in his corner, or angle. Says Ra, in the last judgment, "Let the great one, who is in his angle call the souls of the just, and have them placed in their abodes near

<hr/>

[1] Ch. lxxix.   [2] Ch. cxxv.

the angle." [1] Har-khuti is not only god of the corner, he is personified as "the brilliant Triangle which appears in the shining place." [2] And this god, who rises up victorious on the horizon, spiritualized (Akh) is literally Jack Horner, for Hor is Horus, and "Ner" (Eg.) means victory. Jack's corner has been removed to the place of the solstice, and his victory minimized to the pulling out of plums. One wonders if these plums, like those of snap-dragon, may represent souls snatched from the burning or the abyss. His exclamation, "What a GOOD boy am I," still preserves the title of the youthful god called Nefer (the good, the young), applied to Nefer-Tum, and Khunsu Nefer-hept.

Har-Khuti, god of both horizons, the sum total of the Tum Triad, called "the brilliant triangle which appears in the shining place," seems to be extant in the Lord (Har) CADI, and the triangle to be reproduced in his garland.

The Cadi is a remarkable character among the May mummers in Wales. He is the most active personage, chief marshal, buffoon, and money-collector. He is generally arrayed in a dress of both sexes, male above and female below. The number of the other mummers is thirteen. They are dressed in white decorated skirts worn over black velvet breeches. This dual dress of the Cadi and his followers corresponds to the Two Truths, two birds (light and black) two colours, two origins, and two horizons, of Tum-Har-Khuti, and the mythical personage designated by Taliesin the "two-halved youth." The company carried the May garland, the glorified image of the circle completed and once more renewed at the time of the vernal equinox.

The Cadi, as primus, suggests the god Khuti of Egypt, called Har-Khuti, the sun of both horizons, lord of the two seats or double seat of the equinox, one of whose types was the Sphinx. One name of the Sphinx is Hu, and Hu is the god of the horizon, and the British sun-god. The mixing of sex in the dresses answers to the dual nature of the Sphinx, and the Two Truths. The garland of the Cadi, says the *Every Day Book*,[3] consists of a long staff or pole, to which is affixed a TRIANGULAR or square frame. In the procession the triangular garland is carried next after the Cadi. The god Har-Khuti is pre-monumental. In the record of the divine dynasties, a period of 13,420 years is claimed for the Shus-en-Har or worshippers of the Har, who as Sut-Har (whether Sabean or solar or both) manifested on the double horizon as Har-Makhu and Har-Khuti. Another hieroglyphic of Har-Khuti of the brillant triangle is extant in the three-cornered cake. In the city of Coventry one of the New Year presents given by all classes

---

[1] E. Lefébure. Also, the *Book of Hades*, by Lefébure and Birch. *Records of the Past*, v. x.
[2] *Litany of Ra*, ii. 7 ; *Records*, viii. 116.
[3] Vol. i. p. 562.

of people is the God-cake, invariably made in a triangular shape. The god and triangle meet in one name as Har-Khuti. The cake is the Egyptian symbol of the sun and the horizon ; Har-Khuti of the triangle is god of the horizon, the British Hu. This custom is peculiar to Coventry.[1] The Coventry three-cornered cakes are called god cakes, and the name of God is one etymologically with Khut, the god of the triangle, of which the equinox was the apex, the Khut, the solstices being marked as low down on the horizon, the equinox in the zenith. The cake was a hieroglyphic of the triangle. Coventry is supposed to take its name as Conventry from a priory founded there in 1044, by Earl Leofric and Lady Godiva. But were there such persons as Leofric and Godiva ? It is on Trinity Friday that the Lady Godiva rides naked through the town. The day also agrees by name with the three-cornered cakes and the triangular god, Har-Khuti, who was the manifester of the Trinity. The corner or angle at which the young sun-god was re-born is the Kheb or Kep, *i.e.* Cov. It was in the place of the two times, the " Teriu," where the two became three in one. The Egyptian " Teriu " is expressed by three, and gives us the word, and Kep-en-terui is the corner and the concealed sanctuary of the two times. Terui also denotes the limit, the circumference, and a form of Sesennu, the seat of the eight gods in the lunar birth-place. The Egyptian name is represented by the Welsh DARU or DERYW, an end, and by the TROI, a turn, a circle, the figure of Troy, earlier Trev, the Rep (Eg.) or religious house of Egypt, whence we derive our Trefs, Tres, and Troys. " En " (Eg.) is the preposition " of " or " of the." The name Godiva will resolve into Khutifa, the bearer of this god Khuti, from FA, to bear, carry, be pregnant with Khuti, the child. Thus Godiva, the lady, the patroness of Coventry, apparently becomes a form of the goddess Khet-Mut (Eg.), the British Kêd. Further, IVA or Iua (Eg.) is the boat, the symbol answering to FA, to bear, carry, and Kêd was the bearer whose image is the boat that bore the seed across the " dale of grievous waters, having the forepart stored with corn," a symbol of the mother, great with her child. Khep denotes the secret place, the sanctuary, the Ha-Kheb, in which the god was reborn at the Terui or Troy, to become the young divinity of the double-seat, he being the " brilliant triangle." The same word Kep means hiding, concealing, lying in wait, looking, watching. And Tum in the Kep-en-terui would be in the place of concealment, watching, looking, lying in wait, literally the Peeping Tom of the Coventry mystery. One application of the word Tum actually means to spy, and covet, with the eye for determinative. It would indeed be strange if the Coventry mystery were based on historical characters, for Godiva is just the native goddess who appears on the Monuments as Khatesh or Khen, the bearer, the boat of the waters, first of all personified as the pregnant hippopotamus, Khebt, later Khet,

---

[1] *Notes and Queries*, 2nd ser. ii. 229.

the British Kêd. Kef or Kep is the genitrix ; the word means mystery. Khep is the goddess of mystery, the mystery of fermentation, fermented spirits, and fertilization. The Coventry mysteries were among the most famous in Britain. The word mystery or Mes-terui (Eg.) means the birth, a child of the dual time, born at the spring equinox in Kef-en-Terui. This derivation of the name of Coventry, as opposed to Conventry, is supported by another name, that of Daventry. Tef or Tep is a variant of Kep, and the Tep is likewise the abode of birth at the Terui. Tep was a mythical locality consecrated to Buto or Uati, the goddess of the north, the British Kêd, and it permutes with the Kep or Khab, as the Ha-Khab.

The "try" as a form of the Tref or Tre, Egyptian Rep, Trep, and Terui, our Troy, will not agree with the Convent.

" Curcuddie," says Jamieson, " is a phrase used in Scotland to denote a game played by children, in which they squat down on their hams and hop round in a circular form." The word Curr means to sit in this fashion. It is the Egyptian Kar, to stoop down, bear, carry and be under ; khuti is to make the circuit, go round in a circle. The game is probably an imitation of the lame sun moving round slowly and with difficulty through the lower Kar, belonging to the childhood of the race, and its mimetic mode of enacting ideographic representations. Kar-Cuddie is the hard form of Har-Khuti, and the sun in the Kar-neter is well represented by the English Caddee, a servant employed under another servant ; he is the Kar-Cuddie, the child Har, who was maimed in his lower members.

In the game of " noughts and crosses " there are two players ; one makes the circle and one the cross. It is gained by the one who can first get three marks in a line. Here we find the circle, the cross, and the triad. But when neither of the two players wins the game it is given to " Tom." " Tommy Dodd " is a term also used in tossing, when the odd man goes out. Tum is the god of both horizons, and Hu is his representative of the circle (the Hut); Hak, of the crossing ; when neither Hu nor Hak win the game, it is given to Tum, so that each has it in turn. The cross and four circles or dots of Tit-tat-toe form one of the chief patterns in the artistic designs of the Bronze age.[1] It depends on the particular cult as to which of these three is acknowledged figurehead and primus of the triad. In the Egyptian Ritual Tum is the supreme ; with the British it was Hu, and with the Hebrews it was Jah Iach or מ׳.

The house that Jack built is the solar mansion of the thirty-six gates in the upper half of which was stored the bread and drink of life, both being represented hieroglyphically as grain. Jack is the Akh or Jach who, as Tum, is said to "build the house." The rat that ate the malt is the " abominable rat of the sun," found in the Ritual. The cat-headed goddess Pasht is designated the cat devouring the

[1] Dawkins, *Early Man in Britain*, p. 378.

abominable rat.[1] The dog that worried the cat occurs in the shape of a dog-faced demon, with human eye-brows, that lived off the fallen ones at the angle of the pool of fire in the west, the domain of Athor, the cow-headed goddess, who at this point, having tossed the dog, took the sun (Atum or Tum) between her horns and carried him across to the east. The cow and the cat were both bringers forth of the new sun of spring in the house that Jack built; the house of the two horizons. This was the representation of Egyptian mythology, doubtless the very form in which the facts were taught in the Mysteries.

The Aztecs, at certain religious festivals, as in the feast of Tlaloc, in the sixth month, were accustomed to carry in their arms the images of gods "made of that gum which is black and leaps, called Ulli"; these were named Ulteteu, that is, gods of Ulli.[2] Ulli is india-rubber. And the leaping gods, the Ulteteu, suggest kinship to our Jack-in-the-box, whose progenitors leaped in india-rubber before other springs were invented.

This india-rubber image of deity, a type not yet extinct, is in our day subjected to a great deal of stretching.

A more mystic image of our Jack is the dance of sunbeams on the ceiling, reflected from water in motion, called Jack-a-dandy, or Jack beating his wife with a silver stick. It is emblematical of the two sources illustrated by sun and water; for Jack is a sun-god, and his wife is water, a pail of which he went for with Gula, or Jill, up the hill.

Jack dancing on the water is the same solar image that we find in the *Ritual*—"Oh sun, thou hast lodged dancing";[3] that is, on the waters. The box of Jack is the ark in which the sun lodged dancing, and crossed the water. "The great one crossed in the cabin, capped in the ark."[4] "I saw the sun in the midst of his box when I hailed his disk daily, the living Lord,"[5] says the spirit in crossing from this life to the other.

Atum was the lord of An: lord of the double-seated boat in An. Atum is one with Aeddon (Hu), and this solar god of the Britons appears in one of Taliesin's poems as TEYRN ON, the Sovereign On, or of On, *i.e.* An, usually written Annwn. This is identifiable, because in another poem of Taliesin's, on the "Rod of Moses," he connects the British On or An with Heliopolis. He says of Joseph, "the son of Teyrn On collected treasures from his associates, and the sons of Jacob had those treasures in possession." The title of his poem is "Kadair Teyrn On," the chair of the sovereign of On. In this he sings of the "Person of two origins of the race of Al-Adur, with his divining staff and pervading glance, and his neighing coursers, and his regulator of kings, and his potent number, and his blushing purple, and his vaulting over the boundary, and his appropriate chair:

---

[1] Birch, *Gallery*, p. 18.　　[2] Bancroft, iii. 340.　　[3] Ch. xv.
[4] *Ibid.*　　[5] *Ibid.*

amongst the established train, the sovereign of On, the ancient, the generous feeder," or Heilin Pasgadur, the feeder.[1]

Hu (Eg.) signifies corn, food, and aliment. Tum is the generous feeder. He provides the bread of Tu and the drink of Tep for the Osirian. "My father Tum did it for me; he placed my house above the earth; there are corn and barley in it; unknown is their quantity. I made in it the festival of Tum."[2] Tum is the lord of An, and the feast is in An. The altars in An are piled with plenty. Tum is called Hetu Abi,[3] and hetu means bread. Tum is the ancient god, called Ra in his first sovereignty, and the oldest of the chiefs, who is re-presented as Har-Makhu of the two origins, or horizons. The boundary was that of the horizon, where the seat was established in An.

The poet sings of Teyrn On, "Let him be the conductor of his fleet, then, were the billows to overwhelm beyond the strand, so that of firm land there should indeed remain neither cliff nor defile, hill nor dale, nor the smallest sheltering cover from the wind when its fury is roused, yet the sovereign of On will protect his chair: skilful is he who guards it."

This will appear less remote when we have set forth the typology of the Ark and the Deluge. The writer apparently means that were the deluge to break forth again, there is always one place of safety in the ark of On; that seat of the god will remain secure. This was the seat of Atum in An, the established region; the double-seated boat is there, the ark of Sekari, found with Atum in the procession of the great gods.[4] "There let them be sought; let application be made to Kedig for the men of Kêd, who have been lost."[5] That is, in An, the established region, called Tattu the Eternal.

The Eel was a type peculiar to Atum as sun of the under-world. It took the place of the solar serpent, as the crawler through the waters and mud of the abyss. The eel preserves its divine name, and being a divine type, it was too sacred to be eaten. That was the primitive law of the case. Things forbidden to be eaten were hal-lowed and not abominated. This was the later phase when the theo-logy had changed. At first the Jews did not eat the pig because it was sacred, a form of the multimammalian mother; afterwards because it was degraded and denounced. The later cursing implies previous consecrating. And to this consecration of the eel in Egypt the present writer attributes the yet surviving horror of the eel found in Ireland and in Scotland, where it is invested, rightly too, with the character of the serpent. This repugnance to eating the eel is a superstition; the feeling against eating it was once religiously fostered because it was a divine type, and when the theology changes

---

[1] Davies, p. 528.
[2] Ch. lxxii.
[3] Ch. lxxviii.
[4] Wilk. *Mat. Hierog.* pl. 65.
[5] Davies, pp. 527-532, whom I have here followed. He is not to be compared with Skene as a translator, but was right as to the Barddas being in possession of the ancient mythical matter, although it was not derived from the Hebrew writings.

and the thing is anathematized as unclean, the horror of eating it is there, ready to be set against it.

The superstitions of folk-lore and religion are mainly a deposit of denaturalized mythology, and not until the original types are interpreted and rightly explained can these superstitions be estimated justly.

The eel is a symbol then, extant in our islands, but not understood, which can be interpreted in Egypt, where it belonged to a deity of the dark, worshipped in the remotest times. This accounts for the eel that was seen by a man in Lorn as he was fishing, which was passing from morning until sunset without coming to an end— that was a long eel! Not at all. It was the type of a circle, or the completion of the circle passing through the deep, as the sun-god Atum, whose name denoted a water-type. A namesake of Tum is extant in the Timber, a kind of worm.

Tum was known in Egypt as the living god. That is the Ankh. And we have our divinity of the same name in the god Jingo, whose worship has outlived that of Kêd, Hu, Prydhain, and others of the ancient Pantheon. Jingo is the modified Kingo, the Mentula type of deity. Jingo was a god, also, of the Bask people. "By Jingo" is a common oath, but the more emphatic form is "by the *living* Jingo"; that identifies the Ankh (Eg.) with the living one.

"Ankh," the living, and also the name of the King, was an oath and a covenant, so sacred that it was profane and punishable to use it vulgarly, or to swear by the life (Ankh) of the Pharaoh. Profane swearing consists in making the sacred usage common.

The Irish BEANGAN and Welsh PINCEN, for a sprig or branch, are derivatives from Ankh, the living. The "living Jingo" apparently translates and identifies the Egyptian Ankh, an oath meaning by the living or the life. This sense of life enters into our words jink and "high-jinks." Jink is to be gay and ebullient with life. "High-jinks" are the very festival of frolic life.

Unki (Eg.) is also a god, or the name for God. According to Brugsch Bey, the special Ankh, Unki, or Jingo of Lower Egypt was the god Atum, the only one who is expressly denominated the Ankh or living god. Our Jingo ought therefore to be identified with the Tum Triad, as he may be. Eidin, a form of Aeddon, signifies the living, and both names are identical with Adon and Atum, who is the living god of Pithom, the Ankh, our living Jingo.

Again, the wedding-ring was formerly placed on the thumb. The author of *Hudibras* refers to this:—

> "Others were for abolishing
> That tool of matrimony, a ring,
> With which the unsanctified bridegroom
> Is married only to a thumb." [1]

---

[1] iii. ii. 303.

The Hereford, York, and Salisbury missals direct that the ring shall be first placed on the thumb and left on the fourth finger. But as late as the time of the first George it was a custom to place it on the fourth finger during the ceremony, and afterwards it was worn on the thumb.[1] Here we have the Ankh coupled with Tum, the ring being an Ankh-sign of to pair, to clasp, and to make a covenant.

At Kidlington, in Oxfordshire, the custom was on the Monday after Whitsun week for a fat live lamb to be provided, and the maids of the town used to run after it having their THUMBS tied behind them, and the one who caught it with her mouth was declared Lady of the Lamb.[2] This points to the time when the vernal equinox occurred in the sign of the ram. Possibly the thumbs tied behind may have been symbolical of Tum, the hinder sun, now transformed into Hu, in the sign of the ram. Tum (Eg.) is also a name of the mouth.

Tut is the hieroglyphic hand, and the name of number five or one hand. We have Tum on the hand as the lower member, and Tut as the sign of five in the little finger. In the ancient nursery lore the hand is reckoned up as " Tom Thumkin, Betty Bodkin, Long Gracious, Billy Wilkin, Tutty-Woo." Tutty-Woo, the fifth sign, is number five in two languages, " Tut " in Egyptian and " Wu " in Chinese. There are two versions of the last line ; the little finger is likewise called " Little Tut," and in this version Tut is a hieroglyphic of five, fifth, or a hand. It is this little finger Tut or Tutty that knows and makes known. In Piedmont mothers are accustomed to awe their children by making believe that it reveals everything.[3] Tut (Eg.) is speech, the tongue, the word, the manifester and revealer of the hieroglyphics. The revealer personified is Tut or Tahuti, the lunar deity. In fact, we have two Egyptian deities on one hand in the thumb and tutty-woo.

The first month of the year in Egypt was called the Tat, and this is the Irish name for the first or opening day of harvest. Also the Irish god of harvest was called Tath. Another name of Taht is Takh, and Dagh was a god of the Irish Tuatha-Dadanan ; Deaghd is a name for divinity.

On the Monuments the lunar deity Tahuti, lord of the moon in its first half, is represented by deputy in the second half. One form of this deity is the dog-headed monkey, the Aan, earlier Kan. From this connection of the Cynocephalus with Taht, we derive the well-known man in the moon, who is followed by his dog as Taht was by the dog-headed monkey. These two images of Egyptian mythology have their abiding-place in the moon for ever. One legend makes the man to be Cain, that is, Kan the dog, or Cynocephalus.

The man is supposed to carry a bundle of sticks, said to have been gathered on Sunday, the origin of which has been derived from the

---

[1] *British Apollo*, vol. i. p. 270.
[2] Blount's *Jocular Tenures*, Beckwith's ed. p. 281.
[3] Gubernatis, *Zool. Myth.* vol. i. 166.

Book of Numbers.[1]   The earlier representation may have been coupled with the Hebrew story to point a moral, but the image is sure to be Egyptian.   In our elder poets, Chaucer and Shakspeare, the bundle is a bush of thorns, and a bush is but a branch or tod, and Taht is the bearer of the palm-branch of the Panegyrics ; he is also lord of the date-palm.   Time was reckoned by the palm-branch of the festivals.   The great spring festival was that of our first of May. The branch of May was a sacred sign of this season, and that is the white-THORN bush !   Thus we recover Taht and the Cynocephalus in our man in the moon and his dog, whilst the palm-branch is represented by the bush of thorn or branch of May.

But to return to Hu, the sun-god.   A relic of the Disk-worship apparently survives at Silchester in connection with the onion. Onion-pennies is the name given to Roman coins when found there.   According to tradition, they are so called after a giant whose name was ONION.[2]   The great god, lord of heaven, divinity of the disk, is the Hut, and Hut is the onion.   It seems to follow that the giant Onion is a form of the solar god.   Further Hut, the onion, for the god and the disk, is also the name of silver, and the pennies are the disks of Onion.   The giant is one form of Hu, the great god of the Hut sign and circle, the great solar circle.   Huten (Eg.) is this circle and the name of a ring, and from Huten in the hard form of Khutn comes the Norse Jotun or Eoten, the old English Etin, the giant.   In Egyptian, Khut, Hutn, Aten, all denote the ring or circle of time. The giant was a figure of great extent, a type of the larger course.

In one of our western isles, that of Borera, there was a vast stone, on the hill Criniveal, some twenty-four feet long ; this, the natives said, marked the spot where a giant of a month old was buried.[3]   Of course, when time came to be reckoned by hours and minutes, the lunar period of time looked a giant; that of Hu or Aeddon was a year. This type took one form as the eye of the Cyclop, or giant, the one-eyed monster, the eye being another ideograph of the circle.   The Norse Jotunheimr, the giants' home, is a region of the eternal, or on the way to it, by means of gigantic cycles of time.   The Saxon Eoten for giant is a word unknown in the Teutonic branch of language. Nilsson traces it to a Lap word.   Grimm thought it had been derived from Etan, to eat.   It comes from Katen, an image, a ring, as the representative of a large circle of time.   The eye as a symbol of the cycle was given to Horus, to Taht, and to Hu.   It was likewise assigned to the giant as the Cyclop, and putting out the eye was synonymous with slaying the giant.   The story of Odysseus and his escape from the monster whose eye he had put out has been traced by M. Antoine d'Abbadie among the tribes of Abyssinia.   In this version the hero escapes from the cave by being carried under the belly of the ram.   This gives the thread of a clue to the maze.

[1] Ch. xv.   [2] Wright, *Dictionary*, vol. ii. p. 711.   [3] Martin, p. 59.

Odysseus is a form of the giant-killer or circle-ender, like Khunsu, and to put out the eye is figurative for ending a cycle personified in the Cyclop. Khunsu, the Egyptian Hercules, was the god who represented the full moon, and with the full moon of Easter the cycle of the year, the eye of the Cyclop, ended. This, in the ram calendar, was where the sun entered the sign of Aries. When the vernal colure was in Pisces, the solar hero was conveyed in the belly of the fish.

The onion of Hu was a form of the Ankh, or living, and as an emblem of life the oldest spelling of the name Onion retains the primary significance; it is the INGAN, and ANKH-AN is the repeater of life, who was worshipped as Tum, the living, and Hu, the lord of life. Juvenal satirizes the Egyptian veneration for the leek and onion. He says it is impiety with them to violate and break with the teeth the leek and the onion. "O holy race to whom such deities as these are born in their gardens."[1] The onion of Hu, or Tum, has been given to "Saint" Thomas. Burton, in his *Anatomy of Melancholy*, speaks of a kind of divination with onions laid on the altar on Christmas Eve. In the instructions for divination with onions the buyer is told to be sure to select a shop with two doorways and to go in at one and come out by the other. The onions are to be placed under the pillow on St. Thomas's Eve.[2] St. Thomas takes the place of Tum or Tom, and the double doors correspond to the double horizon, the double house and double-seated boat of Atum.

Here is another meeting-point. The great god Hu was the youthful sun-god, son of the old Atum, and ONION (Hut) is an English name of a young child. Also the ONION is the little one in the sailor's reckoning of so many knots and an ONION.

According to Stukely, the remains of a stone temple at Navestock, in Essex, showed that it had represented a circle with wings. He could not have derived this supposed image from the hieroglyphic Hut, being wholly intent upon the serpent, yet it is the figure of the Hut or celestial sun, the chief sun, the life-giver, the winged disk of the god Hu, called the solar disk spread out. The disk with wings, however, does interchange, as Api, with the disk and serpent. So the British Hu is called the gliding serpent.

The Hut sun is closely connected with our Whit-sun. Hut (Eg.) means white. It was a common superstition that whatsoever was asked on Whitsunday morning at the instant the sun arose and played or danced, God would grant. The god was Hu, and the sun his Hut; the Sunday his White day. Evans, in his *Echo to the Voice of Heaven*,[3] says he went up a hill to see the sun rise betimes on Whitsunday morning, and saw it at its rising "skip, play, dance, and turn about like a wheel." As the Hut sun was the sun of the

---

[1] *Sat.* xv. 1.   [2] *Choice Notes*, p. 244.
[3] 1652. P. 9.

equinox, our Whitsuntide is apparently seven weeks late or nearly 4,000 years behind time.

On one of the British coins the word "Att" accompanies the solar disk or wheel.[1] Aedd was an abbreviated form of Aeddon, the solar god. "Att" is the solar circle in the hieroglyphics, also to fly and soar; be a type of Aten or Aeddon. This circle is known to the bards as the "Barrier of Eidin," the encircling mound,[2] built of stones in the circular temples, and culminating at last in Edin-burgh.

Edinburgh was doubtless one of the seats of Aeddon. Buru (Eg.) is the height, summit, cap. In the hard form this yields the brig or arch over, and the Burgh. "Ethan" is an unknown place, named in the Pictish Chronicle.[3]

The Caer of Eidyn is mentioned in "Cunobeline's Talisman," and in "Gwarchan Maelderw," as well as in the Gododin, and in the sixth song of the latter poem, the "Knights of Eiddyn" are celebrated; they are the equivalent of Arthur's Knights of the Round Table, or Circle of Edin, Aeddon, Aten, or Adonai, the lord. Dun-Edin, another name, is the Tun, elevated seat, throne, mount of Aeddon, the solar god.

Aten and Atum are identical as the youthful god. The Nefer-Atum of a later cult reproduced the Aten or Adon of the earlier. This will be fully explained, meantime Atum (Tum) and Aten are interchangeable names. Hu, the youthful form of Tum, is the earlier Adon or Adonis.

In the Eton Montem it seems to me we have a surviving relic of the worship of Aeddon. Eton itself has the name of the youthful god. The Montem is a peculiar ceremony, said to have been coëval with the foundation of the college. Such foundations as this and those of Cambridge and Oxford were made, so to say, over the crypts of the more ancient cult.

On May the 12th a procession was formed of the boys, who carried standards and were accompanied with music. The scholars were dressed in military or in some fancy costume, and the procession went to a small mount on the south side of the Bath Road, called the "Salt" Hill, supposed to be a British barrow or burial-place of the dead.

The hill was ascended, the grand standard unrolled, when the captain made a speech, and the "Salt" (money) was then collected. The money bags were richly embroidered, the salt-bearers were superbly dressed. Members of the royal family attended at times, and their donation was called the "Royal Salt." The origin of the Montem, as of so many other immemorial customs, is unknown. At one time it was celebrated on the 6th of December, the festival of St. Nicholas, the same day as that on which the boy-bishop was

---

[1] Poste, *Keltic Inscrip.* pl. 7, coin 5. [2] Davies, *Myth.* p. 585.
[3] Jamieson, *Anc. Culdees,* p. 112.

elected at Salisbury and other places from among the children attached to the cathedral.[1]

The boy-bishop was a survival under Christianity of the youthful sun-god, and the ceremony will help to identify the meaning of the Eton Montem. Eton Montem! We use the word immemorial, forgetting what ineffaceable memorials are registered in words! Eton contains the name of Tum, who transformed annually into the child, the Adon or lord. Men-tem (Temau) in Egyptian reads the procession, memorial, dedication, gift of restoration. Eton Mon-tem would thus denote the festival of Aeddon's restoration.

A passage in the "Status Scholæ Etonensis" (A.D. 1560) shows that in the papal times the Eton scholars elected their boy-bishop on St. HUGH'S Day, November 17; St. Hugh being a supposed real boy-bishop at Lincoln, whose day was November 17. St. Hugh is just the papal name of the sun-god Hu, written in the Welsh form, otherwise Aeddon, Egyptian Aten.

At the time of the spring equinox the old god was restored in youthful form, and the event was celebrated in all lands. Eton is still dedicated to the Young god as the especial college for boys. In the same manner on the mount did the Druids unfold the dragon-flag of Aeddon at the time of the vernal resurrection. It was called "MAGNUM SUBLATUM."

"I have devised a huge standard, the mysterious glory of the great field of battle, and its excessive toils. There the victor directs his view over Manon, the luminary, the Arkite with the lofty front, and the red dragon, the Budd of the Pharaon; it shall accompany the Advaön, flying in the breeze."[2]

Salt (Sart) in Egyptian is the name of wisdom and science, and the word has the sense of sowing, planting, distributing, augmenting, and extending, which was no doubt typified by the "Salt" as money-means. Salt is still a recognized emblem of learning and wisdom.

The people of Alnwick formerly celebrated St. Mark's Day in connection with the making of "freemen of the common." The custom is locally attributed to King John, who is said to have once attempted to ride across Alnwick Moor and got stuck in the morass, in commemoration whereof he commanded that all freemen should pass on foot through Freeman's Well. When any new freemen were to be made, a small rill of water which runs through the morass was kept dammed up for a few days before the ceremony was performed. In this way a miry bog chin-deep in mud was made, and through it the freemen passed. King John is here as great an imposter as St. Mark. There was a race for the boundaries, in running which the young freemen were obliged to alight from their horses, in passing an open part of the common, and to place stones on a cairn at intervals

[1] See Dyer, p. 291, for authorities.
[2] Davies, *Ancient Welsh Poem by Gwarchan Maelderw.*

as a mark of boundary. This shows the true Mark, March, or Boundary signified. Then the race was continued to the Town-law or Twin-law Cairns, a high hill, for the honour of arriving first. On this mount the names of the freemen of Alnwick were published. Having competed for the honour of winning the boundaries, the young men returned to the town in triumph, and were met, according to tradition, by women dressed up with ribbons and flowers, playing upon bells, who welcomed them home with dancing and singing. These were called "TIMBER WAITS," a supposed corruption of timbrel-waits.

The celebration may be entirely interpreted by the solar mythos, of which so much has to be written in this work. To begin with, it celebrates the making or becoming free. This freedom was attained by the young deliverer, the sun-god Aeddon (Hu), who crossed the abyss of waters and landed on the mount, the rock of the horizon, where the Hall of the Judgment and of the Twin Truths was located.

Alnwick Moor was anciently called the Forest of Aidon or Aeddon, which identifies the passage with the solar god. King John, who crosses the morass, takes the place of the An or Oan, the manifester, who came up out of the deep, as the sun of the water-signs; the mark or boundary represents the land re-attained and the twin-law the place of the Two Truths, also called the double seat of Atum or Aten. The Timber Waits announced the reappearance of the victors who had won the boundaries.[1]

The "Hill of Aren" is a form of the mount of the horizon and landing-place of the sun. The resting-place of Tydain, the father of the inspiring muse, is in the border of the Mount of Aren; "while the wave makes an overwhelming din, the resting-place of Dylan is in the fane of Beuno, the ox of the ship."[2] The ox of the ship here identifies the landing-place with the colure of the equinox in the sign of the Bull. This "Hill of Aren" is a form of the solar birth place, the Bekh (Eg.), found in An. Renn (Eg.) is the young child, the nursling, and the name-circle; and it is suggested that Aln-wick is one with Arn-wick or hill of Aren, the birthplace and resting-place of Tydain, who, as the British Apollo, is the solar god and a form of Aeddon or Hu. Renn (Eg.) is the typical birthplace, personified as Rannut, and in the Aren was the Bêdd of the youthful god. Bed, in English, is the uterus; the Egyptian But and Pa-t; Hebrew, Beth; Vei, Ba; Keltic-Irish, Beith; Sanskrit, Bheda. This Bedd is the "Pet" of the hieroglyphics, the divine circle of the gods which is synonymous with number nine, the nine months of safety from the Deluge, the nine days associated with the Deluge of Deucalion. Tydain, the father of the Muses, is the progenitor of the nine. Thus the Timber

---

[1] *History of Alnwick*, 1822, pp. 304-309. *Antiquarian Repository*, 1809. Dyer, p. 201.
[2] Davies, *Myth.* p. 194.

Waits may have represented the Muses. Tema (Eg.) means a choir (Temau, choirs), and Ma signifies number nine. The nine were extant in the damsels whose breathings warmed the cauldron of Keridwen. The Gallicenæ of Sena were the Nine. "The tuneful tribe will resort to the magnificent Se of the Sêon," says Taliesin.

Hu, the sun-god, was celebrated by the Barddas for putting an end to the dragon-tyranny. Hu, the bull, was son of the dragon, as in the Bacchic Mysteries the bull was born of the dragon. The dragon was a type of the mother Kêd, Draconis of the sphere. She was the deity of darkness and the night-side; Hu, the god of light, who was her son and consort, became the father who superseded the Sabean mother. Hence we hear of the "Deluge that afflicted the intrepid dragon." [1]

Atum was especially called the Lord of An, which may be rendered Har-An; and there is reason for supposing that "Heron," from whom the city of Heroöpolis was named, was a title of Atum, as lord of the lower world. Champollion considered the analogy between Atum and Heron confirmed by the monumental inscriptions, giving to the kings the title "Born of Atum," since Hermapion, in his rendering of the obelisk of Rameses, calls that monarch the "Son of Heron." In Egypt the An of the Monuments, the Æan of Pliny, is the black land, an appellation of the Heroöpolitan nome. Har-an is Lord of the Black Country; a title of the Pharaohs. In the inscriptions the king is called Lord of the Red-land (Tsher), and Lord of the Black-land, An. Har-An has his likeness in the British Arawn, the solar lord of Annwn, the deep.

Osiris was also a lord of An. Ben Annu is a title of the god in An, which is echoed in a title of Hu, as Pen Annwn, ruler of Annwn. By aid of the Osirian myth with Osiris as Lord of An, and his relation to Horus-Tema, the avenger of his father, we shall be able to correlate the myths of Arawn and his son Pwyll or Pyr. Pwyll, like Horus, the son, changes characters with Arawn the Arkite, who answers to Osiris shut up in the ark by Typhon. Pwyll transforms himself into this character in order that he may become the avenger of Arawn the Arkite, just as Horus is the avenger and defender of Osiris. Arawn is the sovereign lord of the deep. "Behold," he says to Pwyll, "there is a person whose dominion is opposite to mine, who makes war on me continually; this is Havgan," a power also in Annwn; "by delivering me from his invasion, thou shalt secure my friendship." On the day that completes the year Pwyll was to kill the usurper with a single stroke. This was the rôle of Horus, who did battle with Typhon, the "day of the fight between Horus and Typhon," as it is described in the Ritual, as if on a certain day the battle was concentrated into a blow. This was at the time of the spring equinox, and the conflict was in Annu. It is absurd to

[1] Gwalchmai, Welsh Arch. p 202.

render Havgan by summershine.[1] Summershine did not dwell in Annwn, the deep of winter. Hef is a name of the gigantic serpent, the Apophis of the Ritual, and Havgan is the analogue of the Egyptian Apophis.

Har-An, however, is but a title, and it equally applies to Shu as one of the Lords of An. Shu is the Egyptian Mars, god of battles ; and Mars, according to Cæsar, was one of the divinities of Britain. Arawn Pendaran is the Lord of Thunder, and as Master of the Hounds, the dogs of the deep, he partakes of the character of Shu. The Welsh "Cwn Annwn" that appear with Arawn are found in the Ritual as the dogs of shade, that is, Shu or shadow, *ergo*, shadow-dogs. These are hard to lay hold of ; they are thus spoken of : "Oh, leader of the boat, thou goest in the waters. The Osiris shoots through every place in which he has been, through a person who has been to him swifter than the dogs following after Shade." Three times in this chapter (24) the swift shadow-dogs or dogs of shade (Shu) are quoted. The Shadow Dogs of the Ritual are the Echo Dogs of the Welsh myth. When Pwyll is hunting in the Vale of the Boat, GLYN CWCH, listening to the cry of his pack, he hears the cry of another pack of a different tone coming in an opposite direction. These belong to Arawn, the Lord of the Deep, who is here one in person, with Shu, the Lord of the Shadow Dogs of the Deep. The Cwn Annwn, or Echo Dogs, unite both the echo and the shadow character in popular belief to this day, and are supposed to hunt the souls of the dead in shadowy apparition by night, as they do in the Book of the Dead.

It has to be shown that the constellation Kepheus and the star Regulus were two starry types of the god Shu, who was depicted both as the hunter and the shepherd. The star Regulus is in the Babylonian astronomy the shepherd of the heavenly flock.[2] As Kepheus, he is the lawgiver in the North. The shepherd in British mythology is the swineherd. We are told that the first of the mighty swineherds of the island of Britain was Pryderi, the son of Pwyll, chief of Annwn, who kept the swine of his foster-father, Pendaran Dyved, in the Vale of Cwch, in Emlyn, whilst his own father, Pwyll, was in Annwn. Pwyll and Pryderi, called father and son, are the swineherd in two characters agreeing with the two phases of Shu-Anhar. The star Regulus in the Lion was the shepherd or swineherd, the lawgiver and guide when Kepheus was low down in the northern hemisphere. Pwyll is designated Lord of Annwn and Dyved, in which there are seven provinces answering to the seven halls in the house of Osiris, the seven circles of the Troy figure, and the seven caves of the American myths. Dyved is the Tepht (Eg.) or abyss of the beginning, in the region of the north.

Now we are told by Taliesin that it was through Pwyll and Pryderi

---

[1] Davies, *Myth.* p. 421.
[2] *Vide* chapter on the Two Lion Gods.

that the god entered what Davies calls the ark, or the inclosure of Sidi. This, as will be shown, means that these two as a double Regulus were the determiners of a circle of the year. Hence Pwyll is said to govern Annwn, the great deep, the place of the waters of the Deluge, for a whole year, for the solar god of the underworld called Arawn. All this will be vivified later on, at present we must establish our comparison. There is another name of Pyr, called Pyr of the East, supposed to be another character altogether. But we take Pyr to be a local form of Pwyll. Pyr of the East was the son of Llion the Ancient; that is, of the waters called Llion, which burst forth and overwhelmed the world. Llion is the British form of Nun (Eg.), who is the father of Shu, the Egyptian Mars. Nun signifies the celestial water. Pyr, son of Llion, equates with Shu (Kepheus), the son of Nun, and Pwyll is the god of the solar boat, as Shu is in the Egyptian mythos. Thus we identify Pwyll as the British war-god Mars.

Now, to complete the proof that Arawn is the same as Haran (the sun in An), and that Pwyll is Shu, it can be shown that Anhar-Shu-si-Ra-Neb-Khepsh[1] has a character in which he represents or is assimilated to Har-Tema under the style of Har-Tema of Tinis.[2] Har-Tema is the lord who represents justice visibly, whether as the solar Horus or as Shu, and is a representative of the great judge Atum Har-an. HERIAN or HERRAN is likewise a name of the Norse god Odin, the huntsman with the hounds who is the equivalent of Shu and Pwyll with their dogs.

The Cwn Annwn or dogs of the deep are, at times, accompanied by a female fiend named MALT-Y-NOS. This name in the Isle of Man is spelled MAUTHE, where they have the dog of death called the Mauthe dog. Math or Maut was the Hecate of the Britons. She is the Egyptian Mut, to die, Mut, the tomb, underworld, personified as Death. Maut was a form of Mut, the great mother who as Isis was accompanied in her wanderings by the dog. The dog of Mut in Egyptian reads the dog of death. The Druids had Mut in her unfallen form, as the Mother Nature. Math signified kind, Nature, who created out of nine principles or elements. Ma in the hieroglyphics is number nine, and Mat is the mother. A form of "Mat" in Egyptian is fruit, and one title of the Druidic Math was the fruit of the primeval deity, or "FRWYTH Duw Dechrau."

One name or title of the Druidic creatoress is Henwen, the ancient lady. Another divine name of the primordial life-spring or of springing into life at the lowest point of animated existence, out of the chaotic mass of matter in its uttermost stage of disintegration,[3] personified as the deity who was the most ancient and unoriginated ruler, is Ddi-henydd. This can be read by the Egyptian Han or Nun.

---

[1] Harris, *Magic Papyrus*, 2, 3.   [2] Brugsch, *Dict. Geog.* 95.
[3] *Barddhas*, vol. i. p. 218.

Han or Nun is the bringer, called a god. But we shall find the feminine is always first. The NUN is the primordial cause in the negational, passive phase of being, the water, as the factor contrasted with breath; NUN is typical, and water is one of two types, the oldest in the mythical creation. One ideograph of the Han or Nun is the vase, and the vase means the womb, the As. Han, the deity of the heavenly water, is primarily female, as is Hen-wen. Henydd appears to represent the Egyptian "Enti" (Hen-ti), the name of existence, or Hent, the matrix, the water-dam and reversed vase. Hent signifies ruling power, and Ddi-henydd is the unoriginated ruling power. Ti (Eg.) is two or reduplicative; Ti-enti is dual existence; Ti-hent, the plural of rule, in short, the two Truths of all beginning according to Egyptian thought.

Ddi-henydd so rendered is the ancient dual divine being, of which so much has to be said, and then it will be manifest how ancient is this Druidic portrait of cause. We are told in the *Anglia Sacra*[1] that the name of the mother of David (Dyved) was Non. This serves to reproduce the female Nun (Han), the bringer of the hieroglyphics, the Nun of the celestial abime, and the primordial factor of creation, the divinity of the heavenly water.

Among the Irish deities are Krom-Eacha, the god of fire, and Man-a-nan, the divinity of the waters. Akha (Eg.) is fire; Akhu, the furnace. Mena is the wet-nurse, and Nun (Eg.) is the typical primordial water, the inundation. "I have a sword which MAN-A-NAN MacLir (Son of the Sea) gave me," said Naisi of the "Sons of Uisnach."

At Lydney Park, near Chepstow, Gloucestershire, a god was found bearing the Romanized name of "Deus Nodens," who is not known as a Latin divinity. An inscription on one of the votive tablets runs thus: "To the god Nodens. Silvianus has lost a ring; he has made offering (*i.e.* vowed) half its value to Nodens. Amongst all who bear the name of Senecianus, refuse thou to grant health to exist, until he brings back the ring to the Temple of Nodens." Amongst the other relics preserved are certain letters cut out of a thin plate of bronze forming the words "NODENTI SACRUM," which are supposed to have been affixed to the alms-box of the temple wherein those who consulted the oracle deposited their offerings. This may afford a clue to the meaning of the name Nodens.

In Hebrew the Nethen (plural Nethinim) was one who was offered, consecrated, and dedicated to the service of the temple. Nathan (נתן) means to give, to offer, place, set, bestow. It signifies many forms of offering, including the sacrificial; those devoted to the sword[3] or slaughter;[4] having especial relation to blood-sacrifice and offerings of blood. "I have set (נתן) her blood upon the top of a rock."[5]

---

[1] Vol. ii.   [2] *Glen Etive*, p. 105.   [3] Mic. vi. 14.
[4] Jer. xxxiv. 2.   [5] Ezek. xxiv. 8.

"Thou shalt take of the blood and place it (נתן) upon the tip of the right ear of Aaron."[1]

The form Nuden has similar meanings of gifts, offerings, presents, to present, hand over. "Thou givest thy gifts (נדן) to all thy lovers."[2] Nadan permutes with Nadeh (נדה) for "gifts of all whores," in the same verse. Nadeh signifies the wages of prostitution, the images of impurity, uncleanness, the menstruating woman, which suffices to connect Nuden with blood.

Nethen (Heb) means to pour out a blood-offering, and it has been conjectured that a circular opening nine inches in diameter found in the floor of the temple was made use of for receiving drink-offerings of blood as a libation to the god Nodens. Nuden (Heb.) denotes a belly-shaped receptacle, and this terracotta funnel-shaped orifice was ringed round with outer bands of blue and inner bands of red, the two typical colours of flesh (blood) and spirit in relation to the Two Truths of Egypt.

Nat (or Nut) in Egyptian is the name for gifts, offerings, to present tribute, make a collection, to bow, address, hail, help, afflict, punish, save. Enti (Eg.) signifies existence in the invisible form, the lower of the Two Truths, that of blood, the flesh-maker. Enti (Eg.) or Nat is the name of the red crown and the negative form of existence determined by the bleeding flower; Nat, therefore, as in Hebrew, means blood, the lower of the Two Truths, and Nadeh, the flowers,[3] are one with Nat, the flower of blood. The origin of blood-sacrifice will be shown to be related to or suggested by the menstrual purification. So interpreted, Nutenti (Nodenti) indicates Blood-offerings, and "Nodenti sacrum," the sacred place, a mystery of blood-sacrifice; hence the belly-shaped receptacle. When the spirit was offered up to heaven, the blood was poured out in libation to the mother earth the Egyptian Neith, goddess of the lower heaven, that is, earth. Thus Nodens, whether male or female, or both in one, appears to have been a divinity of blood-offerings.

Calves and lambs which happen to be born with a certain natural mark in the ear called the "Nod" or token of Beuno are still chosen as offerings to the Church of Clynnok Vaur, in Carnarvonshire, on Trinity Sunday.[4] The "Nod" is the mark of offering, the blood-sacrifice of Nodens. Beano in English-Gipsy means birth, and the Nod-Beuno is probably the birth-mark. The Bennu (Eg.) was a type of re-birth.

The name written Noddyns has been translated by Keltic scholars god of the abyss. Neith was a Keltic divinity of the mystical water, or blood. The name of Noden also is a well-known English proper name. In the Chronicle of Ethelwerd (A. 508) "Nathan Leod, King of the Britons," was slain by Cerdic. Natan-Leod sounds much as if

---

[1] Ex. xxix. 20.
[2] Ezek. xvi. 33.
[3] Lev. xv. 33.
[4] Dyer, p. 295.

the name had been adopted from the (Romanized) Nodens of Lydney.[1] An inscribed stone found at Lea Mills, on the east side of the River Avon, two miles below Bristol,[2] has on it a bust with sun-like face, which a pair of eardrops proclaims to be feminine. The legend reads "Spes (O) senti" with the circle O broken. Spes might be the Latin for expectation or the resurrection, and this would be corroborated by the cross in the centre ; the dog and cock on either hand corresponding to Anup and the hawk. But these signs only prove the imagery of the equinox, which was pre-Christian and pre-Roman. The rays round the head show the divinity, probably that of the solar goddess known as Sul Minerva. Senti is Egyptian for worship and breathing homage. Spes or Sps is a hieroglyphic variant of the statue As, the sign of the noble, the Great. Here, the one worshipped is feminine. And Spes (Eg.) is the spouse. Seps or Shaps (Eg.) also denotes the bringer forth of the child. Taking the imperfect O to be the hieroglyphic circle, Spes (circle), Senti is the statue erected in the circle of worship to the genitrix, who gave birth to the solar child of the crossing every vernal equinox. This reading would not determine whether the monument be Roman and Mithraic, (there was a feminine Mithras) or ancient British. The inscription, however, contains the leaf-stops that took the place of the ancient papyrus roll of Egyptian punctuation.

The Egyptian Fates or Parcæ are seven in number, called the Seven Hathors, who are in attendance at the birth of children. In the "Tale of the Two Brothers" the Seven Hathors came to see the newly created wife of Bata, and they prophesied with one mouth that she would die a violent death. In the tale of the "Doomed Prince" the Seven Hathors greet him at his birth and predict his fate.[3] They appear in the Ritual in the form of seven cows, with the bull who is the husband of the seven.

The Seven passed into Persia as the Seven Sisters or Wise Women who are present at the birth of children and at other sacred times. They appear in the Rig-Veda as the Seven Sisters who are also Seven Cows like the Hathors. The Chinese have the Seven Sister-Goddesses in connection with the Seven Stars. These seven are found in a diminutive and elfish form among the Manx.

Waldron, in his account of the Isle of Man,[4] relates that a woman, who was great with child and lay in bed waiting for the good hour of deliverance, saw in the night-time seven (or eight) little women of the wee folk come into her chamber, one of them having an infant in her arms. A scene of christening ensued, and they baptized the infant

---

[1] *Roman Antiquities at Lydney Park, Gloucestershire.*   By the late Rev. W. H. Bathurst, with Notes by C. W. King.

[2] *Archæological Journal,* v. xxxi. p. 41.

[3] *Records of the Past,* vol. ii. pp. 145-155.

[4] Works, p. 132 (1731).

by the name of Joan, by which the woman knew she was bearing a girl, as it proved to be a few days after.

The Seven in Waldron's story were accompanied by a male (the bull), who acted as a sort of scribe or minister.

The number seven was continued in divining. Mother Bunch says of the experiment of the Midsummer shift: " My daughters, let seven of you go together on a Midsummer's eve just at sunset into a silent grove, and gather every one of you a sprig of red sage, and return into a private room, with a stool in the middle, each one having a clean shift turned wrongside outwards hanging on a line across the room, and let every one lay their sprig of red sage in a clean basin of rose-water set on the stool; which done place yourselves in a row, and continue until 12 o'clock, saying nothing, be what it will you see ; for, after midnight, each one's sweetheart or husband that shall be shall take each maid's sprig out of the rose-water and sprinkle his love's shift." This too presents a picture of the Seven Hathors.

By aid of the Cauldron of Keridwen or Vessel of Kêd the genitrix we may recover the Egyptian Un, the Goddess of the Hours. The "Pair Keridwen" was a vessel, and the typical name of the whole circle of laws and doctrine of the Druids. Cauldron or Kart-ren is the circle by name. Pair is the Egyptian Per, to go round, surround, be round, and is synonymous with Pail or Pale.

Keridwen, with due attention to the books of astronomy and the hours of the planets, collected plants for the cauldron, which boiled and bubbled for a year and a day, to obtain three " blessed drops of inspiration." These three drops represent the knowledge of the cycles of the sun, moon, and stars.[1]

On a certain day about the end of the year, whilst the ancient mother was muttering to herself and feeding the cauldron with plants, three drops flew out and the cauldron divided in two halves. The two halves typify the two divisions of the circle of the year completed in An, the place where the pool and water of the Two Truths are found in the Egyptian mythology. And this Pair, out of which came the Druidic inspiration, is variously called the Cauldron of Keridwen, of Prydhain, and of Awn. From the Cauldron of Awn came forth the Waters of Truth. The divine drink was brewed in it for a year and a day. It is called the Cauldron of Five Plants, and these represent five planets.

"Manifest is truth when it shines ; more manifest when it speaks, and loud it spoke when it came forth from the Cauldron of Awen, the ardent goddess."[2]

An in the hieroglyphics is speech, and to speak, a form of the Word. An also means repetition, again, to be periodic. An was the place of

[1] *Hanes Taliesin*, ch. ii.
[2] Kadair Teyrn On, *Welsh Arch.* p. 65.

periodicity, the place of re-birth. An and Un interchange, and have one meaning. One form of An is lunar, the Ape-Deity An is a form of Taht. The Ape An was a type of periodic time. And the time-circle or cauldron is the symbol of "the Goddess Awn," the "ardent Goddess," the "ardent Awn," the inspiring Muse whose cauldron is also said to be warmed by the breath of nine damsels. Un (Eg.) is the hour, the repeating period, like An. Un personified is the Goddess of Periodicity, or the Hours, as is Awn the ardent Goddess.

The sum and substance of the earliest science or inspiration was the teaching concerning time and repetition of periods. Awn was the revealer; Un means to show, reveal. Awn burst open the cauldron that divided into two halves. Un is the opener of the circle. Hence it is now claimed that the Druidic goddess Awn, or of the Awn, is identical in nature with Un and the lunar An.

Noë and Eseye are two divinities celebrated by the Druids as presiding over and being worshipped in the vast temple of Stonehenge, called the "great stone fence of their common sanctuary."[1] They are representatives of the genitrix Kêd, whose seat was in Stonehenge. The Great Mother, as we shall see, divides into two other characters called the Two Divine Sisters, who personify the Two Truths, the two heavens, or Heaven and Earth, the two principles called Breath and Water. One form of the Two Sisters in Egypt is Hes (Isis) and Neft (Nephthys). Hes is the cow-headed goddess, the seat, and Hes signifies liquid, hence the vase ideograph. Nef means breath. Hes or As are two readings of her hieroglyphic. Eseye being identified with As, Hes, Iusaas, or Isis, Noë is one with Nef.

Nef means breath and the sailor of the waters. Nawa in Javanese is breath. Neff in Cornish is heaven above, the place of breath. Isis represented the lower heaven and the waters; Neft, the heaven above the horizon, in the total circle imaged by Stonehenge.

Hes, the seat, equates with the goddess of the hind quarter, and Hes (Isis), the cow-headed, was compounded with Ta-Urt, the bearer, in Hes-tareth. Eseye was our Isis in her seat at Stonehenge, and HSA is also an Irish name of the Great Bear.

Nor was Stonehenge the only sanctuary in these islands of Noë and Eseye. In Strathmore there was an extensive Druidic ground, in which numerous monuments have been found. There is a place named Eassie, and a large circular mound, about a mile from the old church of Eassie, and in the "united parish of Eassie and Nevay"[2] we have the names of the two goddesses, united as at Stonehenge, whilst in the form of Nevay we recover the F modified in Noë. But we must go a little further round.

---

[1] Gododin, Song 15.
[2] Stuart, *Sculpt. Stones of Scotland*, pp. 90, 91.

There is, says Herodotus,[1] a large city called Chemmis, situate in the Thebaic district, near Neapolis, in which there is a quadrangular temple, dedicated to Perseus, the son of Danæ. In this enclosure is a temple, and in it is placed a statue of Perseus. The Chemmites affirm that Perseus has often appeared to them on earth and frequently within the temple, and that a sandal worn by him is sometimes found which is two cubits in length, and that after its appearance all Egypt flourishes ; which is delightful when interpreted. Chemmis means the shrine (Khem) of birth and the child. Thebes also has the same meaning. It was the Ap, Apt, or Aft, the quadrangular enclosure and place of birth. Perseus is the appearing star or child, from Per (Eg.), to appear, show, explain ; and Siu, star, a divine son. His reappearance was astronomical. Two cubits are equal to Mati, the Two Truths. Mati is a pair of feet, the pair of feet found on the stones, and the Egyptians were telling Herodotus of the reappearing star in the place of the Two Truths and dual foot in Mat (Mati) or An, the solar birthplace.

The chief corner of this quadrangular enclosure was at the place of the spring equinox, in Apta, called the corner or end of the world ; that is, the place of completion. Here we find a temple within a quadrangular enclosure, and are enabled ·to see that it represented the reappearing son of the mother within *her* temple or Aft of the four corners ; and this was in Chemmis, the shrine of birth.

We find the Egyptian Khi, one of the four supports of heaven, in Gyvylchi, and in the account of the temple at Dwy-Gyvylchi, given in Gibson's *Camden*,[2] we are told that the most remarkable monument in all Snowden, called "Y Meini Hirion," within the parish of Dwy-Gyvylchi, is a circular entrenchment about twenty-six yards in diameter, on the outside whereof are certain rude stone pillars, of which about twelve are now standing, some two yards and others five feet high, and these are again encompassed with a stone wall. It stands upon the plain mountain, as soon as we come to the height, having much even ground about it ; and not far from it there are three other large stones pitched on end, in a triangular form. The triangle with Meini Hirion thus formed a square, a quadrangular relic of the quadrangular Kaer of the genitrix and her son, who were Kêd, and Prydhain, or Ior, the appearing youth. The four corners are the four Khi, the four supports of heaven. This enables us to restore the sanctuary in its dual form, and to understand the meaning of the double figure. The quadrangular Kaer represented the maternal abode, the Aft or Fet of the primary four quarters. Within or near this four-square enclosure was the temple of twelve stones, which number identifies the twelve solar signs, the temple of the young sun-god, whose statue, as Perseus, was placed in the inner enclosure at Chemmis. The· sanctum sanctorum of Stonehenge

---

[1] B. ii. 91.   [2] Col. 805.

was oviform, as were the Adyta of those temples where the fire for ever blazed, because this figure was female, the circle within the circle, the womb.

The Jewish cult was so emphatically feminine in its origin that they sacredly preserved this ovoid form of the circle. Says Rabbi Simon, son of Gamaliel, "When the rent is round it is forbidden, when it is lengthwise it is allowed."[1] That is, when it was the oval shape of the hieroglyphic Ru, the emaning mouth of birth. This was the type of the holy of holies, where stood the statue of the child. "Beloved of the Aditum, come to Kha," say the two divine sisters in their invocation to the child Horus. The Kha is the type of the uterus.

In the "Song of Cuhelyn," the inclosed temple at Stonehenge is called the "precinct of Ior," "in the fair quadrangular area of the great sanctuary of the dominion."[2] The god Ior is our Perseus and Horus. Ior, says Davies, became a title of the supreme God, but is "borrowed from the British mythology, where it seems to have meant the sun, moving within his orbit or circle."[3] The orbit is identified by the twelve pillars of Gyvylchi. Ior is the British Har, the solar divinity, who was the son of the mother before the fatherhood was embodied in Ra. At Stonehenge, then, we had the quadrangular inclosure and the youthful sun-god Ior united with Noë and Eseye, whom we identify as the two divine sisters of Har (Horus) in the well-known Triad of Horus, Isis, and Nephthys; whilst the double nature of Horus as the elder and younger Har is manifested by Ior, who is the renewed and glorified form of Keridwen's recovered son Avagddu, the child of darkness, who was transformed into the radiant lord of light.

In the Osirian mythos Isis, the great mother, has two children, Har the elder and Har the younger. The elder is born deformed, and maimed in his lower members. Plutarch describes him as the cripple deity, who was begotten in the dark. He dies prematurely, or rather he transforms into the second Har. The elder Har is portrayed finger to the mouth, and named Hor-pi-Khart. Khart is the Egyptian word for silence. Hence the Greek Harpocrates was designated the God of Silence. In the astronomical allegory the child Horus was the mystical Word; the second Horus is called Ma-Kheru, the True Word, or the Word made Truth. This is effected when the Silent One is united with him who is the True Voice, and Horus is "transformed into his soul from his two halves."[4] The meaning will be made apparent, but for the present this much is stated on purpose to show that the Druids had the same myth. Keridwen bears a deformed first son, who is hideous to behold, whose misfortune is the grief of his mother. The name of this

---

[1] Mishna, t. xi. ch. ii. [2] Davies, p. 313.
[3] *Ibid.* P. 315. [4] *Rit.* ch. xvii.

child, Avagddu, is said to mean "black accumulation,"[1] and we learn that no change could occur for the relief of both until a certain time appointed, which was set forth as the annual boiling of Keridwen's vessel; then came the change, the "Correcting God" formed the child anew under another name, which indicated the one bursting forth with radiancy. This transformation implies the transition from the elder Horus, the dumb and deformed child of Isis, to the younger Horus, the true light of the world.

There is an Irish word, "Pocrat," signifying, according to Vallancy, "*lame in the foot.*" Po-krat is usually read the child. But Plutarch says he was maimed or lame in his lower members, and here in Irish is "pocrat" for lame in the foot. Vallancy knew nothing of Egyptian.

We have the "Crut" also in English, as the dwarf and the puny child. P-crut is the Crut or Khart, the elder Horus.

In the *British Mythology* we have a character named Gwion the Little. The Welsh Gwion is the Irish GAN, the little one, the diminutive. Gwion is said to be the son of Gwreang, the Herald of Llanvair, the fane of the lady. Gwreang the Herald identifies the impersonation with the Word, or Logos. The Lunar Herald, or Word, is Taht, who is associated with Khunsu, the Victorious Child, or brave boy in the Moon mythos. Gwion was stationed in Caer Emiawn, the City of the Just, in Powys, the land of rest, by Keridwen, to superintend the preparation of the cauldron which boiled for a year and a day to produce the Water of Inspiration and Sciences intended for her son. Three drops only could be obtained. About the end of the time these very three drops chanced to fly out of the vessel, splash the finger of Gwion the Little, and burn him so that he put his finger into his mouth. As soon as he did so, his eyes were opened and all futurity was present to his view. The cauldron divided into two halves, and Gwion the Little fled in mortal fear of the angry goddess, who pursued him and eventually caught and swallowed him.[2]

We may well suspect that Gwion the Little is not only a form of Khunsu, but that his name throws a light on the meaning of Khunsu's name. Khun is depicted with the infantine lock of Harpocrates, the child Horus, and Khunsu as the Child (Su) is Khun the little. Gwion, son of the herald, corresponds to Khunsu, the boy-representative of the moon. Khunsu is depicted as the time-reckoner, holding the palm-branch of the panegyries, and marking the years with a stylus. And the cauldron of Keridwen attended by Gwion, which divided at the end of the year, or at the place of the equinox, represented the time-cycle kept by Keridwen. The drops of the water of life were emblematic of the knowledge whereby future events could be known, that was, astronomical knowledge which afforded real ground for

[1] Davies, *Myth.* pp. 190, 203, 263.
[2] *Hanes Taliesin.* Davies, *Myth.* p. 213.

prophecy.   Gwion was stationed in Powys, the land of rest, for the preparation of the cauldron.   And one title of Khunsu is Nefer-Hept, the Child, or Prince of Peace.   Khunsu is stationed in the zodiac of Denderah, in the sign of the Fishes, figured with the pig and full moon, which is the full moon of our Easter, the sign of the solar resurrection, and the point of renewal for another year.   In the Welsh legend the myth is physiological as well, for when Keridwen pursues, catches, and swallows Gwion, he is again born of her at the end of nine months.   Khunsu is a form of the elder of the two brothers of mythology, and, as such, is a Har-pi-Kart, who is represented with finger pointing to his mouth as the symbol of the mystic Word. Gwion the Little was represented in the same manner, only he put his finger in his mouth, whereupon his eyes were opened, or his transformation came.[1]   Khunsu is a luni-solar form of the son, and Gwion the Little likewise transforms into the solar hero.   "I have," says the Initiate, "been for the space of nine months in the belly of Keridwen.   I was formerly Gwion the Little ; henceforth I am Taliesin."[2]   Taliesin, or radiant front, is a title of the sun.   This is the luni-solar transformation of Khunsu.

Gwion the Little is identical with the Gaelic Con, the son of Cruachan, and hero of a hundred tales, who wields the sword of light against the giants in the underworld of the dead,[3] and who is thus related to the Egyptian Khun, the slayer of the giants according to Macrobius ;[4] the vanquisher of the proud rebels in the Book of the Dead.[5]   Khunsu, the bringer-up of the orb of light from the world of the dead, is figured as Con, who gathers the gold down among the dead, and ascends with it in the giant's creel.   Con-al, or Khun-ar, is the exact equivalent of Khun-su, the brave boy.   But to recover the allegory from the Gaelic tales is somewhat like trying to spoon out the sparks of sunshine from its reflections in the water. Nevertheless, it is shining there.   For instance, in the Tale of the Fine, where Fionn and his heroes are in the house with seven doors, and they sit altogether on the one side to breathe, and the king and people of Danan sit on the other ; "Yonder side of the house be theirs, and this side ours ;" the house is the double solar house, the house of Osiris, with the Seven Halls in the Ritual.   Fionn and his men are the celestial heroes, the Danan are the people of earth.   Ta-nan (Eg.) is the type of earth.   The ensuing battle is that of Horus and Typhon, who is the black dog of the people of Danan.   Fionn slaying the Danan seven by seven with the jawbone of the boar is the same solar or luni-solar hero as Samson slaying the Philistines with the jawbone of an ass.   The deadliest battle of Fionn, when he set his back to the rock on the "longest night that came, or will come,"

[1] *Hanes Taliesin.*   [2] *Welsh Arch.* p. 19.
[3] Campbell, *West Highland Tales,* No. 7.
[4] *Saturn.* i. 20.   [5] Ch. 83.

was the struggle of the sun with the dark power on the longest night of the year.[1]

The common Irish form of Conal's name in O'Connel, and O'Conner adds the word Ner (Eg.), meaning victory. Thus Conner is the victorious Con. Con, as Khun-su, will account for the Gaelic tradition that Conor lived at the time of the Crucifixion. Khun was the king of the crossing, the determiner of the very moment at full moon. It is possible that the stone of the ball in Conor Mac-Nessa's brain may have been derived from the full-moon borne on the head of Khun-su. The legend relates that when he observed the darkness on the day of the Crucifixion, and was told by the seer that the "Innocent One" was then suffering, he got so excited that the ball flew out of his head and he died. In this version of the myth, Conal is designated Conor Mac-Nessa. Nessa appears in Irish legends as the widow with her son Conal; she is said to marry Feargus Mac-Roy, but is as likely to be entirely mythical as Conor who carried the ball in his brain. The only object of introducing the name of Nessa here is to point out that it is an Egyptian feminine name. Nesa means "her," and a daughter of Khu-en-Aten[2] was named Nesa.

Prydhain was a name and character of Hu, the sun-god, the youthful character into which the solar divinity transformed every spring. The same is found in all the mythologies. In the *Mabinogii* he is called the son of Aedd the Great; that is of Aeddon, a name of Hu. He also interchanges names with Beli as the solar son. The young god appears in the British fragments as lord of the seven provinces of Dyved in Annwn the Deep.

These seven provinces answer to the Seven Halls in the house of Osiris in which the young solar god is annually reborn, and from which he emanates. Pwyll also proceeds from the seven provinces and the high place of reappearing in Arberth and from Diarwya, called by Davies the "solemn preparation of the egg." The egg was a symbol of the circle, and this Diarwya looks very like the Egyptian Teruu, the circumference, a name of Sesennu and a form of number eight the expression of the Seven—whether of the Great Bear or planetary Seven—as in the person of Taht. Pwyll or Per read by Egyptian means coming forth, manifestation. With the terminal t this is Pert, to appear, emanate, proceed. Thus Per and Pert, our Pwyll and Pryd, meet in one meaning. Hain (Eg.) is the youth; Prydhain, the appearing, emanating, manifesting youth, or the young solar god of various names. Hu is the God of Corn, and the son and corn (seed) are synonymous. Per is corn, grain, the seed. Pert, the corn or food appearing; Hain, the young. Prydhain is the young seed or corn of Hu, who reappeared at the time of the vernal equinox. We have Pryd personified as Corn.

[1] Campbell, *Tales of the West Highlands*, No. 29.
[2] Amenhept iv.

Martin speaks of a custom in the Western Islands,[1] observed on the second day of February, in which the mistress and servants of each house take a sheaf of oats and dress it up in woman's clothes, put it in a large basket, and lay a wooden club beside it. This they call Briid's Bed. Then the mistress and maids cry three times, "Briid is come; Briid is welcome." This is done just before going to bed, and on rising in the morning they look among the ashes to see if the impression of Briid's Club is visible there; if so, it is a presage of a good harvest and general prosperity. Briid and Pryd or Prydhain are identical, and from Pryd, through Briid the corn, comes our name for bread. The son of Hu, whose name means corn, was the bread of life in person, and Pert in Egyptian is the food made of corn. Briid or Prydhain was the earlier Christ, and when the new theology was adopted in these islands, Briid's bed was made for Christ. It was, as already related, the custom at Tenby, in Wales, for young persons to meet together on Good Friday to "make Christ's Bed." This was done by gathering the long reed-leaves from the river and weaving them into the shape of a man. The image was then laid on a wooden cross, and left in a retired part of some garden or field.[2]

But the son of the great mother is a star-god at first; the solar imagery is latest: one form of the genitrix and son is that of the Bitch Baal and the Dog Baal, the Baali or Baalim of the Hebrews, and Sut-Typhon of Egypt. The dog who accompanies Isis, and is said to be born of Nephthys, is the Dog-star, Bar-Sutekh or Sut-Anubis, the Sabean Son. This divinity reappears in Britain as Cunobelinus, whose name is found on the British coins or amulets. This was the title of a famous prince in the reigns of Augustus and Tiberius, said to have been the father of Caractacus. By his title he is assimilated to the Cynvelyn of the Bards. That is Baal or Belin with the style of Cun, or Cyn, which we identify with the dog as in the Cynocephalus. In the hieroglyphics a headless dog is a Khen; a conductor without exterior vision, therefore a type of interior perception, hence the name given to the Kenners. The priests of Kêd are designated dogs, and she is the bitch. They represented her son. The Cynvelyn of Helvelyn, of Belin's-gate, and of the Druids, is the Dog Baal, in the diminutive from of Belin or Velyn. This form, like that of Sutekh and Saturn, determined the god as the child, the little one who was the son of the mother in the Sabean Cult. Saturn was the planetary type of the male Sut-Anubis of the Dog-star, and in the dialogue between Ugnach and Taliesin we read: "Seven blazing fires will counteract seven battles: the seventh is Cynvelyn, in the front of the mount." Skene renders "Cynvelyn the seventh in every foremost place."[3] The seven are the planets, of which the seventh is Saturn. Therefore Cynvelyn is identified with

---

[1] P. 119.   [2] Mason's *Tales and Traditions of Tenby*, p. 19.
[3] *Battle of Ardderyd*, Skene, ii. 368.

Sut in the form of Saturn, and is one with the Egyptian Bar-Sutekh, the Sabean Baal.

Sut was the great warrior-god ; the dog of battles. And in the Talisman of Cunobeline the Dog Baal plays the part of Bar-Sutekh. "Cunobeline, the indignant, the lofty leader of wrath, pamperer of the birds of prey, and that divine allurer Dirreith, of equal rank with Morien, shall go under the thighs of the liberal warriors. In equal pace shall the Gwyllion proceed with the benign blessing. Amongst the splendid acquisitions of the mystic lore, the most majestic is the Talisman of Cunobeline. It is the shield of the festival, with which the man of fortitude repels the affliction of his country."[1] In this Cunobeline is coupled with Dirreith, who has been shown to be the great Mother Ta-urt, or Rerit. These are the Sabean Mother and Son as Goddess of the Great Bear and the Dog of Sothis, the first-born son of heaven. Now it appears to me that the mythical Arthur is primarily a form of Cynvelyn, the dog of battle.

Arth is the ancient British name of the Great Bear, and this constellation was associated with Arthur. Arth corresponds to Urt, the goddess of the Bear, and we may derive Arthur, the son of Urt or Arth, in one of two ways, Ar-t-ur (Eg.) as son of the old mother, or Arth-ar, the old mother's son. He must have been the solar son in the later myth of the Round Table with the twelve seats for the twelve companions. There is an Egyptian Artaur, rendered by Maspero the flames of God. But the first son of the genitrix was Sabean, not solar ; Sut-Har (Ar) of the Dog-star, Sut-Anubis, the earliest form of Hermes, the heaven-born.

The Vervain plant was used by the Druids in casting lots and foretelling events. It was gathered without being looked on by the sun or moon at the rise of the Dog-star. In digging it up the left hand alone was to be used, and when dug up, it was waved aloft. Leaves, stalks, and roots were dried separately and in the shade.[2] This serves to connect the plant of prophecy with the Egyptian Star of Annunciation, the Dog-star, the son of the great mother, who appears to have been reproduced as Arthur, the son of Arth or Ta-Urt. Arthur, son of the Great Bear, is the equivalent of Sut-Har of the Dog-star, which leads me to conclude that Arthur was the Sabean son before he became the solar representative. The parents of Arthur are the Great Dragon and Eigyr. The Great Dragon is Typhon, the old genitrix.

The British Arthur is primarily represented with the Seven in the Ark who are the only ones that escape from the Deluge in the circle of Caer Sidi. Sidi corresponds to Suti (Sebti, Sothis, the Dog-star, Sut). A poem of Taliesin's called Preiddeu Annwn, the "Spoils of the Deep," contains this Arkite imagery. In the house of Osiris there are seven halls and seven staircases. These seven came

[1] Davies.   [2] Davies, *Myth.* p. 276.

to signify the circles and pathways of the seven planets, but the first seven in mythology are not the planetary seven, they are the seven companions in the constellation of the Bear.   These are the seven Rishis of India, the seven Hohgates of the Californian Indians, the seven sons of Sydik in Phoenicia.   They appear to be the seven companions of Arthur of whom the bard sings in their escape from seven different Kaers, " Thrice the number that would have filled Prydwen we entered into the deep ; excepting seven, none have returned from Caer Sidi."   The subject matter of this mystical representation is the escape of Arthur and the seven companions from the Deluge based on the time and circle-keeping of Arthur's Star, and seven other stars.   Now if Arthur were here considered a solar god, there would be one too many for the seven planetary gods, therefore the seven are those of the Bear, Arthur's constellation, and Arthur is identical with Sydik, the Egyptian Sutekh of the Dog-star.

Again, the Talisman of Cunobeline is a shield, and it is in Arthur's shield Prydwen that he and his seven companions escaped from the Waters, or the so-called Deluge.   Prydwen, the Lady of the established order of things, is a form of the Ark, which also contains eight persons in the Hebrew mythos.

Arthel is a British word, written Arddel in Welsh, to avouch, prove, justify ; a similar meaning to that of Makheru, a title of Horus.   At Exmoor the number eight is called Art ; eighteen is Arteen.   Arthar is Har the prince or lord of the eight, the manifester of the seven. This, however, belongs to an earlier myth than that of the eight great gods of Egypt, in which Taht was the manifester of the seven.   Arthur was the eighth to the seven Kabiri of the Great Bear, the manifester of the seven, or the son of the sevenfold constellation, considered as the great mother.

Arthen (Welsh) is the bear's cub.   Arth-al is to growl as the bear.   Al interchanges with Ar, as the voice, speech, faculty of speech, and Arth-ar is the speech or utterance of the Bear.   This is the doctrinal word or Logos.   So An (Anup), the Anush, is the speech, the announcer of the year of the Bear.   Ar (Eg.) is the earlier Har, from Khar, the speech, to speak, be the Word, the son and Word being identical.   Arthar is thus the Word as son of Arth the Bear.

Arthur, in his first estate, then, we hold to have been the Sabean Mercury, son of the goddess of the Great Bear, and identical with Sydik and Sutekh, who was continued in Egypt as Sut-Har, god of the sun and Sirius-cycle, known as the Negro Sut-Nahsi and Sut Nubti, a Sabean-solar combination to be found in other mythologies, in which a star-god of fire becomes a sun-god.

The series of astronomical legends or myths found on the Assyrian tablets is known to consist of twelve in number, one for each sign of the zodiac.   In the " Fight between Bel and the Dragon," in which appears the sword that turns and flames all round the circle, wielded by the hand of Bel against the Dragon, when the battle is over, it is said "the eleven tribes poured in in great multitudes, coming to see the fallen monster." Evidently the twelve signs were said to be peopled. These correspond to the twelve tribes of Israel, and to the twelve labours of Hercules; the conflict occurs in the twelfth sign, and the people of the other eleven rush in to see the result. The twelfth sign is the last of the old year, and the fight of Bel with the dragon is the same conflict as the battle between Horus and the evil Typhon, the earlier Akhekh—serpent, griphon, or dragon. This battle occurred annually, and specially just before the time of the vernal equinox, and is called "the day of contending of the lion-gods," "the day of the battle between Horus and Sut, when Sut puts forth the ropes against Horus."[1] The contention, being equinoctial, is represented as under the lion gods who kept the level on the horizon, whilst the light and darkness contended in the balance, and each pulled at the ropes of either scale. This belongs to mythology in the latest stage, the solar. Most of the Assyrian matter yet recovered relates to this later stage, although we do get glimpses of earlier things submerged in Akkad. These twelve representations in the twelve signs, the present writer considers to be akin to the twelve battles assigned to Arthur by Nennius (50), the twelfth being a "most severe battle, when Arthur penetrated to the Hill of Badon," or, as we interpret it, to the Bed of Tydain, Tiotan, or Tethin, the solar god, who was reborn in the hill.

Cæsar affirms that the Britons chiefly worshipped the god Mercury; of him they have many images, him they consider as the inventor of all arts, as the guide of ways and journeys, and as possessing the greatest power for obtaining money and merchandise. But we have to reckon with two forms of Mercury; the Sabean and the lunar. Sut was the first form of Mercury; Sut-Anubis is the guide of ways. Taht is the second. This is acknowledged in the Ritual,[2] where we read Taht formerly, or otherwise Sut, when Taht had superseded Sut as the Word, announcer, and reckoner of the gods.

The deity Gwydion has been considered the same character as Mercury, the son of Jove and Hermes, the councillor of Kronus.[3] He is called Gwydion ap Don ; Don being the father of the gods. There is an Egyptian divinity, Tann, both female and male, a type of the earth. Gwydion, the son of Don, they say, by his exquisite art charmed forth a woman composed of flowers, and early did he conduct to the right side as he wanted a protecting rampart, the bold curves and the virtues of the various folds ; and he formed a steed upon the springing plants, with "illustrious trappings."[4] Or, as Skene renders it, "Gwydyon-ap-Don of toiling spirits, enchanted a woman from blossoms, and brought pigs from the south. Since he had no sheltering cots, rapid curves and plaited chains, he made the forms of horses from the

[1] *Rit.* ch. xvii.   [2] Ch. xliv.   [3] Davies, p. 264.   [4] Davies, *Myth.* pp. 263, 264.

springing plants and illustrious saddles."[1] It was Lleu and Gwydion who "changed the form of the elementary trees and sedges."[2] The elementary trees belonged to the ten primary Ystorrynau of Kêd. These were the earliest branches of the "Kat" (Eg.), tree of knowledge. They were so old that Taliesin in the "Battle of the Trees," (Cad Godeu) says, "The mountains have become crooked; the woods have become a kiln, formerly existing in the seas (like bog-oak or a geological stratum), since was heard the shout," the triadic "Beam," or Cyfriu sign. "The tops of the birch," he continues, "covered us with leaves, and transformed us and changed our faded state."

In this poem of the trees, "the head of the line" is described as a female who issued forth altogether alone, and the birch was a much later arrival. Probably the birch, Bedwin, the male emblem, refers to the change in the elementary trees made by Gwydion, and the introduction of the masculine type of the creative power.

But, says the old goddess (or her poet for her), "when the chairs are judged, mine will be the most excelling; my chair, my cauldron, and my laws, and my pervading eloquence meet for the chair."[3]

At the time of the mythological deluge, we learn from a poem by Taliesin, that in the living Gwydion there was a resource of counsel, and when "Aeddon came from the land of Gwydion into Sëon of the strong door," then Gwydion advised him to "impress the front of his shield with a prevailing form, a form irresistible."[4] By this means the "mighty combination of his chosen rank was not overwhelmed" when "Math and Eunydd set the elements at large," which is described as producing a deluge. Gwydion is credited with devising means for saving what Bryant and Davies call "the Patriarch and his family," when the deluge is about to burst forth and overwhelm the world. This he accomplished by forming the "bold curves," and the "virtues of the various folds," and making a "protecting rampart," the shape of a shield or of a circular pattern, a form irresistible. A mode of meeting the coming flood, which is elsewhere figured as building an ark.

Now in the Egyptian Ritual Taht says he built the ark. "I am the great workman who made the ark of Socharis on the stocks."[5] We shall see the gist of this when we come to the Deluge and the Ark.

Gwydion then is here identified with Taht in character as the ark-builder. Taht was the Word, the manifester of the gods, lord of letters or types. The companion given to Gwydion as inventor of an alphabet of sixteen letters is named Lleu, and in the hieroglyphics the Ru is the reed pen, the paint, and the written word of the scribe; the Ru sign accompanying Taht as writer and the lord of letters. Both pen and papyrus were made from the reed.

---

[1] Skene, ii. 296. [2] Skene, ii. 275.

[3] *Chair of Keridwen,* Skene, vol. ii. p. 297.

[4] Davies, *Myth.* 263-4. [5] Ch. i.

Amongst the plants referred to more particularly was the Elestron, the water-lily or flag-flower, the lotus of the Druids. This answered to the papyrus sceptre, the Uat held in the hand of the Egyptian goddesses. The woman composed of flowers is called the rainbow, that is Iris, and the flag-flower is the Iris. Thus, the woman enchanted from blossoms identified by the Iris leads us to see a personification of the reed as the instrument of the written letters invented by Gwydion in Britain and by Taht in Egypt.

The Uat or papyrus sceptre is identical by name with the goddess of the north, Uat, the earlier Kheft, our Kêd. And this goddess of the papyrus reed is replaced by Sefekh as mistress of the writings and consort of Taht. Sefekh reads number seven, which identifies her primarily with the seven stars and with the seven colours of the rainbow, or Iris. It was for this goddess of the rainbow, the seven colours, that Gwydion formed the horse (or horses) on which she was to ride forth as mistress of the writings or as the feminine Word.

The order of mythological sequence is first the Sabean, next the lunar, and lastly the solar. We know the lunar zodiac preceded the solar, and just as Taht claims to have made it or built the Ark, so Gwydion is credited with instructing the solar god how to meet the coming flood of destruction. The eagle of Gwydion takes the place of the ibis of Taht. Thus far Gwydion appears to represent Taht, and the inventor of symbols and memorial types answers to the lord of letters and scribe of the gods.

There is a god on the Monuments, however, named Khetu, of whom little is known. He is called a god of things, it may be of types, as Khet is the seal-ring ; it may be of letters as the temple-scribe or hierogrammat is a Rekhi-khet. Whether a god or only a title, the name supplies a root for Kadmus, the reputed inventer of letters. Khet, Shet, and Set are synonymous in many meanings, and this Khet would seem to be a form of Sut-Anubis, the earliest, the Sabean Mercury. The name of Sut has an earlier form in Khut. Khut is the goer-round, the circle-maker. Khut is a modified form of Kheft (Kêd), the goer-round, as the Great Bear. The first goer-round, as her son, was Sut or Khut, the Dog-star, who bears her name in the modified form. Kuti means the traveller round, the maker of the circuit, the particular god of the Britons. The name was continued in Egypt as that of Har-Khuti, god of the two horizons, who, it will be maintained, was a Sabean Har-Suti before the solar-god, Har-Makhu, assumed the title. And this hard form of Suti is found in the name of Gwydion. Hence, it is argued, he derives from Khut or Sut the Sabean Mercury, who preceded Taht as the scribe of the gods, and that, as in Egypt, the lunar god eclipsed the older star god. But the obscurity of the matter on the Monuments doubles the difficulty with the British mythology. Still, the sixteen letters appear to identify Kadmus, Woden, and Beli, each of whom is credited with introducing that number of types, runes, or letters, into his particular country, and Beli is identical with Bar or Sut, whose name in the earlier form would be Khut, the god of things, and who corresponds to Cæsar's description of the British Mercury, the inventor, the guide of ways, and deity of commerce, with the same passing into the lunar Taht that we find in Egypt.   Another link.   Sut or Sebt deposits the god Seb, whose type is the goose, which in Welsh is the Gwydd. The Welsh Gwydd and Gwyddion, the teachers, sages, seers, men of letters, derive their name from this god of letters.   The first written signs, with the Druids, were cut.   Khet, in Egyptian, means to cut and imprint, or to seal.   Gwyd, in Welsh, signifies the wood that was cut, the letters that were cut, the sage who cut the letters, and the manifestation of knowledge by means of the letters.   Gwyd was the whole science of letters.   The names are countless that come from this root, Khut, to cut, imprint, type, show, or reveal, from which the earliest sacred words were called Ghetas in Sanskrit and Gathas in Zend; Cuth, in English, to be taught, instructed; Kith, knowledge; Guth, Irish, speech; the Gwyddion, who made known; the Godi, or Hofgodi, of Norway, chieftains who in olden time were at once pontiff, judge, and godi, god of the hof or temple in one,[1] doubtless identical as religious rulers with the Druidic Gwyddion, the followers of Gwydion.   Archagetas was, according to Pausanias, a name of Æsculapius, signifying the primeval divinity.   It was the arch-Khutu, who came from Egypt on various routes.   The CADEU-ceus of Hermes was in name and nature a type of Khet, to shut and seal, found to be Egyptian.   Lastly, God is the same word as Khut, although it does not retain a single primal element.   Khut (Eg.) means a spirit traceable to fermentation.   But the earliest Khut or god was the maker and reproducer of a circle, the goer-round in a circle, the opener of the circle.   Hence Ptah and Sut are the openers. The two aspects were those of opening and closing, and here the god is one with the cutter or cleaver.   To cut is to open, to open is to reveal; the primordial god is the opener, and the axe sign is the hieroglyphic of divinity, the type of the cutter, a primitive expression for manifesting and making known.   Kêd (English) is to make known, whence " UN-KED," a word used to describe the horror of the unknown.   Thus, when letters or other signs were cut, the cutter was Khetu, Kadmus, or Gwydion.   Behind the god is the goddess Kêd, and so the god is secondary.   Kêd was Khept, the feminine of Khep, to figure forth, form, and typify.

It is certain that we have Sut, the god of the Dog-star and the inundation in the British Pantheon.[2]   Seithenhin, the diminutive of Sut, has been called the son of Saidi; but is rather Saidi, who is the son.   Seithenhin and Saidi resolve into one and the same character.

---

[1] Mallet's *N. Ant.* Bohn's ed. p. 289.
[2] *Welsh Arch.* vol. ii. pp. 4 and 26.   Davies, p. 198.

Seithenhin or Saidi, the son, has the style of Kadeiriath, the language of the chair. This title rendered in other words is the Word of the genitrix, who was represented by and as the seat or chair; the Word (announcer) who preceded Taht, the lunar form of the manifester in Egypt, as Sut-Anubis or Bar-Sutekh.

Plutarch had heard that about Britain there were many small and desolate islands, and that in one of these the ancient Saturn was detained a prisoner fast asleep in chains. Saturn was the Egyptian Sut, who went out of Egypt in remote times, and was afterwards deposed within it.

Seithwedd is a name implying his sevenfold nature, or having seven courses, which relates Sut to the constellation of seven stars in Ursa Major. Sut, as Sothis, the dog, watched the waters of the inundation, and announced the coming overflow. Han (Eg.) denotes the Bringer of the Waters.

The Welsh Triads preserve the tradition of Sut (Seithwedd or Seithenhin), who was placed in charge over the waters of the deluge, and who upon a certain time was intoxicated, and whilst in liquor let in the inundation over the world, and drowned a district. Seithenhin, sometimes called the son of Seithin, is designated the drunkard. In the "graves (or cities) of the Kymry" one of them is designated the grave of the "weak-minded Seithenhin." From Seithenhin-Sut has been derived the Saint Swithin of the Christian calendar. In him Satan has become a saint. Swithin is called the "Drunken Saint," which identifies him with Seithin the drunkard. Also Swithin's Day, our July 15th, is nearly coincident with the inundation of the Nile, proclaimed by Sut; and if it rains on that day, says tradition, it will continue to do so during forty days. This belongs to mythology, not to meteorology, for, according to the observations at Greenwich, for the twenty years preceding 1861, the greatest number of wet days after St. Swithin's day occurred in the years when the 15th of July was dry. The Christian story which tells how it rained for that length of time on the death of St. Swithin, in the year 865, and prevented the monks of Winchester from removing his body from the churchyard, where he wished to lie, into the choir on the 15th July has been exploded by Mr. Earle,[1] who shows that the weather was most fair and propitious at the time. Further, when it rains on Swithin's Day, the drunken saint is said to be christening his apples. And in the Egyptian zodiac the dog Sothis is stationed in the tree constellation. This tree was the vine in some planispheres; in others the apple-tree on which grew the golden apples in Avallon. Swithin, the drunken saint, is none other than Seithenhin, the drunkard, of the mythos, and the forty days' flow of rain is connected with the overflow of the Nile conducted by Sothis or Sut.

The producer of the inundation became in other skies a meteoro-

---

[1] Saxon MS. published by Rev. John Earle, Prof. of Anglo-Saxon, Oxford.

logical prophecy, and so this was transformed into a rainy saint. In old calendars St. Margaret's Day, the 20th of July, was the one on which it is said " all come to church that are or hope to be with child that year,"[1] and the 20th of July was New Year's Day in Egypt, the day of the inundation announced by Sut, and the overflow of the river which poured its fertility into the lap of Egypt. The 20th of July was considered to be the first of the dog-days in England, and in Egypt it was the first day in the year of the dog. St. Margaret reminds one of her more familiar names of Peg and Page, the same as that of the goddess Pekh, the lioness.

The British Nav and Irish Nevvy is not necessarily the god Khnum of Thebes. Still the name of that deity is found as Knufi,[2] and he became the Gnostic Knuphis. Khnum is lord of the inundation, and Nef means the sailor. Khnum is represented by the bearded he-goat, and Nef is a name for an old goat. This figure of Nef, the old goat found in the Western Isles of Scotland, survived to a late time, as the deity exhibited for worship at the Witches' Sabbath, the last flickering shadow of the ancient mysteries. In the examination of the French witches whose confessions are elaborately recorded by De Lancre, we find the deity or devil often appeared in the form of a bearded goat, at other times as a serpent, and the serpent and goat constitute the biune form of Knum or Nef. The pitcher is a hieroglyphic of this god; with this his name is written. And at the Sabbath the goat issued out of a vast pitcher set in the midst, and began to swell and swell and grow monstrous by inflation until it was fearful to behold. Nevi, in the African Wolof language, means to swell and swell. Evidently this was the god of breath, or Nef, portrayed in a dramatic representation of the breathing source, one male type of which was the goat. At the end of the Sabbath the inflated form subsided and returned again into the pitcher. This is a rendering of the nature of Nef, primitively perfect, identifiably Egyptian by four hieroglyphics, the goat, the breathing, the serpent, and the water jar. It is a masquerade of Egyptian imagery, in which Nef, the lord of breath and dominator of the waters, manifests in his serpent-crowned ideograph of the goat as the image of inflation or breathing life. Marie d'Aspilecute of Handaye deposed that, when she was initiated and introduced to the goat-deity, she had to kiss him on the hinder face, this being the face of a black man, and hidden under a great tail. Khnef was a form of Af, the black sun of the lower regions called the hinder part. The posterior face had not the power of speech, and thus corresponds to the dumb Har-pi-Khart of the dual Horus. Sometimes the deity manifested as something between a tree and a man. This is akin to our Green man and the Jack in the Green, who is the hieroglyphic of leafy life on May Day.

[1] Cited by Grainger, *Biog. Hist. of England*, iii. 54.
[2] Pierret, *Knufi*.

Khnef is pictured on the Monuments as the Green God. At other times the witnesses saw him shaped as a great man enveloped in a cloudiness or a smoke, flambuoyant and red-faced, like iron coming out of a furnace. The sun of the under-world was also the Red Sun and the god of fire. Nefer is the "heat emitted from the mouth of Sekhet," the goddess of fire, a feminine form of divinity.

Neb also signifies gold, golden, to gild; and the great pulpit in which the old goat Nef sat enthroned was gorgeous with gilt, and as they all agreed, glittered very pompously. So the tinsel pageantry of Jack consists of gilding as well as green. This identifies Nef, if not Khnum, on the continent. In celebrating the Egyptian mysteries or the Eleusiniæ in Egypt, Sharpe tells us that "within the temple the hierophant wore the dress and mask of Khneph; the crier, the mask of Taht; the priest of the altar, the emblem of the moon, whilst another with the dress of Ra carried a torch."[1] De Lancre drew a picture from descriptions given by the worshippers, which shows the Triad of the Two Sisters and the Male-God as in the Triad of Isis, Nephthys, and Horus.

Ard-Macha is the ancient sacred name of the city of Armagh in many Irish documents, some of great antiquity.[2] The oldest of these is the Book of Armagh, known to have been transcribed about the year 807; in this the name is translated by Altitudo Machæ, which determines the meaning to be the height of Macha. Ard for height is found in 200 Irish names, and this is the Egyptian Arrt or Ert for the steps, staircase, or ascent. The tract called Dinnsenchas in the *Book of Lecan* professes to give the origin of the name as being derived from some wonderful woman of the name of Macha.

From other sources we learn that Macha, the first of three of that name, came into Ireland as the wife of Nevvy, who led a colony into Ireland 600 years after the deluge. Mythologically interpreted, this suffices to identify Nevvy with the Nav of the Welsh and the Nef who in Egypt is the sailor and lord of the deluge. There is a goddess on the Monuments of unknown office and relationship whom Wilkinson met with but once and copied.[3] He read her name Makha or Makht. Dr. Birch reads it Menka. It reads both ways according as the first sign is taken for an ideographic Men or phonetic M. The goddess is really the wet-nurse Menâ or Menka, as is shown by the two vases held forth in her hands in place of the two breasts or the woman suckling. She typifies the water of life, one of the two factors. Breath or Nef is the other, and Menka or Maka appears in Ireland as the consort of Nevvy. Makh abrades into Mah or Meh, still the wet-nurse as mother, who is called Meh-urt, the meek fulfiller. Mah signifies to be full, complete, covered, filled, satisfaction, also

[1] *Egyptian Mythology*, p. 80.
[2] Joyce, *Origin and History of Irish Names and Places*, first series, p. 77.
[3] Pl. 70, 4.

the number nine connected with gestation, and Mah modifies into Mâ, the goddess of the Two Truths, which are typified by the two vases of Menka (Maka). So that there are three of the name in Egypt —Menka, Mah and Mâ, and the Irish Macha was the first of three of that name in Ireland.

The three, however, may merely mean the Triad of the Great Mother, who becomes the Two Sisters. This was Hathor of the Spotted Cow, who is called the Golden Hathor, and the chief, the second, of the three Machas is known as the Golden-haired Macha.

The Irish Macha is the older form of Meh, Mehi, Maya, and May, recovered from the Monuments as Maka or Menka, the wet-nurse. Macha answers to Meh, a form of Hathor, whose type is the Cow. Hathor or Athor, the habitation and wet-nurse of the child, is extant with us as name of the womb, and she is represented in Ireland by her own cow, that still rises up from the waters in many legends, as did the cow Athor to receive the sun when setting in her own region of the west.

The Irish antiquaries have been cajoled by writers who, like the author of *Rude Stone Monuments*, explain everything by means of "the Danes," and who no sooner come upon a window that opens into a farther past than down they pull the blind, assuring you there is nothing to see.

Dr. Joice supposes the name of the Hill of Howth to be Danish, a form of the word Hoved, or head. But these names go thousands of years deeper than any Danish deposit.

The ancient Irish name of the ground was "SEAN-mhagh-EALTA-Edair," rendered the old plain of the flocks of Edair, Edair being the Hill of Howth. The tradition is that the first leader of a colony, Partalon, took up his residence with his followers on this plain.

Now the names of Edair and Howth may possibly identify the Hill of Hathor or Macha. The Irish Mhagh for plain corresponds to the Egyptian Makha, the level, the scales, the place of the equinox, where Hathor is represented by the cow's head in the Egyptian planisphere close to the Scales.[1] "Shen-Makha" is Egyptian for this level in the orbit of the twin heavens; the flock of Hathor was a herd of seven cows. Wilkinson says this goddess at times wore a peculiar headdress of a hawk, a perch, and an ostrich feather, which denotes that the Lady of Hut is then in the character of the president of the western mountain. "Lady of Hut" was her title at Thebes.[2] This is the Hut for height, the Howth we are in search of, and the Hill of Howth is Edair-Hut, the Western Mountain of the Cow-headed Athor. The "hut" has many forms, a seat, throne, boat, table, shrine, all of which have been found on our hills and heights. HOWTH is a form of HOVED, but that means more than a headland. In Egyptian Hut is modified from Khut, and Khut from Kheft. Kheft is a goddess of the west, the

[1] Drummond, pl. 3.   [2] Wilk. 2nd series, vol. i. p. 391.

lady of that country. Hathor as Lady of Hut is the earlier Lady of Kheft, as the solar west. The Hill of Howth or Hoved implies the Kheft-name of the West, and this form abrades into the " Hib" of Hibernia and the " Ib " of Iberia.

Edar, the Irish name of Howth, is connected by tradition with a female named Edar, said to be the wife of Gann, one of the five Firbolg brothers who divided the land between them.

Ben Edar would be their own mountain of the west to the emigrants as they sailed into the sunset, the Mountain of Hut and of Hathor.

There is a tradition of a colony dwelling on the Mhagh of Edair that perished of the plague.[1] In Hebrew Makkeh (מכה) is the plague or plagues, and in Egyptian Makha-ka is the desert, the desolate land. Outside Eblana there is a small island called " Edri Deserta " on the map, and Edrou Hēremos in the Greek text of Ptolemy, that is, the Desert of Edyros or Edair. Edair connects this with Makha (Mhagh), and suggests that the people who perished of the plague in the desert were mythological, and so helps to identify the imagery.

In a papyrus quoted by Champollion it is said of Hathor in her two characters, " She is called Neith in the east country and Mâ in the Lotus and the water of the West ; "[2] that is, as goddess of the equinoxes. Makha is the name of the equinox or level, and the earlier name of Mâ. Hathor the Golden, whose statues were often gilded, is the lady of the two doors of entrance and egress for the sun. She was the " Goddess of the Lovely Face," of mirth, music, and the dance ; the Venus of Egypt.

The following old lines are sung of the English Hathor, who is the Irish Macha :—-

> " Sing, reign of fair maid,
>   With gold upon her toe—
> Open you the west door,
>   And let the old year go.
> Sing, reign of fair maid,
>   With gold upon her chin—
> Open you the east door
>   And let the new year in."

Hathor was also designated "Daughter of the Water ;" her lute was strung with sunbeams, and her cows were seven in number. This lute—of seven strings—may be represented in another lilt of song, which was formerly sung by the children in South Wales, carrying a jug full of water newly drawn from the well on New Year's morning :—

> " Here we bring the water
>   From the well so clear,
> For to worship God with
>   This happy New Year.
> Sing levez dew, sing levez dew,
>   The water and the wine;
> The seven bright gold wires,
>   And the bugles they do shine."[3]

---

[1] Joyce, 1st series, p. 161.   [2] Wilk. Mat. Hierog.
[3] Choice Notes, pp. 66, 67.

It has now been shown that the Britons worshipped the Great Mother Kêd, who was the Egyptian Kheft (or Taurt) and identical with Kubele. In her lunar form she was the horned Astarte, our Hathor. In her character of Keridwen, the ancient mother was the British Goddess of Wisdom. Gwydion, the Sabean son, is the British Mercury, the Sut-Taht or Hermanubis of the Egyptians. Arthur we parallel with Sut-Har, the Sun-and-Sirius of the Druidic cycle of thirty years, which was the Egyptian Sut-Heb.

The Sun-God Hu is the Solar Hu of the Monuments, called a Son of Tum. His name of Aeddon identifies him with Aten, the son of the mother who became Atum as the divine Father, the Jupiter of the Romans; Atum being Ra in his first sovereignty as the father of the gods and in the dual form, Horus of both horizons, whence the Iu-Pater. Hu retained the character of the son as Aeddon, or Prydhain, the youthful Son of God, corresponding to Tum as Nefer-hept or Iu-em-hept the Son, who comes with peace, the Apollo known to Cæsar.

Pwyll is the Druidic Mars, and Pwyll and Pryderi have been paralleled with Shu, the Egyptian Lion-God in his two characters. Hercules we have identified with Khunsu, as Gwion and Con; the two children of Kêd with the double Horus, and sufficiently shown what Cæsar meant when he said the Great God of the Druids was Mercury, and that after him they also worshipped Apollo, Mars, Jupiter, and Minerva. But there were things in the British mythology indefinitely older than the Roman cult, as known to Cæsar.

When we have collected and correlated the legendary lore of many nations, and can read the symbols in their primal significance, and reconstruct the myths, we shall find, at the head of all, the mythical divinities of Egypt as the oldest things extant; that is, these personifications embody the earliest configurations of human thought, and are proveably of Egyptian origin, and traceable in other lands by their nature and in many instances by name. Words will help us much, but the divinities more. Through them we can get down to firm standing ground on the Old Red Sandstone of the pre-eval world, the primordial pavement of the past on which the footprints of antiquity are fossilized; through them we can get back to the primitive types which culminated in deities, and the dumb symbols of early expression that have been exalted to the status of religious doctrines and revealed dogmas, and prove that these types, the fossilized footprints of the past, are neither Roman, nor Greek, nor Hindu, nor Semitic, but identifiably Egyptian.

SECTION IX.

SOME consciousness of the sacred significance of certain words seems to have yet lingered livingly in the mind of the people of the Western Islands of Scotland when Martin visited them nearly two centuries since. In St. Kilda they had common and sacred words for the same things. They held it absolutely unlawful, he says, to call the island by its proper Irish name of "HIRT," but only designated it the "High Country." St. Kilda is the farthest west of the Scottish Isles; in this, Conachan, the highest point, is 1,450 feet above the sea.

In Egyptian the word "hert" means the high country. Hert is height, above, over, the name for heaven, and either they did not know that "Hirt" was the proper name of the high country, or this was their mode of preserving the fact that it signified the high country, and so they kept the old name as too hallowed for common use, this being one of the most effective means of preserving the mental impress.

Hert, as the height, the upper land of England, would seem to have given the name to Hertfordshire, for it is the summit of the land. The Grand Junction Canal reaches its summit in Hertfordshire, and descends both ways for Middlesex and "the Shires." This is the highest of the counties south called by the name of shires, so that it is the Hert, the land above, in a double sense; highest in altitude and by name as the upper boundary of the shires. "Scarce one county in England," says Camden, "can show more footsteps of antiquity" than Hertfordshire. The highest hill in the county is named Kensworth, and Worth answers to Hert (Eg.), the highest or uppermost, as an inclosure.

The shore, Martin remarks, which in their language is expressed by "Claddach," must be called "Vah." FA, in Egyptian, denotes canals or water inclosed, and the "peh" is the hieroglyphic sign of a water-frontier. Pa is the shore or bank in Maori. These people were preserving their hieroglyphics; V or F being the earlier form of the P.

Tep or tef (Eg.) signifies the point of commencement. This enters into the name of Dover. Tep-ru (Eg.) is the first outlet, gate or port. Tepru is the Egyptian name of Tabor, the sacred hill, a point of commencement in the solar allegory. Dover is our outlet and point of commencement. From Dover starts the great road called Watling Street; this runs northward on its way through the island. It is called the Roman Road, but Uat is a name of the north in Egyptian; Uati is goddess of the north. Uat also means distance, the long, long road, and is determined by three roads arranged lengthwise.[1] So interpreted, the Watling is the long long north road, and the ling does but repeat one meaning of Uat, a common mode of compounding English names.

Uat is both water and way, the water of Nile was the first road in Egypt. This dual meaning survives in English. Watford (Uatford) is the Waterford, and Wat is the name for the ford, so that Watford is the ford of the Uat (water), and the way (Uat) across the water. The north where the three watersigns are placed was the Uat (wet) quarter.

To wattle is to intertwine osiers and make wicker-work, and this WATTLING was an early form of Irish bridge or Uat for crossing over water. Tiling a roof is still called Watling.

The naming of the Isle of Thanet is a curious relic of Egyptian. It is not an island in the ordinary sense, not an isle of the sea, but is insulated by the aid of the river Stour forming two branches, which separate Thanet from the rest of Kent. A thousand years ago the arms of the Stour formed a channel three or four miles in width, named the Wantsume.[2] TNAT means divided in two, cut off, insulated, the river constituting the boundary line and land-mark of the division called Thanet.

Our word Gate is the Egyptian Khet, which, in relation to the water, is a ford, port, or harbour; Khet, a port, to navigate, go, stop, be in closed. This supplies the water-gate as in Margate and Ramsgate. Mer (Eg.) is the sea and also a land-limit, the boundary of a region on the water. Mer-gate (Margate) is the gate at the limit of Thanet at the north-east extremity of the isle. Ruim or Ruym[3] is an ancient British name of Thanet. Ramsgate is the gate of Ruim. Ru (Eg.) is an outlet, water-way. IMA or IM is the sea. RUIM is the mouth or outlet to the sea. The gate in Ramsgate is a repetition of the Ru in Ruim; according to the reduplicative mode of compounding the later names, Ramsgate, the sea-gate of Thanet, is already expressed by the ancient name of Ruim or Ruima. There is a park in Thanet named QUEX Park, still famous for its coursing. KHEKH (Eg.) means to chase, follow, hunt. The QUEX family are probably named as the Hunters. Theirs is a very ancient seat.

---

[1] Prise, *Mon. Egypt*, 15. [2] *Encycl. Brit.* 8th ed., art. " Kent."
[3] Nennius, *Hist.* 31.

Deal, rendered by Ter (Eg.), is an extreme limit of land, a frontier point. Cæsar writes the name DOLA. In "Domesday" it is called Addelam. The corresponding Egyptian is Atr-am, or Atr-ma. Atr is the land-limit, and this modifies into Ter (Deal) am (Eg.), belonging to, also the place of, as in the ham of hamlet. Ultima Thule, the northernmost point known to the Romans, the Thuly of Drayton, the Isle of Thyle (Thylens-el, a name of Shetland), we may derive from Teru (Eg.), a measure of land, the extreme limit of the land, the frontier and boundary. This underlies the Gothic TIULE, the most remote land, and the Greek TELOS, the end; TRO, Cornish, circuit, turn; TORA, Irish, border or boundary; TARA-TARA, Maori, palings. DHAL and TYREE are also found at the extreme end of "the Lewis." Dunnet Head, the Caledonian promontory mentioned by Richard of Cirencester as the extreme northern point of Great Britain, has that meaning in TUN (Eg.), to complete, fill up, determine; and Net, the limit, or end of all.

Ban or Ben (Eg.) means to cap, to tip; the Ben is the extreme point, as the roof; the Ben was a pyramidion; with us it is a mountain. F adds the pronoun It. Ban-f, the extreme point applied to land, describes the promontory or jutting point of Banff. Near Banff is Gamrie Bay. Ka (Eg.) is the lofty, up-lifted earth, the high place, headland, and MERI denotes the limit of both land and sea. On the other side of Gamrie Bay is CROVIE Head. Kherf (Eg.) means to steer, and paddle; and this was the headland by which the deep-sea fishers who left the shell-mounds of Banffshire had to steer or paddle in coming in. Out to sea stands TROUP Head, the home and haunt of multitudes of sea-fowl; "all the birds in the world" are said to come there. In Egyptian TERP is a name of ducks and waterfowl, and also means food. Thus, this breeding-place of the Terp (in America a particular kind of duck is the terapin) is designated in Egyptian as the place of the ducks and food.

The fowlers of Rutlandshire formerly celebrated St. Tibba's Day with great rejoicings. Tibba was their especial patroness. Camden mentions the town of Rihall as particularly addicted to this worship; the passage in which he describes this was ordered to be expunged from his *Britannia*, by the Index Expurgatorius, when the book was printed by Louis Sanchez at Madrid in 1612.[1]

Teb is the Egyptian name of waterfowl; the duck and goose are called Teb, Tef, and Ap. Ap or Af (with the article) denotes the first (Tep and Tef) born of; the duck, goose, and swan were types of the genitrix, who, as the old great mother, was personified as Tep or Typhon, the bringer forth from the waters. Typhon was made a

---

[1] The passage runs thus:—" Rihall, ubi cum majores nostros ita facinasset superstitio, ut deorum multitudine Deum verum propemodum sustulissit, Tibba minorum gentium diva, quasi Diana ab aucupibus utique rei accipitrariæ præses colebatur." —*Brit. Lond.* ed. 1590, p. 419.

saint in Tebba, but the fowlers of Rihall had the pre-Christian form of the lady, and the expurgators knew it.

Ru is another name of the waterfowl, and Rui (Eg.) means mud, marsh, and reeds, hence, perhaps, the name of Rihall.

Caithness is assumed to derive its name from the Catti, of whom Tacitus writes. The Ness is of course equivalent to the nose, or jutting, but we have no such expression as nose of land. In the hieroglyphics the Nes is a tongue, and we have the expression, a tongue of land for the jutting. Moreover, at the base of this NESS, the tongue is still preserved in the town and the Kyle of TONGUE. Hence the Ness is probably the Egyptian tongue. Caithness may be the abraded form of Kheftness, Kêdness, the tongue of Kêd. Kheft is the north, the hind quarter to the north. Kheftness is the northern tongue of the land. This meaning of the north (Kheft, Caith) is corroborated by the south land lying next to it in Sutherlandshire. Sut, or Suten, is the Egyptian south, and the south of Sut (Dog-star) and the north (Kheft) were the two halves of the total land. In Egyptian Khata is the end of land, and the unabraded Khap-ta is the northern end, the Caithness of Egypt. In an ancient poem of the Irish Nennius, " From the region of Cait to Forcu " is synonymous with from north to south. Cait is Caithness, and Forcu the Forth. Caith or Saith also signifies number seven, corresponding to the Egyptian Seb-ti, Hepti, or Khepti, for seven, and Kheft (Kêd) was goddess of the seven stars of the north.

Kaer Gybi (Holyhead) stands on an island at the western extremity of the county of Anglesea. The Kebi were the four genii of the four corners, the watchers over the sarcophagus or the four cardinal points. The Kaf or Hapi was the dog-headed watcher of the road east and west. The especial point of the west is connected with the goddess Khaft, as lady of the west. Khef or Khep (Eg.) means to look, watch, watching, and in Ireland the hill of watching, which preceded and survives the watch-tower, is called a COVADE, Covet, or Kivet, as in Mully Kivet, Fermanagh. Look-out points, says Dr. Joice,[1] intended for places of watching, to guard against surprise, are usually designated by the word Coimhead, pronounced COVADE. The title is generally applied to hills which overlook a wide expanse, and Kaer-Gybi is the inclosure of watching, or the watch-tower. On the mount of Gybi, 700 feet high, are the remains of a circular watch-tower, and on the sides of the mountain traces of extensive British fortifications.

The Island of Sark, says Pomponius, was greatly celebrated on account of the Gallic god.[2] Sarkh or Serkh is in Egyptian the temple, palace, and shrine. This in the parent language gives the name to the island, as the place of the shrine and oracle of the god mentioned by the Roman writer. Also the island divinity is recognized as continental. Serkh, an Egyptian goddess, was a form of Isis-Sothis.

---

[1] *Irish Names*, p. 206.   [2] *De Situ Orbis*, lib. iii. cap. vi.

The Islet of Staffa is named in Egyptian from the action of the water on the rock. Stu is to excavate, to make; FA, channels. Stafu signifies to melt down, with the determinative of water; a twofold description of Staffa. STAVE, in English, is to break, throw, crumble down. SCART is the name of one of the caves, and Skar-t in Egyptian is to be cut, cut out, cut piecemeal. Skart may be read as a picture carved, from Skar, to cut and picture.

Opposite Tenby, in Pembrokeshire, there is a cave called the " Cave of CALDI," containing some marvellous chambers and passages underground, one of which is now designated the " Fairy Chamber." The equivalent, Karti (Eg.), denotes holes, passages, and prisons underground, and as the word also relates to running waters, it may have included the stalactite grotto or cave, as at Caldi.

Some caverns in the chalk beds of Little Thurrock, Essex, are called CUNOBELIN'S GOLD MINES, from the local tradition that Cunobelin hid his gold in them. They are sometimes called DANE-holes, and of course the Danes are brought in, and these are claimed to have been their lurking-places. There is a very deep DANE-hole in the chalk near Tring, Herts, locally called " Dannel's hole." Cuno-Belin's gold was also stored in the chalk of the Dunstable Downs. It is known at Totternhoe as the Giant's Money, which you are supposed to hear ring if you stamp on the ground. Also Money-bury Hill is a part of the chalk range at Ashridge.

This hidden money is known by the name of Crow Gold, one form of which consists of nodular balls of iron pyrites, radiated within, which are frequently found in the white chalk without flint, that is, the mass of soft and pulverulent limestone of this formation.

The earliest gold of mythology is fire. The names afterwards applied to gold as a product of fire were given first to fire itself. The early men, be it remembered, had to mine for fire as diligently as the later dig for gold.

The Egyptian PUR, to manifest, come forth, emanate, appear, is the same word as the Greek name of fire or π ΰ ρ. Pliny says fire was first struck out of flint by PYRODES, the son of Cilix (*i.e.* Silex), and the name of the iron pyrites used with flakes of flint for striking fire points to this origin of fire or Pur.

Among the African names for fire and the sun are the Biafada, FURU, fire; Pepel, BURO, fire; Mose, BURUM, fire; Dselana, BUROM, fire; Galla, BERRU, splendour, glory; Kise-Kise, AFURA, hot; Okuloma, OFERE, heat; Mende, FURO, the sun; Gbese, FURO, sun; Toma, FURO, sun; Bini, UFORE, sun. In Arabic, AFR is sultry, and PIRAH is the sun; in Sanskrit, VIRA is fire, and PERU the sun; BREO, Gaelic, a fire; VER, Garo, fire; VUUR, Dutch, fire. The fullest form of the word is extant in the Maori KAPURA, for fire. This modifies into the Egyptian and Hebrew AFR (אור), and Afr into the Welsh AUR for gold. Gold and fire are identified by name in AUR and

AFR, Afr being first. From Aur comes ore. The first ore sought for was not gold, but the iron pyrites, which, when struck against the flint, yielded the precious element of fire. These were found with the flints in the chalks of our downs. The flint manufactories, as at Cissbury, must have also produced the equivalent of the "steel" for striking fire in some form of the iron pyrites. The Eskimos, at the present day, obtain fire by striking a shard of flint against a piece of iron pyrites. Iron was first extracted from the stone in the shape of fire, long before it was smelted. One name of these iron stones is CROW. An iron bar is still a crow-bar. There is a poor kind of coal called CROW-COAL, which does for furnace-fuel, but is of an inferior kind. Crow means inferior, and is therefore the same as Karu (Eg.), the lower of two ; and crow-gold is inferior gold, not the true gold. The crow stone, then, is a fire stone ; and the fire stones found in the chalk contained Cuno-Belin's gold, *i.e.* fire. The name of fire as Tan or Tek-n has already been traced to an origin in the spark, this being emphatically the fire of Baal.

Another English name for the iron pyrites is Mundic. MUN (Eg.) is stone ; TEK is the spark ; and as Mundic is the equivalent of Muntek, the pyrites is thus named as the spark-stone, the stone of Baal, son of Kar-tek, the old spark-holder of the north. Some of the West Australian tribes still say they derived fire from the north.[1] As already said, an earlier form of TEKA, the spark, is shown by the Bushman T'JIH or T'KIH, for fire, the T of which is a click, and the "JIH" or "KIH" reaches its antecedent in the Swahili CHECHI, a spark, and KOKA, to set on fire with sparks ; KIAOKA, Mantshu Tartar, for a fire made with sparks and dry leaves ; CHIK, Uraon, fire ; KAGH, Persian, fire ; QACO, Fijian, burnt ; and English COKE.

Belin is the little Baal, the child Baal, who in Egypt was Bar-Sut. The name of Sut means fire and limestone, the firestone that fermented. Sut-Nub is both fire and gold. And this identity of fire and gold may be found in the god Sut-Nub, whose name includes both. Cuno-Belin was our Sut-Nub, god of the sun and Sirius combined, and the limestone (Sut) contained the ore, aur, afr, per, or fire, in the iron pyrites called Crow-gold, Cuno-Belin's gold, and the Giant's money. Fire, then, was Cuno-Belin's gold. This was hidden in the chalk as Crow-gold, that is, fire-gold, in search of which the chalk of Dunstable Downs was undermined for miles together, and at one time the Dunstable people, who dwelt a considerable distance apart, could visit each other's houses by passing underground. As the firestones were obtained from the chalk, it follows that the word DANE is the TEIN or TIN for fire. Baal-tein signifies the fire of Baal, and Cuno-Belin's gold is Baal-tein. Tin also means money,

---

[1] Angas, *Savage Life*, vol. i. p. 112.

and both the gold and the money were hidden in the Tin-hole or Dane-hole.

The name of Sut, earlier Sebti, contains Seb, No. 5, and ti, No. 2, and is a form of No. 7, found also in Hepti. At Lambourne, in Berks, there are tumuli at a place known as "STRIKE-A-LIGHT, SEVEN BARROWS."[1] How the old names cling! Sut, the fire-god, our Cuno-Belin, was the embalmer of the dead. His name of Sutekh also means to embalm, and to lie hidden as did the dead in the Barrows, where the fire-and-flint stones were often dug out to strike a light, and replaced by the bodies of the dead.

Kent's Cave or Hole has been called the Bone Cave from the quantity of bones found in it.[2] And if such a place had been named in Egyptian, it would be as the Ken-Kar, or, with the article suffixed, Kent-Kar, signifying a hole underground, having some relation to bone. Ken (Kent) is bone; Kar, the hole, beneath. KEN also means carving in ivory or bone. The KEN is the carving tool, the BURIN, as well as the cartouche in which the name is inscribed. The Kent is the man of the Ken, the sculptor, or literally the scraper. Kenti would be a plural form. In the Stele C, 14, of the Louvre, Iritisen calls himself a Kent, or sculptor. Ur-Kent, the chief sculptor, occurs in another text.[3] It may therefore be conjectured that Kent's Cave was the workshop of the bone-carvers, hence the bone implements discovered there, the bone awl, bodkin, and harpoon, which had been shaped by the rudest flint tools; the philological evidence shows the naming to be Egyptian, and the Kent-Cave, in English Kent's Cave, buried like a ten-thousand-years-older Pompeii, when opened up, reveals the earliest workers in stone and bone, ministering to the simplest human needs as Egyptians. Kent's Cave is in the parish of Tor, whence Torbay. TERU (Eg.) means to work, fabricate, decorate, ornament, and the TERU implement is also the Ken of the carvers. Later, Teru is the name for portraying in colours with the scribe's palette, when the artists who had carved in bone became the men who drew in colours. The word teru enters into the name of Druid, who was doubtless the figurer of other things besides the time-cycles.

The Bone age is the necessary complement of the Stone age; the bone supplied the book for the pen of stone. Stone and bone were the first implements of registering, the primeval KEN of the Kenners, who wrought in the Ken (cave and sanctuary) before temples of learning were built or books were made to bear that name.

The first men of Kent's Hole were Palæolithic. They could not polish stones, but, as may be seen from extant specimens of their work, they attained great excellence in the art of drawing. In the Cresswell Cave the figure of a horse "delicately incised on a fragment

---

[1] Dawkins, *Early Man in Britain*, p. 358.
[2] *Kent's Cavern: its Testimony to the Antiquity of Man.* Pengelly.
[3] Maspero, *Tr. Soc. Bib. Arch.* vol. v. pt. ii. p. 557.

of rib is the first trace of the art of design in this country." [1] But the faculty must have been developed in a high degree among the cave-men of France, where they left their drawings of the Reindeer and whale, their hunting scenes incised on antlers, and, in one instance, the mammoth engraved on ivory.

In the Deruthy Cave,[2] near Sorde, in the Western Pyrenees, a neck-lace was found, formed of the teeth of lions and bears, and on the teeth were drawings of the seal and pike, also a pair of gloves. Altogether there were no less than forty teeth variously engraved. The cavemen cut their pictures on bone, antler, stone, and ivory.[3] Considering that their graving tools were only flakes of flint, the execution of their figures is marvellous. Strangely enough, their art of drawing, engraving, and sculpturing, was indefinitely superior to that of the later Neolithic age. And yet not so strange when we remember that this was the one especial art of the cavemen, of which the Eskimos, Kaffirs, and Hottentots have furnished such remarkable specimens.

The Cave of the Carvers, the Kennu or Kenti, found at Deruthy, is near SORDE, and in Egyptian SURT or SRT means to carve, engrave, and sculpture ; which suggests that Sorde was named as the seat of the sculptors, carvers, and engravers, whose buried work has been found in the Deruthy cave. The word SURT is determined by the KEN graving tool, the sign of bone and ivory.

Tradition tells of the bloody rites of the Druids enacted in gloomy groves. The hallowed grove of the Keltæ was called a Nemet, whence probably the name of Nymet Rowland in Devon. The sacred character assigned to the secluded Nemet is found in Nemet (Eg.), the retreat. The Egyptian Nemet is also the place of execu-tion, the name of the gallows and the block. This throws a light on the dark recesses of the Druidic Nemet, where, no doubt, they put their criminals to death. The Nemet (Eg.), as shown by the Nam symbols, was the scene of judgment and execution. A form of the judge is yet extant in the Nompere, later Umpire. In Gaul the Nemetum had become a temple, but the caves and sacred groves were the earlier temples.

Buchan, in his *Annals of Peterhead*,[4] describes a vast stone, thirty-seven feet in circumference and twenty-seven feet across, which was still in the "Den of Boddam" or Bodun, in the year 1819. Both names are Egyptian. "Batun" means the bad, the criminal, the malefactor ; whilst But-tem signifies the execution, cutting to pieces of the But, the criminal, hateful, evil, infamous, and abominably bad. The Den of Bodun was probably the dungeon of the malefactors ; the Stone of Boddam, the block of their execution.

It was in the Links of Skail that the beetles were found in the stone

[1] Dawkins, *Early Man in Britain*, p. 220.
[2] Explored in 1874 by MM. L. Lartet and Chaplain Duparc. *Matériaux*, 1874.
[3] Dawkins, *Early Man in Britain*, figs. 75, 76, 82, 84.   [4] P. 44.

coffin of one of the ancient barrows. The name of Skail is identical with that of the Island of the Written-Rocks in the Cataract near Khartoom, just where the land of the Inundation begins. Skul (Eg.) denotes not only writing but instruction, counsel, design, picturing, and planning; from which we may fairly infer that SKAIL was a seat of learning named in the most ancient tongue. The root SEKHA (Eg.) means to memorize and remember.

The Cornish GUIRRIMEARS are supposed to have been miracle plays. GUIRIMIR, according to Lhwyd, is a corruption of Guari-mirkle, a miracle play. The word "Guary" is found in English.

> "Thys ys on of Britayne layes,
> That was used by olde dayes,
> Men called Playn *Garye*." [1]

But Lhwyd does not go deep enough, to say nothing of the inevitable "corruption."

Guare, in Cornish, means a play, gware in Welsh, guary in English, and in Egyptian Kher means speech and to speak. But the play was enacted on spacious downs and natural theatres of immense capacity, which were encompassed round with earthen banks and in some places with stone-work. These places, it is now claimed, were the Mirs or Mears. The Mer or Mera (Eg.) is an inclosure of land or water. The Water-Mer is extant in the Mere. The Mer is also a circle, and the Guiri-mir or Kheri-mer is the inclosure or circle where the speeches were made and the play was performed. The size of the Mears shows they were at times beyond speech, hence GUARE means a game, and Kher (Eg.) is also a picture, a representation, that which was acted, the acting drama being earliest. For the Mir is our moor, and in Kirriemuir we probably have the Guirimir by name extant also as a place.

The so-called Anglo-Saxon and German Worth, for an inclosure, is called a test-word, showing the Teutonic settlements. But the Garth, Garter, Garten, and Garden are equally the inclosure. The original of all is the Kart (Eg.), an orbit or circle, that is, the Kar or Caer with the article suffixed. The Kart is the Russian Grod and Polish Grod, a burgh. The modified Hert (Eg.) was the name of the inclosure as a park or paradise. We have it as large as a county in Hertfordshire, and small as the tiny cup of the Blae or blue berry. This is called the whortle-berry, that is, the inclosed berry. But another form of its name is the HERT. In Hertfordshire it is known as the bilberry-hert. Thus we have the Wort and Hert in one. Did the Teutons also carry the Hert into Egypt, together with its earlier form in Kart? The fact is simply that the thing Kart, Garth, Hert, and Worth existed; the W is a later letter, and the later sounds were applied to the earlier names of places.

---

[1] Emaré, 1032.

The Egyptian Kart had to do double duty. The terminal T may denote two, and one Kar (Kart) is the lower; the other, the upper, is the Har (Hert), and the Hert becomes the Art, the ascent, the steep, the height. Kart is downward, and Hert is above. This Hert or Art becomes the Ard, of which there are 200 in Ireland, as the upper place, the height. The Irish Ard is the Welsh Alt, a steep place, and this becomes the Old, as the Old Man of Coniston and Old Man of Hoy. The Art (Eg.), Irish Ard, permutes with Ret, the ascent, and this enters into the ridge or rudge, a back or height. In the Irish Ard, the height, we have the Mount of the Great Mother Macha, whose seat was at Ard-Macha (Armagh), and whose name of Arth is that of the Great Bear.

The Inland Wick, represented by the A.-S. Vic, Irish Fich, Mæso-Gothic Vichs, is with us the homestead, the inclosure of the farm. It is the place of property, of plenty. Feck means plenty, much, most, the greatest part. It is the Egyptian Fek, fulness, reward, abundance. The Fog is a second crop, and the fat of land; allied to the Vic, a marsh or moist land, where plenty of food was grown for cattle. This is the Uakh (Eg.), a marsh, a moist meadow-land.

Cattle were an early form of Fekh, Feh, or Fee. Pekau (Eg.) is fruit or grain. Pekh, as in English, is food. The Fog, Vic, or Wick is the place of food, and becomes at last the inclosure or homestead where the produce is stored, it may be as Fech (vetches), Feh, cattle, Fek (Eg.), the reward, abundance, plenty of food. The Wick is thus finally the inclosure of the VICtuals.

The wick as a creek was derived neither from the Norse nor Saxon Vikings. It is the Uakh (Eg.), an entrance, a road. This Wick is so essentially a corner that in Northumberland the corners of the mouth are called Wikes. It is well known that some of our wicks are places where salt is produced. But these are sometimes far away from any sea-wick, and the wick as bay has no necessary relation to the wick as salt-work. Wick is a sediment and the name of a strainer. The word relates to the salt-making. A dairy is also a Wick in the same sense, with butter for product instead of salt.

Mr. Taylor's suggestion that the name of bay-salt is derived from the evaporation of sea-water in the bay may be doubted when we know that Baa (Eg.) means stone, or rock, solid substance; it may be salt so far as the sign goes, and bay-salt is called rock-salt. Besides which bay is sure to stand for an earlier form of the word. Bab (Eg.) means to exhale; Bak or bake is to encrust. Bekh (Eg.) is the rock, Bakhn being a name of basalt. Also Pakh (Eg.) means the separated; Maori Paka, the dried.

The Wick takes several forms. The A.-S. Wig is a temple, monastery, or convent; the Gaelic Haigh is a tomb, or grave, like the Quiché HUACA. The name goes back to the CHECH or stone chest, and KAK for a church; the KAK (Eg.), a sanctuary; the KHAKHA (Eg.), an altar ; CHAKKA, Hindustani, a circle ; KHOKHEYE, Circassian, circle ; COKOCOKO, Fijian, ring of beads ; KIGWE, Swahili, string of beads ; KEKEE (Ib.), a bracelet ; GIG, Scotch, a charm ; חיג (Heb.), a circle ; IGH, Irish, a ring ; COICHE, Irish, mountain ; KAWEKA, Maori, mountain ridge ; ECA, Portuguese, an empty tomb, in honour of the dead, who are the Egyptian Akh. In Cornish the modified HAY is a name of the churchyard.

The "Ton," says the author of *Words and Places*, is also true Teutonic, although non-extant in Germany. It is a genuine test-word to determine the Anglo-Saxon settlements in the isles, where there are thousands of tons, tuns, and duns, over 600 in Ireland alone, but none to speak of at home. What an amazing anomaly !

In a paper on the " D:stribution of English Place-Names," read by Mr. W. R. Browne, he gave a table of the results obtained by examining 10,492 names in Dugdale's *England and Wales*. Those ending in "TON" formed nearly one-fourth of the whole, being 2,545 in number ; the Hams came next, 702 in number.

Dr. Leo has computed that in the first two volumes of the *Codex Diplomaticus* the proportion of our local names compounded with tun, as Leighton, Hunstanton, is one-eighth of the whole.[1] It is characteristic of Anglo-Saxon cultivation, he says, that their establishments were inclosures (Tuns). No other German race names its settlements Tuns. This fact struck Kemble, who observes "it is very remarkable that the largest proportion of the names of places among the Anglo-Saxons should have been formed with this word, while upon the continent of Europe it is never used for such a purpose."

Mr. Coote sees in it another proof of Roman origin. Our tuns, inclosures, our hedgerows, he affirms, were all Roman. The truth is that the Tun or Tem marks an earlier stage or stratum of society than anything extant with the Germans, Angles, or Romans. They did not possess it, and could not have brought it here. Egyptian will tell us what the Tun was. It is not necessarily the settlement, and consequently the arguments of Mr. Coote founded on its being so are beside the mark and of non-effect. The Tun was not based on the Roman *limitatio agri* and allotment of the land, for it existed before there was any sense of possession in land that could be inclosed. In Egyptian the Tun takes divers forms. The Tun is a region, an elevated seat, a throne. This is extant in our Downs, the high and still most uninclosed of places. In the so-called "Dânes' Graves" found on the Yorkshire wolds, where many tumuli are to be seen, the graves do but repeat the Tun in a plural form, and pervert the old spelling in the name of the Danes. The downs were the judgment-seats of the Druids, like the Tynwald Hill of the Manxmen. The Tun as high place is found on the downs, as are the two Gaddesdens. Tyntagel is the Tun or elevated seat on a rock. Dynas Emrys was a

[1] *Die Angelsächsischen Ortsnamen.*

Druidicial TUN-AS in Snowdon, the lofty seat of the gods. The Zulu Donga (Tun-ka) is a division or cutting in the land, but with no necessary sense of inclosing a property. One of the most primitive forms of the Tun was the Cornish Dynas or fort, a simple entrenchment with stones piled together without cement, and raised some twelve feet high. The Tun is here the high seat, and As (Eg.) is the house, chamber, tomb, the secreting place. Hence the Dynas or fort. So Ab Ithel derives Dinus from DIN and YSU. The barrows and burial-places of the dead are found near these forts, as if the first places of defence were built to protect the dead. To all appearance the first property claimed in land and right of inclosure was on behalf of the dead. We have a possible relic of this in the popular belief that a common right of way may be claimed wherever a corpse has been carried.

The first Tun as an inclosure of land is the tomb. One hieroglyphic Tun is the determinative of a tomb, and Tun in this sense means to be cut off, separated. The TEEN, Chinese, is a grave; THAN is a shroud; TUNA, Zulu Kaffir, a grave; TANU, New Zealand, to bury; DUN, French Romance, a sepulchre; DEN, English, grave. The Den or Tun leads to the Dynas, as the house or general sepulchre of the dead.

The Down, however, is one type-name for the elevated seat, the high place, the burial-place, and doubtless in some of these, now swept bare of all their ancient monuments, there are yet concealed precious proofs of the prehistoric past. The downs were the high places, and the reason why the word "down," came to mean below, is because the Tun, den, or tomb, represented the under-world, where the dead went down at whatever height it opened. The tun, ton, or town, as the inclosure of the living and of property in land, is the final form, not the first; the Roman, not the Egyptian or Druidic. Tun or ton is far older than town, hence the reversionary tendency to the older formation in pronouncing the word town. The ton did not denote a town when it was the Cornish name of a farmyard.

In English, Scotch, Welsh, Irish, Gaelic, Manx, French Romance, Biscayan, Lusatian, Old Persian, Chinese, Coptic, Tonquinese, Phrygian, and other languages, the DUN or TUN is the hill, the summit found in the Egyptian TUN, the elevated seat. Irish philologists understand the Ton (or Thone) to signify the same as the Latin Podex, but the seat is primarily feminine and mystical, the Mons Veneris, the Hes of Isis, the Khep of Khept or Kêd, extant in the Irish CEIDE or Keady, for the hill as the place of sepulture.

Ten and Tem permute; the Tem (dumb, negative) are the dead, and the temple is also the house of the dead. So with us the Tun and Tom are interchangeable as names of the burial-ground. The Tom, Gaelic, is a grave; Tom, Welsh, a tumulus; Tuaim, Irish, a grave; Toma, Mantshu Tartar, a tomb for the dead; Toma, Maori, a place where the dead are laid. The Tema (Eg.) was also a fort, a place of defence. There is a mound or natural fort near Barcaldine old castle, known locally as TOM OSSIAN, or Ossian's Mound. It is a habit of the people roundabout to give many grave-mounds the name of Ossian. In this case it is said to be a place where Ossian sat, according to a local legend.[1] These mounds, being natural forts, were TEMAU. The word TEM (Eg.) also means to announce and pronounce. The Tem as the seat of the singer agrees with the plural Temau (Eg.) for choirs.

Now Ossian was a typical bard, one of the Asi or Hesi, by whom the announcements of the law were made from the Seat. The As is this seat of rule and sovereignty; the As is also a mote or mound (which was the seat of justice) and the resting-place of the dead. Thus the Tom is the tumulus and the tomb, the seat of sanctity, defended as a Tem or fort, used also as a mount of justice or a mote. Another mound named "TOM-NA-H-AIRE," the mound of watching, between Dun Cathich and Connel, further identifies the TOM and the TEM, fort, as the watch tower.

Mr. Taylor describes the syllable "ing as the most important element which enters into Anglo-Saxon names."[2] This is found in more than one-tenth of the total names of English hamlets and villages. In such as Tring, Woking, Barking, it is the suffix merely, but in Paddington, Islington, Kensington, we have the Ton or seat of the Ing belonging to the name prefixed.

The Billings, for example, were a royal race doubtless because they were assimilated to the god Baal; the Thurings are from Thor; the Sulings, of Sullington, in Suffolk, from Sul-Minerva; the Ceafings, of Chevington, in Suffolk, the Cofings, of Covington, in Hunts, and the Jefings, of Jevington, in Suffolk, or of Ivinghoe, Bucks, from the Kef of Kêd. This is merely by way of illustrating the type-name.

The Ing denotes a body of people founded on sonship, human or divine. The mother was the primary parent thus derived from, and afterwards the male. But Kemble's theory, that names ending in "Ing" indicated an original seat of the Angles or English, is apparently negatived by the almost entire absence of "Ings" in South Suffolk.[3] One "Ing" of the Angles is an enclosure. We have it in the far older form of Hank for a body of people confederated (Var. d.), identical with Ankh (Eg.), to covenant. To be at inches with, meaning to be very near together, is an expression belonging to the Ing relationship. The Ingle, a parasite, in a depraved sense, is named from the Ing. Thus we have the Ing as the Hank, and the Ankh was extant in Egypt not only as the living representative,

[1] Angus Smith. *Proceedings of the Society of Antiquaries of Scotland*, vol. ix. 1870-71.
[2] *Words and Places*, p. 82, 6th ed.
[3] Paper by W. R. Browne on *The Distribution of English Places and Names.*

the son, but as the body of people belonging to a certain district, who are designated the Ankh, whilst the topographical inclosure, the Ing, Eng, Inch, Inis, is as old as the naming of the isle, inclosed by the waters. Cheddington, for example, is the Tun, the high place, seat, inclosure of the Ing, which derives its name from Kêd, whose own Tun, or elevated seat, her throne, was higher still at Gad's-den (Kêd's tun).

The Chipping, as in Chipping Norton, Chipping Ongar, Chipping Barnet, or Chippingham, did not originate in Chapping and Cheapening. Cheping Hill and Chepstow take us up to the old high places of Kêd, where we find her cave, cabin, or Kibno, as in the Kibno Kêd, a form of the Cefn or Cefn Bryn or Cefn Coed. This Cefn is the Kafn (Eg.), an oven (a symbol of the Llafdig), and the Kabni (Eg.), a vessel, a ship, which was represented by the boat, the cauldron, and the divine sanctuary of birth and re-birth. The war-chariot of the Britons was a Covine.[1] This too was a kind of Cefn Kibno, or cabin of Kêd, a type of the bearer, who was called Urt, the chariot. The Chevin, in Wharfdale, or on the hill near Derby, or the Cheviot Hills, is not merely the ridge, but the cabin, cave, or Khep—sanctuary on the height, sometimes found in the hill itself, or in the stone-circle on the hill. The Cef, or Cev, is the Cornish Coff, womb, and the wife. Now the Ing community that bears this name are the children who derive that name from the mother's womb, the Coff of Kheft or Kêd, hence the Chip-ing and the Chevening on the great ridge in North Kent. The Kippings were still extant by name, not many years ago, in the neighbourhood of Ivinghoe (Kiv-ing-hoe) and Cheddington.

The Coff being the birthplace, the Coff-ing or Chip-ing is the clan, confederacy, or Hank, named from the feminine abode. The name of the Roman CIVITAS, anciently an association of families, a corporation, and that of CIV-ilization itself comes from the cave and the genitrix Khef. This is a principle of naming direct from nature. The Cefn Cave at the village of Cefn, near Denbigh, is not designated from the village of that name, as shown by the Cefn-caves elsewhere; and as this was only discovered and cleared out some forty or fifty years ago, and had then been filled up with sand from time immemorial, the name of Cefn must have been continued from time immemorial, before the cave was filled up.

The BED is another name of the uterus, and the Bed-Ing is the gens named from the birthplace. The Cwm, or Quim, is another, whence the Cum-ing and the Cwmmwd. In these cases the place of abode has extended to a county, in Bedfordshire and Cumberland. Thus COMBE, according to Ovid,[2] was Mother of the Curetes. The Ken is another form; hence the Ken-ing and the Kennings. The Hem, another, whence the Hemmings. Kêd, the mother and place

---

[1] *Richard of Cirencester*, ch. iii.
[2] *Met.* vii. 383.

in one, supplies one of these type-names, whence Chedding-ton, the seat of the family of Kêd.

This subject will be pursued in the "Typology of Naming." Enough for the present. This alone is origin from the typical birthplace, and such names as Wamden in Bucks, Wambrook in Dorset, Wembury in Devon, Wampool in Cumberland, instead of being corruptions of Wodensburg, are from the living home, Wame, Weem, Uamh, Hem, Cwm itself. This is shown by the pool and the brook, for the Wam was the place and the Pool of the Two Waters and Two Truths of mythology. The Wam as birthplace is identical with woman. The Uamh is extant on a larger scale in the place named Meall na Uamh, South Uist, where the Beehive is still a human habitation.

The Beck and By are said to be Norse or Saxon names. Both are Egyptian; both British. The bi (or bu) is a worn-down form of the Beck. The bu is the feminine birthplace, which, with the terminal T, is the But, or Beth, the abode. With the KH it is the Bekh, the birthplace. Bekha is the land of the birth of the sun; the Bekh is the solar birthplace. Our Beck is applied to the river at its source. The Bekh of the sun was represented by the Hill of the Horizon, the Tser Rock, stationed as a figure of the equinox. The Egyptians placed their equinoxes high up in heaven, in the zenith; this was where the sun was re-born every 25th of March. The Bekh was imaged as the bringer-forth, the earlier Pekh, a form of the genitrix, also named Buto.

The Bekh-Mount had been Sabean first, the Mount of the Seven Stars, and was afterwards made use of as a figure for the initial point of the solar zodiac and the birthplace in the sign of the Fishes. The same hills served in both cults, the worshippers of the Great Mother turning, like the Jews, to the north, the adorers of the solar son to the east.

The mount, throne, royal seat, is the Ten (Eg.), and the word also denotes the division, the birthplace at the equinox, the Bekh. Thus the mount of the Bekh is Ten-bekh, and in the worn-down form Tenby. Now we know the earlier name of Tenby is Tenbich or Den-bigh, and the name is founded on the mount of the Bekh, or solar place of birth. We may further infer the same origin for the town and shire of Denbigh, as the Bekh of the Ten, the birthplace on the mount.

The Peak is another form of the word, also the pike, as in Lang-dale Pikes, the Welsh Pig, the Pyrenean Pic, Italian Bec, and the Puy in Auvergne. The hill behind Bacup is one of our Bekhs. The mountains called "Backs" (as Saddleback) are birthplaces, only these are pre-solar; they typify the mount of Kêd, and of the hinder part. And in this meaning only do we reach the root for the names of our Beacon Hills.

The Bekhn (Eg.) is a fort, tower, fortress, magazine, or stronghold. Bekhn is a name of basalt, the hard, strong stone. The Beacon Hill would thus be the natural stronghold. The Bukan (Eg.) is also an altar with fire burning on it, and that too was a beacon.

These, however, are but applications of the Bekhn or beacon. The origin is in the Bekh as place of birth. Bekhens (Eg.) are called dwellings of the gods, the Bekhen being the Pe, heaven; Khen, sanctuary. Bekhn is the typical birthplace. This may be reckoned in the north, the east, or the south. We have each of these initial points, equinoctial and solstitial. For example, the Ten is the division, and this may be either; at Tenbury we find the solstitial Ten. The 20th of April is the great fair-day of Tenbury, and there is a belief that the cuckoo is never heard till the day of Tenbury Fair, or after Pershore fair-day, which is the 26th of June. The cuckoo is our bird of the cycle, and here the end of his period is the solstice. Buri (Eg.) denotes the highest Ten or division.

The Bekh represented the hill of the resurrection and ascent to heaven. Sinai was one of these as well as Tabor, the Egyptian Tepr. From this top (tep) the sun-god mounted to the upper half of the circle. The rock of the horizon, as it is called, is perfectly portrayed in Blake's picture of the old man entering the rock of the tomb below and the young spirit issuing from it upward. It was the place of burial as the Tser (rock), and the place of re-birth as the Bekh. And this image of the mount of burial and re-birth is the prototype of our Beacon and Back hills, on the top of which the dead were buried in the symbolical birthplace.

On the Palatine Hill in Rome, they show an opening in the rock which is said to be the cave of the she-wolf that suckled the twins Romulus and Remus; this cave also represented the primeval place of birth, the Bekh on the Bekhen Hill. The divine birthplace gives us the names of Buchan, Beckenham, and Buckingham, as the Ham of the celestial place of re-birth, our Heliopolis, and Sinai, for the Egyptian name of this mountain is the Bekh (Bekht). The Bekh as the place of issuing forth may be variously applied to the sun of the resurrection, the infant stream, or the beak of a bird, and the Bacch (bitch), the back of a mineral lode, the bag (womb), and others. But this is perhaps the most curious in its compound condition.

The Port of London extends for legal purposes to a point six miles and a half below London Bridge. This point of egress and entrance to the port is known as "Bugsby's Hole." The current interpretation of names would possibly explain this by asserting that it was derived from the circumstance of a man named Bugsby having "made a hole" in the water at that precise spot.

This is a form of the Bekh, which in one spelling is the Puka, or hole. "Bugsby's Hole" is the Bekh or Puka of entrance and egress by water to the City of London. In the hieroglyphics the Bekhs (or Beks) is the gullet, a passage of entrance and egress. The By repeats the Bekh, and the hole is a third name of the same significance. It is a common mode of continuing the ancient names by a sort of gloss. Beks-by-hole, as the place of passage at the boundary and dividing line of the port, is the Bekh three times repeated.

But for the Teutons it seems we should never have found the English HOME. "This word," says the author of *Words and Places*, "as well as the feeling of which it is the symbol, was brought across the ocean by the Teutonic colonists, and it is the sign of the most precious of all the gifts for which we have to thank them."[1] There was no home in Britain, nor the feeling for it, till the Teuton came! Why, the home is as old as the womb. Word and thing existed as long ago as the Scottish WEEM and the Irish UAMH, when the home was a hole in the ground. As for the particular forms in Ham and Hem, they come from the Egyptian Hem, the seat, abode, dwelling-place, that goes back to the birthplace. Hem is the typical seat, and habitation, the female Ems, the woman, the wife. It was so old that the Hemu, abraded into Amu, are the residents, residing, seated, and inclosed. The Am likewise indicates a residence with a garden, park (" Hert "), or paradise. Nor did the Egyptians bequeathe us the Ham undistinguished.

The Hem sign, which is also the Han, is the symbol of the seat or home on the water, and denotes a water-frontier. The Hemu are the watermen, sailors, and fishers. The Hannu or Hanti are the voyagers to and fro. Both Mu and Nu are the water in Egyptian, hence the interchange of Ham and Han. In the same manner the names of our coast hams and hans interchange, and Ellingham in Hants is represented by Ellinghen in France. On the coast-line of Oldenburg and Hanover the ham takes the shape of um, as the Frisian suffix. The Egyptian Ham or Han being primarily a water-frontier, a place on the coast or river bank, rather upsets the Teutonic derivation of names based on it, whether found in England or France. It makes one feel afresh that the less we know the easier it is to generalize. The Hun (Hunt) is the matrix. This permutes with the Hem or Ham, the Khen, Khem, or Skhem. All have one origin in the earliest place of birth, and were applied to the abodes of the living and the tomb of the dead, as a place of re-birth. How near to nature is the Ham as the seat is manifest in the name of the thighs. The Khem or Ham might be illustrated by a score of types, and each one can be traced to the female, and her type of types, the womb, Khem, Hem, or Ama, the primeval house and home ; the KWAM, which in Khaling denotes the mouth or uterus ; the Quim, כוה, or Khebma, who is the most ancient genitrix of Egypt and the black land.

The SKHEM (Eg.) is the shut place and secret shrine of the child Horus. This form is extant in the African Gura, SAGUMA, and Icelandic SKEMMA, for the house, the abode. One type of

[1] P. 82, 6th Ed.

the genitrix, and therefore of the Khem or Skhema, is the leopard-cat (Pasht), and in Arabic a cheetah kept for hunting is the SHUKM, whilst in the African Bambarra the ZIAKUMA is a kind of cat. The Khem is the feminine shrine, a name of Hathor, the habitation KIMA, Arabic, house, home; KAM, or KIM, in Dumi, the home; CHEM, Tibetan, house; KHEMA, Swahili, a tent; KOMA, Persian, straw hut. The KAM, in Nupe, Susu, Basa, Doai, Ngodsin, and other African languages, is a farm; GAMA, Singhalese, a village; the CHV-MAH (חומה), in Hebrew, is the wall, or the walled inclosure; YUM Magar, a house; UMAH, Javanese, house; UAMI, Uhobo (African), house; CHIM, Zincali, country or kingdom. And it is here we shall find the true meaning of the Combe, the place between the thighs of hills. The Combe answers to the Khem (Eg.), the secret shrine, the Shut-place of Horus, the child, in which he transformed into Horus born again. The Combe is supposed to be the bowl-shaped or crooked formation. The Welsh form of the name, the Cwm, compared with the same word used in vulgar English, the Quim, will sufficiently recover the ancient meaning. It is akin to the home, the Weem, the Cam(ber), for which there is but one prototype in nature.

The underground houses called Weems, the Gaelic Uaimh, a cave, are synonymous with wames or wombs, and represent the womb of the auld wife, the Mother Kêd. Weem or Uaimh answers to Khem (Eg.), a shrine, a secret shut-place, which may be that of the living child, Horus in Khem, or of the dead (Khema). "Can a man enter a second time into his mother's womb," Nicodemus asks. That was exactly what these simple souls symbolically sought to do!

A large cromlech at Baldernock, nine or ten miles from Glasgow, is denominated the "Auld Wives' Lift." The lift is the heaven or sky. The "Auld Wife" is probably the correct rendering. She was Kêd or Kef, whence wife, and in Cornish, Kuf is both the wife and the womb. The Auld Wife's Lift was the Meskhen, or Mastebah, the place of re-birth, to which they looked for a lift into another life in the Lift above. Auld means first and great, the exact equivalent of Urt (Eg.), the first, the great, the old mother, who was the bearer that gave the lift in her chariot, called the Urt, or the womb of the Khebma.

The place of birth being the type of the tomb, the abode of re-birth will account for and explain the hole-stones so frequently found at the circles. Through these apertures children and initiates were passed in the Druidic rites and representations of the mysteries, as a mode of regeneration and re-birth from the womb, the ark, the Cwm, the prison, the cell under the flat stone, the Weem or Khem of Kêd. The root of both Cwm and Cefn is Khef or Khep. MA is the mother or place. The Khef is the Cornish Coff, the womb, or belly. The Kep (Eg.) is the concealed place, a sanctuary; the Khep, or Khepsh, is determined by the hinder thigh, as the feminine abode, and the birthplace in the northern heaven. As cognates we have the Cop, an inclosure with a ditch round it, a heap, and a mound ; the cove and the cave ; the oval, the hop or hoop, an inclosing circle. Khebm modifies into Khâm and Kam. The same root with the terminal N forms the word Khefn, Chivan, or Cefn, and this modifies into the Chûn and Ceann.

One of the cromlechs is called the "Chûn Cromlech." This is a prevalent name for the maternal abode, the KUN of birth and re-birth, the Mes-Khen (Eg.), which the Chûn Cromlech imaged. Chûn is Chiven in Hebrew, the Kymric Cefn, at once the mount and the cave of birth. Now Grimm's law need not be appealed to in paralleling the Gadhaelic CEANN for the mount with the Kymric Pen and Gaelic Ben. It is the reduced form of the Kymric Cefn and the name of the Cevennes. This modification of Cefn occurs in the English Keyntons in Devon, Shropshire, Dorset, and Wilts. The double N of Ceann occurs in Conan, the old name of Conisborough. The Pen and Ben are the Egyptian Ben, the height, cap, roof, top. The Ben was the solar pyramidion ; the obelisk was one of its types. It is masculine, as another application of the Pen will prove. The Cefn is feminine. In this way the TYPES will often take us beyond the region of mere sound-shunting, and give us the definiteness of things in place of verbal vagueness.

The Chûn Cromlech shows the application of the womb-type to the tomb ; the place of birth to that of re-birth. In Glamorganshire there is a circle of stones named Kevn (Cefn) Llechart. Thus the crom-lech and circle of stones are identical with the type of the birthplace, which was first of all found in the feminine nature, then applied to the cave of the hill, and afterwards externalized in the rude stone structures erected outside as the burial-places of the dead.

The ark, pair, vessel, or uterus of Kêd was represented by such stone sanctuaries. The cauldron or cooking-place of the ancient mother was designated the KIBNO-Kêd. In the hieroglyphics the Kabni is also a vessel, a ship, or ark, the English cabin, and the KIBNO-Kêd is the mother-ark. The KAFEN (Eg.) is an oven, and means to bake, and the KIBNO was figured as a cooking vessel, whether for boiling or baking.[1] In one language the belly or womb is the KABIN, and in Welsh the CAFN is a boat and a baker's trough.

The cabin of the ark, the Kafn or oven of the Lady of Bread, the Kibno of Kêd, the Kevn Llechart, the Chûn Cromlech, the Cenn and Cefn of the mountain cave, the Scottish GOVAN are all illustrations of the one original type, the birthplace called the Coff or Kep of Khept, the British Kêd.

The Combe is often found with the Beacon Hill, and in "Cwm Bechan" the birthplace is named twice over.

The beehive-house, which was a human habitation before the type was passed by and left behind for the bees, has two names in Gaelic,

---

[1] By Aneurin, for example, in the *Gododin*, Song 24.

the BOH and the BOTHAN. Boh corresponds to the PEH (Eg.), a form of the lower, hinder part, the Hem, a female type; BOTHAN to the BUT (Eg.), Belly and NU, receptacle; the Hebrew נב, the *receptaculum*, and נבן, the belly, the uterus, and primordial abode.

The primitive borough, burgh, barrow, bur, or bury, is the Bru (Welsh), the mystical residence; BRU, Irish, the womb; Bara, Vei, the womb; Apara, Sanskrit, the womb; Pal, Akkadian, sexual part of woman; Pir, Gond, the same; Por, Armenian, the belly; Bar, Hungarian, and Bayar, Canarese, the belly, and, lastly, the Belly is derived from the same origin.

The Cairn does not mean a mere heap of stones above ground. Mr. Anderson has shown that it is what we might infer by deriving the name from KAR (Eg.), an underground cell or hole, and NU, a receptacle, house, feminine abode. Then it becomes manifest that the Welsh CALON, or GALON, for the womb, is a form of the word cairn. We derive the charnel or carnary from the cairn.

There is an ASCIDIAN simplicity about the beginnings of human thought, as manifested in the earliest typology, which shows the commencement to have been akin to that initial point in evolution, a mere sac, with the dual function of including and excluding water. In the human beginning the sac is the uterus, the abode of two truths of life, those of the water and the breath, feminine pubescence and gestation. All utterance appears to have originated with this primitive utterer.

All human feelings can be traced back to two desires, the one being that of self-preservation, the other of reproduction. These constituted the total stock at starting in the dimmest dawn of human consciousness. And to this early stage we have to look for the first rude mould of thought and expression. Nothing that ever belonged to it has been entirely obliterated, and its evidences are visibly extant, as are those of the Palæolithic age. No origin has ever been wholly lost, any more than spoken language has altogether superseded the clicks. The desire of reproduction by itself alone is sufficient to account for what is termed the Phallic mould of thought and utterance, and the final stage of that desire constitutes religion.

It can and will be shown that the leer of Priapus is an altogether later expression added to the face of the subject commonly called Phallic Worship. There is no lewd grin in the look of the early men; their beginnings were lowly, but their observations were made in a spirit as seriously intent as that of modern science or of childhood. Hence Egyptian art, however near to nature, was pure and unashamed in its nakedness.

The feminine abode of birth was the typical home of the Troglodytes, who dwelt in the caves of the earth and named these after the mother. These caves were afterwards devoted to the dead more freely when the living could defend themselves outside, in the open space, or on the mound. In this way the abodes of the living were named as the habitations of the dead, as in the Tun or the Cleigh.

Cleigh is a Gaelic name for the burying-place. There is a cleigh in Lochnell, identified as a burying ground by its monument, a great cairn some sixty feet in diameter. A stone chest, an urn, and a bronze dagger were found there. Cleigh resolves into KL or KR, the cell; and Akh (Eg.), the dead, the Cleigh being the cell of the dead. The Arabic KALAGH is the stone inclosure of a tomb. The Clach stone is another form of the same word, the stone being the representative sign of the burying-place. The proof may be found in the Clachan. The Cleigh (Clach) is the dwelling of the dead, and around this was formed the Clachan, a small village built round the church which had superseded the Cil or Cleigh of an earlier time. Thus the Clachan of the living has its roots in the Cleigh of the buried dead.

The glebe land and ecclesiastical revenue are not primarily the present made by the people to their god, as Mr. Spencer puts it, for the first possession of the land was taken by the dead, who constituted the earliest form of the landed interest, and instituted the most primitive kind of landed property. The dead were the cause of a sacerdotal class being established in their precincts to protect them, and the church lands as ecclesiastical property are the last result of this ownership, on behalf of the dead, of the soil thus made sacred at the centre, with its surrounding circle devoted to the sustenance of a priesthood.

The type of the tomb-temple becoming the house of the living was preserved in Egypt to a late period. Twelve thousand inhabitants are ascribed to a single temple at An (Heliopolis) by a census taken in the reign of Rameses III. So the tem or tomb became the fort, village, city, and king-DOM.

This origin of the artificial inclosure as the sacred precinct of the buried dead is further corroborated by an Akkadian ideograph. Bat (Akk.) means to die, the ideograph being the portrait of a corpse. Bat is also a fortress, and the ideographic corpse is the sign of an inclosure. The corpse-inclosure was primal, as the Kester, and the corpse remained as a determinative sign of primitive usage when the Kester had become the castle, citadel, or city.

In the "Black Book of Caermarthen" there is a long series of verses on the "Cities of the Kymry." The cities are the graves. Each city is the grave of some mythological or legendary hero, whose name it bears, and these cities originated in the Caers as circles of the dead. Beyond these are the "Long Graves in Gwanas," of which it is said "their history is not to be had; whose they are and what their deeds." We are told, "There has been the family of Oeth and ANOETH, naked are their men and their youth—let him who seeks for them dig in GWANAS."[1]

---

[1] Skene, vol. ii. p. 313.

The Long Graves in Gwanas are the "Long Graves" of the cavemen of the Neolithic age, who turned the natural Cefns into chambered tombs, such as are found in Cefn near St. Asaph, in Denbighshire.[1] Gwanas is Gwan-as, that is, Cefn-as. As (Eg.) is the sepulchre, the chamber of rest, of birth and re-birth, the maternal abode. The cave was this at first, and the chambers were excavated afterwards ; the one being used by the men of the Palæolithic age, the others by those of the Neolithic age. The Cefn was a natural formation ; the Cefn-As (Gwanas) was artificial. Both are apparently recognized in the two burial-places by "Oeth and Anoeth."

The "Long Graves in Gwanas" mean the same as the Long Barrow at West Kennet, Wiltshire, and others found in Somersetshire and Gloucestershire. The name of Kennet likewise identifies the Khen Sanctuary. Khen-net (Eg.) reads the lower-world Khen, and the west was its entrance. The Long Barrow at West Kennet was 350 feet in length. These were made by the men of the Neolithic age.

Cleidh-na-h-Annait is the name of an ancient burial-ground in the west of Scotland with two stone cairns in it. The word Annait is commonly connected with sacred places. ANNOIT, in O'Reilly's *Irish Dictionary*, is explained as "One's Parish Church." In the Highlands the church was at one time synonymous with "the stones." The Annoit, says Skene, is the parent church or monastery which is presided over by the patron saint, or which contains his relics.[2]

The parent church is the mother church. The stone cairn was the earlier Annait, sacred to the dead, and this was built by each person contributing a stone. Nat means an offering, to present tribute, as is done in accumulating the cairn. Annt (Eg.) is tribute, and in English Anne means to give, and Annet signifies firstfruits. Anit (Eg.) also means to anoint, and is the name of incense. But the offering of the stone, An, was a far earlier mode of making sacred, and the Annait was the first stone sanctuary before larger stones were hewn. The Annait can be traced upwards from the cairn to the church, and the stone chest or "sanctuary of the saint's relics." The Welsh ANNEDD is a dwelling-place. In connection with this it is noticeable that the solar birthplace and the soul's place of re-birth in the *Ritual* is An, An being the name for stone, and one especial symbol of An is the stone or obelisk ; also Anit is a name of the genitrix, who was the earliest form of the mother church. The Annait is probably identical with Taliesin's Circle of Anoeth. An-at as Egyptian would also denote the circle of repetition.

Cuhelyn uses the term "Anoeth" for Stonehenge, and speaks of the "study of the circle of Anoeth." Arthur is said to have been imprisoned for three nights in the inclosure of Oeth and Anoeth,[3]

[1] Dawkins, *Cave Hunting*, p. 161.   [2] *History of Ancient Alban*, p. 70.
[3] *Welsh. Arch.* vol. ii. p. 12, Triad 50.

like the other solar heroes who were three days in the fish's belly or in the underworld, the place of transformation and reproduction.

If asked, what is a Hoe? most Englishmen would reply, a hill. So many hills are called hoes. But the hoe as name of a hill is secondary; the hoe is not the hill except that the high place and hoe place are synonymous. The hoe is primarily a circle, and need not be on a hill. The letter o is its symbol. Ho is a boundary; "out of all ho ' is out of all bounds. Our hoe is the hieroglyphic Heh, the cycle with the sign of the circle. The hoes were stone-inclosures of a circular form, whether on the hill or in the plain. True to the primordial type, these circles have perpetuated the primitive idea even in their names. In the Orkney Isles they are called Ork-hows, that is ark-circles. "Much fee was found in the ork-hows," says an inscription in the Orkneys.[1] The primary form of hoe is Kak or Khekh.

The Hay, Haigh, or Hak, as in the Cornish Hay, a churchyard, and the Hak-pen at Avebury, is derived from Kak, an old local name for the church or stones. The Kak is neither derived from the German Hag, a town, nor the Dutch Haâg, an inclosure, nor the Sanskrit Kaksha, a fence or bush. It exists as the root of all in Kak (Eg.), a sanctuary, an inclosure, and KAKUI, a coffin. The Kak or Khekh may be manifold. One of the earliest is the Kak, a boat, a CAIQUE, Welsh CWCH; another is the English CEGE, a seat. It may be the keg or cask, the Whiche or chest, the Kymric Gwic or the Norse Haugr, a sepulchral mound. The stone-chest or Kistvaen is also called a CHECH by Camden. The Kak is an extant provincial name for church. The Kak (Eg.) is a boat and a sanctuary. This boat is the Welsh Cwch, the Coracle of the goddess Kêd. Hence the hoe or how is an ark, and the Ork-hows are the arks of the dead.

The name of the Orkney Isles is undoubtedly derived from the old Kymric word Orch, which means a border, a limit. This renders the Egyptian Ark, an end, limit, to cease, be perfected, finis. They are named in Egyptian as the extremity or end of the isles. Nun (Nnui) signifies countries in relation to water and fellows of the same type, as we say the Orkneys. Nnui (Eg.) is the name of water, and Ark-nnui is both the land and water limit. The isle is also an ark of the water, especially chosen in ancient times as a place of sepulture. The ARACH in Gaelic is a bier; the ORK, Icelandic, a sarcophagus, and in Irish the womb.

The writer is fully aware that the repetition of certain words and names used so frequently by the Arkite triad, Bryant, Faber, and Davies, will be to many as the offering of water in hydrophobia. Nevertheless the dreary Arkite and Druidic subjects have to be gone over again with the expectation of seeing a winged transformation of the grub long buried underground, and stamped underfoot, as if for ever, by many a passer-by.

[1] Farrer, *Inscriptions*, p. 37.

The hieroglyphic sign of land and orbit, called the cake, occurs four times on a stone found in the Rose Hill tumulus at Aspatria, near St Bees.[1] This, like the hoe, is the symbol of a completed period. That period was fulfilled when the sun had passed through the three water signs, and entered the first of the nine dry signs. The cake signifies land and horizon, the place of landing from the waters. The circles represented the ark generally on the hill-top, out of the waters. Our cake is synonymous with the Egyptian Khekh, to check.

The hoe or howe goes back to the Khekh (Eg.), the horizon, collar, the round. Khekh, the balance or level, denotes the circle completed at the equinox. The Khekh collar worn by Neith has nine symbolic beads, corresponding to the nine maiden stones and to the nine nobs on the Scottish Beltein cakes. Many of the circles consisted of nine stones. The relation of this number to land and a completed course will be amply illustrated. Enough for the present to point out that in Egyptian Meh means to fill full and fulfil, to complete. Meh is the number 9, and a water line, the same in significance as the cake symbol. Meh-urt is a form of the cow-headed goddess Hathor, and Mehi a name of the lunar deity Taht. Meh is likewise the north. When we are told that Maes means a field and Magh a plain, that explains nothing. They mean much more than that for the present purpose. The Magh, as plain, is based on the level of the equinox, the Makhu (Eg.), level or balance. The ancient name of Dunstable was Magintum, and it is a lofty table-land. Ard-Macha (Armagh) is the level aloft. Hence the place Makhu interchanges with Mat, the mid-way; and the Swiss Mat, the plain, level, or meadow, is the Magh. Both meet in Egyptian, where Mat is an old name of the Makhu in An. Makh and Meh denote the place of fulfilment.

There is no proof extant of the original number of stones in Maes How, which bears the same relation, however, to the standing stones of Stennis, in Orkney, that Maes Knoll does to the circles at Stanton Drew,[2] showing a likeness in the nature of these monuments, as well as in the name. And we know, by the Nine Maidens of Cornwall, and the Nine Stone Rig, that some of the stones were nine in number, and that number would in Egyptian denote Meh's How, or the circle of Macha.

Kemp How is a tumulus in front of a circle among the remains at Shap. Shap, in the hieroglyphics, signifies time, epoch, period. The Shebu is a collar forming three-fourths of a circle with nine points. Shebu means a certain quantity of flesh. Shap is to shape, figure, image; bring forth, evacuate. The root of the matter is the measure of time, nine solar months, that it took to clothe a soul in flesh or

---

[1] Ferguson, *Rude Stone Monuments*, p. 157.
[2] *Ibid.*, p. 153.

shape it and bring to birth. The Shapt were persons belonging to religious houses, such as we infer gave the name to Shap.

Kemp, in English, is a champion ; Kemb, a stronghold. In Egyptian, Khem is the champion, and the Khem is a shrine of the dead, with a circle for determinative. Khem-p-how is a circle of the dead. Khenf is bread or food offered to the dead, and the Shebti are sepulchral figures and images of the dead.

Pomponius Mela speaks of the Island of Sena in the British Seas, where the nine priestesses ministered in a round temple, which they unroofed annually and covered again in one day, before sunset.[1] He relates that if in the process any one of the women dropped or lost the portion she was carrying to complete the work, she was torn in pieces by the rest, and the limbs were carried round the temple in triumph, until the Bacchic fury had abated. Strabo affirms that there always happened some instance of this cruel rite at the annual solemnity of uncovering the temple.[2] The same thing is alluded to by Taliesin as the metaphor of a hopeless calamity, "a doleful tale, like the concussion, like the fall of a SE, like the Deluge."[3] It was most probably a representation in the mysteries. The nine "Se"s were the nine months of child-bearing impersonated. If one of these let fall the burden, it was fatal to all; the eight were depicted as turning on her and rending her piecemeal. Such was the drama of mythology. In the same sense the Gallicenæ are said to have turned themselves into whatsoever animals they pleased. So the sun's passage through Aries and Taurus was his transformation into the Ram and the Bull.

The name "SEON" is not necessarily that of an island, although Strabo mentions an island of Sena.[4] The root meaning enters into senate, sennet, a total or round, and is the Egyptian "Shen," a circle, orbit, round, circuit, period. The Druidic Caer-Sëons were the primitive type of these, and they were stone circles. The Caer-Sëon, or Sëon with the strong door, typified the landing-place of Hu after the Deluge, the station of the sun on his ascending out of the three water signs into the circle of the nine land signs. Whether an island or a Caer, the Sëon was the circle emblematic of the divine circle of the gods, the Put of the hieroglyphics, signifying number nine. And the nine maids or priestesses were one with the nine muses of Greece, the nine that danced about the violet-hued fountain as described by Hesiod.[5] Taliesin says, "The tuneful tribe will resort to the magnificent SE of the Sëon."[6] "Sua," in Egyptian, is loud singing ; Shen, the circle.

The vessel or cauldron of Keridwen, the symbol of this circle, was said to be warmed by the breath of nine damsels ; in Taliesin's "Spoils of the Deep," it is the cauldron of the ruler of the abyss.[7]

[1] Lib. iii. c. 8.   [2] Lib. iv.   [3] Gwalchmai.
[4] Lib. iii. ch. viii.   [5] Theogony, 1.   [6] Davies, *Myth*. p. 67.
[7] Davies, *Myth*. app. iii. p. 518.

These were the nine muses of Britain, and of greater antiquity than those of Greece.

The nine personify the nine months of gestation and of giving breath to the child; in the eschatological phase they performed the rites of the dead, and represented the "wake," the resuscitation, and re-birth of the soul of the deceased, as did the nine in attendance upon Osiris. Hence the nine maidens of memorial in the nine stones.

The accented ë in Sëon shows the elided consonant. This is recoverable in Segon (Caer Seiont, from Segont), and Segon is the Sekhen (Eg.), the inclosure, place of settling, of rest, a breathing place, from Skhen, to give breath to. And in Caer-Sëon the cauldron of Keridwen was warmed by the breath of nine damsels or muses, the Gwyllian of the Sekhen who become the nine Gallicenæ of Mela's account. The Sëon or Sekhen is found in several forms.

In the year 1843 seven urns were exhumed at Swinkie Hill; these were inverted and imbedded in an artificial mound. Near at hand is a monument called the Standing Stone of Sauchope. Sau-Khep (Eg.) denotes the sanctuary or place of transformation for the mummy or dead body. Of course the Sau may only have denoted the deceased, but doubtless they preserved the dead to the best of their ability.

Swin in Swinkie answers to Skhen (Eg.). Ki (Eg.) signifies the land, earth, interior region. Thus Swinkie is the domain of the Sekhen, sanctuary, resting-place, where the dead were gathered together, literally, as the hieroglyphics show, to be embraced in the arms and inclosed in the womb of the mother earth in the Sekhen or Khen shrine, as at Swinkie.

We are now able to show that Scone in Scotland is another Sëon or Segon. The Moot Hill of Scone preserved the original, that is Egyptian, meaning of the name, as the Sekhen.[1] It was designated the "Collis Credulitatis" or Mount of Belief. It is called the "Caislen Credhi" by Tighernac (728), which is rendered the "Castellum Credi" in the Annals of Ulster. The Pictish Chronicle, in recording the assembly in 906, says from this day the hill merited its name, viz. the "Mount of Belief."[2] Now the Egyptian name for belief is "Skhen," which also means to sustain and give rest. Thus the Scone Mount is the Sekhen Hill, in a double sense.

The nine maids of the Segon or circular temple have bequeathed their name and number to some stones standing on the downs leading from Wadebridge to St. Columb, which are generally called the "Nine Maids."[3] The legend relates that the nine maidens were turned into stone because they would otherwise keep dancing on Sunday, which riddle is easily read when we know the nature of the nine, and that the birth depends upon their established fixity. Other circles of nine

---

[1] Stuart, *Sculp. Stones of Scotland*, pl. 59.
[2] *The Coronation Stone*, p. 30. Skene.
[3] Borlase, *Ant. of Cornwall*, p. 189, pl. 17, fig. 1.

stones in Cornwall are known as the "Nine Maidens." In Scotland we find the Maidin stone or stones.

We have also the Rekh or Rig of Nine Stones. In "Barthram's Dirge," "They shot him at the Nine-Stane Rig, Beside the headless cross."[1] Near this "nine-stane rig," in the vicinity of Hermitage Castle, was the "Nine-Stane Burn." Also there was the Lady-Well. A most precious preserve of the ancient imagery this of the nine stones, the waters, the feminine fount, the pre-Christian cross; we shall see, directly, the relationship of nine stones to the waters, and the cross without a head.

We are told in the poem on the Graves of the Kymry that they also buried their dead on the shore where "the ninth wave breaks," and here we can arrest the symbol just where it passes into false belief. The ninth wave and earlier tenth does not mean the sea-wave, but relates to the reckoning by nine and ten in the time of ten moons or nine months and a three months' inundation, still manifest in the three water-signs. The water side of the circle was one quarter, and the nine waves, nine stones or nine maids, represented the nine dry months of bringing forth. The ninth wave and the tenth, the nine pins and the ten have their prototypes in the two Collars of Isis, the Gestator who wears nine Bubu or Beads, whereas the Collar of the wet-nurse called MENÂT implies the reckoning by ten water periods of twenty-eight days each, as Ment (or Mêt, Coptic), signifies number ten, and Men-t means liquid measure. The cross without a head is an equivalent symbol of three quarters out of four. So the Put circle of the nine gods contains three quarters filled in and one quarter left hollow, ◐. The horse-shoe images and the head-dress of Hathor like-wise typifies the same three quarters of the circle as the nine stones or the headless cross; the zodiac, minus three water-signs.

The "Nine-Stone Burn" was also represented near Dunstable (Beds). There is an earthwork near the town called the "Maiden Bower" and the "Maidening Burn." The "Maiden" identifies the nine stones when interpreted. It may be noticed that Dunstable stands on chalk hills that have been turned into catacombs by enormous excavations which were made with the most primitive implements of the Bone and Stone age.

The "Maidens" do not derive directly from the word maid, but from the nine, which is both Meh and Mâ in Egyptian. The Egyptian Meh, to fulfil, and Meht, earlier Makht, to be fulfilled, represent the German MAGD, for the maid, in MÄDCHEN and in the Gaelic MAIGH-DEAN, as the one whose period is fulfilled. Makha, to measure, is the earlier form of Meh and Mâ; and the Makht of the Equinox was the Meht of fulfilment in the north quarter. The Ten is the terminus; and the Meh-ten, the terminus of the nine, is equivalent to the name of the maiden, Makh-ten or Maighdean. These circles were

---

[1] Scott's *Minstrelsy*, "Barthram's Dirge."

the seat of the nine whether as the Meh-t or the Put. Taliesin calls himself the "Bard of Budd" who conversed much with men, and as Budd is the Egyptian Put, the divine circle of the nine, Bard of Budd is identical with the poet inspired by the nine, the nine muses or maidens of the Budd circles formed of nine stones. In the Gododin [1] the bard celebrates the fame of the "established inclosure of the band of the harmonious Budd," that is, the Put in the hieroglyphics, the nine. This circle of the nine called Put and Mâ (Meh) is established by Ptah and Mâ in the Egyptian mythology. Ptah is the framer and Mâ the fulfiller. The circle of nine, it is repeated, is based on the nine months of fulfilment in gestation, and on the nine dry months which in Egypt with an inundation made a year.

The "Maid" stones were probably limited to that number, and Meht is the number nine fulfilled. This name is extant in Maidstone. The Maiden Stone in Scotland, and the Maiden Castle, possibly mean the "Ten" (Eg.), throne seat of Meh, the nine. Bridget had her nine maidens, and in her legend as the Virgin Martyr it is affirmed that the castle of Edinburgh was called the Maiden after her. But there were "Nine Maidens," as at Boscawen-ûn, and three other places mentioned by Borlase, consisting of nineteen stones, which have been mixed up with the nine maids. Also the inner elliptical compartment of Stonehenge, within which stood the stone of astronomical observation, consisted of nineteen granite blocks. Now we shall see the further use of the root Meh for nine. Meh (or Mâ) is the number, and "Ten" has different meanings, as Ten, the throne, seat, place, division of the nine. Ten is also our English number, 10; Ten, a weight of 10 Kat, an unity of weight, the ideographic Ten, or sign, formed of two hands or ten digits.

Meh-ten may be read either the nine total or 9—10, our 19, the exact number of Maiden-stones at Boscawen-ûn. Now when we remember that the Metonic Cycle is a period of nineteen years, at the end of which the new moons fall on the same days of the year and the eclipses recur, it is exceedingly strange if it was left for a Greek astronomer of the name of Meton to discover this cycle, B.C. 432. The nineteen Maiden Stones in Cornwall, and the nineteen at Stonehenge, already figured and stood for the Cycle of Meton, or possibly of Mehten, meaning the number nine-ten.

The stones varied in number according to the nature of the circle or Caer. The Caer was sometimes a quadrangular inclosure, then it symbolised the circle with four corners, like that of Yima in the Avesta which had four cardinal points, and was a four-cornered circle. Two of these Caers with four corners, but left open, would be the two houses of the sun, the lower and upper Caers or courses, and these would equate with Sesennu the region of the eight great gods. The circle of nine, whether called a Bedd or Maes How, represented the

---

[1] Song 22.   [2] *Vendidad,* Fargard 2.

nine months of childbirth, and the sun in the nine non-water signs. There may have been a circle of ten stones, which number, as in the ten pins and tenth wave, was superseded by the solar nine. Twelve stones stood for the total of the solar signs, and nineteen for the Metonic or Maiden Cycle. They range at least up to seventy-two, the one-seventh of a Phœnix period of five hundred years. The dead were buried in or around them, but they served the purpose of the living registers and rolls, and were the figures of the astronomical chronology.

The reader will gather from this that the Men-an-tols of Cornwall meant something more than merely holed stones. Ter, the circle, round, to encircle, of course includes a hole, the Cornish Tol, but is more than that. Ter, in the simplest form, is time, the mover in circles, tide, season, limit.

The Men-an-tols were gnomons and dials of time. Max Müller has observed that a Men-an-tol stands in a field near Lanyon, flanked by two stones standing erect on each side. Let any one go there, he says, to watch a sunset about the time of the autumn equinox, and he will see that the shadow thrown by the erect stone would fall straight through the hole of the Men-an-tol.[1]

The name of Carnac, in Brittany, is the same as Karnak at Thebes, and resolves, as Egyptian, into Kar-en-akh, the circle of the dead. It comes to the same thing if we read Carn-Akh, as the Cairn in English ; CRWN, Welsh ; CRUINN, Gaelic ; CERN, Cornish, and CREN, Armoric, denote the Cairn-Circle. Kar is the underworld, underground ; Kar, a chest, sarcophagus or coffin ; Karas, a place of embalmment, a chamber for the mummy. This is the origin of our Kar-stones, from which so many places are named. It is not that Carragh (Ir.) merely means a rock. The Car stone is a rock, but the full form of the rock, as Craig or Carragh, includes the Car (Kar) of the Akh (Eg.) or dead. In that case the Car-akh and Car-rekh have the same signification, as both the Akh and Rekh denote the dead. The Rock is a worn down Caraig or Cleig-stone of the dead. At Carrowmore, in Ireland, a large number of sepulchral remains have been found. The unabraded form of the word is Car-raighea-mora. Mora is a region, land. Kar is the underworld, the sarcophagus, the hole or passage. But it may be questioned whether Raighea does not mean more than rock. In Egyptian Ruka is to hide, to stow away in safe secrecy. We have the form ruck, to crouch down out of sight. LLYCH, Welsh, a hiding-place, and LLECH, to lie flat or horizontal, apply equally to the dead and the flat-stone. So interpreted, Carraighea-mora is the region limited to the sarcophaguses or mummies of the hidden—that is, buried, dead. The part of Arthur's Seat called Salisbury Craig was doubtless a Car-akh Hill.

[1] *Chips*, vol. iii. p. 296.

There is a stone in Aberdeen designated the Craba Stone, and if we apply this principle of formation to its name, Craba becomes Car-akh-ba. Ba in the hieroglyphics is the stone or place of the hidden corse, and " Car-akh-ba " reads the stone or place of the hidden—that is, buried, dead, the final form of which is the Grave-stone, GRAVE being a form of Craba, and Craba an abraded Kar-akh-ba. With the Ba rendered stone there are many Crabas known as Cra-stones. And as Cra alternates with Crow, other stones are called Crow-stones, or Clow-stones. In this transformation of Car-raigh into Crow, we come upon the meeting-place of Rook and Crow, two names of the black long-lived bird of renewal, adopted in our islands, and named after the Egyptian Rekh.

In Cornwall the stones with a circular hole, made use of to pass the children through as a type of new birth, or some kind of cove-nanting, are called CRICK-stones. Crick-stones, they maintain, were also used for dragging people through to cure them of various diseases.[1] This offers us another Car-rekh stone. And we must beware of supposing a compound word like this has but one meaning. In the Crick-stone the Kar (Eg.) is the circle, the hole, and Rekh (Eg.) signi-fies to whiten and purify, therefore to heal. A feminine Rekh (Eg.) is a laundress. The Crick-stone, then, as the Kar-rekh stone, becomes the hole-stone made use of for purification and healing. As the Car, Crow, or Craba-stone it was a type of re-birth ; the grave itself was but a hole of passage, the emaning womb of another life.

Kirkcaldy in the full form is probably the Kar-rekh-Caldy, the circle of the Rekh, who were the Magi, known in Scotland as the Culdees, or, as Kar-rekh becomes the Kirk, known in the same country as the stones, and then the kirk, Kar-rekh-Caldy is the stone circle of the Culdees. Many of the stones are called Leckerstones, as those near Abernethy, the Liggarstone in Aber-deenshire, the Lykerstone at Kirkness. This is the reverse form of Kar-raig, with the L instead of R. Here the name is identical with that of LECKERBAD, the place of the purifying sulphur baths.

Rekh (Eg.), to whiten, wash, purify, in connection with the Crick-stones used for healing, makes it appear probable that the Rocking-stones were employed as Rekh-ing-stones—that is, stones of purification. Roke (Eng.) is to cleanse. Mineral ore is ROCKED in cleansing. The Rocking-stone, says the Arch-Druid Myffyr Mor-ganwy, was the Yoni-stone ; it typified the womb of Kêd, and was called the Ark-stone. In the Mysteries the initiated entered the womb of the mother, were cradled and rocked in it, renewed and born again from it. Rekh (Eg.) means to reckon, calculate, know, and the oscillating or rocking-stone was also used for purposes of divination.

Mr. Botterell, a Cornishman, wrote to one of the papers some time ago, and informed the public that a few years before there was a rock

---

[1] Botterell. Max Müller's *Chips*, iii. 292.

in the town-place of Sawah, in the parish of St. Levan, known by the name of Garrack-zans. This is a dialect form of the Crick and Carraig stones. The word zans is a valuable addition. Sans or Snes (Eg.) signifies to salute, adore, invoke. Sens is to breathe, to breathe the earth, that is, begin to breathe. Ssen, to breathe, pass, begin, has for determinative the slug or snail, an image of the lowliest beginning to breathe the earth. San is also to heal, prepare, preserve, and save. We have it as saine, to bless, and save. SAU in Cornish means health, and denotes healthy. The U and W imply an earlier F, as in save. Sefa (Eg.) is to purify, and Sawah was the place of healing. San-su (Eg.) would signify preserve, heal, charm, save the child, as was done in the process of regeneration and re-birth by passing it through the Kar-rekh or circle of purification.

In the parish of Lansannan, Denbighshire, there was, according to Stow, a circular plain cut out of the solid rock on the side of a stony hill which contained twenty-four seats, and was called Arthur's Round Table. Twenty-four, as the four-and-twenty elders, was a solar number as well as twelve. The Welsh Llan is a shrine, a sacred inclosure. Ren (Eg.) is a symbol of inclosing. San (Eg.) means to preserve and save, also to heal. Nen may be the type and likeness.

TAOURSANAN is the Gaelic name given to the circles of standing stones. It is read "Mournful Circles," or supposed places of sacrifice. The Sanan is one with the Welsh Sannan, and the Llan and Taour, or Ter, interchange. The dead were buried in these "Mournful Circles," and the mournful is extant in the Ter (Eg.), the layer out and mourner.

The conclusion we arrive at here is that the circle of the Sannan or Sanan was the place of preserving the dead, and on that account the other circle through the stone was the symbol of salvation and renewal in the doctrinal sense. The transformation and regeneration postulated for the mummy laid in the womb of earth were applied to the child and the initiate in the Mysteries, and they were re-born from the Crick or cloven stone, the Yoni-stone, connected with the circle of the dead.

Our ancient Menhirs or high stones are named from Men, a fixed stone memorial or monument, and "Her," high, over, above. Mena also means the dead, whence the Minnying-day, or anniversary in which prayers were offered for the dead. According to the Egyptian language, the "Menhir" signifies the stone erected over the dead. The Menhir was a symbol that conveyed a profound meaning. Men (Eg.) is a name of heaven. The Her (her-t), means the image of heaven and of hereafter. Her is also the way, the road, to fly away, leave, go out, ascend. The Menhir was a fixed and lofty memorial of the higher life.

The Men-Ambers, as they are called, through the modification of the K sound, were originally Men-Kam-bers, and the word is commonly spelt Mencamber, or Mincamber, by the Cornish people. In this form the name explains itself. Men is the fixed memorial. Khem (Eg.) is a shrine, and the dead ; ber (Eg.) is the top of the obelisk, the roof of the house. Cam is the name for the ancient earthen mounds and ridges which the Khem (Eg.) as shrine of the dead (Khema) identifies. The Cam-ber, or roofstone over the dead, is our first form of the chamber. Camber is also an English name for a harbour. The Mencambers were harbours of the dead. The oldest chambers, cambers, shrines, are the Cams, mere ridges, mounds, burrows, tumuli on the downs. The Egyptians made some of their cambers and sarcophagi of obsidian, that stone being named Kamu. The greatest weight, of hugest size, of hardest stone, lifted to the fullest height, was the fittest embodiment of their type of Eternal, and this they expressed with tremendous toil in quarrying, hewing, and heaving heavenward their monuments, menhirs, mencambers, and piles vast as Stonehenge or the Great Pyramid.

This meaning of Kam and Khem will account for a place like Camelot, near South Cadbury Hill, in Somerset. As described by Drayton,[1] it was a hill of a mile in compass at the top. Four deep trenches with the steepest of earthen walls enclosed about twenty acres of ground. Egyptian will tell us what for, in the name of Camelot. Kham is a shrine for the dead, and Ret (lot) signifies to retain the form. The Ret is also the ascent or STEPS. Camelot was the shrine in which the dead could best and longest be preserved. Cadbury tells the same tale. It is the bûry, barrow, burial-place. Khat (Eg.) is the corpse, dead body. Khet means shut and sealed ; Khat, the womb, personated by the goddess Kêd. One of the Men-Kambers is described as being a rock of infinite weight, laid roofwise on other great stones, so equally poised that a child could move it, but no man remove it.[2] This would be rocked in the Mysteries. Another enormous stone in Gower was calculated to have weighed thirty tons, erected as the primary type of permanence. Such was the longing for life to be continued, as may be read in the various types of permanence, when we can see through the symbol, whether this be the mummy type perfectly preserved, or thirty tons of millstone grit elevated and suspended, or only a shin-bone split and painted red and buried in a mound of shells.

The immense flat stone was called Arthur's Table. The table of Egypt is the Hept, the sign of peace, offering, plenty, welcome, sunset ; the table of heaven and of the sun, heaped with food.[3] This was Arthur's Table, and Camelot, the lofty shrine of the dead, was but this table on a larger scale, round which the gods were figured sitting at the eternal feast.

[1] *Poly-olbion*, Notes to Song 3.   [2] *Chips*, vol. iii. p. 289
[3] Herodotus, iii. 18,

The stone monuments of Britain are none the less Druidic because their likeness is found in other lands. They are some of the scattered remains of the primitive cult, relating to the keeping of time (Teru) and the preservation of the dead. They are the dumb witnesses to the human desire for continuity, which attained such profound and persistent expression in the Egyptian art of symbolizing the mummy as the type of self-continuity.

In England the grave was formerly called the "pytte," and the same name was given to a well with an intermittent spring; over this well the enormous flat stone of Arthur was elevated and suspended as at Kefn Bryn in Gower, where a vast unwrought stone, from twenty to thirty tons in weight, was supported by six or seven others over a well which had a flux and reflux with the sea. Here the well and grave were one in the pytte, as they were in the Great Pyramid or the Masteba of Egypt.

The interior of each tomb consisted of three parts, typical of the vault and void of the two heavens, and the middle earth or passage between the two, called by excavators the Serdab. The void was the well containing the mummies in the underworld. The open chamber typified the upper world of the future life, where the deceased sat at the celestial feast surrounded by his friends in his eternal home. When the friends in the earth-life come to visit their dead and bring their offerings, these are representative of contributions to the feast; the life above being the reflex image of the life below. In the passage between, or the Serdab, was placed the sepulchral image called the Shebt or double, the type of transformation from the one life to the other. This had the same significance as the scarab emblem of Khepr, the beetle, that went underground to make his change, and to issue forth once·more in the shape of his own seed. The Serdab was the place of Sem-sem or the re-genesis, and the only communication between it and the rest of the tomb was a small hole scarcely large enough for the hand to pass through. This usually opened toward the north, like the entrance to the Great Pyramid. It was the place of egress from the womb, the Mest of the Masteba, and has its analogue in the hole-stone of our far ruder and far older structures. Mariette describes the Masteba as a "sort of truncated pyramid built of enormous stones and covering, as with a massive lid, the well at the bottom of which was the mummy."[1]

Our primitive sepulchres were open to the passers-by, as were the Egyptian Mastebas, in which the friends of the deceased deposited their offerings or came at times to pray and hold their feasts of the dead on the anniversary day. The Masteba was the chapel over the grave or pit, representing the underworld. It contained the table on which the contributions were deposited.

---

[1] *Monuments of Upper Egypt*, p. 73.

In the case of Arthur's Stone, the slab was the table, and the large stones still bear evidence of the offerings that were made as well as the mode of offering.

At Bonnington Mains, near Ratho, there is a cromlech with cups, bowls, and basins in the cap-stone. The cap-stone is a reminder that the cap, roof, top, is the Ben in Egyptian, the cap or roof of a monument. Benen (Eg.) is also a surname of the Horus of Resurrection. The Benn is the Phœnix, another type of re-arising. The cups were hollowed on the outside of the covering, the cap-stone, so that, if no longer filled by friendly hands, they would still catch the rain, a type of the water of life besought by the builders of these monuments uplifted towards heaven as their petrified prayer.

Arthur's Flat-stone laid over seven others with the well beneath corresponds to the most colossal Masteba of Egypt. For the Great Pyramid is an enormous Masteba, and it contains seven chambers with the deep well underground. The oldest form of the pyramid known in Egypt is found at Sakkarah, which has seven steps like the Babylonian towers. In this form the seven steps correspond to the seven chambers of the Great Pyramid, which has the mystical number within instead of without. Arthur's Stone was supported on six or seven other stones. We may be sure the correct number was seven.

In the hieroglyphics the number seven is Hept, and the same word signifies the table of offerings, the HEAP of food, the shrine, the ark, and peace. The earlier form of Hept is Khept, the Goddess of the Seven Stars, and it is here claimed that the Seven-Stone, or stone supported by seven, or the seven-tiered tower, the seven-stepped or seven-chambered pyramid, represents the birthplace personified by the genitrix who was Khept in Egypt and Kêd in Britain. From this it follows that the British Masteba is of an earlier type than the Great Pyramid of Ghizeh or the more ancient Pyramid of Sakkarah. The number seven is also connected with the name of Arthur in the form of seven companions in an ark. One of the Druidic stones is known as the Seven-Stone. The monument in Llan Beudy parish, or the house of the ox (sign of the bull), shows that Arthur's Table was identified with "Gwal y Vilast," the couch of the greyhound bitch, that is, the couch or lying-in chamber of Kêd. In this place the flat-stone or table supported by other stones is only about two and a half feet high.[1] This then was a burial-place that represented a birthplace, the birthplace of the divine child Arthur, and abode of re-birth, variously called the Cell of Kêd, Maen Llog, Llogel Byd, Maen Ketti, the Ark-Stone, and the Stone of Keridwen, known to-day as representing the womb of the Great Mother.[2]

The AFT, couch and name of the goddess Aft or Fet, is repeated in the Cornish VETH for the grave, and Gaelic FUADH, the bier.

[1] Gibson's *Camden*, col. 752.
[2] *Letter of Myffyr Morganwy*.

The Khet or Kat, seat of the mother and her child, became our Cat-stone, often supposed to denote a place of battle. The cat-stone is the stone of Kêd, the genitrix, and marks the birthplace of her child, whether Sabean or solar. Cat, the French Chat, is the Egyptian Kat. This seat was the mount of the Great Bear in the earliest time; afterwards it was turned into the Bekh or birthplace in the rock of the horizon when the zodiac was formed.

The seat in Egyptian is the Khet, with steps denoting an ascent, and the Kat, a seat or throne. The latter is a conventionalized lioness, which was used as a palanquin or portable throne, with considerable likeness to Arthur's Seat. The seat is the feminine abode; the same words signify the womb, the seat, or Kat of the child. Thus Arthur's Seat is synonymous with Arthur's Stone at Kevn Bryn or Arthur's Table, or Arthur's Quoit, as the symbol of the mother, who was the habitation (Kat or Hat) of the child. Hence the lioness or the lioness-shaped portable throne was a type of the bearer.

At the foot of Arthur's Seat lies Duddingstone. Tut (Eg.) is the throne, image, or region of the eternal. Tattu was the established region in this sense. And in Tattu was the rock, the Tser Hill, the Hebrew Tzer. Duddingstone may be named from the stone of establishing, the type of the eternal identified as Arthur's Seat.

Our word "dole" is the same as the Egyptian Ter. Dole is to divide or separate, portion, tell, mark out. The Dole-stone is a land-mark or bourne. Dole is to lay out and grieve. Ter (Eg.) is an extreme limit, boundary; ter, to indicate; ter, a quantity; ter, erect a limit; ter, a layer out, or mourner.

Men (Eg.) is a monument, a stone of memorial; Men-a, death, or the dead; Men-t, a bier; Men, to arrive and rest. These sufficiently identify our Dolmens as places of burial, but whereas the cromlechs may have been cemeteries, the dolmens seem to have been marked off as more especially individual tombs. The dolmen is, however, the same word as the Irish TERMON, and the Toda Dermane, a god's house or residence of gods. Inside the inclosure or Dermane there was a round tower called a boath, a kind of Pictish tower or conical temple. "Round about the Boath," says Marshall,[1] "there was a kromlech, and numerous stone kairns dotted about with the outlines of stone walls on a large scale surrounding all." The Dermane was also named a Gudi, i.e. temple. Kudi, in Sanskrit, is a house, and to curve round. Kudu or Godu, in Toda, is to collect together; Kattu (Tamil), build, bind, bond; KETUI (Eg.), a building, a circle.

In Ireland a small piece of ground fenced off round the church was in some places called a TERMON. It was land belonging by sacred right to the church, and to this Termon the criminal and other fugitives could flee for refuge, and were held in safety for a time when once within the prescribed boundary. The phrase "Termon lands" is

[1] *A Phrenologist among the Todas.*

common in Anglo-Irish writings. The Termon of course agrees with the Latin terminus, but that does not explain the right of refuge. The full significance of the Termon lands and sacred boundary can alone be found in the fact that it was the dead who protected the living within their own domain, and that MENA (Eg.), denotes the dead, and the TER (Eg.) is the limit, boundary, the word also meaning to hinder. The Termon was the boundary-limit within which the dead were allowed to hinder the further pursuit of those who sought sanctuary from justice or from their foes. It was the dead who conferred a right of refuge, and formed an asylum of their sanctuary to the criminal or debtor who fled to them for protection from the living, and in this sense the precincts of Holyrood House were a termon-refuge for the debtor on Sunday. The Termon is extant in Termon Castle, an ancient residence of the Magraths, also called "Termon Magrath" in the "Four Masters." The Magraths were hereditary wardens of the Termon, and in this we have another allusion to a Termon, founded on the charge of the dead, the sanctuary of the dead and living, like that deduced from the name and customs of Caistor church. Dr. Joyce says the Termon in several places shows the former existence of a sanctuary. The O'Morgans were the wardens of Termonomorgan in the West of Tyrone. Mer (Eg.) is a superintendent, and the Khen (Eg.), would signify the sanctuary of the dead. The Termon suggests that the tors of Devon, the rock-towers, the natural round towers or Turagans, may have been early places of sepulture.

Mis Tor is in Devon, and Mes (Eg.) denotes the birth or re-birth of the dead in the Meskhen and, it may be, in the Mes-Tor. Yes-Tor and Hessary-Tor (Devon) possibly represent a Kes (Eg.) tor; this being a burial and embalmment, at which point the Kestor and Kester would meet, the Tor being the natural mound and type of the later Kester, Castra, and castle, when the sanctuary and defence of the dead was turned into a place of defence and offence for the living.

Ketui, in Egyptian, is an orbit, circle, with determinative of house and plural sign. It is literally Ketui-house built circularly, our "Ket's Coity Houses," Khet meaning in Egyptian shut and sealed. The Ketui is the Gudi of the Todas, the enclosed temple and place of burial, exactly as our churches stand in an enclosure amongst the dead. The Toda enclosure was crowned and typified by the Boath, the shape of which, as of the Picts' towers, is preserved in the extinguisher. This Boath, God's house or residence of the gods, is the same word as the Assyrian Bit, Hebrew Beth, Scottish Bothie, Egyptian Paut and Pauti, lastly Put, the circle and the company of nine gods; the hieroglyphic being a circle three-fourths or nine-twelfths filled in. Some of the stones were called Coits; this name is preserved in the Quoit or ring. Ket's Koity House is the Koity, Coit, or Quoit, as the circle of the goddess Kêd. This circle of the goddess Kêd was a reality in spite of the Arkite lunacy of Bryant, Faber, and Davies, and had its physiological and astronomical prototypes.[1] Khet, in Egyptian, is the secret, intimate abode. Khat is the womb, the secret, intimate abode of the creative powers on the physiological plane of the myth, and in the astronomical or eschatological stage, the ark, the circle, called by the name of Kêd. Koity fairly represents the Egyptian Ketui, the circle, orbit, or quadrangular Caer. The circle Ketui or Coity was the same as the KIBNO-Kêd, the Kafn (Eg.), or oven, the baking-place of the mother of corn or bread, and of the " Pair Keridwen " of the Barddas. But, whereas the earliest type was the cave, a natural formation, the stone circles and enclosures had to be erected, and Ketui (Eg.) means built. Raising the stone of the Ketti was one of the three mighty labours of Britain.

Our " Ket's Koity" is Kêd's Ketui. The Welsh GWAITH (as in Gwaith Emrys) means work, labour, workmanship, identical with KAUTI (Eg.), work, labour, especially to carry and to build. Gwaith Emrys (Stonehenge) is thus an enormous koity-house of Kêd, the bearer. Also GWAITH (Welsh) signifies the course, turn, or time, and this is the Egyptian KETUI, an orbit, circle, or course of time, showing the relationship of the building to Time as well as to the dead. Excavations made in the neighbourhood of Ket's Koity House showed that it was a burial-ground full of sepulchral chambers in groups, each single group being generally surrounded by a circle of stones.[2]

About five hundred yards from the particular stones called Ket's Koity House is another monument, named the Countless Stones, and there are indications that the stones in this neighbourhood were countless. Ket's Koity House is but a perverted form of Kêd's Koity Hows, the hows or circles of Kêd, the Great Mother. Even without the name of the goddess, the words " Khet," to be shut and sealed, " Ketui," a circle of stones, an orbit, still suffice to identify the HOWS as the enclosures of the dead.

Khent, in the hieroglyphics, is a garden, and the English Kent is still called the Garden of England. Kent is our south land, and Khent is the name of an unknown part of Egypt, but it was ob-viously one with the south, the way of the Inundation and source of fertility. Horus, as Lord of the South, is designated the Lord of Khent.

In the Annals of Rameses III., the king, in an address to Ammon, says, " I made thee a grand house in the Land of Khent."[3] This is mentioned as one of the four quarters along with the north, east, and west. The Grand House in the south erected by the Kymry appears to have been represented by " Ket's Coity."

Both in Egyptian and Welsh, Kêd or Khet signifies the enclosure. And this is applied also to Emrys, as Gwath Emrys or the enclosure of Emrys, which is Stonehenge. The name of Emrys is yet extant

[1] Davies, *Myth.* 402.   [2] Wright, *Wanderings of an Antiquary,* 175.
[3] *Records of the Past,* vol. vi. p. 26.

on the spot, though transformed into Ambrose in the re-christening. It is also known as the Circle of Sidi and the structure of the revolution, that is, of the celestial bodies. Res, in the hieroglyphics, means raise up, watch, with the ideograph of the heavens. Am (Eg.) signifies to discover, find out. Am indicates a residence with a park or paradise, that is, an enclosure. So interpreted, Gwath Emrys may be the enclosure of the watch-tower, observatory, or the stone of astronomical observation.

Hor-Apollo [1] tells us that the scribes of Egypt have a sacred book called AMBRES, by which they decide respecting any one who is lying sick whether he will live and rise up again, ascertaining it from the recumbent posture of the patient. In Egyptian, AM-(P)-RES would read "to discover the rising up," and this would equally apply to the celestial bodies. One wonders whether our Emrys, Ambres, or Kambers, may not include the rocking-stones raised up (res) for purposes of divination or discovery. Am, to find out, discover, has an earlier form in Kem, with the same meaning.

Another of the three mighty labours of the Island of Britain was building the work of Emrys, later Ambres. Dinas Emrys was the sacred place in Snowdon. Emrys is said to have been a sovereign at the time when Seithenhin the drunkard let in the Deluge. A character in the British mythology, a supposed prince, who fought with Hengist, was Emrys or Ambrose, called the president and defender of the Ambrosial Stones.

Stone-Henge is the Stone-Ankh, the living-stone. As the vocabulary shows, we have the English equivalent of the Egyptian Ankh. Ankh, to clasp, to double, is imaged in our hank of thread, a double loop tied or crossed in the middle. Hank is to tie. A Hanger is a fringed loop appended to the girdle for the small sword, and the Egyptian Ankh was used as the buckle of a girdle. The Ankh symbol was the ideograph of life, and united in one form the cross and circle.

This Ankh sign is the original of Stonehenge; every upright and horizontal stone made the figure of the cross all round the circle itself: that was the Ankh. It was built of stones: that was the stone Ankh. The stones were of the hugest size and the most enduring that could be found: this made the stone Ankh a colossal image of eternal life. Ankh, the living, was also preëminently applied to the departed. Such is the signification of Stone-Henge, read by Egyptian. The fact that hang also means to suspend, and these stones were partly suspended, may be thrown in. Stone-Henge was a topographical and typical form of the Ing, enclosure.

In Welsh, ANG denotes the open capacious place for holding and containing, it may be embracing, which agrees with Ankh (Eg.), to clasp. The stone-hank has its analogues in the Persian KANK, or temple, and YANICK, a grave; YINGE, Zulu Kaffir, a circle; Chinese

[1] i. 38.

Ying, a sepulchre; Italian Conca, a tomb or burial-place; Ying, Chinese, a kind of bracelet; YING, Chinese, a kind of necklace; INGU, African Ako, a circlet of beads; KUNK, African Dselana, a bracelet; KHEUNG, Chinese, a stone bracelet; CINGO, Latin, to environ or surround.

It is quite possible that the horseshoe and circle of foreign stones within the outer circle of Stonehenge represented the earlier temple belonging to the Great Mother and her starry son. If the triliths surrounding the inner ellipse were, as some authorities affirm, seven in number, they would form the perfect figure. If there were only five of them, the ten uprights would still illustrate the Sabean-lunar reckoning, which was superseded by the solar nine. The outer range would represent the temple of the sun. Thus we have the Emrys, or Stone of Observation; the nineteen stones of the luni-solar cycle, the seven triliths (or ten uprights) corresponding to the seven stars, or the planetary seven, with the outer circle representing the addition of the later solar reckoning. The development of the Cult will account for the two periods apparent without implying two different races of builders. We may take the disk-shaped barrows of the Bronze age, for instance, to be typical of the solar circle, the latest of three, stellar, lunar, and solar, corresponding to the Palæolithic, Neolithic, and Bronze periods.

Stone-Henge or the Stone-Ankh was the great national tomb-temple. Sir Richard Colt Hoare counted 300 tombs round Stone-henge, within twelve square miles, and in Stukeley's time 128 were to be seen from a hill close by.[1]

The Cursus or course at Stonehenge into which one of the avenues leads is called the " YSTRE "; it is half a mile from the temple itself, and consists of a course ten thousand feet or two miles long, enclosed by two ditches three hundred feet apart.[2] The YSTRE is mentioned in the "Gododin," a poem ascribed to the bard Aneurin. It has been already shown that the STER (Eg.) is the couch of the dead. The word means the dead-and-laid-out, to lie on the back, be laid out together, and is determined by the lion-couch of the dead. The Yster is either the Ster uncompounded or a worn down form of the Khi-ster or Kester. It has been assumed that the Sters, of which there are many in Caithness, as in Stemster, Shebster, Lybster, Ulbster, Seister, Scrabster, Thurster, are derived from the Scandinavian " Saetr," the name for a farm. The Egyptian " Ster," however, has now to be taken into account.

This meaning of Ster (Eg.), to lay out, the place of laying out, and of the dead laid out, renders unnecessary the assumption that three out of the four provinces of all Ireland, Ulster, Munster, and Leinster, were named as settlements of the Norsemen, from the seat or dwelling called a Saeter, as a farm or homestead. They were

---

[1] *Archæologia.* xliii. p. 305.　　[2] Maurice, *Ind. Ant.* vol. vi. p. 120.

neither laid out nor settled nor named by the Scandinavians. The dead were the first laid out, and their burial-place was the primitive Ster. The first minister was probably the Mena-ster as the layer out of the dead, the Minster being the later place of laying out on the couch. Munster may derive its name from the place of the dead, the commonest starting-point of the living. Leinster would thus be the Llan-ster, the enclosure of the laid-out dead, which afterwards became the church as the Llan. This, of course, is not the only possible mode of naming the province. Ster is to lay out. Set is the Egyptian name of a nome, and the r (ru) means a mark of division, which in the Stour is a boundary river, and still the three Sters are independent of Norse naming. The oldest spelling of the name of LEICESTER shows that the place was the KESTER of the LEIC, or laid-out dead. "MANCHESTER" is probably a kester of MENA (Eg.), as in MINSTER, the ster of the dead.

The Stool, the lowly seat or rest for the feet, is an extant form of the Ster, couch. The redstart is the redtail, which is long and stretched out, as it is in "Start point." From this Ster, latter end, comes the stern of the vessel. In one instance it is the tail of the bird, in another of the vessel, and in another it applies to the end of life. And from Ster, to lay out, extend, &c., we probably derive the ster terminal in maltster, seamster and webster.

The "Ster," as the act or place of stretching out the dead in burial, has particular significance when we call to mind that the men of the Stone age, Palæolithic and Neolithic, did not lay out their dead, but buried them in a sitting and contracted posture, with bent thighs, their heads resting on their arms, and faces turned towards the daylight world beyond the mouth of the cave. Instead of laying out the dead, the cavemen folded them somewhat in the manner of Peruvian mummies, and left them in an attitude the exact opposite to those of the Ster. The tomb being founded on the womb, this will at once suggest that the contracted crouching posture was adopted in imitation of the fœtus, and the dead sitting in their caves were arranged according to the likeness of the child in the womb.

The reader has but to refer to the ground-plan of the chamber in the round cairn at Camster, Caithness,[1] to see the likeness to the uterine type. The figure is that of the vagina and womb, which exists in a more conventionalized form in the hieroglyphic Kha ⟺, the ideograph of the Khat, the belly and womb, and Kha was the name of the Adytum of Isis, formed on the feminine model. Khem (Eg.) is the shrine, and Ster means laid out, dead.

The "Ster" of Caithness alternates with the name of CAS or KEISS, as Sinclair Cas, Dunbeath Cas, Berriedale Cas. Kas (Eg.) is the burial-place, the coffin, and denotes embalmment and burial, and in Berriedale Cas we seem to have the proof that the Cas has this

---

[1] *The Past in the Present.* Mitchel, Fig. 51; see also figs. 48 and 49.

meaning. The Welsh Cas occurs in Cas-Llychwr (Loughor), where there is a Roman altar. The Gaelic "COS" is a hollow scooped out of the hillside for a kind of dwelling, a very primitive habitation, as it may also be made in a tree.

The tree was an early kind of coffin. This type of the Great Mother, who personified the tree of life that bore the child as the branch, was likewise made use of in death and burial, and a scooped-out tree, a "Cos," would be the Kas (Eg.) coffin. The Kas is the lowly dwelling-place of many languages, always traceable, like the Khem or Khen, to the birthplace. It is the KHEPSH (Eg.); GUSA, M'barike; KOSOA, Guresa; the QUISSE or COISSE of the French euphemized as the thigh, and as the hip in the Gaelic CEOS and Latin COSSA. The Kas is represented by the Latin Casa or hut-house, as in the Casa Santa at Loretto; the COSH, English, a cottage; CHEZ, French house, home; also CHOSE, peculiarly applied; QUESSA, Quiché, a nest; GAZA, Persian, small hut; KHUSS, Arabic, house of reeds; SAS, Romany, nest; SOZ, French Romance, an enclosure. The KAS, as burial-place, supplies the names of Egyptian cities, as in KAS-verver and KAS-kam, opposite to Antæopolis, therefore on the western side of the Nile, the side of the tombs. Kas-khem denotes the funeral shrine. Kesslerloch is the name of a cavern of the cavemen near Thayingen, Switzerland. Cayster was a name of the ancient plain upon which Ephesus was built. That is Keph-ster, Kak-ster, or Kas-ster, the Ster of the Sanctuary. Keswick is a KAS re-named as a wick; there was formerly an oval at this place containing forty stones. At Cissbury, on the South Downs, near Worthing, there is an ancient British camp which was also used by the Romans. It is excavated with regular shafts and galleries. There is another at Chisbury, in Wiltshire. These have nothing to do with the Saxon Cissa. The Bury as in Mena-bury Hill (Herts), near Aldbury, does but repeat the Ciss or Kes, the burial-place. No doubt the excavating for flints and iron-stone led to the formation of some of the chambered tombs.

The CHEESE-wring at Liskeard is a Kas-ring, or circle of the dead. The Wring is a place where cider is made, and not inappropriate for the place of the dead who were transformed into spirits. So the Egyptian name of the sanctuary Kep also means to ferment and turn into spirit. The Cheese-wring is a mass of eight huge stones, rising to the height of thirty-two feet. They have now the appearance of nature's handiwork alone, like the rocks at Brimham, in Yorkshire, probably on account of their extreme age. Sufficient time has never yet been allowed for a true judgment in the matter.

In the language of Wordsworth :—

> "Among these rocks and stones methinks I see
> More than the heedless impress that belongs
> To lovely Nature's casual work ; they bear
> A semblance strange of power intelligent,
> And of design not wholly worn away."

Also, at times, the names of these stones are very arresting. One of these groups, supposed to be the effect of some convulsion of the earth, is named "Kilmarth" Rocks (Scotland). Of course the marth may denote the old word mart, for wonderful. But the stones erected or hewn by human hands belong to the dead, who, in Egyptian, are the Merti. Kar-Merti signifies the circle or underworld of the dead, and this was kept by the dog Dor-MARTH, the British KERBERUS.

From Kas (Eg.), the funeral, to embalm and bury, comes Kast (Eg.), the coffin, the enclosure of the body. This is our Kist, and Kistvaen. FENNU (Eg.) is dirt or earth ; English fen, mud, mire. The Kist-vaen would thus be the burial-place underground, or the earth-coffin.

Considering the importance of the burial-place as the point of impinging on the earth, the centre of the living group from the Llan up to the city, it is extremely likely that the Russian Gostinoi-Dvor of every large town is derived from Kas (Eg.), to embalm and bury, and Kast, the coffin or burial-place. This would account for its universal character as the bazaar, the meeting-place, analogous to the church amid the dead, the sacred place of meeting. We have the "Cos" as the tree-coffin; the Kistvaen as the earth-coffin ; the Cas-Llychwr of the Welsh burial-mound, the Casses of Caithness, and in the Mount of Belief at Scone, the "Caislen Credhi," where the word "Caislen" includes the Llan, enclosure, of the Kas, coffin (Eg.), funeral and burial, identified with the Mount of Belief.

It was at a place named Keiss, in Caithness, that the burial-mound was discovered near the harbour, containing the implements of stone and bone belonging to the Palæolithic age. Rude sepulture had there been given to human bones supposed to have been previously split to obtain the marrow for eating.[1] We now claim the mound at Keiss as a most primitive form of the Kas (Eg.) or Kester, a place of preservation for the buried dead.

Castallack Round was an ancient circle, destroyed of late years, like so many others yet to be grieved for in vain. It stood in the parish of St. Paul's, Cornwall. Kes-ter-rekh, or Kes-ter-akh, the Egyptian equivalent, shows the Kester of the dead, and as Lack denotes stone, the Castallack is the stone circle of the laid-out dead.

Roskestal is another name containing the Kes-ter, the circle of the dead. At Roskestal was one of the Garrack-zans, as at Sawah. Ross adds another Egyptian element to the rest. Res means to raise up, to watch ; Ras, the south. The Castallack Round opened with a doorway to the south. And there in the south, the place of the summer solstice, where Khepr made his transformation in the sign of the Crab, the Egyptians had located the land of eternal birth, which the sun reached on the 30th of Epiphi, our Midsummer, the year began anew, and the spirit was "at peace in its place, full at the fourth

[1] *Prehistoric Remains of Caithness,* S. Laing and Professor Huxley.

hour of the earth, complete on the 30th of Epiphi," and the person of the spirit (Eg.) was then in presence of the gods.[1] " He has his star, or shade (or soul) established to him, says Isis, in heaven at the place where the goddess Sothis is. He serves Horus in Sothis. He becomes as a shade, as a god among men. He has engraved a palm on his knee, says Menka [or Maka, Irish Macha]. He is as a god for ever, reinvigorating his limbs in Hades." [2]

This theology was known to the Kes-tel builders. Ros-Kes-tal is the raised circle of the embalmed or buried dead. The burial-place was lifted up, as it were, in the arms of the mother Earth, and the outlook turned south to the land of eternal birth. The pathos expressed on the face of these early ideas, when we have lifted or seen through the veil of symbol, makes the heart ache.

One thinks the divine consciousness must surely feel a parental love for this our world and all its creatures in it, if only for the upward yearning of humanity in its infancy, the touching appeal of these primitive ideas and emblems in which the early men portrayed their deep unquenchable desire to nestle nigh and nigher to the ever living heart of all! And, as death was one of the first, profoundest teachers of man, it would be ghastly strange indeed if it had nothing to reveal after all, as the unknowers assume to know and assert, but a death's-head horribly agrin, as the type of the eternal, and this universal abode of life, were but a vast, hollow, eyeless skull, with no sensorium of consciousness within.

The prose Edda also says, " At the southern end of heaven stands the palace of Gimli, the most beautiful of all, and more brilliant than the sun," possibly because it was pre-solar.

One name is frequently repeated in connection with the stones in the forms of Rath, Roth, and Rut. Rath-Kenney, Meath, is the seat of a cromlech. There is one also at Ratho in Midlothian. At Rothiemay, in Banffshire, there are remains of a stone circle. In Rudstone churchyard, there is a fallen monolith which once stood twenty-four feet above ground, and has been calculated to have weighed forty tons. Ruthven in Forfarshire, Ruthin in Denbighshire near which is the "hill of graves," Ruthwell, and many others of the same name, are all places where the stone monuments are found. With the interchange of the letters R and L, it still holds good as at Lethham Grange, and Linlathen. Indeed, the Lothian Hills themselves, with the numerous remains and hut circles on their summits, appear to derive their name from the same origin.

Rat (Eg.) is a stone, a hard stone, a carved stone ; the word means to engrave, cut, plant, to RETAIN THE FORM. To retain the form was the object of the stone hewers and carvers. Mummifying was a mode of retaining the form. Burial in high places, in dry ground, in stone coffins and beneath stone covers, was intended to preserve the form.

---

[1] *Ritual*, ch. cxl. Birch.　　　　[2] Rubric to ch. ci., *Ritual*, Birch.

The Rath-mounds were chosen or made artificially, and circumvallated for the purpose of protecting and retaining the forms of the dead. Also the writing of the name of the deceased on the gravestone is an individualized mode of doing what was formerly done *en gros*.

This naming may be followed by the aid of Ren or Lin (Eg.), to name. Thus a name like Linlathen indicates the place of the stones (Rat or Lath), which retained the form of the dead in the mounds and the tumuli, or their memory in the mass, ages before the individual was separately distinguished by the name cut on his own tombstone.

One of the largest carved rocks found in Northumberland is called the Rowtin-Linn Rock. It contains fifty or sixty ring-cuttings and over thirty cup-cuttings—to quote the phraseology of Sir James Simpson. Rowtin-linn as Rat(en)-Renn—the linn here retains the double N, and represents the form of renn, to call by name—denotes the carven stone of naming. The mode of naming is of course symbolical or hieroglyphical, and is ancient in proportion to its rudeness. If they aspired to an individual record, they had not in those times the means of securing it, but there was a general record at the centre of each group of people, or appointed place of burial.

Some of the stone buildings of our goddess of the north were of the same simple, rude, massive type as was the temple of Buto or Uati. There was a Druidic stone at Locmariaker reputed to weigh 260 tons. These enormous stones were raised up and supported on other stones, and one of them in Cardiganshire was called the flat stone of the Giantess. The " Maen Ketti " shows that the one of the "three mighty labours of the island of Britain," called " lifting the stone of Ketti," refers to these suspended stones.

Herodotus observes, " Of the oracle that is in Egypt, I have already made frequent mention ; and I shall now give an account of it, as well deserving notice. This oracle in Egypt is a temple sacred to Latona, situated in a larger city, near that which is called the Sebennytic mouth of the Nile, as one sails upwards from the sea. The name of this city, where the oracle is, is Buto, as I have already mentioned. There is also in this Buto a precinct sacred to Apollo and Diana : and the temple of Latona, in which the oracle is, is spacious, and has a portico ten orgyæ in height. But of all the things I saw there, I will describe that which occasioned most astonishment. There is in this enclosure a temple of Latona made from one stone both in height and length ; and each wall is equal to them ; each of these measures forty cubits : for the roof, another stone is laid over it, having a cornice four cubits deep. This temple, then, is the most wonderful thing about this precinct." [1] The temple of Latona made from one stone is the type of the ark of Kêd ; and as the one was an oracle, so doubtless was the other. It represented the birthplace, and the place of new birth, and was consequently used by the Druids

[1] B. ii. 155.

and diviners as the place of consultation and for the utterance of their teachings.

The next most wonderful thing to the oracle of Buto seen in Egypt by Herodotus was, he tells us, the "Island of Chemmis," situated in a deep and broad lake near the precinct in Buto. "This is said by the Egyptians to be a floating island, but I myself saw it neither floating nor moving, and I was astonished when I heard that there really was a floating island. In this, then, is a spacious temple of Apollo, and in it three altars are placed; and there grow in it great numbers of palms, and many other trees, both such as produce fruit and such as do not. The Egyptians, when they affirm that it floats, add the following story. They say that in this island, which before did not float, Latona, who was one of the eight primary deities dwelling in Buto, where this oracle of hers now is, received Apollo as a deposit from the hands of Isis, and saved him, by concealing him in this which is now called the floating island, when Typhon arrived, searching everywhere, and hoping to find the son of Osiris. For they say that Apollo and Diana are the offsprings of Bacchus and Isis, and that Latona was their nurse and preserver; in the language of Egypt, Apollo is called Orus; Keres, Isis; and Diana, Bubastis. Now, from this account, and no other, Æschylus, the son of Euphorion, alone among the earlier poets, derived the tradition that I will mention, for he made Diana to be the daughter of Keres. For this reason they say that the island was made to float. Such is the account they give."[1]

We also have an island of Buto, and the account furnished by Herodotus affords us the means of comparison and identification of the island in the north which was described by Hecatæus and reported for us by Diodorus Siculus in his chapter on the Hyperboreans.[2] He tells us there is a British island opposite the coast of Keltica, lying to the north, "which those who are called Hyperboreans do inhabit. They say that this island is exceedingly good and fertile, bearing fruit twice a year. They feign also that Latona was born in this island, in regard whereof Apollo is adored above all other gods. The men of the island are, as it were, the priests of Apollo, daily singing his hymns and prayers, and highly honouring him. They say moreover that in it there is a great grove or precinct, and a goodly temple of Apollo, which is round and beautiful with many rich gifts and ornaments, as also a city sacred to him, whereof the most part of the inhabitants are harpers, on which instrument they play continually in the temple, chanting forth hymns to the praise of Apollo, and magnifying his acts in their songs. These Hyperboreans use the proper language of the Greeks, but they are especially joined in league of friendship with the Athenians and Delians, for they say that certain Greeks came in times past to them,

---

[1] Herodotus, ii. 155.   [2] B. iii. ch. xv.

and in their temple presented divers sumptuous gifts inscribed with Greek letters, whereupon one of them, named Abaris, passed into Greece and confirmed the amity which a long time before was contracted with those of Delos. Now they which command in this city and preside in the temple are Boreades, the progeny of Boreas, who hold the principality by succession."

The name of the Boreades would seem to have travelled still further north and to be extant in the Hebrides. It has been erroneously supposed that the island was England, but it is self-identified by name and the mythological scheme as Bute, one of the seven isles of Buteshire, the namesake of Buto, both being sacred to Latona and Apollo. Bute lies off the Keltic coast of Scotland, as the Kelts or Gaels were then reckoned. Moreover, it has in Arran the twin island, which was called Chemmis in Egypt, and was known as the floating island. Aren is an Ark-Island, Aren being a name for the ark, therefore it represents the same floating island of the ancient symbolism. Also the seven isles of Bute are a form of the sevenfold seat of the goddess of the north and the seven stars.

The Irish goddess of wet or moisture is one with Uat by nature, and as the divinity of Buta-faun, the temple of Buta, the present Buttavant, in the county of Cork, she is likewise identical with Buto. Butafane is the temple of Buto; the goddess was known to the Irish as Be-Baiste, and Peht, a form of Buto, was the divinity of Bubastes. Moreover, BITH or PEITHO is a name of Venus in Gaelic, and Buto is the Egyptian Uati, Goddess of the North, a humanized form of Khept, British Kèd, whose name of Wen or Ven, in Keridwen and Ogyrven, represents that of the Greek Venus, and Irish OINE.

A floating island was an early form of the ark, a means of crossing the waters mentally or actually before boats were launched or bridges built. This constituted the land of life in the deep, the Ankh-land or Inch. Herodotus describes the floating island called Chemmis (the shrine of birth) in the lake at Buto, in which Latona saved Apollo when pursued by Typhon. That island was the Ankh-land. It was on account of this origin that the natural floating islands of the lakes were objects of great reverence and religious regard.

The tree-coffin, the boat scooped out of the tree, the Wrn (Aren) Cwch, Coracle, or ark, the cave in the mount, the beacon hill, the couch of Kèd, the bed of Tydain, the seat or quoit of Arthur, the ship of the earth, the Kak sanctuary or Skhen shrine, the Kas and Kester, Tom and Tun, stone cell and cromlech, Kistvan and Ket's Coity House, the Roundago, Mencamber, Kibno, the circle of the nine maidens, of Anoeth, of Sidin, Cor-Kyvoeth (Stonehenge), or Camelot, were each and all types of the mother to whose bosom the dead were committed for burial and re-birth; to these may be added the Island of Bute.

The Druid Bedds, circular sanctuaries, sacred to Tydain and Kêd, were cemeteries, as Beddau are graves. In those formed of nine stones, the tomb was just the womb. The bed in English is the uterus. This was the Egyptian Put, the divine circle of the gods ; and the bed of nine stones was its ideograph. Thus the dead were returned to the place of birth to await their transformation. This was why they were the enclosures of Kêd, the great mother, who took them to her bosom again as the nursing mother of eternal life.

A remarkable cluster of names occurs in the Duke of Hamilton's grounds in the Barony of Mawchane, in Lanarkshire, with their Lands of Carsbaskat, the Cross of Netherton, and the Moat-hill or Seat of justice in the Haugh. Lan-ark is the Ark-enclosure. Ark (Eg.), Orch (Welsh), denote the end. This was the enclosure of the dead. Nuter (Eg.) is divine. Tun, the lofty seat. The Makhen or Makhennu (Eg.) is the bark (ark) of the dead. The Kars (Eg.) is the place of embalmment and burial, bas (Eg.) means to hide and protect, transfer or transfigure, and Kat (Eg.) is the womb or the circle of reproduction.

The Haugh in the Norse Haugr, the Hag-pen, the Hogh, Hawk-law, How, and Hoe, were funereal mounds and enclosures of the dead. The Hag in Northumberland is the womb, prototype of the Hag-tomb. The Kak is the old church. The Moat Hill is a most ancient form of the Egyptian Hall of the Two Truths or Maāt. The goddess Mâ presided in the Maāt-Hall. Her name in the hard form is Makh, the Irish Macha. Now, there is a great mound in Westmeath, the Mound of Moate, called Moategranoge, a name derived by tradition from the young Grace or Graine, who was said to be a Munster lady. Dr. Joyce refers her ladyship to the same origin as the Milesian princess, who, according to the legends, took on herself the office of Brehon, and from this moat adjudicated causes and delivered her oral laws to the people.[1] This Moate we claim as the Irish Maāt or Macha, who was goddess of justice and lawgiver in the Maāt-Hall of the Two Truths in Egypt. The various moat hills were her seats, one being in the Hamilton grounds. The Ham (Hem) is the feminine seat and abode, and the original tenure of the Hamiltons, it may be inferred, was based on guardianship of the sacred ground belonging to the dead, the same as that of the wardens of the Irish Termons and the lord of the manor of Hundon in Caistor.

The Irish Sidh is an abode, habitation, cave in the hill, and sub-terranean palace of the spirits as fairies. The "Wee folk, good folk," the supernatural beings are called Men of the Sidh. The Banshu is the Bean-Sidh. The Sidhean is a fairy mount. The ancient name of the Rock of Cashel, and of several other fairy haunts, was Sidh-Dhruim. Rocks, mounts, and mounds wherein the dead were buried, are especial forms of the Sidh. There is an ever-famous Sidh at Ballyshannon, Donegal, where William Allingham enshrined the

[1] *Tour in Connaught,* by Cæsar Otway, p. 55.

"Wee folk, good folk," in an immortal lyric. The "airy mountain" is the Sidh Aodha Ruaidh, a great resort of the fairies. It is a hill now called Mullagh-shee, the hill of the Sidh or fairy palace. It was lately found to have been a sepulchral mound; recent excavations have shown that it contains subterranean chambers. This was the burial-place of Aedh-Ruadh, father of Macha of the golden hair, his ark of the waters.

Sidh is also applied to the spirits themselves, who are called THE Sidh. SIDHEÓG means a fairy spirit. This, however, may be the spirit (Akh, Eg., is a spirit and the dead) of the Sidh. But the immediate point is this. In Egyptian the Irish Sidh is represented by Shet, a name of the chest, box, sarcophagus, another hiding-place of the dead. The Shet is also a space, closed, secret, and sacred; a void, the tomb; all that is mystical and mysterious in relation to burial is expressed by the word Shet, English shut. Shetu also denotes a kind of spirits, the spirits of wine. One sees how the hill of the dead would be transformed into a primitive kind of spirit-world, the home and haunt of mysterious beings, the palaces and mansions of the glorified dead.

On the sculptured stones of Scotland there is a representation of some fragments[1] of stone coffins from Govan, of which no account is given. Two of these are tortoise-shaped, and one especially is marked in a manner to suggest that it is a symbolical or conventionalized tortoise in stone. The tortoise is Shet (Eg.), an ideograph of the mystery and secrecy expressed by the word. There is a "Chapter of Stopping the Tortoise" (36) in the *Ritual.* It had then become an emblem of evil in the world of the dead.

If we are right respecting the direct Egyptian origin of our institutions and ideas, it is certain that our teachers, say in the second stage, that of the Keltæ, must have inculcated their horror of the body's returning to the elements by the way of the worms, which amounts to an agony at thought of it, as expressed in the Book of the Dead.

At Chysauster, in Cornwall, there were a series of caves and excavated passages, which have been destroyed within living memory. The name of these tells us in the old tongue that they were places in which the mummy was preserved. Ki (Eg.) is the ground-plan of an abode, and means an inner region; Khi is to screen, cover, protect; Sau is the mummy; Ster is laid out together, laid on the back, with the image of the mummy laid out on the lion-couch of the embalmed dead.

In almost every case where excavations have been made, it has been proved that the stone circles were places of sepulture. Knockmany Hill at Clogher, Tyrone, when opened, was found to contain sepulchres chambered in the rock. This may perhaps account for the name of the numerous Irish KNOCKS, as the gathering-places of

[1] Stuart, pl. 134.

the MENA or Dead. CNUCH in Welsh means to join together, and represents the Egyptian ANKH. In English the CNAG is a knot, or cluster; KNOGS are nine-pins; the KNOCKING-place is one of general resort. The KANK, Persian, is a temple; the YING, Chinese, a sepulchre; Italian, CONCA, the tomb. The KANK or KNOCK is an earlier form of ANKH or HENGE, applied to the hill before the stones were erected on the plain.

There is a hill in Renfrewshire out of which issues the River Kart; the "Kart Waters," a synonym of the "Black Kart." Kart in Egyptian means the silent, stealthy, black as night. This makes it feasible that the name of the hill, the "Staick," is likewise Egyptian. "Stekh" signifies to embalm, hide, to escape notice, lie hidden, make invisible. This, therefore, looks like a burial-ground. Hills were, of course, the dry places in our climate. Also this meaning of Stekh, the concealed place, may perhaps identify the origin of our "Stocks" as places hidden in nooks or by greenery. Woodstock was the place of the famous maze or labyrinth which may have been a primitive Stekh, as the place of concealment that secured the sanctity of the dead.

In a Charter of King Athelstan, dated in 939, printed by Kemble,[1] there is a description referring to Avebury, one portion of which is called "COLLAS Barrow." This, in Egyptian, would be "Karas"; where we find Karas is the place of embalmment; "Karas," the funeral and embalmment. The Karast is the mummy, the preserved body, our corse. The meaning of Kars or Karas lives in our kerse, to cover a wall with slate; clize, a covered drain; and a close, Cornish clush. COLLAS Barrow answers to the Egyptian Karas, the place of preservation for the dead. The same description of Avebury mentions the Hack-pen, taken by Stukely to mean the serpent's head. But if this be Karas Barrow, the burial mound, then the Hag-pen is the Hag, How or Kak, the sanctuary, and Pen is the mount; Ben (Eg.), the height.

In Hebrew the KARAS is the belly or paunch, used as a vulgar expression for the בטן or womb. In the Mishna it signifies the pregnant womb, and the mummy of the dead in the Karas was the image of the child in the womb; a fœtus of the future life. In another spelling CHARAS (חרס) is identical with the Egyptian Karas, as the clay-place; also the sense of earth, earthy, plaster, to be sticky, agrees with Karas as the term for embalming the mummy and embedding it in the earth.

The coating of the body with ochre, which preceded the Egyptian Mum or pitch-plaster, is implied in the Hebrew Charas. Still another variant of the same word, as קרש, yields the boards of the Tabernacle, which was an image of the womb and tomb in one; the COFFIN, as the final form of the CEFN, KAFN, CABIN, or KIBNO of the bringer-forth.

[1] *Codex Diplomaticus Ævi Saxonici*, v. p. 238, No. 1120.

The temple of Classerness, which stood in the Western Isles of Scotland, contains the Karas (Eg.), the place of burial and embalmment, in its name. Ser (Eg.) means the holy place; Ness is the promontory or jutting of land. The "Roundago" says the same thing more briefly. Ren, to name, is to ring round, whence round (ren-t), enclosed; and the Akhu are the dead. There was a Roundago at Kerries, and Karas in Egyptian again identifies the place of embalmment or burial. Kerries corresponds to Collas Barrow at Avebury, and to Classerness. CRESSwell Cave, where the oldest traces of design and drawing on bone have been found in Britain, is probably a form of the Karas, the place of embalmment and burial. The carved bones, reindeer horns, and ivory, like the Jade stones, were early forms of the Fé and the inscribed tablet or papyrus buried with the dead; these are now represented by the tombstone erected over the dead.

The "Kaer of the Gyvylchi," in Snowdon, was a form of the enclosure of Kêd. The Initiate, speaking of the mysteries, exclaims— "I shall long for the proud-wrought Kaer of the Gyvylchi, till my exulting person has gained admittance. It is the chosen place of Llywy, with her splendid endowments. Bright-gleaming she ascends from the margin of the sea. And the lady shines this present year in the desert of Arvon, in Eryri."[1] Llywy was a form of Kêd; the branch and token of the egg belonged to her, she presided over the mystical transformation.

Gavr-Inis is the name of a cromlech. Inis means an island, and the dead of Memphis were conveyed to the Island of Tattu, in the Nile, there to await their change and transformation, whereby they were established for ever. This change is called after Khepr. And in the cromlech of Gavr-Inis we have a form of the island, the ark amid the waters, in which the dead awaited their resurrection.

In Egypt the beetle (Khepr) was the type of transformation and resurrection, as were the Druidic egg and branch in Britain; both egg and beetle showed the same change, and the beetle is found in the barrows. In Egypt the beetle was observed to settle on the banks of the Nile just before the Inundation, where the soil was moist and doughy. On this its eggs were laid in a pile and the earth heaped over them in a round mound; then it excavated and dug out the earth beneath, and thus shaped a sphere or ball of mould, with its eggs enclosed. Now the waters were beginning to rise, and it was a long way from the place of safety at the rim of the desert sand. But Khepr was equal to the emergency. Turning round and fixing the inward-curving hind-legs to the two sides of the ball, somewhat like the ironwork of the garden-roller, except that Khepr was both handle and operator in one, the rolling began by the beetle pushing backwards the ball revolving on the axis of his legs. At the edge of the sand and beyond high-water mark of

[1] *Song of Hywell*, Davies, p. 284.

the COMING tide, Khepr ceased to be a roller, and turned sexton. He dug down half a yard or more into the dry, pushed in his little world of future life, and then buried himself along with his seed to wait the transformation of the chrysalis. In inscriptions at Bab-el-Muluk and Abydus, Khepr is distinguished as the scarabeus which enters life as its own son; a type that dispensed with paternity, and belonged to the time when there were only the mother and son, and the son was established in the place of the mother, as he was in the person of Khepr-Ptah. "They say," observes Clement Alexander,[1] "that the beetle lives six months underground and six above." This is the type of the sun in the six upper and six lower signs. Watching the works and ways of Khepr the Egyptians conferred on the beetle the honour of being the symbol of transformation into new life. In Egypt they could bury beneath the soil without fear of damp. But in the north they learned that the chief dry places for the dead whom they desired to preserve would be the high places.

The first KHEP, or KOFF, of KHEPR, where the transformation occurred, was the womb; next the cave or CEFN, then the caer of GYVYLCHI, and the cromlech of GAVR-Inis. The final form is the CHAPEL, the lady-chapel, as it is still designated, which, in Cornish, is the female Cheber. The French CIBOIRE, is the pyx; the Hebrew QEBORAH, Hindustani KABR, Swahili KABURI, Arabic KABR, and Malayan KUBR, are names of the sepulchre. The CAFELL, Welsh, is the choir or chancel; the Gaelic and Irish CAIBEAL is a burial-place; the Latin CAPULI, a bier; the Hindustani and Turkish KIBLA, a shrine, and a quarter of the heaven. Womb and tomb are synonymous, and in Irish KOBAILLE means pregnancy; the KEBIL is a midwife, and in Gothic KIPURT signifies birth.

Gyvylchi, in Wales, is identical with Kabal, or Gebail, names of Biblos, where the genitrix had her Kep or sanctuary. The myth identifies the scenery, and Gyvylchi is the high earth of Gebail or Kabal, and one of the four supports of heaven. Khibur, the Egyptian name of Hebron, is the same mount in mythology. Cyverthwch is the name of another place in Eryri, the Cliff of Cyverthwch. The Druidic Kyvri-Vol, near Gower, is the ark or chest (Vol) of Kyvri. The Egyptian imagery shall be identified past doubt.

The Kep (Eg.), a concealed place, sanctuary, abode of birth, is our Cave. The Messiah is born in a cave of the rock or mountain. The cave of the Peak in Derbyshire is likewise called the Keb, a name for the Peak. The Kep or cave was the type of the birthplace, the feminine abode. Hence the cave of the mountain is the sanctuary of the Great Mother, in the Keb of the Peak, as well as in Gebail, Khibur, or Hebron. Kep (Kêd) or Kheft, the Typhonian genitrix, was represented by the Khepsh, or hinder thigh, the thigh constellation. Now when Typhon was degraded in this country, as in Egypt, it was

[1] *Strom.* 5.

the devil, and to show how definitely the Egyptian imagery was imprinted on our land, the Keb or cave of the Peak, the symbol of the Khep, as hinder part, hinder thigh, is known at this day as the " Devil's Arse."

Avebury or Abury was a form of the mount, but reared by human hands. It is certain to be a type of the Kep, the image of Kêd, and therefore the earliest form will be Kaf-bury. Af (Eg.) has an earlier form in Kaf. The bury in this shape is explained by burui (Eg.), the cap, tip, roof, supreme height, which has the same meaning as Ben, determined by the pile, obelisk, or pyramid ; the Hag-PEN being a part of Avebury. Af and Kab (Eg.) mean born of. Av-bury is the lofty birthplace. The Barddas call it the Pile of KYVR-ANGAN. It was also a form of the Ankh, that is, a symbolic image of life associated with the idea of transformation or transfiguring, a typical place of re-birth for the dead laid out together (Cyvr), also used in the Mysteries for the enacting of the doctrinal drama. The builders were imitating the beetle in burying their dead as the seed of future life, waiting in a dry place for the resurrection, and the receptacle was representative of the Kept or Kêd, the Egyptian Meskhen.

Cor-CYFOETH was a name of Stone-Henge, and in Welsh CYFAWD means to rise up ; CYFODI may be rendered the Resurrection. Abury being a work of the builders, the name can be glossed by GOBER, Welsh, a work, operation, deed ; GOBERU, to work ; GEPHURA, Greek, a mound of earth ; KEBER, Cornish, CABIR, Welsh, a rafter, roof-work ; CEIBRAW, to joist, lay on rafters ; CIVERY, English, a compartment in a vaulted ceiling ; KABARA, Persian, a beehive.

There are writers, who like Goldziher in his *Mythology of the Hebrews*, have imprudently characterized the system of British Druidism as a modern imposture and perversion of Christianity. But the truth is there is far more in it even than has ever been claimed by the Barddas. When the matter is tested by the comparative method, this will be proved.

The chair of the bards was a great symbolic institution, the chair of Keridwen. This is an identifiable type. The chair was the Hes or As of the genitrix in Egypt. Hes, the chair, is likewise the Egyptian name of the singer, the bard, and means to praise, applaud, celebrate. Tut is to unite together, a ceremony, typical, and Put is the divine circle of the gods ; the Put circle of nine in number. An earlier form of the circle is that of FUT, the four corners, the quadrangular Kaer. HES-TUT-FUT, the celebration of the singers in the quadrangular Kaer, the circle of Kêd, gives us the EISTEDDFOD continued, in keeping with its original character, to the present time, as an annual gathering of singers and reciters with the seat (Hes) in the circle. The Eisteddfod is a living link with Stonehenge, the Stone of Eseye, and with Egypt. Further, as SILL is an old English name for the seat and throne, equated by the Egyptian Tser for the temple or palace; Ser the seat and rock of the horizon; it is probable that Silbury Hill is a form of the "Seat of the Throned Bards," who were likewise the lawgivers.

The language of the chair was personified in Kadeirath, the son of Saidi. Kadeir is chair, and in Egyptian att or "uti," is a name for speech, utterance, language, the Word.

The typical teacher of Druidic lore, Taliesin, characterizes his mystical utterances by the name of "Dawn y Derwyddon."[1] Dawn, in Welsh, is the Lore; Dawn y Derwyddon, the lore of the Druids. The Tan or tannu, in Egyptian, further identifies the kind of learning; Tan, measure, extent, complete, fill up, terminate, determine: Tennu, lunar eclipses; Tennu, reckon, each and every amount. Thus the Druidic lore consisted in reckoning up each and every one of the circles and cycles of time. This is described as "The study of the Circle, the Circle of Anoeth."

"I know," sings Taliesin, "what foundations there are beneath the sea. I mark their counterpart each in its sloping plane,"[2] that is in the lower signs, the nether part of the circle of Anoeth. This circle as Solar was called the precinct of Iôr, or the year. "Iôr, the fair quadrangular area of the great sanctuary,"[3] is the equivalent of the four-cornered circle of the Zend Avesta, made by Yima.

The stones of the circles are sometimes called DAWNS-MEN, and this title was perverted into dance-men, and the dancing men of legendary lore. Finally the dawns-men and dance-men were converted into Danish men, and the Danes take the place of the Dawn made plural in Dawns.

This Dawn is Taliesin's Dawn y Derwyddon, the Druidic lore. The Dawn-Men are the Stone-Memorials of the Druidic lore, the knowledge of the time-circles registered in the stones. That they localized the circle of Anoeth in these islands is shown by the name of the parish in Scotland where the Stone of Kirckclauch was found,[4] which is Anwoth. An (Eg.) also means to speak, hear, listen. WOTHE (Eng.) means eloquence. Anat (Eg.) is the stone-circle. An-at is the circle of repetition. "I require men," says the god Hu, "to be born again," "RY ANNET."

The heaven was divided in two halves, sometimes represented by Nupe above and Neith below. Nupe (or Pe) bends over the earth and rests on her hands and feet in the form of a half-square, equivalent to the half-circle, and this figure was conventionalized. A stone in the Edinburgh Museum of Antiquities shows a figure that corresponds to the upper half of the heavens, represented by Nupe as the upper hemisphere, or by the human figure conventionalized into

[1] Davies, *Myth.* p. 117.
[2] *Song of Cuhelyn*, 6th century. Davies, *Myth.* p. 314. *Ib.* p. 52.
[3] Davies, 315.
[4] Stuart, *Sculptured Stones*, pl. 123.

mere line.   A cross within the enclosure intimates the place of
the equinox, the division of the two heavens, where the sun entered
the upper one.   Here was the "Hall of the Two Truths," whose
duality takes so many forms.   Here was the region of Tattu, the
eternal.   One sign of this was the wheel or cake symbol of the orbit,
which became the ancient wheel-shaped Theta of the Greeks.   Thus
the letter Theta with cross and circle combined repeats Tattu or
Teta, the established region in the zenith.   This same sign is found
on the Scottish stones.   It is the especial emblem of the equinox as
the place where the circle of the year was completed and renewed.
Two such cakes or wheels denote the double equinox, as in
the Hebrew בת־דבלים (Beth Diblaim), the house of the double
cakes or circle, and other forms of doubling.   Har-Makhu was the
solar god of this double horizon, with its station at the place of the
equinox.

Now it is claimed by the present writer after long study, that the
little house of the double-cakes, disks, or circles found on the sculptured stones of Scotland,[1] is the Hall of the Two Truths in An
or Tattu, the solar birthplace, and that the image of the Two
Truths and dual circle is what is commonly termed the "spectacles
ornament."

This is sometimes represented across the little house of the two
circles as in plates 15, 17, 33 (Stuart), and at others by the double
disk.   In either case the double circle is crossed by the crooked
serpent or the Z-shaped figure.   The Egyptians placed their equinoxes
up in the zenith and their solstices low down on the horizon.   The
place of the equinoxes was a mount, and if we imagine an enormous
plank laid across the top, on the ends of which two figures ride up
and down at see-saw, we shall be able to realize their scales as they
ascended and descended north and south.   This see-saw of the solstices in the scales or balance of the equinox is necessitated by the
one being in the zenith, the others on the two horizons.   The see-
saw on our stones is the serpent or Z-figure oscillating across the
double disk of the Hall of Two Truths.   This can be shown.   One
name of the figure of the double horizon is Tset.   Tset is the serpent (Tet), and this serpent Tset becomes our Zed.   Thus the serpent
and the Z are equivalents as on the stones.   The Zed or serpent, then,
belongs to the double horizon north and south, its head and tail are
solstitial; these go up and down across the dual disk, which is therefore the Egyptian equinox in the zenith.

The serpent depicted in plate 37 of Stuart's *Sculptured Stones*
from the monument at Newtown is the basilisk, the goggle-eyed or
spectacled serpent, which is the especial warder of the gateway of the
path of the sun.[2]   In keeping with this character it is portrayed

---

[1] Stuart, *The Sculptured Stones of Scotland*, 2 vols.
[2] Sarcophagus in Soane Museum.

with four wings, which represent the four corners of the earth. It is also depicted under the name of Hapu with four heads. And again, on the same sarcophagus, it appears in a fourfold form as Apt, having four figures on it. Apt is the name of the four corners, and the basilisk is the serpent of the four cardinal points, that is, of the solstices and equinoxes, therefore, of the circle of the year.

Another basilisk on the same monument is three-headed, and it represents the trinity of father, mother, and son, or Osiris, Isis, and Horus which is perfected and completed in the conjunction at the time of the vernal equinox. The typical serpent of the Egyptian monuments has the same signification on the Scottish stones.

The Sweno stone, supposed to commemorate the defeat of Sweno, is to me the Shennu stone. Shennu (Eg.) is the circle of time consisting of two halves (Shen or Sen). Shen also means the brother and sister, the male and female halves. These figures are portrayed on Sweno's stone (plate 18, right hand), and on plate 20 the two figures are bending over the child born at the place where Osiris, Isis, and Horus met in " Shennu " at the crossing.

The hall of the double disk is found on stones at Tyrie and Arndilly. Both Dilly and Tyrie correspond to Terui (Eg.), our Troy, the circumference and limit of the whole, consisting of the two times called Terui ; which is also a form of Sesennu and number eight, the total as the Ogdoad, like the eight in the ark, here represented by ARN.

At Bourtie (plate 142, Stuart) there are two stone-circles, the two disks of the drawings. There is also an eminence called the Hawk-Law. Two cairns were opened about fifty years ago. In each there was a stone coffin enclosing two urns of hard baked clay.[1] The name Bourtie answers to Per-ti (Eg.), the dual circle and double house in An.

A rock on Trusty's Hill, near Anworth, Galloway, has the double disk and Z-sceptre. The equinox is in line with a conventionalized fish, and there is a sign pointing expressly at the fish. The Worth is an enclosure, and An we claim as the solar birthplace, the celestial Heliopolis. An also means a fish in Egyptian, and here the equinox is in An ; the monument in Anworth.[2]

When the solar birthplace was in the Fishes, it was represented by the genitrix in the shape of a mermaid who brought forth the child.[3] Now the well-known symbols of the Mermaid are the comb and the glass. These are frequent on the Scottish stones. The comb and mirror are depicted on the " Maiden Stone," [4] which thus becomes the Stone of the Mermaiden Goddess, half woman, half fish, the Derketo, Atergatis and Semiramis, who was represented in Britain as the mother Kêd, our Keto.

---

[1] Stuart, p. 42, and plate 132, fig. 3.
[2] Plate 97, Stuart.   [3] Hermean Zodiac.   [4] Plate 2, Stuart.

The comb represents puberty, the first of the Two Truths in the mystical sense. At this period the maiden bound up her hair for the first time with the comb, plaited, knotted, and snooded it, according to Egyptian usage. The mirror is the type of reproduction, like the eye, which is likewise figured at the place of the vernal equinox. This was the symbol of the other of the Two Truths. Both were united in the mermaid or fish-goddess, or yet earlier water-horse. But where is the mermaid herself? She is represented by the elephantine monster of these drawings. This figure accompanies the equinoctial imagery of the double-disk in plates 2, 22, 24, 34, 39, 47, and 67 (Stuart).

The same figure accompanies the crescent or semi-circle in plates 4, 10, 40, 47. It represents the goddess of the Great Bear, whose type in Egypt was a monster compounded of hippopotamus, crocodile, the Kaf-ape, and lioness.

The monster of the stones is the same ideograph as the Mare with feet fettered fast around the cake-type of Tattu, the eternal, or depicted full gallop on other of the coins or amulets of Cuno-Belin. Hippa, the mare, is but a more European form of Kefa, the female water-horse.

The monster is the Scottish version of the conventionalized Bear, portrayed by the Welsh as mare and boat and bird in one image.

In either case the object was not to imitate nature, but to compound an ideographic symbol. It happens that the spectacles-shaped double disk is found on the Assyrian monuments, as a form of yoke, and is said to denote A FOUR-FOOTED ANIMAL TRAINED TO THE YOKE. Our word yoke and the Latin Jugum are forms of the Egyptian Khekh, the balance and the place of the equinox. The Roman Jugum appears as a kind of cross.[1] Thus the cross, the balance, and yoke, are types of the equinoctial level, the crossing, and the word Khekh names all three. The Jugum as the top or ridge of a mountain also corresponds to the Kekh of the horizon or height. In Eskimo, Kek is the boundary; Kakoi, in Japanese, means to enclose, clasp, fence round, and the four-footed animal trained to the yoke is our mare with fettered feet, and the monster whose tethered turnings round denoted the earliest year, that of the Great Bear. It was by aid of the Great Bear that the early observers determined the equinoxes and solstices. The Chinese say when the tail of the Great Bear points to the east, it is spring; to the south, it is summer; to the west, it is autumn; and to the north, it is winter. This was the constellation of the bringer-forth of the child as Sut, the Dog-star, in the pre-solar time.

The bird is often found on the stones, and on one of them[2] there is a form of the boat, with a paddle in the forepart.

[1] See pp. 481, 482, *Trans. Bib. Arch.* vol. vi. part ii.
[2] Plate 85, Stuart, St. Orland's Stone, at Cossius.
[3] Plate 15, Stuart.

The fish appears on the Edderton stone,[3] and again on the Golspie stone,[1] accompanying the symbols of the equinox. This can only indicate the colure in the sign of Pisces.

On the Mortlach stone,[2] two fishes are portrayed, and they are joined together like the two of the zodiac. There is a figure of the Ram beneath, as if superseded by the fishes. Further, plate 118 shows a ram-headed figure over the fishes, or twin-fish, also an inverted human figure. This read hieroglyphically—the inverted figure is among the hieroglyphics—signifies the reversal of the signs, and says the colure has left, or is leaving, the sign of the Ram for that of the Fishes. The imagery is on the cross of Netherton, which, in Egyptian, means the divine seat; this seat was denoted first and foremost by the cross of and at the crossing. It was at this point the hero Horus overcame the Akhekh dragon of darkness, the Typhonian type of evil. And on the Golspie stone,[3] the hero is portrayed fighting the battle of Horus against Typhon, which terminated at the spring equinox.

At the place of the equinox was the double holy house devoted to Anubis, the double Anubis who may be seen biformis, back to back, at the crossing in the planisphere of Denderah. We know the Druids made use of the ape in their imagery, and this was one form of Anubis. This double Anubis as dual ape appears in plate 63.[4] The duality is curiously expressed in the way they are twined and intertwined together. The same twins are apparently intended in plate 45, from the Kirriemuir stone.[5]

When the Great Mother was first typified by the bear, or water-horse, Typhon, Sut was her son, and his type was the Dog-star. As Apt she is expressly called the Great One who gave birth to the boy.[6] The boy in Britain was Beli, the star-god, and Belin, the solar Baal.

And in one of the archaic sculpturings,[7] the so-called Z-sceptre is drawn with the double disk in a boat-shape figure, like that of the Hindu Meru, with the seven heavens at one end, and the seven hells at the other, on the north and south poles. The dog's head is appended to one end of the balance. It is repeated in fig. 34.[8] The dog is obviously at the head end, that is, in front, the south; the north being the hinder part, represented by the loop or tie of Typhon, and this points to the Dog-star, the announcer of the solstitial year. Thus we have the mother and son, Sut-Typhon, as Great Bear and Dog, among the earliest of all the Sabean types figured in the heavens.

Every type found in cluster on the stones connected with the cross ideograph of the equinox shows the astronomical imagery in the eschatological phase. The great mother, the sun-bird, the mirror, comb, serpent, and hall of the double disk, all denote the resurrection

---

[1] Plate 34, Stuart.   [2] Plate 14, fig. 1, Stuart.   [3] Plate 31, Stuart.
[4] Stuart.   [5] Stuart.   [6] Birch, Gallery, p. 41.
[7] Simpson, p. 170.   [8] P. 171.

or reproduction of the sun-god and the soul, and so proclaim and prove the monuments to be memorials of the buried dead.

Evidence of what may yet be called the Druidical cult, maddening as is the name to some, is not limited to the monuments, but survives in the names of places where the stones have been destroyed. So long as they stand, our hills will talk in the primeval tongue, and while Helvellyn lasts, its name will prove it to have been the seat and scene of the worship of Kynvelyn, the British Belin.

The present work has been partly written on ancient Druidical ground. The author was born in its neighbourhood, and has lived in the heart of it for many years ; born in the shrine of Belin, at GAMBLE, which may be rendered the Khem of Baal. This is shown by the Bulbourne river, and the ancient city of that name. An old distich of the district says :—

> " When St. Alban's was a wood,
> The ancient city of Bulbourne stood."

Bulbourne was the boundary of Baal.

If it be objected that the word Gamble is an English name for the leg, my reply is the leg (hinder) is the especial hieroglyphic of the genitrix, who was herself the shrine of Baal.

The Druidic ground is chiefly on the Chiltern Hills, at the corners of three counties, Herts, Bucks, and Beds.

There are three Hoes, Ivinghoe, Totternhoe, and Asthoe. The seat or throne of Kêd is extant as the Ten or Den of Gad, divided into the larger and lesser Gaddesdens in accordance with the dual mapping out. The Den or Dun is a division as well as a seat, following the Egyptian Tena, to divide, separate in two, and Nettle-den is the lower den, the nether of two (so a jakes is a nettle-house), like the Neter-Kar. The Dunstable crows are both black and white.

At Dunstable we have the Maiden-ing burn or bourne, possibly both in one. PYTTE-STONE implies the stone of the intermittent well.

Ashridge Park was anciently in two divisions, and one of these, the south-eastern, was always stocked with fallow deer, the northern with red deer.[1] These were as true symbols of the two halves of the solar circle as the white and red crowns of Egypt.

There is a tradition that the San Grael was at Ashridge, the house of the Bonhommes. Skelton, in his *Crown of Laurel*, speaks of " Ashridge beside Barkanstede, that goodly place to Skelton most kind, where the Sang Royall is, Christ's blode so red."

Ashridge House has its legend of the cross, because it stands at the crossing. It stands in two counties, and is so completely divided that during the time the present writer dwelt in its neighbourhood a sudden death occurred, and a coroner's inquest ensued. The doctor

---

[1] Lipscombe, *Hist. of Bucks*, vol. iii. 447.

chanced to mention that the man had died in another room; this was in the next county, and another coroner demanded.

Here, then, was the crossing, the topographical and symbolical analogue of the astronomical crossing of the equinox, with the Hall of the Two Truths, an image adopted by the religion of the Cross. Hence followed the token of the crucifixion in the presence of the San Grael.

My conclusion respecting the meaning of "Ashridge," which has nothing to do with the ash tree, is that it represents the mountain-ridge which is preëminent in the Welsh Esgair; Gaelic, Eisgir; Irish Aisgeir or Eiscir, that is, the ridge of hills and mountains; the rocky ridge, Ysger in Welsh, being the rock or stone, which is repeated in Asgr-ridge, or Ashridge. This is the Egyptian Skaru, the name of a fort; the Assyrian Ziggurrat, a tower; the Hindustani ZIARAT or Shrine. The fort or place of defence is recoverable in the Welsh YSGOR, a circular entrenchment. As the Greek ESCHARA, it is an altar for fire-offerings, and in the Hebrew אזכרה (Azkerah) it also relates to sacrifice and memorizing. In Egyptian, As is the seat, throne, sepulchre, sacrifice; Kheri means the victim bound for sacrifice. The mountain was the first altar, and its caves supplied the earliest tombs. The word as Eskar has also been adopted by geologists for the isolated heap left by the ice or water at the foot of the hills or on the plains. In accordance with a common principle of compounding and, it may be, of interpreting names, the Ridge does but translate the Aisgr.

The name of the ancient mother Kêd is extant at Cheddington, earlier Kettington, the ten (high seat) of the Ing of Kêd. Kêd-tide is an old name for Shrovetide. We have also the river Gad. Wad's Combe (vulgarized into Ward's) and Wad-Hurst (as it is written in the old maps) are probably forms of Gad, or Kêd, and in the Hurst or wood of Kêd the greater part of this work was prepared.

Monybury Hill is a portion of the table-land of Gad (Gaddesden), and thus we may have our Gad and Meni on the same ground. The "Gallows" Hill, Welsh GWALAS, means the Couch or Bed on the high table-land, which was the lair or birthplace of Kêd. One of the hills next to Gallows Hill is still called Steps Hill. Cadr-Idris is said to have had 365 steps cut in it. The hill at Cheddington is cut in three vast coronal-like tiers, most distinct still, although it has been ploughed over for ages. The early men cut indelibly whether they worked in adamant, sienite, limestone, or only in the earth itself. Their seals and impressions were worthy of the divinity whose name, Khet, means to cut and to seal.

These three steps are ranged towards the sunrise, a triad of tiers corresponding to the three solar regions, the upper, mid, and lower, found in all the mythologies, and with the three ranks and symbolical colours which were apparently disposed in the same way as in Egypt;

blue for the highest rank, white for the second, and green for the lowermost; blue for Ammon in the height, green for Num in the deep, and white for Khem on the horizon (to judge by white as the colour of Khem-Horus); and these three are the chord of colour found on the Druidic glains. "We find some of them blue, some white, a third sort green, and a fourth regularly variegated with all these colours."[1] On these three tiers or steps of gigantic heaven-ward stride, we may conjecture the Druids stood in their triple ranks at sunrise, and on gala days in front of a stone temple, crowning the Hill of Kêd, toward which these steps ascended. In the hill named "Steps" Hill, close to Gallows Hill, there is a vast ravine or gorge, not a natural formation, not the work of the elements, and unaccounted for as the work of human hands. But the name of the hill itself suggests the clue; it is the STEPS Hill, and at the head of the ravine is the place, scope, and height required for an ascent of steps as at Cadr Idris. This hill leads up to Gallows Hill, the highest of all, otherwise known as the Beacon Hill. With the strange ravine and ascent of the Steps Hill we may compare the Egyptian Feast of the Dead, which was a festival of the steps and of the valley, or of the ascent from the valley.

We cannot but associate the Gallows Hill with the Gwal of Ast, the couch or seat of Kêd; Ast being one of her names, as Gaddesden the ten or seat of Gad (Kêd) is close at hand, and not far off is Asthoe, the hoe of Ast, and near by is Aston; also with the Gallicenæ and the Gwyllion of the Druidic mysteries, a plural of Gwyll or Gwely, the Gwal or Gwalas being the couch, ark, circle of Kêd, formed of the nine stones; the Nine Maids that stood in a circle, still represented by the circle trodden indelibly round the top of the Gallows Hill; the nine damsels who warmed with their breath the cauldron of Kêd, and the Nine Gallicenæ, the Galli of the Shen (Eg.), circle, the Welsh Sêon. These were the British muses, identical with the daughters of Mnemosyne in Greece, and the nine who attended Osiris in Egypt. And at Dunstable—an ancient seat of the CATIEUCHLANI—we have the Maiden-Bourne or Maiden-ing-burn.

There are no stone remains in all this region now, nothing but the cloven and circle-showing hills, and the imperishable records of the past preserved in names. Hills and ridges like these are not so easily carted away. There are three Commons on this portion of the Chilterns, and commons are religious remains; the last relics of general property in land under the Druidic system of government, which was primally the land of the dead, the Khem-mena (Eg.) a common place of the dead, being earlier than commons for provisions.

In one of the songs of poor old Merddin, the Caledonian Druid, who uttered the death-wail of the ancient cult, he exclaims, "How great my sorrow! How woful has been the treatment of Kedey"—a familiar name of the Mother Kêd. "They"—the opponents—"land

[1] Davies, *Myth.* p. 211.

in the Celestial Circle, before the passing form and the fixed form, over the pale white boundary. THE GREY STONES THEY ACTUALLY REMOVE,"—as if the mournful fact were too pitiful[1] for credence! Now let us turn to the lands of the living.

An old distich says,

> "By Tre, Ros, Pol, Llan, Caer, and Pen,
> You may know most Cornish men."

These are prominent names of places after which the family or community were named. The Llan is an enclosure in Cornish, also a church, the latest form of the sacred enclosure. In Persian the Lân is a yard. There are close upon one hundred Llans extant in the village names of Wales. And Dr. Bannister has collected 300 proper names in Cornwall based on Llan. This, in the hieroglyphics, is the Ren, the name and to name. The ideograph is the Ren, ring, an enclosure, a cartouche, for the royal names of the Pharaohs. We have the far more primitive Ren-enclosure as a Ran, the noose or band of a string, and in Ren, to tie up. With the participial terminal ren is ren-t, the enclosed and named, and that is the formation of the enclosure named, a primitive mode of getting on the land. One form of land is the ground between the furrows in the ploughed field. Land is that which is enclosed and named or Ren-t. The Run-ring for cattle was an early Llan, and the Ren sign is a noose for holding cattle by the foot. The orbit of that run was a primitive llan, and the payment made for it was RENT.

The same antiquarian has collected 500 Pens, named from the headlands, the Scottish Bens. Ben in Egyptian is the height, the point, cap, tip, roof; the Ben-ben is a pyramidion. In the same list of names there are 400 "Ros"'s; the Ros is a rock or headland, a natural elevation, which would be seized upon first for its position. It is the same at root as the Irish Lis, and English Rise.

There are 1,400 townlands and villages in Ireland having names beginning with Lis. The Lis is a raised place; it may be the natural or made mound turned into an earthwork. In the *Book of Ballymote* the Rath is used to denote the entrenchment of the circle, and the Lis is the space of ground enclosed. The Lis was sometimes enclosed within several raths or entrenchments. The Egyptian Res, to be elevated, raised up, to watch, be vigilant, best explains the nature and meaning of the Lis, as place of outlook within the protecting circle, before towers and fortifications could be erected.

The Welsh and Cornish Trevs, Trefs, Troys, or Tres, are probably the Egyptian Rep or Erpe. The Tre is understood to mean a homestead. The Erpe or Rep (Eg.) was a temple, a sacred house. With the article prefixed, this is T-rep, answering to Trep, and many of the Treps were certainly religious foundations. In Egyptian

---

[1] Davies, *Welsh Arch.* vii. p. 48.

we find the Taru, a college; Terp, the rites of Taht, a name for literature; and Teru, for the circumference, the Troy. The Teru is a modified form of the Tref.

Dr. Bannister has collected 2,400 Cornish proper names beginning with Tre, and there are a thousand Tres as places. This is the Egyptian Ter, Teru, and T-erp. Ter signifies all the people, the whole of a community dwelling together. The dwelling may be beneath the family roof-tree, whence the Tref (Tre) as the homestead, or it may be a village, as in the Dutch Dorp and English Thorpe. The habitation may be added to the Ter by the Pa (Eg.), a house, abode, place, or city, whence the Terp, Tref, Thorpe. Without the T (the article The) the Rep or Erp is an Egyptian temple, the house of a religious community. Thus we have Ter (Eg.), the community, and in Craven, Trip denotes the family and the herd, while the worn down form of Tre in Cornish means the homestead, dwelling-place, enclosure. The Erp (Terp) is the religious house. In Holstein the Tref or Thorpe is called the Rup without the prefix.

In the Scilly Isles there were vast monumental remains in Borlase's time, especially in an Island named "Trescaw," from whence, according to Davies,[1] a graduate in the Druidical school was styled Bardd Caw, one of the associates. Cuhelyn ab Caw was a British Bard of the sixth century. The songs of Keridwen were sung by the chanters of Caw. The plural Caw is found in Kaui (Eg.), a herd or band.

Trescaw, then, was a foundation of learning. The Caw is the Egyptian Khau or Kaf. The Khau as a scholar is implied by the Khauit being a school, a hall of learning with cloisters or colonnades. The Khau (Eg.) is a dog, and the priests of Kêd were dogs, *i.e.* kenners or knowers; the dog being a symbol of the knower with the Druids as well as in Egypt. The full form of the Khau is the Kaf ape, the Cynocephalus, a type of Taht, the Divine Scribe; also of the priests and of letters.[2] With us, too, the shepherd's dog, the knower, is designated a Cap; a Cap being synonymous with a master or head. Hence the symbolic cap of the scholar. In Egyptian the Skhau is the scholar and scribe, from Skhau, to write, writing, letters. So in English for the Caw we have the scholars, and Trescaw, otherwise Ynis Caw, was the island of the scholars. This was one of the Scilly Isles. The name of Scilly identifies the school. Skill means to know, to understand. The Scilly Isles repeat the name of Ynis Caw, the island of the scholars; which suggests that the Trê in Trêscaw is a modified Tref or Trep, as T-rep (Eg.) the temple or sacred house, and that scaw may represent the Skhau (Eg.), to write, writing, letters, the scribes and scholars. The Ys in Welsh was added to augment and intensify words, and this would make Caw Yscaw. Thus Ynis Tre(f) scaw would be the island of the Druidic "Erp," a temple of the scholars; the school implied by the later name of the Scilly Isles. The Tref as family became the Irish Treabh and English Tribe.

We have a group of counties, or hundreds, anciently known as Sokes, in Essex, Sussex, Middlesex, and Wessex. Our Soke is the Egyptian Sekh, a division, to cut out, incise, to memorize, remember, depict, represent, rule, protect. The Sekh is a division mapped out, marked off, cut out. The British Soke was the territory on which the tenants of a lordship were bound to attend the court. Also the SOKE of a mill was the range of territory within which the tenants were bound to bring their corn to be ground. The word "Sekh" has many meanings. It is a variant of Uskh for water, the earliest of all natural boundaries and divisions of the land. Sekh, to cut out and divide, has the meaning of share. The right of socage is the right to a share, held in later ages on varying terms. For example, in the Manor of Sevechampe, *Domesday*[1] records that there were four sokemen; one of these held half a hide, and might sell it; another held one Virgate, and could not sell it without leave of his lord (Elmer); the third and fourth had right of sale. King Edward had sac and soke over the manor. In Egyptian "Suskh" means free to go, have the liberty. As sock and suck have the same meaning, the Soke is a companionship, the basis of the Soke (guild), and the primeval socage was the freedom to graze cattle in a certain division, still extant in the right of common pasture, accorded to the company who held the land on the communal system. The earliest socage was so simple that it may be described as a right of suck or succour at the natural fount of life, the breast of the great mother of all, from which the children were not yet forcibly weaned, as they had not parted from their birthright and heritage. The socage then became a franchise, the parent of that liberty, freedom, frank-pledge, or what-not, now conferred by the honour called the freedom of the city. The primitive socage belonged to common ownership, the later to lordship, when the ownership was made special and several, with the right to levy soken, that is, toll. Port-Soken Ward, in the City of London, means a municipal district having the privilege of levying soken or toll in the shape of port-duties. Applied to territorial division on the large scale, the Sekh gives us the plural Sex, our four counties. In Essex, Sussex, Middlesex, and Wessex, we have a complete system of the territorial sokes, arranged according to the four cardinal points, and named in Egyptian. Uas is the west, a name of western Thebes. Wes-sex is Uas-sokes, the west divisions. Wessex was Hampshire. Robert of Gloucester calls Hampshire Suthamtshire, and Sut-amt in Egyptian is south-western. Both Sut and Su signify the south, and in Sussex, Wessex, and Essex the English follows the parent language in dropping the terminal T. Sussex is the south sokes, and on the same principle Essex is the east sokes.

[1] *Lib. Domesday*, fo. 141, No. 36.

Ast, to be light, answers to our east. In this chart Middlesex is to the north. The northern boundary of the zodiac as well as of Egypt was called Mat in the oldest records. Mat signifies the mid-middle division, which was the north-east quarter of the compass. Thus we have a circle of the sokes, with London seated on the water in the right position to represent the solar birthplace in Mat or An, the celestial Heliopolis. It will bear repeating that Sussex county was divided into six parts called Rapes, each of which had its river and castle. Now as the castle is but a later Kester, it looks as if the original Rape may have been the Egyptian religious house called the Rep or Erp, just as the Sekh or Uskh was also the Hall of the Two Truths. Sus (Eg.) means six, and whether intended or not, Sussex reads the Six Sokes. A religious foundation connected with the dead is at the base of all our living institutions that are deep-rooted in the past.

Our Sters, as before shown, are the resting-places of the dead. The hieroglyphic ster is variously compounded in the Min-ster, the ster of the dead; the Kester (Ke-ster) a house, region, land, inside place for the stretched-out dead; and with the Kaer or enclosure of the dead. The Chesters are also known as the Kaers. Portcestre was formerly called Kaerperis, Gloucester was likewise Kaerglou. Winchester was formerly called Kaerguen, which shows that Win is the modified Guen. Guen answers to Khen or Khennu (Eg.), the sacred house, hall, or sanctuary. Thus Guen-chester is the sanctuary of the buried dead, who were shielded and sheltered in the Chester. Khen, the sanctuary, also signifies to alight and rest, and Khen-khe-ster (Eg.) is the protected resting-place of the laid-out dead. The Glou in Gloucester takes the place of the sanctuary in Guen-chester. Its equivalent Kheru (Eg.) means a shrine, house, sanctuary, or cell, so that the significance is the same in both. Kher and Khen are determined by the typical quadrangular enclosure, and the Kaers were called quadrangles as well as circles. GLOU has the V sound in GLEVUM, and Kheru (Eg.) has the equivalent in Kherf, a first form, the model figure, or type of the Kher; it denotes the chief, excelling, surpassing, sacred. The Egyptian Kar is a hole under-ground, and with the terminal F for "it," we may obtain the grave as the equivalent of Kherf, a first form, a model figure, whilst Glev (Glevum) in Gloucester is really synonymous with Kherf, and grave, the inner place of the dead. In Cirencester both names are united, and Kar-en-khe-ster (Eg.) is the enclosure of the Chester or protected place of the buried dead, unless we read the word Chester as com-pounded from KAS (Eg.), burial, and TAR, the circle or to encircle. We have both forms in Caistor (church) and Ros-Kestal.

As burial-places, the Kaers, Khesters, and Minsters acquired their greatest sanctity, and for that reason were adopted and continued as places of Christian worship and rites; for churches and cathedrals.

Deep digging beneath and round some of the Chesters and Minsters would reveal many a glimpse of our pre-Christian, pre-Roman, pre-eval past, buried alive and still calling dumbly for rescue.

The Caers preceded the shires. And Nennius enumerates thirty-three Kaers as the names of ancient British cities, and as Kaer is the hard form of Shaer, it is evident these Kaers became our Shires. Kart (Eg.) means dwelling in. The Karrt is a name applied to the dwellings of the damned in Hades. With us the s forms the plural instead of the TI in Egyptian. The Egyptian Kars were the lower places from the south as they were in Wales, and in the mapping out of England the shires, or kars, are the lower counties. We have the meaning preserved in another way. The lower is also the left hand, and the Car-hand is an English name for the left hand. When the Druids plucked the magical plant with the left hand, that was on the night side, and the transaction belonged to the lower world.

We owe the words weal, wealth, weald, to this same origin in the Kar or orbit, the enclosure. Wealhcyn is not derived from the word Welsh as a name of race. That had a common origin in the Kar, gower, gale, or weal. For example, hemp, the halter, is called Welsh parsley, and the cuckoo is the Welsh ambassador, because the one makes the noose round the neck, the other makes the annual circle, each being a form of the Kar or weal. In the same way the whelk is named from its spiral circles. To welke is to wax round like the circle of the moon, and the Ring-dove is also called the Wrekin-dove, Wrek and Welk being synonymous. Wales and Corn-wales are on the borders of the land ; they are the outermost counties lying where they look as if conscious of being the first Kars enclosed from the common waste. Next comes what used to be known as the Wealhcyn or the Wreakin, as the word is found, in Shropshire. Wealhcyn does not mean Welsh-kin ; it is applied to the land as in the Wreakin, not to the folk. Cornwall, was one of the two Wales. Somersetshire and Devon were the Wealhcyn. Khen (Eg.) means within, inner, interior. The Wealhcyn are the interior or more inward of the Kars, Shires, or Weals, *i.e.* an inland Wales. The people may change, but names are ineffaceable.

The inner Wales leads to the suggestion that the name of Cornwall is derived from Kar-nu-wale. Nu (Eg.) signifies within, and Kar-nu-kar reads " Kar within Kar," or the inner of the two Kars called Wales. Cornwall was formerly Cornwales. Thus we begin with Wales the Kars the lowermost counties, the west being the way to the underworld, and Cornwall was anciently known as one of the two Wales. Kar-nu-wale is Wales within, and the Wealhcyn is a still more interior Wales. In this way we see the advance inland from what looks like a point of commencement in Wales.

One name of Wales known to the Barddas, is Demetia. Seith-wedd or Seithin Saidi is represented as being the king of Demetia or Dyved.[1] Dyved, later David, is a typical name of Wales, the land of Taffy. Temti (Eg.) is the total of two halves, the plural of Tem, a place corresponding to the dual Wales. In the old maps Demetia is called Dyved. This, in Egyptian, indicated a figured point of commencement, from Tef (Tep), the first point of beginning. Tep, however, as commencing point, would by itself apply equally to Dover. But the Tepht (Eg.) is the opening, gate, abyss of source. The Tepht answers to the lower Kars.

This name of DYVED as the Tepht is illustrated by the "Davy's locker" of our sailors, the bottom of the sea, which is the mythical Dyved or Tepht, the place of the waters of source, the pit or hole of the serpent, where the evil Deva or Typhon lies lurking. The Druids figured this underworld, or Nether-Kar, as the place from whence the visible world ascended, and as the place of the evil GWAR-THAWN. Cornwall, formerly called West Wales, was also known as Defenset, and its people were the Defæsetas. Tef-nu (Eg.) is Dyved within, the secondary form of Dyved or Wales. Here is a double Tef as point of commencement analogous to Demetia and Wales.

In a map of Britain (597) carefully collated from local maps and from Dr. Guest's researches by the author of the *Norman Conquest*,[2] we find four counties named Sets; these are Defenset, Dorset, Somerset, and Wiltset (later Wiltshire). These four counties should constitute a land once inhabited, mapped out, and named by Egyptians, for the "Set" is the old Egyptian name of the Nome, a portion of land measured off, divided, and named, *i.e.* nomed. These are the only four Egyptian Nomes named as Sets in the island.

Defenset, in accordance with its name, comes first after Dyved or Wales. Dor-set (in Egyptian, Tur-set) means an extreme limit of the land, the frontier, the very heel of the foot or foothold. Dorset is the frontier nome at an extremity of the land. Somerset is the Water-Nome. Su is they, them, or it. Mer (Eg.) is the sea. Somerset is the Sea-Nome. Wil-set, when equivalented in Egyptian, will be Hir-set, the upper nome. Hir is upper, over, above, high, uppermost boundary. The full form of "hir" as a place-name is hirt, and this may account for the T in Wiltshire. Hert was afterwards applied to the shire of the uppermost boundary of our shires. This goes to show that Wilts was once the uppermost limit of Egypt in England, as the highest of four nomes or Sets.

Our Set is the Egyptian Set or Sat, from Sa, ground, which, with the participial T denoting the Sa is measured or cut off, becomes the Sat (as we say, *sawed* off).

The Sa (Sa-t) has the meaning, in measure, of one-eighth of a quantity of land.[3] Now, if our "Sets" were divided and named on this principle, they would correspond also in number, and there ought

[1] Davies, *Myth.* p. 197.   [2] Vol. i.
[3] *Denkmäler*, iv. 43 ; iv 54, *a.*

to have been eight. There are four Sa-t or Sets in England and "DYVED" in Wales. Now Dyved signifies a measure of four. We have it in the English tofet, tovet, and tobit, a measure of four gallons. Four gallons to one tofet is equal to four divisions of Dyved. Moreover, the Egyptian Aft denotes the four corners, and Teb is a quarter, a place. Dyved was as surely the other four divisions as that four gallons make the tofet, and although they are not extant by name as the other four "Sets" they may have been four Kars, which they were. In the old map we still find Gower, Caeradigion, and two Caerleons. These are four Cars, answering to the four nomes, called Sets. Moreover, four Kars survive as counties in Wales, Cardigan, Carnarvon, Glamorgan, and Carmarthen. "Four Caers there are, stationary in Britain ; their governors are agitators of fire." [1]

The Egyptians divided the circle of the heavens into upper and lower. The lower contained the Kars. The lower half was to the north, the Kar-Neter, the Kar divided from the upper half by the equinoctial line running east and west. Set was the south in Egyptian ; the south was the upper country, and our four Sets are in the upper country towards the south.

On the monuments these two halves or houses of the sun are figured as two quadrangular enclosures with an opening, as two houses named "Iu." And in the Druidic writings, the Caer is sometimes designated a quadrangular enclosure. Two four-cornered enclosures give us the eight regions of Sesennu, as well as the twofold division of the total, Temt, Demetia. The map shows this scheme made geographical on British ground. The four Sets are the southern and upper half of the whole. At the edge of Dyved, close to the dividing water, is Gower, answering to the Egyptian Kar, the lower and divided Kar-neter, our nether Gower. This Kar is denoted in the hieroglyphics by the sign of a half-heaven, because the Kar-neter was but the sun's course for half the round, the lower, northernmost half that begins with Gower.

The Kar or Kart is a course in Egyptian, an orbit or measure ; in this case the sun's course through the lower half of the divided heaven. Two Kars in the hieroglyphics read Kar-ti ; the ti duplicates the Kar, and the determinative of Kar-ti is two half-heavens. Karti, then, abrades into Kart, the total orb, in English the garth, girth, garter, or quart. The Egyptian Kar-ti, the plural of Kar, have various forms as orbits, holes, passages, enclosures, prisons, showing they were enclosures of whatever kind, and the Welsh Caers were known as fenced enclosures. Karti is the exact equivalent of Wales. Four Kars in Dyved would complete the eight required to make the unit of the Sat of eight "Sa"s. Four "Sa"s or Sets and four divisions as Kars, make the total of Dyved, as in Egyptian Tebt, the measure, which in one form is equal to our bushel, in another it is a table, with which we may compare the

1 *Kadair Teyrn On,* 4.

Round Table, in another a sarcophagus. The Teb or Teb-t, as an unit of measure, was variously applied as dry, liquid, and land measure. Also we find the "Sa," divided into one-sixteenth of a measure of land, as in England the Tobit is subdivided and differs in different counties.

The division of eight, however, is primary, and the look of the whole thing is that the land of Dyved was the twin-total, afterwards divided into eight nomes, four Caers in Dyved, and four in Defenset, Dorset, Somerset, and Wilset considered at the time to be the TWO LANDS of Wales; Devonshire being called West Wales. Wales is Gales, Kars, Gowers, the plural of a course or total. That total being Egyptian was twinned, the lower and upper Kars, the two Kars, Gowers—Gales, Wales.

The two Tebs in Dyved and Defenset, if designated in Egyptian, would be Teb-ti, the dual Teb, as Teb-ti, a pair of sandals; and we find that the "Tebti-pehu" was an Egyptian name of the 12th nome of Upper Egypt, meaning the Water Nome [1] of the double division.

Tibn-ti, the double Dyfen, appears on the Monuments.[2] Our two Tebs or Tebn, Dyved and Defu, form the double division of the water nome just as does the Tebti-pehu of Egypt. Also, Dyfen as the one-half of the whole, is extant in the Welsh Dobyn, a half-pint measure. This total, these two halves, these eight nomes, four to the south and four to the north, yield the eight regions of Egyptian mythology, and an Egyptologist would expect to come upon the Sesennu or eight great gods of Egypt. These also were known to the Druids; they were the eight persons in the ark, assumed by Bryant to be Noah and his family.

Taliesin sings:—"A song of secret significance was composed by the distinguished Ogdoad, who assembled on the day of the moon," that is, on Monday, the day of Taht, the lunar deity, lord of Sesennu. They assembled, and "went in open procession; on the day of Mars, they allotted wrath to their adversaries; on the day of Mercury, they enjoyed their full pomp; on the day of Jove, they were delivered from their detested usurpers; on the day of Venus, the day of the influx, they swam in blood; on the day of Saturn (lacuna); on the day of the sun, there truly assemble fine ships."[3] Skene's version is somewhat different, still the EIGHT are there.

In the *Ritual*, where the solar imagery has become eschatological, and has to be read backward to recover the primary meaning, the solar (or spiritual) place of re-birth is in An, the On of the Hebrew writings. In this region we find the Hall of Two Truths in which "a soul is separated from its sins." One name of the hall is the Uskh, the water-place, the limit, the division. The Uskh Hall has for determinative the three feathers, corresponding to the three feathers

---

[1] Pierret, *Tebti*.   [2] Maspero.
[3] Ancient Poem, *Welsh Archæology*, p. 74.

of Wales, and Lazamon, in his *Brut*, tells us that, when the good Belin had made the burgh of Caerleon, he called it "Caer-usk." The ex and usk of our water-names sometimes permute, as do husk and huck, for a pod ; and as before suggested, Oxford with its Uskhs (halls) is not merely the water-ford, but represents that crossing of the boundary where we find the Uskh Hall in An.

The crossing is preserved by name in the Ex, X, or cross sign. Exan is a name of Cross-wort and the Ex-ford is the ford of the crossing where the water and the Hall of the Two Truths are found in the solar circle. The Uskh-Hall is extant in the Esking, a name of·the pentice or sloping roof.

Caer-leon, which had belonged to the Sabean naming, was changed by the sun-god Belin the Good, *i.e.* Nefer-Baal, into Caer-Usk. The quadrangular Caer of the Kymry is the four-cornered Kher of Egypt. This was the shrine of religion, the cell of the priest, the oracle of the divine word. The Kymric Caer or Car passed out as the Gadhaelic Kil and English Cell. There are 1400 Kils in Ireland, a considerable number in Scotland, and some in Wales. These were not founded, although they were adopted, by the Christian missionaries, the cuckoos who did not build their own nests. The Kirbys are the places of the ancient Kirs and Kils, which were there ready to be re-named.

The suggestion now to be made is that the four Sets and four Kars of the double Dyved were a localization of Sesennu, and that this region was the probable place of the first landing, colonization, and naming of the Egyptians in Britain.

In British fable, Devon is one of the heroes who came into the island with Brute, our Pryd. He is famous for chasing a giant to a vast pit EIGHT lugs across; the monster, in trying to leap the chasm, fell backwards and lost his life. The giant is a type of the vast, the unmeasured; Devon is the mapper-out and measurer; hence, when Brute portioned out the island, this fell to Devon's-share.

> "And eke that ample pit, yet far renowned
> For the great leap which Devon did compel
> Coulin to make, being eight lugs of ground,
> Into the which returning back, he fell." [1]

A lug is a measure of land, as is a league, it is the log or reckoning, Egyptian Lekh, of various lengths, as a pole, a sea-mile, or three miles. The mythical pit represents the Kar (Eg.), and it is the pit of eight "lugs" across. Devon, according to Spenser, is followed by Corin, who gave the name to Cornwall. These answer to the double Kars or Wales. Devon, being a mythical name, applies equally to Dyved, and the eight lugs correspond to the eight Sets and Kars.

But we can bring this naming of the two lands, according to the

---

[1] *Faery Queen*, ii. 10, 11.

Egyptian imagery and mode of expression, to a yet finer point than in the double Caers of Wales and Cornwales with the four Sets and four Caers on either side of the water. It will be suggested that the landing-place was in Menevia, now called St. David's. In Dyved we find the seven provinces of Sut-Typhon. Dyved from the Ap or Af is primal. This Ap enters into Menapia as the primordial, ancestral district. Not far from this point and place of landing is Cardigan Bay. Into this runs the Tefi, named like the land, as the first of the rivers of Dyved. Its water is the line of division between North and South Wales. Here then is the lesser and prior form of the dual circle of two halves; in Egyptian, this is Karti, and Karti-gan is Cardigan. Khen (Eg.) means to alight, rest, a sanctuary, and a CENTRAL apartment, or dwelling-place. And the central dwelling-place in the double orbit of north and south, the Karti, still bears the name in Cardigan.

We may venture a little further inland. The first of the shires distinguished from the Caers and Sets is Shropshire. Shrop, Scrob, or Salop, are all derived from Kherp (Eg.), the first form, model figure. The first division, called a Shire instead of a Caer, would be Kherp- (Eg.) Shire, or Kherf-Shire, and in this county the name is extant as that of the river *Corve*. Moreover we see the people of inner Wales pushing farther in, as the first inhabitants of east Shropshire known in the pre-Roman times were the Cornavii.

The Romans called Salisbury, or Sarum, Sorbidunum, *i.e.* Kherp-dun, and the name in connection with Stonehenge on the Plain shows that here was the sovereign sanctuary the Kherp, the first, consecrated, excelling, surpassing, ruling seat (dun) of worship. So in Coptic the Egyptian Kherp becomes Sorb. The same root is represented by the royal name of *Corfe*, and the *Glev* of Gloucester.

This word Kherp is the most probable original of the name of Europe, answering to the first quarter named in the north. This important root will be elaborately treated in the "Typology of Naming." Meantime it may be reiterated that Kherp means first in form or any other condition of being. The Kherp is the king as first person; the prow of the vessel as fore-part; the paddle as primary means of propelling. It is the first castle as Corfe, the first shire as Shrop, and will equally apply to Europe as the north land discovered by the Kymry or Khafitic race.

Kherp meant to paddle and steer, at a time when both were one, and Europe, the isles of the Gevi, were the first lands steered for, therefore the Kherp, whence Europe. This also is the most probable origin of Albion. Aristotle mentions the islands of Albion and Iërne four hundred years before Julius Cæsar is supposed to have named the land in Latin. "Beyond the Pillars of Hercules is the ocean that flows round the earth. In it are two very large islands called

Britannia; these are Albion and Iërne."[1] The name is not derived from Albus (Lat.), the White. The ancient inhabitants are called Albionës. Uni is the Egyptian name for inhabitants. The Kherpi-uni (Albioni) would be the first people of the isle, as the Kherp. Such a derivation may be followed farther north to the land of the Lap (Kherp), the first who prowled or paddled to that region.

The various names of Ireland, Eiri, Eri, Eriu, Heriu, Ieriu, Iveriu, Iberiu, Greek Ierna, Ptolemy's Iouerna, Mela's Iuverna, and the still earlier Hibernia, all point to a typical name corresponding to the form Iberia, and Ib, Iv, and Hib all meet to unify at last in Kheb or Khef, the name of the genitrix. This name, first applied to the north by the Sabeans to denote the hinder part of the heavens, the cave of production, when the dog-star determined the south to be the front, was extended to the west, the Ament in the Solar reckoning, and the Kheb, or Sabean north, became the solar west. Hence there is a goddess Kheft, who is lady of the country, or heaven, the lady of the west, the place of going down of the sun and hinder part to the east, the front, reckoning by sunrise. Now the persistence of the "Iu" in the variants Eriu, Ieriu, Heriu, Iveriu, and Iberiu (the N in Erin and Hibernia, is later) leads me to think it may be the Egyptian "Iu," which is dual and duplicates. Thus Kheb-er-iu would be the Twin, secondary or duplicated division (er) of the Kheb quarter, in short, the western Kheb, and secondary to the north in accordance with the solar reckoning. KHEB-ER-IU read as Egyptian is the secondary Kheb, which was the western the solar Kheb, whereas the northern was stellar,[2] and Ireland is still the typical "Land of the West."

With the restored readings (no primitive word begins with a vowel), Kherp-ion (Albion) and Kheb-eriu (Ireland) will also yield the first and second in another sense, and in the order of Albion and Iërne (Aristotle), the final Great Britain and Ireland.

Romana was one of the native names of the island of Britain. Rumena in Egyptian signifies the extent of, extending as far as, the limit, or thus far. So read, ROMANA would be named as the farthest point of land. THULE is another name, which read as Egyptian corroborates that of Romana. Tur (Eg.) is the extremity, boundary, frontier, land's end, as in Ultima Thule, or the DHUR of the Butt of Lewis.

In the accounts preserved by the Triads one of the three names given to Britain is "Glas Merddyn," or the green spot defended by water; that is, the green island. Mer or Meru (Eg.) is an isle. Mer and mer-t are names of the sea, the water-circle. Ten means to be cut off, divided, made separate; or Mert is the water, and ten the seat, an early form of the Tun. Mert-tun (Welsh Dyn) yields the island as the sea-surrounded tun.

[1] De Mundo, s. 3.   [2] *Vide* section i. pp. 14, 15, 16.

England, we are assured, is named from the Angles. But one begins, not without reason, to doubt everything currently taught concerning our past. To the people of Brittany this country was their ANCOU-land; the land of souls, to which the spirits of the dead crossed over by night on the ANCOU-car, as the souls of the Norse heroes passed to Britinia, the White Island of their mythology. Ankow in Cornish is death; but in Brittany the Carr-au-Ancou is the soul-car. The Egyptian word "ANKHIU" is a name often used for the departed, and in the inscription of Una the coffin is called the HEN EN ANKHIU, or chest of the living. In the German mythology and folklore England is a land of spirits, and when the REVENANT visits her mortal lover, nothing is more common than for her to hear the bells ringing, or the spirit-voices calling for her in England. But this could hardly be because some people called Angles once landed in the isle. Of course it is not the land we know, that is meant, but the name of England the island and England the spirit-land have a common origin. They are identical, because in Egyptian Ankh is the word for life. Ankh-land was the land of life in mythology localized by name in England. And for the people on the mainland the white island beyond the waters was blended with the Ankh-land that lay on the other side of the waters crossed by the souls in death.

England is thus treated as the land of life, or souls, and a similar thing occurs when Homer sends Ulysses to consult the dead in the north, the country of the Kimmeroi.[1]

Khema is Egyptian for the dead, and RUI, the isles. These were astronomical, and belonged to the underworld in the north, where the sun travelled in passing from the west to the east, and the Isles of the Kymry are located geographically in the same direction. There is another cause for this confusion or interfusion. England, according to the Roman report, was looked upon from the continental side as the supreme fount of Druidic lore. If, as is more than probable, the Egyptians made this their earliest seat and permanent centre, if this was the island first lighted up, the beacon first kindled to shine across the waters as an intellectual pharos to the mainland in the dark night of the past, the fame of the geographical England would also help to blend it with the mythical Ankh-land. Moreover, there are reasons for thinking that this was literally the land of the dead (or spiritual living), used as such for the burial of those who belonged to the Druidical religion, and that to cross the waters for burial was a typical custom, a symbolical ceremony, whilst our island was the favourite funeral ground, an ark amid the waters, the Ankh-land that was the Ark-land.

Ankh-land is an Egyptian compound as Ankh-Ta, the name of a quarter in Memphis. Ankh-taui is the double land of life, or the land of death and new life. Between the two lay the water that was

---

[1] *Od.* b. x. and xi.

crossed in death, and this passage was represented in the ferrying of the mummy over the River Nile. Britain and Brittany were the two halves of this water-divided land of life. And according to Egyptian ideas, the dead would be carried to the other side for the resting-place across the water. THIS would be the Ankh-land to THAT, and Brittany to Britain. Thus we find the Ankh-land there in Anjou and Angevin. The name of England as the typical land of life is illustrated by the mummers or guisers of Derbyshire, who perform a play of St. George. The opponent of the hero is Slasher, a type-name for the fighter. The equivalent of Slash is found in Sersh, an Egyptian name of a military standard. Slasher is slain, and it is the part of the king of England to restore the fallen Slasher to life again.[1] The monarch explains that he is the king of England, the greatest man ALIVE (Ankh).

"When Hempe is spun, England's done," says the ancient distich. Bacon interpreted this as a prophecy signifying that with the end of the reigns of Henry, Edward, Mary, Philip, and Elizabeth, whose initials form the word Hempe, England would be merged in Great Britain. Such prophecies belong to the hieroglyphics. Hemp is synonymous with the hank as the hangman's noose. The noose is the Ankh. The goddess Ank wears the hemp on her head; the Ankh (hank), loop of twisted hemp or flax, was the sign of living; when this (as hemp) is spun out, the Ankh-land is done. This seems to be an allusion to the living and to the land of life.

When Bede calls his countrymen the Angli, it does not seem probable he should mean that the people of the island were Angles because of three boat-loads of Norse pirates having landed in Thanet, who were followed by hordes of Jutes, Saxons, and Angles. The British people could not have become the Angli in that sense any more than they had become Romans. Procopius, in the sixth century, mentions the ANGILI of Brittia, opposite to the mouth of the Rhine. Had Britain then received its type-name from the continental ANGLES? The ANGILI, INCH-ILI, ENG-ILI, were the islanders. Ankh is an ethnological or topographical name in the Texts as "Ankh, native of a district."[2] That district would therefore be Ankh-land. The dead of Memphis rested in Ankh-ta, the land of life. The eternal region was represented by an island, the Island of Tattu amid the waters of the Nile. Ankh-ta is Ankh-land, and as an island or Inch-land that was England in Egypt. Lastly, England has been the Ankh-land ever since it was named Inis-Prydhain by the Kymry. Inis and Inch (as in Inchkeith) are identical with Ing, Eng, or Ankh, and the island is the Ankh-land, the Inch-land, Ynis-land, or England, because it was the island and the land of the INGS, which name was afterwards turned into Angleland.

It has been suggested that the Euskarian or Iberic ETAN, as in

---

[1] Dyer, p. 469.   [2] *Select Papyri*, 108, 1.

Maur-ETAN-ia, Lus-ETAN-ia, Ed-ETAN-i, Cos-ETAN-i, Lac-ETANI, Carp-ETANI, Or-ETANI, Turd-ETANI, and many others, is contained in the name of BRITAIN. The present writer sees in the ETAN a form of the Tun, as circle or enclosure. Aten or Uten (Eg.) means to form the circle, and Huten is the circle. The exact equivalent of Etan is UTAN (Eg.), later Etan, the name of a consecration, sacrifice, offering, and libation. These were made in the Tun, as the seat and circle of the dead. Uti (Eg.) is the name of the coffin and embalmment. HUDUN in Arabic is burying; and as all the chief type-names for the dwelling-place are derived from the place of sepulture, the Etan is not likely to be an exception.

The ancient Britons also called the country Inis-Prydhain, the Isle of Prydhain. Nennius derives the name of Britain from Brute, whom we identify with Pryd or Prydhain, the youthful sun-god of the Britons. But it appears certain that Britain was inhabited by the men of the River-drift type in the Palæolithic, if not the Pleistocene age, before Britain was broken off from the mainland to become an island, and it happens that an English word BRITTENE means to divide, to break off, divide into fragments. In Egyptian Pri or Prt signifies the thing or act in process, visibly appearing, bearing off, and running away; Tna is to divide, separate in two halves. At one time the water-way was a mere FRITH, and PRIT, PART, or BRIT is equivalent to Frith; TEN, as in tine and tint for one-half bushel, is the Egyptian TENA, to be made separate or TWAIN. As we have seen, this principle of naming the land visibly divided and made separate was applied to the Isle of Thanet; and the Brittany on one side of the Channel and Britain on the other are geologically known to have been divided in two; the names are there in accordance with the fact as if to register it, and prove that they had been one, whilst Brittene in English and Prit-tena in Egyptian agree in showing they were named as the land that was known to be, was manifestly, even visibly broken and separated in twain. Britain and Brittany, then, we take to have been named as the broken and divided land; as the visibly-divided land, or as the land in the process of visibly dividing, separating, and becoming two.

So in a thousand ways and things, myths, rites, customs, folklore, superstitions, words, names of places, and persons, dead Egypt, so called, is yet living in Britain, and has but undergone her own typical transformation which the rest of the world considers to be death.

SECTION X.

TYPE-NAMES OF THE PEOPLE.

HERE, it is submitted, is direct positive evidence of a remote pre-historic time as interesting to us as deciphering the cuneiform or hieroglyphic inscriptions or exploring Palestine. Speculative dreaming over a far-off past which never had a present has nothing to do with these facts of language ; these names applied to things, places, persons; this total system of mythology. The present writer did not begin as one of those poor pitiable "Keltomaniacs" who had been poring till purblind over their reliquary remains of a past which they could not prove, still holding fast to their faith in the preciousness of what they clasped in their hands or enclosed in their heart of hearts, and who, when they shyly showed their treasure in the light of the present, were told their diamonds were but charcoal, and the look of faith and wonder in whose yearning, dreamy eyes was met with scorn. or the simper of superior knowledge, until they felt the increasing light of to-day did but serve to make their folly all the more definite. Such a one was MYFYR MORGANWY, who lately died as Arch-Druid of Wales. He was certainly in possession of the ancient cult more or less, which has never been altogether extinct in the country. He adored the sungod Hu as his saviour, and assembled the brethren at the time of the winter solstice to celebrate the coming of his Christ to bruise the serpent of Annwn, that seed time and harvest might not fail. He maintained to the last that Jesus was Hu, and the Christian system a corruption of Bardism.[1] Not as one of these did the present writer begin, and not as with them is the matter going to end.

We shall now turn with increased interest to the Roman and Bardic reports concerning the learning of Britain. Those stern Roman eyes hard as granite, out of which the British battle-onset had so often struck the fire-flashes, like the granite broken all a-sparkle, have in

---

[1] Letter in the *Western Mail*, March 12, 1874. See also the letter at the end of this volume.

them an arresting lingering look almost of wonder as the writers turn
to speak of the barbarians into whose faces they had peered so often
under the battle-shield, and whose souls they had never penetrated ;
whose past history they had never fathomed up to the time of leaving
the island in a last retreat.

"The Gauls," says Pomponius Mela, "have a species of eloquence
peculiar to themselves, and the Druids are its teachers. These profess to know the size and form of the earth, and the universe, the
motions of the heavens and of the stars, and the intentions of the
immortal gods. They take the young nobles of their tribe under their
tuition, and teach them many things in secret. Their studies last a
long time, as much as twenty years, in caves, or the depths of the
forests. One of their tenets which has transpired is the immortality of
the soul and the existence of a future state ; which inspires them with
much additional courage in war. As a result of this doctrine, they
burn and bury with their dead all those things which were adapted for
them when living. In former times they carried their accounts with
them to the grave, and their claims for debts ; some of them would
even burn themselves on the same funeral pyre with their friends, that
they might be with them in a future life." [1]

"Bardism," say the Barddas,[2] "originated in the Isle of Britain.
No other country ever obtained a proper comprehension of Bardism.
Three nations corrupted what they had learned of the Bardism of the
Isle of Britain, blending it with heterogeneous principles, by which
means they lost it : the Irish ; the Kymry of Armorica, and the
Germans." Beyond the Barddas are the Druids. "This institution,"
says Cæsar, "is thought to have originated in Britain, and to have been
thence introduced into Gaul ; and even now those who wish to become
more accurately acquainted with it generally repair thither for the
sake of learning it."

It is not necessary to notice the customary explanations of ancient
names as Roman or Norse, because, if the present reading of facts be
true, they are to a great extent superseded. Our land was mapped
out and named and trodden all over ages before the Romans and
Norsemen came, and their bloody hoofs did but little to obliterate
the deeper footprints of the earlier men of a peaceful invasion.

It is beginning to be felt more and more that the effects of military
conquests on the life of the land have been vastly exaggerated. Such
conquest does not sink very deep ; although it makes a great show on
the surface, it melts into the earth like a snowfall and passes away.
The re-conquest by the conquered begins at once. This is especially
illustrated by the conquests of the Turk. It was so more or less with
the Romans in Britain. No such Romanization occurred as that
which is advocated by one class of writers, except in the Codification
of the laws. No *tabula rasa* was ever made by the Romans, or

---

[1] Pomponius Mela, iii. 2.   [2] *The Barddas*, Williams, pref. p. 27.

they would have remained; nor by the Norsemen, for they were incorporated and absorbed. Both fertilized the race that fed on them and flourished.

Arnold of Rugby gave utterance to a false cry in English literature on the subject of Kelt and Saxon; he was unwearying in his glorification of the Saxon and depreciation of the Kelt. This cry was lustily echoed by his followers, and has often been re-echoed by the present writer in the most frequently demanded of all his lectures, one on the Old Sea-Kings. That cry has been a common bond even between the historians Froude and Freeman. Nevertheless we have been falsely infected with a shallow enthusiasm respecting the Saxon element, and were almost entirely ignorant of what might be signified by the words " Keltic " and " Kymric."

The Kymry and the Keltæ clung to the soil which their names had covered on the surface, and their roots had ramified below. The race was as ineffaceable as the names. The conquerors brought a fresh infusion of life and a wash of new words and later letter-sounds, but the older elements remained. Men might come and men might go, the race went on for ever. The Loegrians of England coalesced with the Saxons from the Humber to the Thames, and must have mainly supplied them with wives, as mothers of the amalgam.

Of course the present mode of diagnosis does not enable us to get beyond the namers, or to distinguish between the Cave-men of the Palæolithic tribes and the men of the River-drift. These have to be left in the lump as the Kymry, the race of Kâm or Khebma, the ancient genitrix of the north first named in Æthiopia. If there had been a pre-lingual race that crawled out over Europe from the warm African birthplace, language could not tell us. At present, however, there is no reason to suppose there was. The Cave-men answer to the Kafruti, whose representatives in Herodotus are the Æthiopian troglodytes.

The Kep (Kef) in Egyptian is the concealed place or place for concealment, the Kafruti of Africa were Cave-men, and language reproduces in the Isles the Kep, Coff, or cave, whether as the womb of the mother or the earth, which was primally personified in Africa by the Kheb-Ma or Mother Kheb, the hippopotamus. And on the other line the Khebm abrades into the Kam type, as in the Cwm, Coomb, Quim, Camster, or Camelot. The men of the Neolithic age as stone-polishers can be identified with the Karti (Eg.) or Keltæ; also the men of the Hut-circles, Weems, Picts-houses and holes in the ground, the Karti, are doubly identified, because Karti (Eg.) means holes underground as well as other forms of the Kar, Caer, or circle, including the dual Wales and Corn-Wales.

Their representatives are still extant in the interior of Africa, where Stanley found them living in the subterranean habitations of Southern

¹ B. iv. 183.

Unyoro, described by him as "deep pits with small circular mouths, which proved on examination to lead to several passages from the mouth of the pit to more roomy excavations like so many apartments." [1] The nearest approach to a Hottentot village is still to be found in a group of beehive houses in the shealing of the Garry of Aird Mhor, Uig, Lewis.[2]

The Egyptian Kar is a hole underground, the Kil. The hole becomes a cell, and the cell a shrine, in the KHER, that is, the KHA-RU or uterine outlet. With the P suffixed, this makes the word KHERP, a first formation which on one line is the CRIB, on another the grave. The entrance and circle of the Cair at CLAVA constitute the womb-shape, and CLAVA represents the Kherp, that is, *the* Kha-ru or feminine cell, which becomes the GRAVE, or, in another type of the abode, the Crib.

The Karti, or men of the huts and holes, are known to have been spinners and potters, weavers and corn-men. A spindle-whorl, fragments of pottery, and a weaving-comb have been found among their relics. Dr. Blackmore discovered the cast of a grain of wheat in the clay which had formed a part of the cover of one of their pits. Also, two concave stones for crushing corn and making meal have been found.[3]

The earliest beings who issued forth from the dark land with Egypt for the MEST-RU (Mitzr), the outlet from the birthplace, were doubtless black and pigmean people. They left their nearest likeness with the Akka and the Bushmen, and these have their fellows, more or less, in the little black or very dark people of various lands. They are extant in the short-statured type of the North. The anthropologists bear witness to the primary pigmean people of the Isles who preceded the Keltæ. The name of the Kymry testifies to the black complexion. Also the Irish preserve two appellations which have been traced back from territorial to tribal names ; one of these is the Corca DUIBNE, the other the Corca OIDCHE. Both were black people at first; the one dates from darkness, the other from night. So, in the African Mandingo, DIBI is the dark ; TOBON, in Mantshu Tartar, and TUFAN, in Arabic, signify the dark night. Not only does DUIBNE denote black, it also identifies the Typhonians, the children of Tef, goddess of the Great Bear, and the celestial black country of Kush, whose star is yet extant by name as Dubhe in that constellation. Also, as the first Goddess of the North was followed by Uati, so the Oidche or night people seem to echo her name ; Uat being a modified form of Khebt, and the Corca Oidche are the people of night.

In Scottish folklore the Picts (Pechs) are the little men, on their

---

[1] *Through the Dark Continent*, v. i. p. 432.
[2] *The Past in the Present*, p. 64. Dr. Arthur Mitchel.
[3] Dawkins, *Early Man in Britain*, p. 268.

way to become Pixies or wee-folk altogether. This tends to connect them with the small dark people of the Palæolithic age. Here language may have preserved an ethnical note, for the name supplies a type used for things minified and small as in the Pixy and Pigmy; Pigwiggin the Dwarf, and the Pykle or Pightle, the small enclosure.

The seven holes underground (Karti), or seven caves; seven provinces, or seven nights, or seven stars, may not be of much avail ethnologically or topographically, but they have their measurable value in the astronomical allegory, as will be seen when all is put together again, and then we shall find that the heavens are a mirror to the pre-historic past of men.

The British beginning was pre-solar and pre-lunar. It was the Sabean beginning on the night side, and the dating from the dark, the mythical abyss common to the oldest races in the world. Cæsar says: "All the Gauls assert that they are descended from the god Dis, and affirm that this tradition has been handed down by the Druids. For that reason they compute the divisions of every season, not by the number of days, but of nights; they keep birthdays and the beginnings of months and years in such an order that the day follows the night."[1] The British in like manner kept the same reckoning, according to the first form of time in the Jewish Genesis. They reckoned by star-time and dated from the darkness. They were the children of Seb, who in the older and dual form is Sebt, and in the oldest Khebt. In the hieroglyphics the star, as Sef (Seb), is the sign of yesterday and the morrow, the star of the evening and the morning that constituted the first time. They were Sabeans by birth.

Hence their claim to be the children of Dis. The Latin Dis is the Egyptian Tes. The meaning of wealth, as that which is derived from the depths of the earth, will not help us. But the Tes is the Hades, the depth, the abyss itself. This is the British Dyved (Tepht) of the seven regions. In the Druidic mythology there are seven provinces of Dyved called the patrimony of Pwyll, over which Seithwedd Saidi was the king. Seithwedd implies septiform, and these seven provinces are synonymous with the seven caves of the Mexicans, Quichés, and other American races, from whence the migration started in the dawn of creation.

Latterly we hear a good deal of the Eusks or Euskarians, a black-haired pre-Keltic race, short of stature, supposed to have left remains in the Basque, the Laps, and among the earliest people of Wales and Ireland. The Basque call themselves the Euscaldunac. The Eusk name, whatever its origin, is perpetuated on all the waters of Europe, and this in all its forms is traceable to Egyptian.

Sekh or Uskh is an Egyptian name for water. Uskh and Sekh interchange, and the Uskh-ti or Sekh-ti are the mariners of Egypt. The Sekht is a barque of the gods, very archaic, as it represents the

[1] B. vi. 18.

lotus, one of the earliest arks of the waters; the still earlier one being the Irish ORC, the uterus. The Uskh was a large broad boat of burthen, on which the Egyptians moved their armies by water. Uskh also means to range out far and wide. Thus we have the water and the bark, the mariners and the voyagers, all named in Egyptian; also the Uskh people who went out on the Uskh, in the Uskh, rowed by the Uskh, in the Uskh range as far as the migration extended.

In Cormac's *Glossary* the ancient form of the name "Scot" is Scuit, and signifies the wandering. Sailors are the wanderers of the waters. In Egyptian Khet means to navigate. With the S, the causative prefix to verbs, we have the word Skhet, the vessel, ark, or boat. The Skute is an English name for a wherry, and the Scuit, or the wanderer, or sailor, corresponds to the Egyptian Sekht, the mariner. The word Scot, therefore, may render the Uskht, as the wanderer of the waters, as well as being the name of Sekht, who was the ark personified, the primitive bearer of gods and men. Us (Eg.) has the same meaning of a large, extensive range. Ukha is a name of the bark, and means to seek and follow. These are variants of Uskh, and will therefore include the Ugrian name, the Esquimo, the Ostiak, Uzbek, Osage, Oscan, and others, according to the ethnological data. If to this name we add the masculine article (Eg.), P, P-uskh yields the Uskh, or Basque. Also, just as Uskh and Sekh interchange in Egyptian, so another name of the Basque people is SIKani. And if the Egyptian feminine article be added to Uskh or Sekh, we have the TOSK, and the TSHEK, the native name of the Bohemian tongue, whilst THE Oscan becomes Tuscan.

The Cangia is a native name of an Egyptian vessel. And the Cangiani once inhabited the promontory on the Minevian shore opposite Mona.[1] Menevia or Menapia denotes a primal place of anchorage, and here the Cangiani name preserves that of the Egyptian vessel, the Cangia, Chinese Junk, as the oldest form of the Kennu or canoe, also surviving in the SEGON (S-khen) of Segontium, near Carnarvon, the chief city of the Cangiani.

The Finns call themselves the Quains. In the hieroglyphics the Khent sign also reads Fent, and the Khennu or Khenit are the navigators, sailors, pilots; the men who paddle a canoe. Khen means to navigate, transport, convey by water. The Khenit are equivalent to our Khenti of Kent, who were called the Kymri. The Khen as seafarers may also have had an especial territory (Tir) in Cantyre, as well as in Kent and Segont.

PHANES is said to have been found as the supreme divinity of the Finns.[2] If so, this suggests the derivation of their name from the same original as that of Pan and Fion, or the Fenians. That

---

[1] *Richard of Cirencester*, b. i. ch. vi. 25.
[2] *Recherches sur l'anciens Peuples Finois*, p. 66.

original, according to the present reading, being An, the Egyptian Anup or Sut-Anubis, God of the Dog-star, our Baal. An, Åan, and Khan (as in the Cynocephalus) are interchangeable; also the Ben and the Fenek are types of the dog-headed divinity,.and FAN or FION, Pan or Phanes, are forms of the same name. The Finns also worshipped the Great Bear as a goddess named OTAVA, Tef being the Egyptian Typhon represented by the Great Bear. They have the great hill Kipumaki answering to the mount of the north, the Khep of Khebt. On its summit was the large flat stone of Kêd, surrounded by other large stones; in the middle one there were nine holes for the burial of diseases. The goddess who collected the diseases and cooked them did so in a vessel corresponding to the Pair or Cauldron of Kêd. Her name of KIVUTar seems to be a developed form of Kheft.

The goddess of healing is named SUONETAR, and the invocations addressed to her are called the Runas of SYNTY, *i.e.* of regeneration.[1] San (Eg.) means to charm, heal, restore, and save.

The Finns belong to the so-called Chudic or Shudic races, and the name identifies them with Sut of the Dog-star, the son of the Typhonian genitrix and first form of the male god in heaven. In Egyptian the name of Sut, Suti, Sebti, is a deposit from Kheft, the name of the north and its goddess. The Shudic races are also those who went northward, but Shudic is secondary to the Khafetic or Japhetic name, and if the modification had taken place before the migration occurred, then the people so named would be secondary also, following the Khafetic or Kymric race. Still, the modification of Kheft into Chud or Shud may have been wrought out in Europe and the race may belong to the original people of the north.

The early inhabitants called Britain the Island of Beli, that is Baal, who, as Sutekh, the child Sut, is identifiable with the Saturn of Plutarch, who was bound in one of the British Isles. The people of the star-god Beli are one in mythology with the Finns of Phanes, Fenek, An, or Anup. The "Chronicles of Eri" assert that long before the Kelts left Spain, the god "Baal had sent the blessed stone Liafail" to their ancestors with the instruction as to its proper use. This points to the Keltæ coming by land from Egypt to Spain, and thence to Ireland, and at the same time distinguishes between the Keltæ of Keltiberia and the earlier Kymry. According to the present view, Baal is Bar, the son of Typhon, or Sut-Anubis, who took shape in these islands as the Sabean Arthur, son of the Great Bear. The stone is the seven-stone of the Druids,[2] the stone of the seven stars. One of the stones, the Syth-stone, bears the name of Sut, *i.e.* the Sabean Baal.

There was a people known to the geographer Ptolemy whom he calls the Epidii. He mentions the island of Epidium, lying between

---

[1] *Kalevala.*   [2] Myfyr Morganwy.

Scotland and Ireland, and designates the Mull of Cantyre Epidion Akron. The Epidii may very well represent the Khefti. Also, in Egyptian, Kheft passed into the later Buto (Uati), and the isles of Bute may also represent the modified form of Kheft. They are seven in number, and as Hepta is seven in Greek, and Hept in Egyptian, it seems probable that the seven isles of Bute were the Epidium of Ptolemy, and the Epidii an extant relic of the Khefti, named after the Goddess of the North and the seven stars of her constellation.

A circular enclosure at Dunagoil, in the island of Bute, is called the Devil's Cauldron, in which rites of penance were performed. One part of the purgatorial pains consisted of sleeplessness; the penitents being threatened by the priests with eternal punishment if any one of them went to sleep. To prevent somnolency, they were provided with sharp instruments for the watchful to keep the unwary awake. Also in the burial-ground of the church the sexes were not allowed to mingle, but were interred apart until after the time of the Reformation.[1] These facts show the Devil's Cauldron belonged to the goddess Kêd, and that her rites survived to a late period. Her cauldron had to be watched for a year and a day under the strict injunction that the boiling was uninterrupted for a moment.[2] The penitents were keeping Keridwen's watch, and the dividing of the cauldron in two halves was also imitated in the separation of the sexes in burial.

Kheft (Eg.) came to signify the devil in Egypt, and here the cauldron of Kêd had become the devil's.

The Epidii are in the next stage of naming to the Japhetic race, descending from Kheft or Kêd. The same name apparently enters into that of the Menapii, a people of Ireland also mentioned by Ptolemy. Dublin is supposed to be on the site of the ancient Menapia. Pliny likewise designates the island of Mona or Man by the name of Menapia, by mistake apparently, as Menapia was the name of the mainland-point of the promontory of Segont. Also Menevia is the old name of St. David's. Apia and Evia are modifications of Khefi or Kêd. The old genitrix as Teb seems to have retained that form of her name in Dublin. In Egyptian Men signifies to warp to shore, arrive, and anchor; it also names the harbour. Api (Eg.) means the first, chief, ancestral. So derived, Menapia or Menavia names the place of the first arrival, anchorage, and harbour, on whichever coast it may be found, and the Menapii would be the primordial inhabitants, whether in Wales, Ireland, or elsewhere.

The tradition of the Bards, now to be listened to with more respect, is that the first colonies came forth seeking a place where they could live in peace, and that they fled from a land which they could not possess without warfare and persecution, whereas they desired to do justly and dwell at peace amongst themselves. So they came across

---

[1] *Encyclop. Brit.* 8th ed., article " Bute."   [2] Hanes Taliesin.

the "hazy sea," from DEFROBANI. Defrobani agrees with Tapro-bane, a name of Ceylon. Did they mean they came across the hazy sea from Ceylon?

Here we have to distinguish between the celestial and geographical naming. Tep is a particular point of all commencement in the my-thological astronomy, the beginning of movement in a circle, the starting-point. Ru is the outlet, gate, mouth. Tepru means oral commencement. Tepru is also a name of Tabor, a mount of the birthplace. Ben is the supreme height, the roof. Tep-ru-bani was an initial point in the solar circle, without going back for the moment to the earlier circle of Tep, the Great Bear.

The old writers, in their stories of voyages and the strange crea-tures to be met with in the East, often speak of the mermaid; a being half fish, half woman, that was to be met with off the coast of Tapro-bane. The Mermaid of the zodiac is the original of this, and is still to be found in the north-east or Taprobane, the sign of the Fishes, the lofty outlet or Bekh of the beginning. That is the celestial Tapro-bane, which may have various geographical applications. It happens that we have another name of Ceylon amongst us. The island is likewise known as Serendib. Serendib is the place where the Hindus locate Paradise, the place of beginning. Hence Adam's peak is found in the island. When Adam was cast out of Paradise, say the legends, both Jewish and Arabic, he fell and found footing on the island of Serendib; Eve on Djidda.[1]

Adam is Atum (Eg.), who was the lord of this place and point of commencement in An (the fish) or Serendib. In the Samaritan version of the Pentateuch the name of Serendib (סרנדיב) replaces that of Ararat.[2] These can be identified as one according to the mythos. The Teb or tep is the point of commencement in the circle for Noah or for Adam, and it is the Tap in Taprobane, the Def in Defrobani, the Teve in Teve-Lanka, another name of Ceylon, and the Dib in Serendib. Tep (Eg.) denotes the upper heaven, the top, and the Tepht is the lower. The Ser or Tser was the rock of the horizon; another name for the Tep Hill. This rock, the Ser-en-Tep, was at the initial point, where the solar ark rested in the birthplace of the beginning. From this exalted height we must descend to note a most trivial application of the word and signification of Ser-en-Tep or Serendib. It is a well-known threat, the meaning of which is entirely ·unknown, for our peasantry to promise a boy a "SERENDIBLE good drubbing," and this, which has been perverted at times into a "seven-devil good drubbing," is sup-posed to attain the highest point in thrashing. This is Serendib in England. It is also hieroglyphical. One ideograph of "Ser" is the arm with the sceptre of rule grasped in the hand, typical of the arm of the Lord (Osiris), put forth in the person of Horus, the

[1] Weil, *Legends*, p. 14.   [2] Gen. viii. 4.

Messiah son, at "Ser-en-Tep," each time the sun crossed the equinox of spring. The lifted arm and the topmost point are to be realized in our "serendible good drubbing." This descent of the mythical imagery to such common use serves as a kind of gauge for the length of time demanded for its transformation.

The Ser or typical rock of Egypt, and of the Hebrew writings, whence came the living stream, was likewise the place where precious metals were found, and the word "Ser" is determined by the Tam, that is, gold sceptre; Ser signifying things of a golden hue, and this Ser, the typical rock of old, still survives with us by name. The Tzer becomes the rock, "Sela," for the first time in the Hebrew writings, Num. xx. 8, and in the lead mining districts of Cumberland the beds of rock which contain the ore are called "Sills." The Sil proves that we had the Ser rock of the beginning.

Taprobani or Defrobani may have also been a type-name of Egypt. There was, according to the poet Dyonysius, a Taprobane situated in the Erythrean, that is, the Red Sea. Tep was a city consecrated to Buto (still earlier Tep), goddess of the north. Tep-ru-Benn identifies the point of commencement in the land of the pyramid and palm, if it was meant for Hav. They say they came across the "hazy sea" from the land of Hav. Hav in Welsh and Kheb in Egyptian is the name of corn: Khebu a crop of corn; and they were led forth from the land of corn by Hu, the god of corn. Kheb is the name of Lower Egypt, and Kep of the inundation. Hence we may infer that Defrobani and Hav, if geographical localities, were names of Egypt. Atum (Adam) of the peak in Taprobane or Serendib was the great god of On or An (Heliopolis) in Lower Egypt, and Irish legends assert that the migration proceeded from ANALD, or An, the old.

A people deriving from Taprobane claim to come from the first and loftiest point of commencement. In the celestial allegory this means the circle of Tep or Khep, the Great Bear. If topographical, it is neither Taprobane of Ceylon nor the Red Sea, any more than the heights (Ben) of Dover. They would claim to derive from the high lands in the country of Kam and Kush.

The divine names belonging to the myths are easily traced and identified in almost any language or land. But the names which have become ethnological and topographical for us are now more difficult to determine. Emigrants from Africa would not forthwith become Kymry and Kelt, Gadhael and Pict. The nearest to an ethnological link between Egypt and the isles appears in the name of the Kymry preserved by the Welsh. Kam is a name of Egypt and of the black or dark people. A relic of the "darkness" of Kam may be traced in the meaning of "Gammy" which the tramps apply to their ARGOT as the dark language, or lingo used for keeping dark. The sons of Kam were doubtless very dark when first they came.

Martin, in his *Itinerary through the Western Islands of Scotland*, says the inhabitants of the island of Skye were at that time for the most part black.[1] Doubtless that is over-coloured, but all who have travelled in the isles and in the remotest parts of Wales and Ireland have met with the old dark type, which has been greatly modified by admixture, but is not yet extinct.

On the Egyptian monuments the dark people are commonly called the "evil race of Kush," but when the Æthiopian element dominates, the dark people retort by calling the light complexions the pale degraded race of Arvad. And, in the ancient poem called "Gwawd Lludd y Mawr"[2] the detestation of the dark race for the light breaks out in a similar manner.

"The Kymry, flying in equal pace with ruin, are launching their wooden steeds (ships, the "HORSES of Tree") upon the waters. The North has been poisoned by depredatory rovers of pale disgusting hue and hateful form, of the race of Adam the Ancient, whom the flight of ravens has thrice compelled to change their abode and leave the exalted society of SEITHIN."[3] Here it is claimed for the Kymry that they belong to the dark race, the pre- or ante-Adamic race; the children of Sut, the Druidic Saturn. It is the conflict of the dark and light races, such as is said to be found in the cuneiform legends of Creation. The meaning can only be measured by the mythos which will show that Adam the ancient is the red man, so to say, *versus* the sons of Kam; and whereas the lighter complexions, and later Solarites, the world over, prided themselves on being the children of light, the Kymry date from the dark, and prefer it to the hue of the white men who come to proclaim that they belong to the fallen race of Adam. This preference was expressed long after the Roman period of occupation.

Indeed, it would seem that in the England of our day either the growth of beard is visibly changing the face of the people or else the ruddy, fair-haired, light-complexioned Norsemen, who once came to the surface in these islands and floated for some centuries as the crown and flower of our race, are being gradually absorbed by the primeval dark type just as the foam of a troubled sea flashes white awhile and then merges once more in the dark depths; as if the more ancient hue had included theirs for colouring-matter and still asserted its supremacy.

Kam is a typical name for Egypt, and in the hieroglyphics "ruui" means islands. The Kam-ruui might thus be Egyptians of the isles. But we can do better than this on behalf of the Kymry. And first of the word itself. We are told nowadays by the Latinizers of language that the Kymry are the CIMBRI and CAMBROGES. Why not Comrogues? This can certainly be honestly derived from "Taffy was a Welshman, Taffy was a thief," *ergo* the Kymry were

<hr>

[1] P. 194.   [2] *Welsh Arch.* p. 74.   [3] Davies, *Myth.* p. 569.

the Comrogues or Combrogues, whose language, we may infer, is the Combrogue!

Zeuss[1] gives the oldest recorded form of the word as that in the *Codex Legum Venedotianus*, where it appears as Kymry, plural,—Ruui (Eg.) is plural for the isles,—Kymru, for the country, and Kymraeg, for the language. Kymry corresponds to the Kimmerians of the isles, the north, and the darkness, or the black skin, the Κιμμέριοι, Kimmerii, of which the ground form is said to be Kymr. But there is another possible reading, according to which the Kymry need not have named themselves from Kam or Egypt. Gomer (גמר) is the discoverer of the isles as the first. The Welsh philologists understand the word Kym to mean the first. The Scottish Kimmer, a young girl, is a first form, and Chimp (Dorset) means a young shoot. Kimr, in Keltic, denotes the first in might, as the warrior, and the word appears to enter into the name of the Druidic god known among the Gauls as Camulus. This agrees with Kem (Eg.), to discover; Kem, an instant, first in time; Khem, the shrine of the *first* Horus; Kham, matter, body; Kam, the black; all forms of the first. Moreover, the Egyptians came in the course of time to call the barbarians the ignorant, the savages, the aborigines, KAM-RUTI, and these preceded the cultivated Rut of Kam, or Egypt. Kam or Kem (Eg.) means to discover; rui (Eg.) denotes the isles; and the Kem-rui would be the discoverers of the isles. Here we might utilize the typical Gomer. Gomer, in the book of the generations of Noah, was the son of Japhet, whom we identify with Kheft, the north, and the first of those amongst whom the isles were divided in their lands.[2] So far, so good; but we have further to find the meeting-point of the Kymry and the Kabiri. According to Stephanius of Byzantium,[3] the Cimbri or Cimmerii were called Abroi. Ap (Ab) is the first, the ancestral, head, and with the rui (Eg.) for the isles, the Ab-rui are the first islanders. Abrui is an abraded Kabrui, and Kabrui (Eg.) means the island-born. The Hebrew word Gevi (גוי), rendered gentiles, relates Gomer to the ancient mother Khef and to the north, and the children of Khef were the sailors, the Kabiri, the later Abroi, hence their oneness with the discoverers of the isles, the Kymry. Abaris, the Hyperborean mentioned by Herodotus,[4] was the hero of a story that the writer declined to relate as belonging to the Hypernotians, or one of those tales which are told only to the marines. Abaris was said to have carried an arrow round the earth without eating anything. Abaris personifies the Abaroi or Kabiri, the sailors, and his arrow may have been the mariner's compass. The story was probably told of the Kabiri, who sailed round the world by the aid of the arrow that pointed north. In Egypt only can we find the starting-point of both the Kymry and Kabiri as one people.

[1] Vol. i. p. 226.　　[2] Gen. x.　　[3] *De Urb.*　　[4] Book iv. 36.

The ancient genitrix of the human race in Africa is named Kheb-ma, as before explained, and this word modifies into Kheb and Kam, as names of Northern Egypt. The Ari (Eg.) are the children, sons (a plural of Ar, the son), companions who take one form as the seven Kabiri, answering to the seven sons of Mitzraim, *i.e.* Kheb or Kam, originally Khebma, the place or the mother, Kheb. The letter Y in Kymry is not a primate ; it represents the F, as guilty is the earlier guiltif; so that the original form of the word is KFM-ry, and KFM corresponds to the Hebrew חום, which we find at last in KHEVMA or KHEBMA, the hippopotamus and the type of the land of Kam, the Typhonian genitrix, whose children are the Kamari or the Kabiri. They are the Ari, or children of Kam, the black Arians. The earliest records of the past were entrusted by the primitive people to the keeping of the heavens, and it is there they have yet to be deciphered and read in the primal hieroglyphics. The Kamari or Kabiri in India, Wales, or Scotland, are the children (Ari) of Kam, Kfm, or Khebm, the old hippopotamus goddess of the north pole, who bore the Kabiric seven. These, wherever found, date from the first formation. If the Y at the end of the word implies the K, then we have the Ari in an earlier form as the rekh (Eg.) or race instead of sons, and the Kymraig are the race of Kam, Kvm, or Khebm, the hippopotamus-goddess, the typical abode of birth, still extant in the CWM. The Laps, who call themselves the SABME, are the same people by name, as the ethnologists are now beginning to suspect them to be by race. In Egyptian the Ruti are the race or rekh, and, as we have seen, Khem-ruti is the Egyptian type-name for the uncivilized and savage race.

These then are the representatives of the earliest Ruti of Africa, the Kafruti, Kamruti, Kvm-ruti, or Khebm-ruti, who went out to become the Sabme-ruti of Lapland, and the Kamruti, Kamari, Kam-rekh, or KYMRAIG, or Kymry of Wales.

The traditions of the British bards may be precipitated into some form of solid fact when we can once evaporize the mythic matter which held them in solution. These tell us that the colonizers came across the " hazy sea," considered to have been appropriately named, from the land of Hav. Hav in the old Welsh, as in Irish, would be Ham. Ham is modified from Kam. In Egyptian we have the origin of this permutation of M and V, and there it is M and B, as in Num and Nub, these permutations are made visible in the hieroglyphics by which more than one phonetic value is obtained from a reduced ideograph. If we take the permutation of the V in Hav with B, this gives us the land of Hab.

A modified form of Kheb is found in the name of a land that lay below the second cataract, called the land of Heb. " I entered the land of Heb," says Su-Hathor, " visited its water places and opened its harbours." This was in the time of the Second

Amenemhat, Twelfth Dynasty. That may have been the Welsh "Land of Hav," with the name so ancient as to have then become thus modified from Kheb!

The land of Hav we know as Kab, and it has been shown how the two names of Kam and Kab are interchangeable on account of an original name Khebm or Kvm (בןח), Welsh Cwm. We can illustrate the modification in Welsh by the name of the Gipsy. There is no need to assume that the name of Gipsy is a corruption of Egyptian. Further research will teach philologists that language never has been tampered with in the past as it is in the present. Kheb is the hieroglyphic form of Egypt. Si is a son or child. Khebsi, the child or descendant of Kheb, Kab, Kab-t, Egypt, is our Gipsy. And in Welsh the Kab has not only been modified into Hav, it has lost the H in AF or Aiftess, a female Gipsy, whilst keeping the meaning, and Egypt is known as Yr Aipht.

We begin with Kab, the inundation, that which is poured out on the ground. Kheb is to be laid low, be low ; Kheb is Lower Egypt. Khept is to be extended on the ground, to crawl, sit, or be prostrate on the ground. Kaf and Hef also signify the same thing, to squat or crawl. Hef is the name of things that go, squat, or crawl on the ground, as the viper, snake, caterpillar, worm, and a squatting woman. Af is the sun in the lower hemisphere, the squatter and crawler. See how this survives in English. The Eft crawls on the ground. The Havel is the slough of a snake, cast on the ground. The Hoof and Hoff (hock) are parts next the ground. Ivy or Alehoof creeps on the ground. Haver is the part of the barndoor close to the earth, whilst the Hovel squats near to the ground, and in Welsh the Gipsy is still the Hav or Af, as the squatter on the ground ; this proves how closely language has likewise kept to the rootage in the primary stratum. Aftess or Aiphtess is in Welsh the female Gipsy,—T being the sign of the feminine gender in Egyptian— whilst in the Scotch name for Gipsy the earlier "Hfa" has modified into "Faa."

The Gipsy, as the child of Kheb, is named from Lower Egypt, and from the hippopotamus goddess who became the British Kêd, the Mother of the Ketti. We are only using the Gipsy name, however, as a *camera obscura* in which we can see the backward past. Kheft is modified on the one hand into Aft and on the other into Kêd, from which it is proposed to derive various names of the people of Britain. Whatsoever after-applications the names derived from Kêd may have, the Khefti or Japheti are the people of the north, the hinder part of the whole circle. These are named from the north and the Goddess of the North, who was Kheft, or Aft, Welsh Aipht. The original name is extant among the Finns in their goddess Kivutar, whilst the accent on the E in the name of the British genitrix Kêd denotes the missing consonant.

Javan was one of the sons of Japhet,[1] and the earlier name of the Akhaioi who became Hellenized, is Achivi ; also Koivy is a name of the Hellenistic languages. These, too, were the sons of Kheft, the genitrix, and mother of the Gevi or gentiles. In the modified form we have the Khuti, the Chudic or Tschudic races in general, who were the first to cross Europe, the quarter of Kheft or Japhet. These issued forth as the Khefti or Gevim of Genesis. The Gutium of the Assyrian astrological tablets, who are synonymous with the Gevim, must have belonged by name to the Khefti or Kheftim, whence the Gutim. In the modified form of the name we have the Ketti, Coti, Catini, Gadeni, Cotani, Catieuchlani, and others. With these must be classed the name of the Goth.

This secondary stage of the name corresponds to that of Gwydion, the British Mercury, Sut-Anubis, the son of the Mother Kheft or Kêd, as if these were the children. The first of the sons of Kheft were the Kymry (Gomer) of the isles. In Nennius the descendants of Gomer (or the Kymry) are the Galli, and he derives the Scythi and Gothi from Magog, the second son of Gomer, as the Kymry in the second degree. Kheft then, modified into Khêd, gives us a type-name for the Ketti of these isles. Khêd, the Great Mother, is for ever identified with the ark of the waters represented by the revolving Bear. The seven stars of that constellation are the seven Kabiri, the primordial navigators. The people who came into the isles could only come as navigators in post-geological time. These named in Egyptian are the Ketti. Khet means to navigate, sail, go by water, or with the current towards the north. Thus the Ketti are the sailors, the sons of Kheft, goddess of the north and the Great Bear, the Mother Kêd. The isles must have been discovered by the Kymry, and these therefore were the Kheti of the north, the Chudic people in general, an early branch of whom came into Britain.

We are told by the Barddas that one of the three mighty labours of the island of Britain was lifting the stone of Ketti. This was in building the circles or arks of Kêd, which again relates them to the Ketti as the builders. Callimachus,[2] identifies the Keltæ as the Keaton.

The Atti-Cotti are mentioned by Ammianus.[3] Jerome also speaks of the Atti-Cottos. Concerning two ancient Keltic tribes, the Scoti and Atticoti, he says : "The nation of the Scoti have no wives (each man for himself), but, like cattle, they wanton with any woman as their desires may prompt. I, as a young man, have seen the Atticoti eating human flesh."

Atai (Eg.) is the chief, superior, noble. The "Ara-Cotii, famed for linen gear," are celebrated among the first great founders of the world who ploughed their pathways through the seas in the dim

---

[1] Gen. x.　　　[2] *H. in Del.* 172.　　　[3] xxvi. 4 ; xxvii. 8.

starlight of the past, called the "distributors of men and countries when there arose the great diversity." The Kheti, in Egyptian, are also the weavers, from Khet, the loom, and to weave. Thus the Coti may be the sailors, builders, and weavers, under the one name derived from Kheft.

Moreover, Kep or Kef, the goddess of the Great Bear, the ark of the seven Kabiri, survives in the Kef, the cabin, skip, skiff, and ship, each a name of the water-vessel called after the water-horse. The Coti or Ketti, then, we identify with the Abroi or Kabari, the sailors, and both with the discoverers of the isles, the Kymry. Their fame as weavers agrees with Khet (Eg.), net, or woof, the later form of knitting being to weave.

Kêd or Kud abrades into Hud. In Egyptian Khep is mystery. Khept was the goddess of mystery, or the mysteries. Cibddar, in Welsh, is the mystic, or son of mystery; and the land of Hûd is a name of Dyved. The cell of Kêd is found under the flat stone of Echemaint; and this cell under the flat stone is called elsewhere Carchar Hûd, the prison of Hûd, the mystery. Carchar, the plural of Car, is Karti (Eg.), a name of prisons, the Kars, below. The cell of Kêd and Hûd identifies Hûd as the modified form of Kêd or Kûd, and Hûd, the mystery, is the equivalent of Kep (Kheft), the mystery, in Egyptian, and the Kars of Hûd in the mysteries correspond to the Kars of Dyved in Wales, called the land of Hûd. Hûd as a name of Kêd gives us the feminine Egyptian form of Hu, and we recover the genitrix and her son in Hu and Hûd. Hudol, in Cornish, is a magician, as the son of Hûd. Now, another Welsh name for the Gipsies is "Tula Abram Hood"; that is, the people of Abram Hood. In Egyptian Tula (Turu) means a whole people, a community. In Ab-ram we have another form of Kab, Hab, Af, Hef, the word for squatting on the ground. Rem, in Egyptian or English, is to rise up and remove. Af-rem is squat and go, as is the Gipsy fashion.

Another reading is possible however without going to this root. The Gipsies call themselves the Roma, their language the Romany, and in Egyptian, Rema is the express name for the natives of a country, and a people who called themselves Rema in Egyptian would mean natives of Egypt. If for Ap we read Kab, then Tula-Kab-Rem-Hut is the Hut people who were natives of Egypt. The Gipsy name of Hood, again preserves that derived from Kêd, in the form of Hud. Hu was the son of Kêd, and the people of Hu are the Hut or Hood, the children of Kêd. So Kheft became Uati in Egypt. The son Hut (Eg.) is corn, or the seed of Kêd (cf. the kid), and the priests of Kêd were called Hodigion, bearers of ears of corn. Hod is corn in Welsh, and Hut in the hieroglyphics.

A still more curious connecting-link in relation to the word "Hut" is supplied by the statement of the Lee tribe of Gipsies, who told

Barrow that their name in Egypt signified an onion.   In this country they had identified it with the Leek, which modified into Leigh, and lastly Lee.   Now in Egyptian the onion is " Hut," and they were of the " TULA-APH-REM-HUT," who were the Hut by name, which, as the Lees said rightly, is the Egyptian name of the onion.   Hut, the onion, also means the hot ; Rekh (Eg.), the equivalent of Leek, signifies heat, and Leek modifies into Lee.   Moreover, the onion is worn in the hat by the Welsh, and the Egyptian Hut is both onion and hat.

This forms a further link in the chain which connects the Gipsies, Egyptians and the primitive people of our islands, by name.

The Hodigion were an order of priests.   The magic wand of Mathonwy, which grew on the bank of the River of Ghosts, was called the Hudlath by Taliesin.   This was the staff of the Druidic priests, also named the Hud-wydd.   Wydd is wood cut from the tree of life, the Egyptian and English ash.   The Hut staff is a mace in the hieroglyphics, and the determinative of the onion ; it is apparently headed with an onion.   This sign would therefore be equivalented by the onion worn on the head, in the hat, of the Hut.

From time immemorial there has been a Gipsy encampment near Kelso, at a place called Yetholm, where the Gipsy king Lee died a few years ago.   This was the holm, the land deposited at the confluence of two waters, the location of the Hut.   Yeth and Heth permute in the names of the Cwn Annwn, or hell-hounds.

The Gevi, or gentiles, of Genesis are the heathen, and the Hut give us the name of the Huthen, or Hethen, the worshippers of Hu and the children of Kêd, who were our ancestors.

We have now got the Khut, Coti, or Ketti, in the modified form of the Hut or Huti.   Meantime this had occurred.   The beginning was with the Great Mother, who bore her Sabean son as Khut (or Sut, the Dog-star) ; in later ages she bore the solar child, whose name is Hu.   Kêd and Hu are the Great Bear and Sun.   The moon was also adopted as a type of the genitrix.   Huti (Eg.) is the sun and moon joined together in the Hut or winged disk.   Thus Kêd and Arthur, or in Egypt Sut-Typhon, were first—as will be demonstrated ; the moon was second, and the sun was last.   Now with the Hut and Huti we have the blending of lunar and solar astronomical mythology, and the modification of Khuti into Huti is the replica of the Ketti people becoming the Hut.

When we first hear of Somerset, a tribe of the Hedui are dwelling in the east of the shire.   This word has the Egyptian terminal " Ui," which also means the proper, good, genuine.   The Hedui are the proper or true Hed, *i.e.* Hud or Hut.   They are also the people of the upper half of the circle, the south-eastern part.   Hut (Eg.) is the name of this upper region, and of the white crown.   Hutui is the upper and right-hand half of the whole.

Hut is white and light. On the western side are the Cimbri or Kymry; they are on the dark side of the circle, and Kam (Eg.) means black; the west is the Ament. The Hedui thus claim to belong to the primitive people who derived from Kêd, Kheft, or Japhet. In either sense the nomenclature is Egyptian.

The Hedui of England became the Gaulish Aedui and the Irish Aedh. This is traceable. The Gaulish Druids looked to England as the chief seat of their religion. Both Hedui and Aedh meet in one divine name—and these divine names are alone primordial—that of the god Hu, whose title, as the son of Kêd, was Aeddon, the British Adonis. The priest of Aeddon was named after him as the Aedd.

" I have been Aedd," says Taliesin, describing the transformation of the deity from one character to the other, as it was represented in the Mysteries. At (Eg.) is the lad, and in our English, " Jack's the lad," the meaning is enshrined. Jack (Kak) was the god in the dark, Hu in the light, and Taliesin, having just passed through the change, says, " I am now Taliesin," or the radiant one, *i.e.* Hu.

The origin of the Aedui and Hedui and their relation to the sun-god, who, as the child, was the still earlier Sabean fire-god, are shown by the word Aedha, Greek AITHOS, Latin AEDES, and its connection with warmth, fire, or light; also with the Hindustani ID, the solar festival at Easter, and with OD, the Akkadian name of Shamas, the sun-god.

The Irish ecclesiastics of this name are countless, which shows it was of a sacred type. It is rendered Aidus by the Latinists, and by the English it takes a return curve, and is always rendered Hugh, the hard form of Hu. Mac-Hugh, or Magee, is the final shape of the son of Hu. Cathair Aedhas is the stone or circle of Hu, and the son of Hu is an Aed in the English church, as Bishop Magee.

The god Hu and his followers, the Hut and Hedui, are solely responsible for the " Jews in Cornwall," of whom tradition tells. The Hut with the s terminal instead of the t would become the Hus, as children of Hu. Hu or Hiu becomes Iu. Iu (Eg.) means to go forth, as did the son of the mother called Hu and Iu. The Ius are the Jews. The story is that Jews migrated to Cornwall, and worked as slaves in the mines; this is continually repeated in Cornish books. But if the Jews were ethnological, they would still be the Hut of Wales, who were called Hus or Ius. The Hebrews only got there by an edict of language, through the sound of the letter J. The name is not ethnological. Old smelting-houses are still called " Jews-houses" by the Cornish people. This offers the right clue.

Hut, in Egyptian, is the white, whether metal or other substance. Hut is the name for silver, as the white metal. The white metal, Hut, rendered silver, most probably includes tin, so inseparable is Hut and the white thing, the image of light and of the white god Hu. But that does not matter here. Huel is the Cornish name of a tin-mine. This relates the tin to Hu, the white god, and Hu with the terminal is Hut. Hut being white is also that which is whitened. Iua (Eg.), for instance, means to wash, that is, to purify and whiten. This process applied to smelting and refining ore, making the black tin white, is Iua in Egyptian, and the refinery is the Iua-house, or Jew-house, singular; Jews-houses, plural.

This derivation of the Jew-house from Iua, to wash and whiten, is corroborated by the fact that the smelting-house is also called the white-house, and white is Hut (Eg.). White represents an earlier Quhite in Gaelic; and Chiwidden, in Cornish, is the white or Jew-house. The Egyptian Hut, for white, had an earlier form, as Khu is light, and Akhu is white, and with the terminal these are Khut and Akhut. Chi-wid-den, however, rendered Khi-hut-ten (Eg.), might mean the white house as the place of beating and spreading out thin, to whiten or make the White, and Khi-hut-tahn is, literally, the house of white tin.

The Chiwidden, or white house, is the smelting, refining, beating, whiting house; and for discovering this process Chiwidden was placed in the Christian calendar of saints. Chiwidden, as person, can also be derived from the white god Hu. The white god and the white substance have one name as Tahn, the Repa, and Tin. Chi-hut-tahn is the young sun-god, the heir-apparent, who is the white ruler, the Iu or manifester, who was Hu in the character of Aeddon or Prydhain.

This appearing youth also turns up in Cornish hagiography as a companion of Chiwidden, called St. Perran, who landed in Cornwall at Perran-Zabuloe with a millstone tied round his neck.[1] Per-ran (Eg.) signifies the appearing, manifesting youth; it has the same meaning as Per-t-Han (Prydhain) and again as St. Picrous (of the crossing, where the eye of Horus or the Tahn is found in the Egyptian planisphere), who was also credited with the discovery of Tin or Tahn. Kiran in the legends equates with Piran.

The spelling of the name of Kiran called the saint will enable us to clinch the meaning. Kir or Kar (Eg.) is the course, and An (Eg.) means to repeat, appear the second or another time. Kiran is the great saint who in his old age went to Cornwall to die. Cor-an (Eg.) is the repeated, secondary circle or Cor, and he is undoubtedly the Corin or Corineus who was fabled to divide the island of Albion with Brut in the time of the giants.[2]

Market-Jew is one of the many names of Marazion connected with the legend of the Jews in Cornwall. We have identified the Jew with Iua, to whiten and purify; the market hyphened with it shows its relation in Egyptian, but not by making Market-Jew the Jew's Market. Khat (Eg.) is the name of a mine and a quarry. Mar is a region,

---

[1] Hunt, *Popular Romances of Cornwall.*
[2] Geoffrey's *History*, b. i. ch. xvi.

a limit, a street. Mar-Khat-Iua is the region, the limit, of the mine and smelting-houses. Market-Jew, then, is a name of the place solely associated with the mining and smelting of the famous tin.

Different names of Marazion tell the same tale.

Mara-Shen (Eg.) is the water-limit and the land's end, as place of turning-back.

Mar-Kes-Iu, the form used by Leland, Mar-Kys-Yoo used by William of Worcester, and the other spellings, Markesiow and Marghasiewe, correlate with the Iua or Jew in this way. Mar is the region, land-limit, bank, as it is likewise the water-limit; Khes is to ram, beat, pound, found, and if we keep to our determinative Iua to whiten, the Mar-Khes-Iua is the water-region or land-limit of the foundry and refinery of Tin.

Norris[1] says it is a rule without exception for words ending with T or D in Welsh or Briton, if they exist in Cornish, to turn the T or D into S. Thus the Hut become the Hus, Ius, or Jews in Cornwall. The Jews vanish and leave us the Hus or Hut, the worshippers of Hu and children of Kêd.

Mother and son are typified by the mine and metal. Kheft, the hinder quarter, the north, the underworld, the well of source, is also the mine. Kheft was the mine itself abraded in Khaût and Khât (the Eagle or accented A having been an earlier FA), the mine, the quarry; and the white metal, the Hu, was the child of the womb, the Kêd, Kheft, Khaut. Kheft, the mine, still survives in the SHAFT.

The Rut in Egypt were the race, the people, mankind. And the word lives in the English Lede for people and mankind. It is a type-word in various languages of the same value as Rut. In the names of places we have the land of the Rut, in Rutlandshire, Redruth, Ruthin, Rutchester, Ludershal; the City of Lud, in London, and Ludlow. Then there are the Laths and Ridings.

The county of Kent is divided into five "laths" for civil purposes, and these are subdivided into hundreds. The lath is supposed to come from the Saxon gelathian, to assemble. But the Kymry and the "Rut," as the Egyptians called themselves, are more likely to have brought the name of the lath, especially as the Rut is a symbolical figure of five—the sign of five steps corresponding to the five laths of Kent; and as the Rut image is the determinative of the word Khent, the south, so the country of Kent is in the south of England.

We have the earlier orthography of the lath and leet in our rides, ruts, and ridings for divisions; rod-knights, or reding-kings; rate, to govern, rule; rede, to counsel or advise, and the Irish raths. The laths and leets were relics of the Rut, the race *par excellence* as the people of Egypt, whose name of the Rut, the first, from the root of all, is retained in our word rathe, for the earliest. The Triads

[1] *Cornish Drama*, Vol. ii. p. 237.

speak of the Brython from Lydaw (Britanny), who were of the original stock of the Kymry. Both the name of Brython and Lydaw tend to show the relationship. Gwydion is called the great purifier of the Brython.[1] In later times the Brython from abroad are distinguished from the primitive Kymry as one of the cruel races that afflict them.[2]

Now an early name of Britanny is Lydaw. Au (Eg.) is the place or country. Lydaw or Rut-au then is the land of the Rut, the ancient race. Rut and Lud are interchangeable, but Rut is the oldest form. So the Rutuli of Italy are probably older than the Latin name. Also Rhodez in Rovergne was the place of the ancient Ruteni; Rennes, of the Rhedonis; Rouen, of the Rothomagi.

The Scotian Fir da Leith, rendered the men of two halves, is from the Egyptian Rut, to be divided, the men who separated from the parent stock. Rut (Eg.) means to be several, repeated. These were the Fir da Leith. In the coal country a Leite is a joint or division in the coal, Scottish LITH and Gaelic LUTH for a joint; that is the sense of the divided Leith. The lathe is a division of a county, and the lath is made by splitting.

We learn from the 10th chapter of Genesis that Mizraim (Egypt) begat Ludim. There were various branches from the Rut in Egypt which spread in Africa, Asia, and Europe under the one name. In a sepulchral inscription at Thebes the Ruten or Ludim are described as the people of the "Northern Lands behind the Great Sea." This name may have included the people of Lydaw, Britanny, and Britain. The range of Tahtmes III., Seti I., and Rameses II. was immense both by land and sea. "I have given thee to smite the extremities of the waters; the circuit of the great sea is grasped in thy fist," is said by the god to Tahtmes. "I have given thee to smite the Tahennu, the isles of the Tena"—the divided, cut-off, remote people. "The chiefs of all countries are clasped together in thy fist."[3]

The oldest names of the children are derived from the motherhood. The Ruti bear the name of Urt, the greatest, chief, first, and oldest; the Egyptians of Kheb. The Auritæ or Kafritæ included two names of the Typhonian genitrix. The Ketti we derive from Kêd, who, as Kheft, named the Japheti. This naturally enough suggests a divine origin for the Picts and Scots, who were the Kymry in Scotland. The Welsh bear ample testimony to the fact that the Picts were of the race of the Gwyddyl (Gadhael), whom we call the Kymry of the second degree, that of sonship. It is well known that Ireland was, so to say, the first Scotland, or country of the Scot. In ancient records Scotia means Ireland. In the fabulous history of Scotland it is said that a daughter of Pharaoh named Scota was wedded to a Keltic prince, and these were the progenitors of the

---

[1] Taliesin, "Cad Goddeu."   [2] Taliesin, "The Sons of Llyr."
[3] *Tablet of Tahtmes III.*

Scots. We are told by Hesychius that Venus was worshipped in Egypt under the name of Scotia; by which he means Sekhet (or Bast), called the goddess of drink and pleasure. Venus is a general name for the genitrix, and Sekhet, also identified with the element of heat, is one of the two characters of the Great Mother. The Irish Bridget will help us to recover Scotia in the isles, in Ireland first and Scotland afterwards.

One part of the process in converting the Irish was to take their ancient deities, the devil included, and transform them into Christian saints, and as saints they have figured in the calendar ever since. Thus the mythical Patrick appears to be identifiable with the god Ptah ; not but that there may have been a priest of Ptah named Patrick. The Rekh (Eg.) is the Mage, wise man, priest, and there may have been one or many Ptah-rekhi, the priests of Ptah. Patrick, or Ptah-rekh is probably the hard form of the Paterah known as a Druidical title. Attius Paterah, the friend of Ausonius, was a Druidical Paterah or Patrekh in this sense ; he who was said to have been STIRPE SATUS DRUIDUEN GENTIS ARMORICÆ,[1] and the companion of Dyved.

When the original Patrick landed in Ireland, the country was ruled by Niul of the Nine Hostages, who may possibly be a form of the Welsh Ner, Budd Ner. Budd answers to the Put circle of the nine gods, which was represented by the Bed of Tydain, in Tad Awen, the father of the nine British muses. Ptah was the framer of the Put circle of nine in the solar reckoning, and division of the dry period. It is not necessary to claim the sacred title of Budd as denoting a form of Ptah, but we certainly have Ptah as the "old Puth." The title of Bûdd is written Vytud and Vedud by Cuhelyn, and it is also assigned to the goddess Kêd. Now, in the hieroglyphics, Fut, the earlier form of Put, signifies the circle of four quarters which preceded the Put circle of the nine gods.

Nial of the Nine denotes the second of these two, and this was the order of things when Patrick came to Ireland. Also, there is a character that figures in "Hanes Taliesin" as "Bald Serenity," a form of Ptah, who is pictured as bald or wearing the close-fitting skull-cap, the sign of baldness. Bald Serenity had his abode in the middle of a lake or mere. Ptah carries in his hands the Nilometer sign, the emblem of serenity, as lord of the waters. Ptah was the former of the egg, and a daughter of "Bald Serenity" was named Crierwy, the "token of the egg," also the manifestation of the egg.

The Pudduck is a frog, and Ptah was in one form the frog-headed deity, the biune being whose likeness was reflected in the two-fold nature of frog on land and tadpole in the water. Budd seems to be the British Buddha, and both are probably derived from Putha, so that Pat may be the living representative of Ptah. Patrick is said to have been christened Succath or Socher by his

---

[1] Auson. *Prof.* iv. and x.　　　　[2] Davies, *Myth.* pp. 116, 118, 189.

parents. The Sekht (Eg.) is the ark of Sekari, and Sekari is a title of Ptah; Pt h-Sekari is the Ptah of the Sekhet or ark, the silent or mummy form of the god. Paterah was the companion of Dyved, or Hu, the solar Tevhut, who was the son of Ptah.

Patrick was accompanied by St. Bridget, she whose fire was in the keeping of nine maids, and surrounded with a fence of a circular form, the Put circle of the hieroglyphics, the divine circle of the nine gods; " Put " being the name of number nine.

Nial of the nine hostages looks like the ruler of the Put circle of nine gods who was Ptah in Egypt. If historical, he would be affiliated to that system of reckoning by the solar nine, which includes the partition of China into the nine divisions of Yu.

Sheelah-na-Gig is sometimes called Patrick's mother, sometimes his wife, and is an Irish form of the Great Mother Sheelah's day is March 18th. Sheelah as the mythical mother is known by the figures called "Sheelah-na-Gigs," very primitive and plain in their meaning. These were portrayed over the entrance of ancient churches, and formed in one particular part a hieroglyphic Ru, the typical mouth or CTEIS which Sheelah pointed to as the door of life. Kekh (Eg.) means a sanctuary, in English a church; the Chech is a stone chest, kist-vaen or cromlech, and the birthplace supplied the type of the burial-place. This gave appropriateness to the Sheelah-na-Gigs being portrayed as the way of life over the church-door.

Sherah (Eg.) means source, the waters of source. Serah (Eg.) is to reveal and exhibit. Sheelah-na-Gig may mean the revealer of source in the sanctuary and birthplace. Sheelah is a goddess of drink and pleasure, as was Bast (Eg.), who exists as the Bebaste of the Irish mythology; therefore it may be inferred that Sheelah the revealer and Bebaste are identical. Sher (Eg.) means to breathe with joy, which the Irish do on Sheelah's day, and the shamrock worn on Patrick's, the previous day, is drowned in the last glass of whisky on Sheelah's night to her immortal memory. Further, Sheelah-na-Gig ought to enable us to read the name of Bridget. Ket is the womb, the birthplace personified in Kêd. The Sheelah-na-Gig exhibits the Ket. Brid is a name of Prydhain, the child. Brid-get may therefore be a personification of the mother of Brid, Brute, or Prid, the manifester as the solar son.

In Scotland Bridget was known as Scota; she is called Scota by various writers.[1] Scota, the wife of Patrick, adds to the likelihood of his being the god Ptah, for Sekhet was the consort of Ptah, the mother of the solar son,—the Egyptian Pridhain or reappearing youth,—and she, like Bridget, was the goddess of fire.[2]

The month of Choiak was sacred to Sekhet and her festival. In the sacred year this answers to our October-November. And in

---

[1] Jameson, *History of the Ancient Culdees*, p. 109. See also Vallancy, *Coll. Hib.* pp. 200-251; and *Nimrod*, vol. ii. pp. 639-645. [2] *Ib.* 189.

the Roman calendar the principal festival of Bridget is celebrated on the 7th of October.[1]

The Scottish Scota, Bridget, once identified through Patrick with the Egyptian Sekhet, ought to help us still farther north. For Sekhet was a name of the lioness-headed goddess Pasht, Pekht, or Peht. The consort of Ptah takes two forms as goddess of the Two Truths of the north and the south. As goddess of fire in the south she is lioness-headed; as goddess of moisture in the north she is cat-headed; the one is named Sekhet, the other Pasht, Bast, Pekht, Peht, or Buto. Buto, in the north, is the cat-headed form of the genitrix, and BAUDRONS, the Scotch name for a cat, retains that of Peht or Buto, the cat-headed goddess. Butha is also an Irish name of the moon.

The Irish goddess of moisture is called Be-Baste; that is, Bast or Pasht of Egypt; and Scota (Bridget) is the fire-goddess, that is, Sekhet. These are the two divine sisters in the Ptah Triad, the plural form of the genitrix. The gist of all this is that the names of the Picts and the Scots meet in this· goddess, who is Pekht in one form as the divinity of the northern frontier and Sekht as the mistress of fire in the south.

If we were to derive the Sgiot or Scot from Sekhet the goddess, then the Sgiot-ach of the Scotch probably comes from the Akh (Eg.), meaning the illustrious, noble, honourable sons of Sekhet (Scota), who is recognized on the monuments as the goddess of pleasure and drink, ergo of whiskey, the fire-water; spirit and fire being synonymous in the symbolism of the Two Truths.

The Scotch were anciently called Cruitnich, the Corn-men, the cultivators of corn, and the word Cruitne is extant as the name of a place called Cruden. Sekhet is an Egyptian name of corn. We hear of the Picts and Scots as the painted men, and Sekht signifies the painted. The Gaelic Sgod, a dweller in woods and forests, answers to Sekhet (Eg.), field and forest. The Picts and Scots are generally coupled together; Pekht and Sekhet will enable us to make a geographical distinction.

Pekht denotes the hinder, the northern part. Peh (Eg.) is the rump. The Pukha was the infernal locality of the under-world. Pick-a-back is a pleonasm of hindwardness. Pest (Eg.) is the Pes, the back, and in *Gammer Gurton's Needle* we read, " My Gammer sat down on her Pes." Pes, Peh, Pekh are forms of one word, back. The Picts therefore were the people of the hinder part of the land, the Egyptian symbol for the north. Khebt has the same meaning, and in Kent a Gipsy (Kheb-si) is still called a Pikey. The Peak in Derbyshire has been shown to be a type of the hinder part, and in the *Saxon Chronicle* the men of the Peak are called Pecsætan. They, too, were Picts according to the naming from the north, the genitrix Pekht being a later form of the Goddess of the North who was the earlier Kheft

---

[1] *Roman Martyrologie according to the Reformed Calendar*, by G. K., 1627.

Suâ, the south, is an abraded Suka, and with the terminal T, Sukat is synonymous with south. The Scot was localized as the southerner of the two.

Pasht and Sekhet are a form of the twin-lions of the equinox facing the north and the south. Our Pash or Pasch of Easter is named after the goddess who presides at the place of the equinox, the Pekha, in her dual form. The Pekh or Bekh was the solar birthplace at the time of the vernal equinox.

The western equinox is represented also by her name, although the date is now belated. The Monday after the 10th of October is called Pack-Monday; on this day bears were baited, and dogs were whipped. Formerly a number of cats used to be burned to death on the Place de Grève, Paris, in the Midsummer fire of St. John. The cat-headed Sekhet was the Egyptian goddess of fire. The bear was a type of the ancient outcast mother as well as the dog, both having been symbols of Sut-Typhon. Poke-day in Suffolk is when food is divided and portioned out among the labourers. Pekh (Eg.), is food, and to divide.

Pekh appears to have left her name in the county of Buchan. Her image survives in the red lion of Scotland, red being the colour of the northern, the lower crown, and of the female lion represented by Pekht. The wonderful temple at Bubastis was made of red granite. Bede wrote the name of the Picts as Pehtas, and in Egyptian Pekht became Peht.

The twin lioness (or the lioness and cat), as Pekhti, represents double might and vigilance; Pehti means vigilant and foreseeing. This meaning is modified in our word peke, to peep, pry, and peer into. Pekht was the watcher, and in one of her two forms she must have watched from "Arthur's Seat," Edinburgh, where the dim outline of the lion is not solely drawn from imagination, although the work of man has been almost effaced by the work of time. It is obvious too that this was Sekhet, the goddess of fire, known as Scota (or Bridget), who tended the fire with her nine maidens, at Edinburgh, the maiden castle.

MAI (Eg.) is the cat-lion, ten is the elevated seat, the throne, which in the hieroglyphics is lion-shaped. And Edinburgh was the maiden city, *ergo* not only the circle of the nine maidens of Scota, but also the seat of the cat-lion.

The MAI is also the lion rampant of Scotland, and has its tail between its legs and thrown over its back.

Lastly, in the arms of England the lion and the unicorn are united in a common support, and the unicorn is a type of Typhon, the one-horned, the Ramakh (hippopotamus or rhinoceros), called the mythical unicorn, the ancient Kheb or Kheft of Egypt and the Kêd of Britain.

In the eastern part of the territory of the Rhobogdii, in Ireland, says Richard of Cirencester,[1] was situated the promontory of the same name ; their metropolis was Rhobogdium. Rru (Eg.) is a name for the children. May not these Rhobogdii have been the Rru as children of Pekht, the dual lioness ? Pekht in the form of the twin lions, also named the Rhiu, was seated on the rock of the horizon called the Ru as the place of the lions or Pekhti.

The Rru-Pekhti should be the children of the goddess Pekht, unless they called themselves descendants of the Pekhti. Either way the promontory of Rhobogdium repeats the Egyptian imagery.

Bast, the goddess of drink, has given us some names connected with drinking. To booze is to drink deeply. A bussard is a great drinker. A basking is a drenching. The bush was a symbol of drink at the alehouse door. Afterwards the wooden frame of the signboard was termed the bush. Grose says buzz, to buzzer one, signifies to challenge a person to pour out all the wine from the bottle into his glass, and to drink it, should it prove more than the glass would hold. It is said to a person who hesitates to empty a bottle that is nearly out. To buzz is to empty the bottle, and the buzzard is the coward who refuses. Bes (Eg.) is the inundator or swiller.

Bast as goddess of pleasure is our divinity of bussing. To buss is to kiss, conjoin; to baste is to tack together. To Bast we owe the bastard. To bask is pleasure. Baskefysyke[2] is a name of *fututio*. Bagford mentions an image that once stood at Billingsgate. The porters used to ask the passers-by to kiss the same, and if they refused, they were bumped on the seat against it. He calls it a post, and intimates that it represented some old image that formerly stood there, perhaps of Belin. Bagford adds, " Somewhat of the like post or rather stump was near St. Paul's, and is at this day called St. Paul's Stump." This is probably alluded to as the BOSSE of Billingsgate in " Good News and Bad News."

" The Water-workes, huge Paul's, old Charing Crosse,
Strong London bridge, at Billingsgate the BOSSE."[3]

The BOSH is a figure, the Egyptian PESH a statue.

We can recover the BOSSE as the seat of Bast.

Besa (Eg.) is a name of the seat or of seats. Pest (Eg.) is the back, spine, seat. The Bessy who was Bast made a special symbol of the cow's tail. The gate of Belin, the birthplace of the son, was represented by the BOSSE or Bast, the hinder part, and those who would not BUS her in front were bumped behind.

Pekh, the goddess of pleasure, is possibly invoked in the expression, "an' it please the Pix," *i.e.* if Pekh please, for she was terrible at times ; in the character of Sekhet she was the punisher of the

---

[1] B. i. ch. viii. 11.   [2] Cokwold's *Daunce*, i. 116.
[3] By S. R. 1622 ; Brand, " Kissing the Post."

wicked. Her name is sacred as that of the Pyx, the ve..sel in which the host, mass, or consecrated wafer, is kept. The pyx is an emblem of the womb, and Pekh, later Bekh, is the birthplace, where the bread was BAKED. Mesi (Eg.) is a cake, Mes-t is the womb, and Mes means to engender and generate the child ; the pyx with the bread in it represents the mother and child ; but the mother signified by the name of Pekh, the goddess of Pasche, of bussing, of pleasure and drinking, is the old cat-headed Puss or beast. Kissing the PAX after the service, according to the Ritual of Rome, is a survival of kissing the Bosse-Image of PEKH as the goddess of the Easter crossing.

The fairy is a diminutive of early deity, and Pekh also survives in the form of a fairy, as the Piskey or Pixy of Devonshire and Wales. A man who is glamoured, it may be with drink, is still said to be Piskey-led ; the goddess of drink thus protects her followers, or plays the devil with them, as the Puck or Pouke.

The lioness-headed Pekh (Pasht), who was both Pekht and Sekht, mother of the Pict and Scot, was one of the very ancient deities of Egypt, but not the oldest.

The Goddess of the North and of the Bear is the most ancient form of the genitrix, first, oldest, and chief, as her names declare. She, as the present writer contends, passed into the form of Uati as God- dess of the North, who had her sanctuary at Dep (Tep, the first), at the extremity of the Rosetta branch of the Nile, and who, as Uati the dual one, bifurcates into Pasht and Sekht of the solar myth. Sekhet also has the title of Urt, the great goddess, who had been Ta-Urt.

The secondary stage of the lioness goddesses is well shown by the two lions that draw the car of Kubele, the Great Mother, the Kheb of Egypt.

There may be some sort of gauge in this naming. If the ethno- logical titles follow the order of the divine dynasties, then those who claim from Kêd are primeval, whilst those who date from Pekht and Sekht of the lioness type are in a second or a third degree of descent from the beginning. This assimilation to the divine order is likely to afford some guidance in seeing that the Pict and the Scot were off- shoots from the Kymry of Britain, who claimed to be the children of Kêd. Thus, when we meet with the Atta-Coti in Scotland, and know that Atta (Eg.) is the father and typical ancestor, it looks as though the Atta-Coti still claimed in another country to belong to the original race of the Ketti. The Catini, or Gadini, may likewise have claimed descent from Kêd.

Here for example is an illustration by which the tentative ten- tacles appear to hold fast in another bit of anchorage in the heavens. We are told there was a Scottish tribe during the Pictish period named the SELGOVÆ, in the West of Scotland, their western boundary being the river Dee, and their southern limit the Solway Frith.[1]

---

[1] *Encyclop. Brit.* 8th ed., article, Scotland.

Selk also is an Egyptian goddess who was connected with one of the four quarters, as she is found conjoined with the four genii; which quarter is shown by the scorpion borne on her head. Selk means the scorpion, once the sign of the western equinox; and Serkh is the hole, the opening of the Ament in the west. The Selgovæ were the people of the western quarter. Afa (Eg.) means born of. Selk-afa is born of Selk. Her name is also found in Sark. Selk (or Serk) is a goddess of Dakheh, in Nubia,[1] the Dog-star was sacred to her, which takes her back to the primal motherhood. The name of Dakheh reminds one that Bridget was said to be a daughter of Dakha.[2]

The Picts disappeared in so mysterious a manner because they never existed in the distinct ethnological sense supposed.

The Kymry were the first known people in Scotland; they preceded the Pict, Scot, and Gael; the appearance of the Picts in the north of the country is strictly connected with the disappearance of the Kymry, and will account for it. The disappearance is but the submergence of the name of the Kymry in that of the Picts and Scots. The Kymry being the original race that inhabited Scotland, bifurcated into the Picts and Scots. The Picts vanished, and the Scots' name spread over the land. In India we find the same thing occurs under the same names. Dr. Bellew[3] has recently called attention to the fact that in Afghanistan large sections of the Afridi people and the family of the Khan of Kelat are called the KAMARI, whilst the inhabitants of the "Pukhtún-khwá country" are designated Pacts and Scyths, the equivalent for our Picts and Scots. The name of Pukhtún-khwá, rendered by Egyptian from Pukht, divided in two, and Ka, for country, recognizes that division north and south of the bifurcating race, as in Scotland, and as in the land of Egypt. Here, then, are the Kymry in India, as in Scotland, dividing into the people of the two lands (of Egypt) called North and South, whence the Picts and Scots. Ari (Eg.) denotes the companions, fellows, the family, as in Kabiri, the Seven "Ari" of Kheb, the goddess of the Great Bear. The Pict and Scot, or Pact and Scyth, follow and correspond to the division of the heavens by north and south when the solstices only were marked and reckoned by, and the Great Mother was represented by the two divine sisters, who in one system were Pekht and Sekht of the lioness type of the genitrix. We need not, therefore, be surprised to meet with the Logari of Logar in India answering to the Logrians of ancient Britain. For Dr. Bellew also finds the Logari, whom he parallels with the British LOEGRWYS. The Logari (Rekhari)

---

[1] Wilk, pl. 55. [2] Vallancy.
[3] *The Races of Afghanistan: being a Brief Account of the Principal Nations inhabiting that Country.* By Surgeon-Major H. W. Bellew, C.S.I., late on special political duty at Kabul. (Calcutta: Thacker, Spink, and Co.)

are the Ari, children, fellows; and Rekh denotes the race, or the people of a district. The race of KHEFT or Kêd may also be found in India in the CHEDI, a very ancient people who lived in Bundelkhand, and were famous for their close attachment to ancient customs, laws, and religion. In both directions the naming must have been carried out from one source, which was the land of Kam, םןה, or Khebma, as only by the Egyptian mythology and language can the facts be interpreted. It is this unity of origin in Egypt that has created the glamour concerning the lost ten tribes who have been discovered in so many lands, including Afghanistan. There is a common source for the people, the mythology, and the naming. The AFRIDI folk, found in India at the head of the Kamari, the Pacts and Scyths, may now be claimed as a form of the Auritæ, Afruti, the primal Kafruti of Africa, and the brotherhood of the Kamari, Kabiri, Abroi, and Kymry goes back to the unity of the black race.

Before Cæsar landed in Britain, the people of Dorset were known as the Morini and the Durotriges. The Morini (in Egyptian Mer-uni) are the inhabitants of the water-region and the land-limit. Ter (Eg.) is the extremity of the land, the frontier. The T adds the article the, and Rekhi signifies the people of a district. The Teru-trekhi or Durotriges are the people of the district on the water-frontier at the limit of the land. Rekh (Eg.) gives us a type-name, but one somewhat difficult to utilize. For the Rekhi may be the architects, builders, metallurgists, refiners, time-keepers (Druids), teachers of various arts and sciences. They are the magi, the learned, the knowers, and experts.

Richborough, at the navigable estuary of the river forming the Isle of Thanet, yields the Borough of the Rekh. Buru (Eg.) is the high place. Here was the Roman haven of Rutupia and the Urbs Rutupiæ of Ptolemy, described as one of the chief cities of Kent, and commonly supposed to have been one of the first Roman stations in England. The name proves it was one of the first Egyptian stations. Ru is the gate, road, way. Tepi means the first, the point of commencement, and from this initial point proceeded the Watling Street or Way, the Ru which as primary is in Egyptian Rutupi. The castle of Rutupi on Rich-borough Hill is said to exhibit a more perfect specimen of Roman military architecture than is found elsewhere in Britain. But is not that owing to the Rekh (builders) who preceded the Romans?

Rutupi is not only the point of commencement for the great road, as Tep is the sacred hill, also written Tepr, *i.e.* Thabor, our Dover. The hill of Rutupi stood by the bay,—whence the Romans procured the famous oysters mentioned by Juvenal,[1]—which is now mere marsh-land. As Teb (Eg.) is fish, and Ru a pool, RUTUPI may also denote the oyster-bay. It is not necessary to prove too much, but such is the fundamental nature of Egyptian naming.

[1] " Rutupinove edita fundo."

Ragæ, the chief city of the Coitani, was a name of Leicester; thus Ragæ and Leic permute in the Ster or Kester of the Leic or Rekh. The Logi also inhabited the north-east part of Sutherlandshire and the south-east of Strathnearn. A people named the Regnai, Ptolemy's Pηγνοι, were in possession of Sussex and Surrey before the Romans came and were powerful enough to maintain a kind of independence afterwards. Tacitus mentions their king Cogidubnus as a faithful ally of Rome. Regnum, the modern Chichester, was the capital of the Regnai. Rekh-uni (Eg.) are the native Rekh. As the Rekh were, amongst other things, masters of the art of smelting, and as the Weald of Sussex was famed from time immemorial for its ironworks, it is possible the Rekh in this instance signifies the workers in metal, the Regnai being of the old race (Rekh) of Knowers.

But we can get in a little closer. The Rekh is the brazier and smelting-furnace. Nai is the Egyptian plural. The Rekhnai, the plural Rekh, signifies the smelters, and furnace-men. Khekh-tebn (Eg.), the equivalent of Gogi-dubnus, is the ruler of the division or circle.

Creklade or Greeklade, not far from Oxford, is famed in ancient traditions as the place to which certain Greek philosophers came with Brute. Drayton remarks that the History of Oxford in the Proctor's Book and certain old verses [1] affirm this, and that both Creklade and Lechlade (the physicians' lake) in this shire derived their title from these Greeks and Lechs or Leeches. Once we get the Greeks and Brutus out of the way, we can better interpret the obscurities of tradition. Our Lechs are the Egyptian Rekhi, the learned, the wise men, the mages, the men of science, physicians and astronomers, builders and metal workers. Rat, the equivalent of our lade, a place, means a place, to plant, make fast. Lechlade, i.e. Rekh-rat, is the settlement of the Rekh. The Roll-rich stones, in the same county, mark another place of the Rekh. Rer is a circuit, circle, to go round. The stones formed a circle of the Rekhi, who were also masons, builders, and religious teachers in Egypt.

The name of Crek classicized into Greek is probably derived from Kha, a book-place, an Altar, and Rekh. Kharekh would naturally abrade into Crek, and be as naturally read Greek. Creklade is thus the college of the Rekh. The magi of Egypt are the pure wise spirits and intelligences. And the Rekh exists with us as the Rook, a name for the parson. The Rekh, the pure wise spirit, has for determinative the Rekh bird, the Arabic Roc, a mythical bird. The Rekh is the Phœnix, the ancient bird-type of transformation. This Rekh becomes our ancient and wise bird, the rook. Thus we have the Egyptian Rekh in several forms.

At Caiplie, near Fifeness, in the parish of Kilrenny, Scotland, there is a cave with sculptured walls. The name Caiplie is modernized from

[1] *Leland. ad Cyg. cant. in Iside.*

Caip-lawchy. The Egyptian equivalent will be KEP-REKHI, and Kep is the concealed place, the hidden sanctuary; Rekhi, the wise men, pure spirits, knowers, the mages. "Kar-rennui" signifies the astronomers as namers of the star-courses, from Kar, a course, an orbit, and renn, to name. Another of these caves is found at Dysart, which name in the same language reads Teb-sart, the cave or place of the sowers of wisdom and science. In the name of Rugby we probably have the Bi (Eg.), a place of the Rekh, the learned. Leighton-Buzzard is a town near Ashridge. Leigh-ton denotes the "ten" division, seat of the Rekh. The Buau (Eg.) were chiefs, heads, archons. Ser is to arrange, organize, dispose, distribute, amplify, augment, conduct, console. Sart is to plant, grow, sow the seeds of wisdom and science. The Buau-sart were the chief ministers and officers of the Rekh, Leighton being the centre of a settlement, the boundary of a shire. Rickmersworth introduces another name of the Rekh of Egypt in the Mer, plural Mert. The Mer was a monk, a person attached to a temple. Mars-worth is possibly the worth, a nook of land belonging to the Mert, in the English plural the Mers. Thus Rickmersworth would be the enclosure of the Rekh, the wise men, mages, pure spirits, who lived the monkish or Mer life, and whose sanctuary, park, pavilion, a retired nook, was at Marsworth, and Rickmersworth. In Domesday, however, the name appears with the Egyptian masculine article prefixed as Prick-mare word, that is, "The-Rick-Mer-worth." P-rick-mer-worth was held by the abbot of St. Alban's. St. Alban's held the manor in demesne, and this may give another meaning to the Mer. The Mer is a superintendent, overseer, governor; Morien, a name given to the builder of Stonehenge by the Welsh Barddas may have been the superintending Mer in this sense. The "Fomorians" is one of the Irish typal names of the race. Fu-meri-uni (Eg.) are a large number of inhabitants inclosed and governed, equal to a claim of being civilized. Moor Park is at Rick-mersworth, and Park in Egyptian is Hert, our worth, a nook of land enclosed. Rekh-Mer might signify the prefect, superintendent of the Rekh, who was modernized into the abbot of St. Alban's, and if so, the Hert would be his place, as park, garden, pavilion, whence Moor Park.

The time came when the Welsh specially distinguished England as the land of the Rekh. Lloeger is the name for England known to the Barddas. The Britons divided the island into Lloeger, Cymru, ag Alban. When shut up in Wales, that district constituted three regions, called Gwynedd, Pywys, and Dehenbarth, which were distributed into a number of Cwmmwds, Trevs, and Cantrevs. But in the first Triadic division England is Lloeger. "Lloegyr," says Lady Charlotte Guest, "is the term used by the Welsh to designate England. The writers of the middle ages derive the name from the son of the Trojan Brutus, Locryn, whose brother, Camber, bequeathed his name to the principality. But from another authority, that of the Triads, we learn that the name was given to the country by an ancient British tribe, called the LLOEGRWYS.[1] Lloegrwys is a form of the Logrians of Lloeger.

This brings us to the original type-name of the Rekh or Lekh, which may afterwards indicate the calling, but primarily it means the Race (it is the same word), as the descendants. Just as the Irish Laegh is posterity, or LUCH, offspring; LUCHD, Gaelic, people.

Cormac Mac Cullman[2] derives the OrbRAIGE as the descendants (raige) of Orb or Orbh, RAIGE meaning posterity or race. O'Donovan, in his commentary on this, says:—"Orbh was the ancestor of the people called Orbraigh, who were descended from Fereidhech, son of Fergus Mac Roigh, King of Ulster in the first century." The Irish Orb here represents the Egyptian Kherp, to be first, chief, principal, royal, consecrated. The Orbraigh were the royal race.

The Lloegrian race of England were the posterity of the Kymry, corresponding to the Logari of Afghanistan as the race, the descendants of the Kamari and Afridi. The primeval unity of the Lloegrians and Kymry in race and religion is acknowledged in the *Gododin*[3] by an address to the universal goddess Kêd, "O fair Kêd! Thou ruler of the Lloegrian tribes." The Lloegrians were one in religion after they were divided by new ethnological lines of demarcation in Britain. Lloeger rendered by Rekh-r means the division of the Rekh. All we can say of them further is to repeat that they include the magi, the metal-workers, the builders, the doctors.

They gave us the Rike, the master, the governor; the Leech, the Laghe-man (lawyer), the Lake (player and actor). The Lace-maker and Lacking (fulling), Lake and Lockram (kinds of linen), Leach or purified brine in salt-making, Leches for cakes, and the Loker or carpenter's plane, are named after them.

The Scottish Picts' towers are called Brochs, and as these are built in contradistinction to the caves and holes, it may be the Rekh (Eg.) enters into their name. The Rekh is an architect and builder. The Pa (or B) denotes a house. The P-rekh would be the house that was built, and the Brick would be named in the same way as the means of building. Further the Rekh is the furnace, and the brick is burnt. So read, the PA-REKH would be the burnt house, and the Brock may not only have been the erected house but also the vitrified fort.

The Rekh (Eg.) as people of a district apparently supply the English Ric as in Cyneric. This became a kingdom in the German König-reich, but the earlier Cyneric was derived from the Rekh or people of a district who were of kin and formed the gens. The Bishopric is now the only word with this ending in ric, meaning rule and sway. But is the ric compounded with bishop identical with the

[1] *The Mabinogion*, p. 210.   [2] *Glossary*, ninth century.   [3] Song 25.

ric in Cyneric? Not necessarily. The Rekhi (Eg.) are various. They
are certain people of a district, and may be a religious order, as the
magi, the knowers or men of learning. The ric of the bishop may
therefore denote his sway and dominion over the Rekhi or the rooks.
But as the Rekh or people of a district are also the race, mankind,
at first in the gens, and then generally, the ric becomes secularized,
and Cyneric and Bishopric are two distinct forms of the Rekh.

The origin of the SHIP, as in fellowship, can be traced far beyond
the verb SCAPAN, to shape. Ship is no doubt akin to shape, but there
is an earlier source for the word. The Sep (Eg.) were a body of
persons belonging to a religious house. The Sepa was a district, a
country. This in the earlier form is our scape in landscape; SHAAB,
Arabic, a large and noble tribe; SZCZEP, Polish, family, tribe, race;
SIBBO, A.-S., race or relations; SUBA, Hindustani, a province.

Skab (Eg.) means double; to Kab is to double and redouble.
The Sheb image of the second life is the double. Whence the Kab
(Kab-t) is a family. With the S prefixed, this is the Skab, Shab, or
Ship, as a companionship or fellowship, from the redoubled fellow or
companion; this is the German Schaft, equivalent to Kabt, a family.

In Hebrew the tribes are Shebt. This was the Sept, as the English
clan, race, or family, that proceeded like the seven stars from the
common genitrix. SHEFT (Eg.) means order, and a section; ZEPPET,
Circassian, foundation and groundwork; SHEBT, Hebrew, tribe or
tribes; SAFEDD, Welsh, the fixed state; SAFT, Scotch, a rest, peaceful;
HEPT, Eg., peace, number seven, and the ship; SEPTUM, Latin, an
enclosure; SAPTI (Eg.) to construct, wall in.

The Sept in Swahili is the Kabila, like the Arab tribes of that
name; in Turkish the Kabile is a clan, tribe, or Sept. The Gbalai
in Gbandi is a farm. The Zincali Chival is a village, the Latin
Cubile, a bed, a place of repose. The Welsh Gefail is a smithy; the
Swahili Cofila, a caravan; the Cornish Ceible and English Coble, a
barge or flat-bottomed boat; Malayan, Kapal; Irish, Cabal, a ship;
Gaelic, Cabail; Kuff, German, a kind of vessel, with a main and mizen
mast; Evu, Adampe, canoe; Uba, Ibu, a canoe; Abies, Latin,
ship; Kufe, German, tub; Keeve, Scotch; Cuve, French; Kufa,
Polish. An old tub is a nick-name for the sailing vessel.

The root of the matter in relation to the ship and number seven
is that the first ship in a double sense was the constellation of the
seven stars, the Sept or Kabt of the Kabiri, and all forms of the
name lead back to them, and to the genitrix Khebt or Khepsh. Seb-ti
(Eg.) reads five-two, or seven, and this companionship of the seven in
heaven was copied on the earth in the Sept, Schaft, and the Ship of
the companions, or companionship.

MACCU, MAQVI, and MACWY are different forms of a word frequently found in the Ogham inscriptions of Wales, as in MACCU-
Decceti, MACU-Treni, MAQVI-Treni. In the Irish and Gaelic Mac

it has come to signify the child, the son of. But the individual is not primary; the gens, tribe, or clan is named first, and the word ending as Mac can be traced backwards. Maga, Cornish, is to feed and nourish. MAG, Welsh, to breed, nurture, rear, and bring up; MAEG, English, a kinsman or blood-relation; MAIKA, Hindustani, kindred, relations, the mother's family. And in the Irish *Book of Armagh* [1] Maccu or Mocu has the force of gens or clan, and interchanges with the word Corca, people, in MOCU-Dalon and Corcadallan; MOCU-themne and Corca-temne; MOCU-runtir is rendered by the phrase, "*de genere runtir.*" Magad (Welsh), a brood, follows Magu, to breed, and Magad is identical with the so-called Anglo-Saxon MAEGETH for the family, tribe, people, which is the Egyptian Mâhaut, earlier Makhaut, meaning the clan, family, COGNATE, with especial reference therefore to the motherhood. The Magad, brood, and Makhaut clan of blood-relations, belong to the earliest sociological types, hence in the Quiché language Machu means very old. In Zulu Kaffir, Makade denotes a very ancient thing, and in Arabic MAHKID is root-origin. The Maccu or Makhaut is the clan, gens, family, or community, and the Mocu-Druidi or Maccu-Druidi were the Druidic clan, family, community. The MOCU-runtir was the clan that had their cattle (run, Eg.) in common; ter being the whole people, all. Dallan seems to be the same word as runtir, reversely compounded. In Ar-magh the name is localized, and as it is on the hill, the Ard may be the Egyptian ascent, and as Art is also the ceremonial type, Armagh may denote the place of a religious family, the Makhau, on the height. Possibly the Vi in Maqvi may stand for the Egyptian Fu or Fi, meaning numerous, and as the Welsh QV passes into or comes to be equivalented by the letter P, Magvy is the later mop and mob; the mop being an assembly of servants at a fair waiting to be hired; the mob, an indistinguishable multitude, yet represents the many, the common family or Makh. The fact and relationship of the Makh is shown by the "MAGBOTE," a fine for murdering a kinsman. It also exists in the Magdalen, an asylum, whilst its hieroglyphic sign is extant by name in the mace. Every early name shed by the human being has been applied elsewhere in the animal kingdom or in other domains. Thus the societary phase represented by the Maegeth in English and the Mâhaut (Makhaut) for the clan in Egyptian is now represented by the MAGGOT, which bears the name and continues the type of the Ma'khaut as an indistinguishable swarm or multitude, that bred like maggots, and were massed under a clan-name, a Mac-something or other. Moreover, the human Makhaut or maggots worshipped a god of the same status as their own, a maggot deity—the same process of naming applies to the divinities as well as to animals, reptiles, or insects—and in the CROM-CRUACH the Irish adored the maggot-god. Crom in Irish

---

[1] 9, *a*, 2; 13, *b*, 2; 14, *a*, 1.

means a maggot, and O'Curry says the name of the celebrated idol of the Gadhael signifies literally the "Bloody Maggot." But if we render it the maggot of blood, we obtain a kind of synonym for the MAKHAUT, which was a family, or clan of cognates founded solely on the tie of blood, originally on the mother's side. The motherhood of the family under this name was doubtless represented by the Irish goddess Macha, the Egyptian Makha, who holds the two vases in her hands, or Meh, a name of Hathor, the habitation. This naming of the family, the place, and the person had been applied in various lands. The mother and her emblem typify the abode. In Kaffir the Mâi or MAKI is the womb ; MAGA, Tasmanian, the *mons Veneris*; MAGA, Fijian, the same ; MAKAU, Maori, the female ; MACHHA, Khaling, the old woman ; MAKU, Timbuktu, the womb : MKE, Swahili, the female as producer ; English Make, German MAGD, and Egyptian MEHT. In Swahili the Makao is the abode ; in Hindustani, the Macka is the MATERNAL mansion; Magha, Sanskrit, a typical house ; MUK, Akkadian, a building; MOGHA, Sanskrit, an enclosure or fence ; MAHA or MAKHA (Eg.), a sepulchre, an enclosure ; MUKHDAA, Arabic, a storehouse ; MOK, African Penin, a town ; MOKI, African Murundo, a town. MÁS or MAUCE, in Irish, euphemized as the thigh, in the name of Más-reagh, in Sligo, is a form of the Maccu, abode. Más-a'-Riaghna, near Antrim, the thigh of the queen, identifies the Más with the Egyptian MES-T, the place of birth, the womb, and with the imagery of the Bear and Thigh constellation. From the Maccu as the family or clan is then derived the Mac now prefixed to names with the understood meaning "son of," whereas primarily it signified the clan. The CLAN and MAC are acknowledged to be synonymous by the sacred keepers of the Clan-Stone in Arran, whose family name was CLAN-Chattons alias MACK-Intosh.[1] Thus Mac- Donald was the Maccu or Mâhau, the family or clan of Donald, and this is still indicated by the style of *the* Macdonald, who is Donald of the Maccu named from the uterine abode itself, whence the Magad-brood and the Makhaut-clan ; hence also the personal title, as in the Swahili MKUU, a chief, a great man ; the MEK, in African Senaar, a king ; MAKH, Akkadian, supreme ; the MAKHT (Eg.), a mason and the explorer of mines; the Irish MOGHAIDE, a husbandman ; the Japanese MIKADO, an emperor, and the MAC in the Highlands, who is still the great man.

Typology explains the nature and cause of conservatism in the matter of names. The " O " prefix is still the sign of the mother-circle, whether extant as the Irish O' of O'Brien, the Japanese " O " of titular honour ; the Maori "Ouou ;" Egyptian Uau ; Fijian AH, a prefix and title of respect ; and various others. The O circle is the hieroglyphic deposit of the Hoe, Hoh, Haigh, and Khekh, the circles

---

[1] Martin, *Western Islands*, p. 225.

of many languages, as the Ega, Akkadian, crown; Aukhu (Eg.), diadem; Hak (Eng.), enclosure; Khekh (Eg.), a collar; Kak, sanctuary; Chakka, Hindustani, circle; Khokheye, Circassian, circle; Kekee, Swahili, a bracelet; Cokocoko, Fijian, beads; GOGY, Craven, and Kok, Basque, an egg; a name which on every line of language may be traced to that circle represented by the Irish OG, the virgin (mother) of the beginning. The O', like the Mac, is an ideographic sign of descent on the mother's side, from the time when the children did not know their own fathers.

The name of the Gadhael is derivable from Kêd, the mother. Hel, Har, or Ar (Eg.), is the son. This can be corroborated by the symbolic branch. The mother and son were represented by the tree of life and the branch, the Welsh Pren and Egyptian Renpu. One form of the tree of life is the Ash; an ancient name of this tree is the Kit. Ash-keys are called Kit-keys. Kit-Mut is a name of the Egyptian Great Mother, our Kêd. The ash is one of the few trees on which the mistletoe branch can be found, but very rarely. The mistletoe was the Druidic Pren or branch, and Guidhel is one of the names of the mistletoe, which was especially venerated at the time of the winter solstice. Guidhel, the branch, is Hel or Har, the child of Kêd whence Al, Welsh, for the race. Thus the Guidhel is the son or offspring of Kêd; if we apply this ethnologically, the Gael is a branch race of the Ketti (or Kymry), acknowledged to be so by the name, and whereas the Ketti are directly named from Kêd, the tree herself, the Gadhael are named after the branch as the offspring. The earliest form of the name Guidhel is Gwyddyl. And Gwydd is the wood or tree of the branch race, that became the Gael of Ireland, and afterwards of Scotland.

The W in Welsh often has to represent an earlier F or V; consequently the name of the Gwyddyl, in the Welsh form, preserves that of the Al, the race (in Egyptian the sonship) of Kheft. Gwydd, for example, is the Welsh goose, and Kheft (Eg.) is the typical goose or duck of the genitrix, whose names (Kheft and Aft) it bears. The Gwyddyl are also identified and denounced in later writings as the distillers and devils, and in Egyptian the name of the old mother denoted the mystery of fermentation, and Kheft supplied a type-word for the evil one, the devil, the English QUEDE.

Tradition affirms that five Firbolg brothers divided Ireland between them. The name of the Firbolg people is usually said to mean the big-bellies, and Bolg is an Irish name for the bellows, which are called after the belly. It is possible that the early people were pot-bellied like the soko and gorilla, on account of their diet. It is also conceivable that the term big-bellies might be a nickname employed by the men of a later cult to describe the worshippers of the GREAT or *enceinte* mother. In Hebrew, פלג (Pelg) denotes an image of fulness and inflation, or bulge. But this is not a primary appellation applied to a race.   The Irish Bolg are identical with the Hebrew Peleg in another sense.   The Fir-bolg are the five brothers who divided Ireland between them, and are probably a form of the five kings in the astronomical allegory who came between the flood of Noah and the conquest of Abraham, in which interval we find Peleg. In Akkadian and Assyrian PULAGU and BULUGU mean division, but, as we see in the Hebrew Pelg, this may be a family division, the tribe, kindred, or gens.[1] The dividers of the earth made the divisions into which the first families were formed.   Thus the Swedish BOLAG signifies partnership and the Icelandic FELAG, the MANN-FELAG was an association, a community of men.   In Arabic we have the BALAK for a crowd ; in Turkish, the BULUK is a number of persons, as a military company ; also the German, Russian, and Polish PULK : Persian, FAY-LAK ; Lithuanic, PULKAS ; Greek, PHALAGGOS ; Gaelic, BURACH ; Sanskrit, VARGA.   The one root supplies the English BULK ; German VOLK, a mass of men ; Danish FLOK, a company ; English FLOCK ; Norse, FLOKK ; A.-S. FYLC, a company ; Hindustani, FIRKA ; Latin, VULGI. the common herd ; Spanish, VULGO ; Italian, VOLGO ; and English, FOLK.   All these have one origin and retain the meaning of the division as a distinct body of men associated, so that the flag is the type as a military sign, and the name of it preserves the same word.   These, then, it is contended were the Fir-Bolg.   They may have worshipped the great mother, and so supplied a nick-name, for the prostitute is still known as the BULKER.   Oldham, in his *Poems*, speaks of the " BULK-ridden strumpet," who is so-called from being used by the BOLG or BULK, the whole body of men. The Bulker is the namesake still of the Greek Παλλᾶκίς[2] the concubine who, in Egyptian temples, was the divine consort, the Latin PALLACA or PELLEX, a harlot, a strumpet, and Hebrew פילגש (Pilegsh), the concubine, woman of the court, or prostitute. She represents the earliest status of the woman as consort of the company, the MUT and MORT, who was once the great mother.   The Hebrew פלג likewise denotes this intercourse in common.

BELAWG (Welsh) means apt to be ravaging ; BOLOCH is disquiet ; the Scottish PILK, English FILCH, is to pilfer, BILK to defraud, and BILEIGH, to bely.   Words often conspire, so to say, to give a bad character to the oldest names of the people.

In the African Galla, BULGU is the name of the cannibal, and in Irish Folgha meant bloody.   The Scottish BELGH, or BILCH, is the monster.   And so the Firbolg got a bad name.   But they have left us the BILLY for a brother, a term of endearment because the Bolg was a brotherhood or family, gens, or division ; also the FELLOW, a companion, the W representing a G, as the felly or felloe of a wheel stands for FELLICK.   Moreover, the burgh is a deposit of the Bolg, as the general enclosure of the clan, also the Park for animals

---

[1] Judges v. 15, 16, the LXX, Pesh, Targum.   [2] *Herod.* i. 84.

and the Net of that name for enclosing the fish,[1] and the Paroch, later Parish. The Burgh or Borough is related to the tithing by ten over which the Borsholder presided, as Borow in Old English is tithing, and the tithe is a tenth. The No. 10 is Per in Akkadian, Fer in Dselana; Fura, Yula; Fura, Kasm; Fulu, Malagasi; Fulu, Batta; Pulu, Atshin; Blawue, Basa; Puluh, Malayan and others; the Fir-BOLGS were probably founded on the division by Ten, which was the base of the hundreds and counties.

The Hebrew type-word is apparently compounded from Pu (Eg.) to divide and rekh (lekh), relations, people of a district, the race. The Pu, Fu, or Bu (Eg.) also signifies the belly, *i.e.* the womb, as birth-place. From this, the Bu-lekh or Bolg are the people of the womb, who descended on the mother's side when the individual fatherhood was necessarily unknown.

As an Ethnic name springing out of the fellowship, the Bolg of Ire-land is the same as that of the Belgæ and the Bulgars. All three represent the Folk-name of those who date from the division and discreting of the undistinguishable herd, as well as the earth on which they settled, no matter where they dwelt at first, before the BULK was as yet subdivided into individual families and whilst the woman still remained the Bulker. The existence of a Belgian tribe or people with a language nearly akin to the Irish has been recently proved by the GLOSSA MALPERGA, disinterred by Leo, which may, perhaps, supply an Ethnic link between the Belgæ and the Firbolgs.

The Irish traditions report that the Round Towers were erected by the "Tuath-dadanan," who were opposed in their work by the priesthood of the Firbolg, which led to a religious war.[2] They have been described as an early race of conquerors from the NORTH of Europe. It happens, however, that TUAITH is an Irish name of the north, and for the left hand. TIEVET is also a form of Tuath, as in Tievetory the hill-side north. In the hieroglyphics, the north, the lower of the two hemispheres, is first the Tepht (Eg.) and next the Tuaut (Eg.), and this was the Welsh Dyved of the seven provinces in the north. Tena (Eg.) means to divide, turn away, and become separate; ani (uni) are the inhabitants. This tends to show that the Tuath-da-Dana-an were emigrants from the north, whichever country it may have been. It happens also that the Tuaut of the north is the domain of Ptah the builder and establisher, and one of his titles is Tatanen. Ptah-Tatanen typifies the eternal with a round pillar, which, as a fourfold cross is the emblem of stability, and of the four corners. Tatanen means the Nen, type, image, statue; and tata, the head of ways, or, as we say, the cross-roads. The round tower, with its four windows, one to each quarter, is a similar ideo-graph, and it raises a suspicion whether the Tuath may not have been named from the north quarter in connection with the word

---

[1] Hollyband, 1593.   [2] *Vullancy Coll.* vol. iv. preface, p. 51.

Tatanen or Dadanan, as indicator of the four of which the Tuath was one, just as we have the Tuath in CROSS-THWAITE, in the Vale of Keswick. And if Patrick was a Mage of Ptah, it would explain why he was also called COTH-raige, rendered "four-families," not because he was the property of four masters, but because of the four corners, the Irish Ceather, which in the Manx Kiare (number four) equates with the quadrangular Caer of the Welsh. Cothraige denotes the four-cornered circle, as do the Tat and Round-Tower.

The Tuath-dadanan were reported to have descended from Nemethus, a native of Nem-thor, or Nemethi Turris; to have learned magic in Thebes, and waged war successfully against the Assyrians. They were led into Ireland by Nuadhah, the silver-handed.[1] Now Ptah was the son of Nem, the god of Thebes. Nem-tes is Nem himself; the type of Nem; Nem-ter, the bourne or boundary of Nem, which was that of Thebes. Nnu is Antennæ, feelers, and may have been the hand; huta is silver, Nnu-huta, the silver-antennæd or handed. Thus the descent claimed from Nem himself by the Tuath-Dadanan is exactly that of Ptah-Tatanen in the divine dynasties of Egypt, and no record of this descent of Ptah from Num was above ground out of Ireland until the present century. This increases the likelihood that Path-rick, the priest of Ptah, was the Mage of the god himself, whose name is thus pronounced in the Irish Path-rick.

If so, this will be the British Budd called by the Welsh the Victorious, who as Ptah is the Establisher, the Budd of the "Established enclosure,"[2] and who becomes in English the "Old PUTH." Ptah-Ur the old Ptah is one of the characters of the god.

The Egyptian " Ka " type, cultus, kind, supplies the terminal " Ch " to our adjectives, by which nationality is designated as Welsh, Scotch, Dutch; this as in the Irish " Corca " and in Greek was hard originally. Strange as it may look, the word Welsh is the same as Corca, or Kar-Ka, the type of the Kar, whatsoever that may mean. But unless we know what it does mean we cannot know what " people " may be included under that name.

Corca for people is Egyptian. Kar is native inhabitant and Ka the type of person or function. But Kar is essentially the lower, the nethermost, that is really the earliest. Corca as a type-name then belongs to the Kar, and in this case it may be from Wales. The Corca and the Welsh are synonymous in this sense.

Possibly the Irish type-name of CORCA for the people is derived from the Kar type in another way. Kar (Eg.) denotes the cultivator of the soil, as in the name of the GARdener, corn, and the crop. KARKA (Eg.) means to prepare, and the word will also read Kar, food ; Ka, to create, or Kar, to cultivate, and Ka, the food. Thus CORCA may denote a form of the corn-men, the Cruitneacht of Scotland, the

<hr />

[1] O'Flaharty. See Wood's *Inquiry respecting the Primitive Inhabitants of Ireland.* [2] *Gododin, Song* 22.

Scoti (Skhet, Eg., corn), the Hut or Hud of the Land of Hûd, in Wales. Coirca, in Irish, Kerch, in Armorican, and Ceirch, in Welsh, are the names of oats, and the Corca may have been the oat-men, who probably preceded the wheat-men, the oat being a cereal nearer to the grasses. The Keltæ as the Karti were already known to the Monuments. Karut signifies natives, inhabitants, aborigines, and Karti is but a variant of Karuti, which as previously shown is a modification of Kafruti, and the Karti or Keltæ are finally the Kafruti of Africa.

An ancient name of Babylon was Kar and the people are the Karti or Chaldees. The Egyptian name of Babylon is Kar. Kir is a supposed Hebrew name of Babylon.[1] The meaning of Kir or Qir as that which embraces and encloses, hence a wall or a circle, also agrees with that of Bab (Eg.) the circle, enclosure, and to go round. Inscriptions of Assyrian Monarchs designate the whole land as Kaldi, and the inhabitants as the Kaldiai[2] They also are the Keltæ from the original Kafruti on another line of migration and development.

So old was the word Kart for stone-polishing in Egypt that it was reduced to "Herut,"[3] meaning to polish stone, and as an illustration of the Kelt axes, hatchets, and knives, it may be pointed out that the word herut also means to arm for war. It is noticeable that these stones are confounded or mixed up with thunderbolts. In Shetland the Kelt stones are called thunder-bolts. But is not this because the lightning also kills? The Kelt-stone was polished for killing, and the supposed safety on the spot where the bolt has fallen may arise from the Kelt being a type of protection and defence, for which reason it was interred with the dead.

The Taffies are among the first people in the world. Tefi or tepi (Eg.) means the First. Tef is the divine progenitor, who came to be called the Father, but the female Tep is first. Af the flesh, the one born of, the Eve, Mother of all flesh, is primordial. With the article prefixed Af or Eve is Tep, the Typhonian genitrix, and the Taffies are her children. The British have their Duw Dofydd, a male god like Tef the Divine Father, and through him the Jewish David has been foisted upon them. But their descent is from Teft, the Egyptian Tepht, the source of all, the mother of all, who as place was Dyved of the seven provinces, and as person was the Goddess of the Seven Stars. Teft and Kheft are interchangeable names of this old genitrix. The rhyme of

> "Taffy was a Welshman,
> Taffy was a Thief,"

recovers an Egyptian word Tafi, worn down to Tai, and Taui, meaning to steal, run off with. And shall the ancient mother Tef, Teft, Tâui, be confounded with and merged into the Hebrew David? Let language forbid. This shall not be so long as a child has its toffy, or

---

[1] Amos, i. 5, and ix. 7. [2] Strabo, p. 735.
[3] *Select Pap.* 35, 8.

the sailor his tafia, and the ship its taff-rail, or the caged skylark its divit, or lovers have their tiffs, and married folk their duffle ; so long as daffin and tiffin continue and taff-cakes are made with Taffy—Taf (Eg.) is to carry, bear—riding on his goat, and there is a Dobbin to carry a sack of corn, or a painter to daub, or a dove to croon, or a devil to frighten, or a Davy Jones's locker for a seaman to send to, the name of the ancient mother thus memorialized in English must never be lost in that of the Jewish David or the Christian saint.

It has been already suggested that the Welsh David was identical with Tebhut (Eg.), in which we have the Hut, corresponding with the name of the solar race, of Teb. We shall find the chief type-names of the people are derived from the Great Mother ; the Ketti and Gadhael from Kèd ; the Fenians from Fen, Ven, Wen, Oine, the Irish Venus ; Taffy, from Tef (the same as Kêd), which makes it all the more probable that the Manx people are the children of the genitrix Menka, or Menâ.

Menka (Eg.) is the goddess whose name may be read Mâka, and who is the Irish Macha who bore the twins of mythology. The Manx arms consist of three legs in a circle doing the wheel, each one being a counterpoise to the others. In the hieroglyphics a counterpoise to a collar is called a MAANK ; the counterpoise is also the determinative of the nurse's collar, named the Menâ. The accent shows an abraded sign, which we restore by reading Menka. Taliesin, speaking of one of the mystical characters, says, " I have been MYNAWG, wearing the collar. I hold the splendid chair of the eloquent, the ardent Awen." The goddess Menka represents the Two Truths, implied by the collar and counterpoise which bear her name. She is a rare form of the genitrix, and must be most ancient. Menka means to form, fabricate, create ; and she was the feminine creator to whom the work of creation was assigned before the fatherhood was impersonated. The Manx counterpoise tends to identify the Manxmen, as named after Menka. We have the manger, a horse-collar. Menka, to create and form, is extant in the saying, " We were all MUNG up in the same trough," *i.e.* in creation. The word mingled is a deposit. Manx is modified into Man, as the name of the island. One of its names is Inys Mon, the island of the cow. The cow Mon will identify the wet-nurse of mythology, as Menâ or Menka, the goddess of the Menâ-t, the outcast shepherds of Egypt.

The Menka deposited the Monk, the Isle of Man as Monæ was an especial religious sanctuary. The word Ma-ank also reads Motherking, or, I, the Ruler. Now the Chickasaws at one time had a king whom they called the Minko, and the succession was hereditary on the mother's side.[2] This points to the female Rex, or mother-king (Ank). One particular clan whose hereditary king was on the

<hr/>

[1] Davies, p. 419.   [2] Schoolcraft, *Res.* i. 311.

mother's side, called themselves the Minkos. The same meaning may have entered into the Manx name as the children of Menka for their right of primogeniture still extends to females as well as to males. So the Guanches of the Canary Isles, when discovered by the Spaniards, were found to be living under the government of " Menceys," who were chiefs subordinate to one head.[1]

The Chickasaws were divided into six clans, namely, the Minko, Sho-wa, Co-ish-to, Oush-peh-ne, Min-ne, and Hus-co-na. And when the chiefs thought it necessary to hold a council they requested the king to call one. The king then sent out runners to inform the people of the time and place at which the council was convened. This council of the six clans summoned by the Minko is identical with that of the six sheadings summoned by the Manx. The island is divided into six districts called sheadings (sheth is a separate division, in Egyptian, and in Hebrew, sheth is Number 6) under the jurisdiction of the Court of Tynwald.[2] This court sits for the summary trial of offences for breaches of the peace and misdemeanours. It is still called holding a council. The Court of Tyn-wald sits in Douglas every fortnight; and " tna " is a fortnight in Egyptian.

Wald, by permutation, is wart, ward, or word, and in Egyptian, Uart means the foot, leg, go, fly; that is, word sent with all speed by means of the foot-messenger or runner. Thus the Tynwald was the council summoned fortnightly by the runner. The Manx judges are still designated Deemsters, and, in the hieroglyphics, " tem " or " tma " means distribute justice, make truth visible, satisfy. Anciently the Manx coroner was termed a " moar." The Egyptian " Mer " is a superintendent, and to die. Further, the " mer " is superintendent amongst five as the " Mer-tut,"[3] and the Manx coroner was also the sheriff's officer of a court which had four bailiffs, and he, too, was a superintendent of the five, named the Moar. In the North of England a farm-bailiff is a Moor, whilst the superintendent of the town-walls in Chester was a Murenger. The Mayor is perhaps derived from the Egyptian Maharu, the hero, to whom the collar of gold was awarded.[4] The collar of gold is a part of the insignia of the English Mayor. Still mayors are superintendents, as the French MAIRE DU PALAIS.

All lands, Dutch and Welsh, is a phrase used to express the whole world; and the whole world thus resolved in two halves is according to the Egyptian pattern. But Theod, a people, and Welsh, strange or outlandish, are not primitives of speech. Dutch is the Egyptian Tut with the type-terminal (KH, Eg. type), and tut, tat, tata, denote the upper of the two, the head, chiefs, princes. Tut is the mountain, our Tot-(hill). The Welsh are the people of the Kar, or Well, the lower half of the whole. This symbolical and philological distinction was then assumed by those who bore the name of the Upper, the

---

[1] *Ethnol. Journ.* N. S. vii. p. 107.   [2] *Encycl. Brit.*, Isle of Man.
[3] *Lepsius Denk.* ii. 22.   [4] *Rec. of Past*, vi. p. 7.

Teuts, Teutons, or Dutch, who looked down on the Welsh people of the underworld, the Kars, as the lower. Ethnologically, the loftiest means the latest, nearest the surface, the upper stratum ; and the lower is the earliest. It should be "all lands, Welsh and Dutch."

The Irish Aithech-Tuatha, who are usually identified with the Atti-Cotti of the Roman writers, constantly found to be warring against the dominant races during the first centuries of our era, are rendered by Dr. O'Connor as the *giganteam gentem,* and were looked upon by O'Curry as the rent-paying people.[1] The name read by Egyptian is Aat-hek-tata ; aat, is house, abode; hek, to rule ; tata, heads, chiefs, heads of ways; the established heads of houses, or the equivalent of rent-paying people. Atai means chief, superior.

The name of the Pheryllt often occurs in the ancient writings. They are the legendary wise men of Wales. In Hanes Taliesin, the goddess Keridwen prepares her cauldron of inspiration and the sciences according to the books of the Pheryllt. An old chronicle quoted by Dr. Thomas Williams affirms that the Pheryllt had an establishment at Oxford, where we find the Uskh Halls, the temples of learning, prior to the University founded there by Alfred.

The Pheryllt have the reputation of being the first scientific men and teachers of arts, more particularly those connected with fire. They are supposed to have been the earliest metallurgists and chemists. The name of Pheryllt, in its earlier form, is Phergyllt, and Gyllt answers to Karrt, the Egyptian furnace ; the double R even is repeated in the double L. Per or Pra (Eg.) means to show, cause to appear, exhibit, reveal, manifest, explain, and is a form of the word fire. Those who did this by means of fire might very well be the Perkarr-t, Phergyllt, or Pheryllt, the fire-furnace men, or metallurgists.

It is not necessary to derive the name of the CULDEES, who formerly occupied the CILS or CELLS in Scotland, Wales, and Ireland, from that of the Keltæ race, or Karti, the stone-polishers. They were a kind of monkish priesthood, having especial relationship to the dead. This primarily applied to the covering, burial, and preservation of the mortal remains. The Gaelic CEALL is both a cell and a church ; the CILL is a burial-ground or a churchyard. The English GOALE is a sepulchral tumulus or barrow, and the Cell is a religious house. The CEL, in Welsh, is the corpse ; CIL, Irish, and CIALL, Gaelic, mean death. CLEITH, in Gaelic, is to conceal, hide, keep secret ; CLEITHE, Irish, is concealing and keeping secret. How early this is applied to covering over may be seen by the African Padsade name of the loin-cloth, the KULOTO, the Irish CEALT, and Scottish KILT ; Welsh GOLAWD, a covering or envelopment; Hebrew GALAD, and Javanese KULIT, a skin ; an early form of covering and clothing; Irish CLOTH, a veil or covering ; English

[1] *Manners and Customs of the Ancient Irish,* v. i. Intro. p. 23.

CLOTH and CLOTHING. The clergy are still called the CLOTH; they were the first coverers of the dead, as the Mena-ster is the layer-out of the corpse. The French COLLET is a clergyman. The CULDEE was a COLLET, Hebrew קהלּה, the preacher, but in especial relationship to the covering and concealing of the dead, the stone-circles and the stones. CLETHY or CLUDDU in Cornish is to bury. The Irish CLUDH is a burying-ground; the Gaelic CLADH a churchyard; the CALLAID, Gaelic, a funeral wail; the A.-S. GILD, service, worship; Greek KELADEO, to invoke, call, celebrate; French CULTE, worship; Hindustani, Arabic, and Turkish KHULD, being perpetual, in perpetuity, or eternal. The Egyptian KART was the chest of the preserved mummy. The KARAS was the burial-place; the KARAT, a key, bolt, lock. Our KARAT or CULDEE kept the enclosure of the dead, as at Kirkcaldy. In Egyptian the KA is the religious minister, and RAT means to retain the form, plant, and to beseech, which describes the office of the CULDEE. Teru, to invoke and adore, is also the name of the layer-out and mourner of the dead. This is one of the most important meanings of the root TERU, the Cornish DERUW, and Irish Draoi, the Druid. Also Teru, to mourn, is the English DARE, to grieve; Gaelic TUIR, to rehearse with mournful cadence. The Gaelic TOIR is a churchyard, and in Cornish DAIAROU means to bury; the English THURH is a grave. The earliest religious service was rendered to the dead, and this was the office of the Culdee, who continued the work of the Druid, which was taken up and carried on by the Christian priesthood.

The Irish Corca-Tened were a form of the fire-people or workers in fire. The sons of USNAGH or UISNEACH, also called Usnoth, were another. It was at Usnagh (Uisneach), in Ireland, that the Council of all the provinces met annually and the new fire was lighted that was sent over all Ireland. This identifies the name with fire. As a people, or clan, the sons of Uisneach are especially dear to Gaelic tradition and song, for the sake of the hero Naisi and the woman Deirdre, the fatal beauty of mythology.

The Three Sons of Uisneach are possibly a form of the solar Triad, or Trimurti. Hence their three domains and their booths of three divisions. If so, Naisi the hero and eldest of the three, will be the same representative of the sun in the lower regions as the Nasi, whom we shall find in the Hebrew mythology; one of three Adonim. NAS (Eg.) signifies fire and flame. The NASR is the fiery Phlegethon, and Naisi, in connection with the three brothers Uisneach, would be the solar god of fire.

It seems more than probable that the DEIRDRE or DARTHULA, who was the Keltic Helen, the fatal BEAUTY of the sons of Uisneach, was likewise the Direte of the British coins or talismans; a name of Kêd, who has been identified with Taurt of Egypt. Darth-ula as TAURT-URA is

the first, oldest Bearer.   TA-URT-RRE would be the first, oldest, chief
Bearer of the children as Goddess of the Great Bear, the Spark-holder
or Mother of Fire, and DEIRDRE is connected with the vitrified forts.
The ancient Spark-bearer is named KAR-tek, and GAR-drei is a
local variant of Deirdre's name at Ballycastle, opposite RAGHERY
Island, Ireland, where there is a rock still called CARRAIGH UIS-
NEACH.   This appears to connect the CARRAIGH name with the
smolten rock from REKH (Eg.) the furnace, fire, heat.   If we divide
the name of Deirdre as DEIR-dre (cf. Kar-drei) and DEIR is taken to
represent Taur (used without the terminal T), DREI as synonymous
with TERUI,—the Egyptian form of Troy of the Seven-circled Centre
or Sesennu, found in the Kef of Troy at Cov-en-try, the place of the
Goddess of the Seven Stars and her son Sut-Baal, before the Solar
Triad of the sons of Uisneach was established,—Deir-dre is then
TAUR, the Great Mother of TROY.   Hellen is the Renn (t) nurse of
Har the Renn, and in one version Naisi and his two brothers are
called the sons of Uslinn ; Us (Èg.) meaning to produce, become
large, swelling, and Lenn (renn), to nurse, be the gestator as the Lenn,
or producer of the LENN.   Mythologically the matter is so ancient
that to all appearance it must have gone from Wales to Ireland in the
stellar form, and from thence to the West of Scotland in the solar
stage as a comparatively modern reproduction.

So much for the mythical aspect.   As an ethnological entity the
clan of the sons of Uisneach are credited with being an offshoot of
the Irish Gadhael, who migrated from Ireland into the West of Scotland, and settled in the region round Loch Etive, where they left a
number of vitrified forts or mounds.   They were rock-smelters.
Many rocks are easily melted into vitrified masses, but of course the
time was when it was a thing of wonder.   The workers in fire,
whether as vitrifiers, metallurgists, or makers of fire-water, were looked
upon as magicians, and we find the direct descendan s of the Kymry
denouncing the Gadhaelic fire-people as smelters, distillers, and devils.
The "Rocks of Naisi" and the "Wood of Naisi," still found in the
neighbourhood of Loch Etive, become significant in relation to the
smelters when we know that Nasi (Eg.) is fire.   For the Wood
(birch-trees) was famous fuel for the furnace, and the iron made with
its charcoal still fetches, in Glasgow, four times the price of coal-iron.[1]
This wood of Naisi, the COILLE NAOIS, in Muckairn, was the earlier
coal.

Dun Mac-Uisneachan, on Loch Etive, shows the seat of the sons
of Uisneach, whose names are connected with the vitrified forts.
Us (Eg.) means to create and produce, large, extended ; EN is by,
and Akh is fire.   Us-en akh reads the creators and producers by means
of fire.   Further, the Egyptian has the terminal T in Akht, fire.   This
would allow for the variant found in Usnoth or Usnath as a modified

---

[1] *Loch Etive*, p. 75.

form of Usnakht. But there is more than the vitrified mounds concerned in the name and nature of the Uisneachan. We know that in the time of the Bronze age in Britain cremation was first introdued by the Solarites, the men of the worship under which the barrows were made in the shape of a disk.[1] The vitrifiers were also workers in bronze, and they burned their dead and placed the ashes in urns. Now, in Egyptian, USH means to consume, destroy by fire; YSU, Welsh, to consume; YSSU, Cornish, to burn. ASH, in Hebrew, is fire; OZO, in the African Hwida; UZO, Mahi; WOZI, Pika, and WASHA, Swahili, to set on fire; USH, Sanskrit, to burn; USHĀ, burning, scorching. The word is likewise applied to reducing by means of fire, as USHM (Eg.) signifies a decoction and the essence; USHM is to grind down and devour. This USH would explain the name of the UIS or Us-neach as the vitrifiers and producers of the forts, by means of fire, with another meaning for the Akh, or Akht. These are the dead. Ush-en-Akh would denote the cremationists. Us, then, meaning to consume or reduce by fire, applies to cremating, vitrifying, and distilling (Ushm, Eg., being the essence, the decoction). The Akh or Akht are the dead; EN (Eg.) is "of the," and the Ush-en-akh (or Ush-en-akht, whence Usnath) would be the destroyers or consumers of the dead by fire. In the name of Dunstaffnage, a castle, built in later times in the neighbourhood of Loch Etive, we may find corroboration for this reading. Dun is the seat, the royal seat, of the Gadhaelic kings and chieftains, and STEF (Eg.) means to melt or smelt, to purify and refine with fire, whilst NAGE also represents the "EN-AKH" as in UISNEACH. The name of Uisneach derived from Ush, to destroy by fire, would be most perfectly realized if the people were so designated because they burned their dead; a horrible practice in the sight of the men of the older faith. This would be sufficient cause for giving a bad character to fire itself, whether in the fire-water (whisky) or the element which consumed the dead, hence the "distillers and devils" applied to the Gadhael.

If we may judge by the only living representatives of the Palæolithic men in Europe, the Eskimos, the sole surviving people known to show a total lack of reverence for their dead, the earliest emigrants into Europe took little, if any, care of their dead. No interments have yet been proved to belong to the Palæolithic period. Thus they would not have the very first stake in the soil afforded by the feeling for their buried friends. They drifted over the earth's surface without any permanent ground ties. They continued the life of nomads up into the Arctic Circle, and wandered about the world all over Europe, through Asia Minor and India, in a far larger range than that of their successors, whose increased culture and acquirements enabled them to add agriculture and fishing to their hunting, as

[1] Dawkins, *Early Man in Britain.*

means of subsistence, whilst the respect for their dead helped to tether them down to the soil where they took root and founded on the spot their earliest civilization.   The first emigrants from the primeval home were of necessity wanderers over the earth, and that was how the earth was peopled at first.   But with the burial of the dead and the cleaving to them still, came the more or less permanent settlements grouped about the graves as a fixed centre of civilization.   The caves, kasses, caers, long barrows or disk-shaped tumuli, mounds, cromlechs, tombs, and temples, all tell the same tale in unbroken continuity. The difference in the range does not therefore imply a difference in the race.   Thus it appears probable that the prime incentive to form a settlement and take a fixed rootage in the earth originated in a cling-ing to the burial-place, as if the nomads were first laid hold of by the dead hand, and retained by the skirts, in the locality which made an irresistible appeal to them with the voice of their loved and lost crying "Stay with us and protect us in our helplessness."   One seems to see this in the Skhen or Khen (Eg.).   The Skhen is a shrine, and the name signifies to alight and rest, place, dwell, be in sanctuary, a sanctuary, to institute and establish.   If the meaning of Sekhen be summed up in a word, it means to settle; the determinative being a water-bird.   Khen, Khenen, and Sekhen are interchangeable. We have the Egyptian Khenn, the religious sanctuary in Caer-Conan, the ancient name of Conisborow, situated upon the high lands of the river Don ; here the sacred place on the rock, the Kester, was con-verted into a later fortress or castle.

Khennu (Eg.) is to believe, and the word means to lean on, rest, be sustained.   It may be divided into Khen, to carry, transport, image, and nu, divine, or heaven.   So Nahb, another name of belief, means to sustain.   Thus Khen-nu is divine sustenance and heavenly trans-port in the sense of carriage and conveyance, rest and settlement. The Cangiani are thus traceable to the Kenners, whether we call them believers, knowers, or primitive settlers.   Khen and Sekhen permute, and the people of Caer Segont become the Cangiani.   Ki (Eg.) denotes the land, region, or abode ; uni or ani, the inhabitants.   Thus the Cangiani are the inhabitants who first settle in the Sekhen enclosure, which was a religious foundation, because a sanctuary of the dead, also found at Scone as the "MOUNT OF BELIEF," and the word Khenn acquires another relationship to the dead as if through them the primitive men first laid hold of the other world as well as made fast their earliest foundations in this.   Caer Segont, the Segontium of the Cangiani, answers to the Egyptian Skhent, that is, the Sekhen, shrine, or sanctuary.   The Skhent is the double crown, emblem of the Two Truths, and of the established circle.   Caer-Segont is the resting-place, the established Khen or Sekhen ; a foundation in the Caer or sacred circle, hence a sanctuary, one of those circles, for example, to which the living debtor can still flee and find protection with the dead, although he does not know that it was on their account the right of refuge was conferred.

We might cross-examine other witnesses to the Egyptian origin of the Kymry in ages incredibly remote. For example, "Taht was formerly Sut," we are told, of the two gods. And this conversion of Sut into Tet has its analogue in the hieroglyphics, where Tset is the earlier form of Tet. The Deep is Tset or Set, and the rock is Tser or Ser. A syllabic Tes passes into the phonetics as both T and S, and thus divided, we obtain Set or Tet as the bifurcation of Tset. The snake Tet must have been a Tset. This double sound of TS or TZ is lettered in the Hebrew Tzade, צ, which looks as if it figured the snake with a dual head, or T and S in one. In Hebrew it expresses the strongest sibilant, the hiss of the snake. The snake Tet, then, was an earlier Tset, and survives as the Hebrew Tzade and English Zed. This double sound of the TS also divides in the same way in Æthiopic and Arabic. It is very rare on the monuments, where it leaves an accented T in the snake, supposed to have a similar value to the Coptic Djandja, Ϫ or Dj. This latter letter, however, introduces another phonetic link. It is identical with the hieratic form of the crocodile's tail used as a Ka in Kam. Ka modifies into Sa with the same sign. Thus we have the Sa, Tsa, and Ka, the Ka being first of all. At present, however, we are concerned only with the Tsa, Tes, or Tzade, which divides visibly into the letters T and S. This Tes survives in Assyrian, Hebrew, Æthiopic, and Aramaic, as a representative of the S. Also, DZ in Mpongwe, TS in Setshuana, DZ in Ki-Swahili, DZ in Ki-nika, represent the S. The Semitic Z is convertible at times with D, or rather the Z and D interchange; the Hebrew Zeh (this) is Da in Chaldee; the male, Zakar, is Dekar. The Egyptian form in Tzer appears to explain that the root of both is a compound, and this bifurcates again, and offers a choice in the process of sound-shunting. By this process the Egyptian Tser or Ser, the rock, becomes Tser, Tor, or Tyre. CH in various languages of Eastern Asia is the modern equivalent of T. Ts, in Chinese, sometimes stands for the European S, and the Chinese S is T in Cochin-Chinese. The T often precedes the S, and is aspirated. In several languages of Eastern Asia the CH is equivalent to T. Before the "E," in Latin, G changes to "DJ," whilst in French we get back to ZH, and in Russian we find the primitive Ts. Tzar is purely the Egyptian Tser.

The Egyptian Tes interchanges with Tsh, which is probably the earlier sound, as in that we have the ground tone of the Mongol and Arabic DJ and English J, French ZH, Latin DJ, and G, the English and Sanskrit CH, where we pass back again into the K. The origin of all this lies at the foundation of language, and has to be sought for in the clicks, where we cannot follow it at present. But the point is that this TS, TZ, TSH, or, DJ is a living sound to-day with the Welsh, and one they have been unable to exchange for the later representatives of this primitive of speech. Mr. John Rhys, in a report as inspector of schools in the counties of Denbigh and Flintshire, remarks on the total inability of the Welsh children to master the sounds of J and CH. He says their inability to pronounce these leads them to read a sentence like "Charles and James got a shilling each for finishing the job which they had begun," " Tsyarles and Dsyames got a silling eats for finicing the dsyob whits they had begun." In some parts also the Carnarvonshire habit of giving a sputtering pronunciation to a final dental is not unusual, while the U of North Wales, which resembles the German Ü, is frequently substituted for the English I. Thus the sound which had almost died out of monumental Egypt, the TS or TSH of the snake and the bolt hieroglyphics, is perfectly preserved in Wales, still claiming kinship with the clicks of Africa. According to this pronunciation, the cherry would be called the Tsherry, and that is the Egyptian name for the red. It is not known that they had the cherry-tree; but TSHRU is the red crown, red calf, red land, gore, red blood, red tree. The substitution of the U for the I is also notable, as in the hieroglyphics the I has in it an inherent U, and interchanges with U. The Egyptian U is the English I. Tef and Kef are interchangeable names of the same goddess who became our Kêd, and the equivalent TV and QV both pass into the later P, (B) of Welsh. Also, as already pointed out, the QV of the Ogham is composed of five digits or one hand, and in the hieroglyphics Kef is the hand. The hieroglyphic language must be studied before the " glossic " of the English can be perfected. For instance, the Cornish and other pronunciations of tin make use of an aspirate, which the Egyptian Tahn for tin, will explain. Also the aspirated P of the Gaelic still represents the ideographic PEH, which in one form is the lioness, in another a water-fowl making the visible " PEH " with the open bill, both denoting the expulsion of breath, or aspirate.

The S (Eg.) is the causative prefix to verbs, and in Welsh, words are augmented and intensified in meaning by help of a prefix, YS. The sounds of the "I" in English are related to the Egyptian I, which has the U inherent in it ; and as the U became E, the I also implies an IE, as well as IU. Sounds representing diphthongs tell us of the ideographic stage of language and its complex types, out of which the simpler phonetics were finally evolved.

Some of the Scottish antiquarians seem inclined to consider, in direct antithesis to the reputed wise man, that there is nothing OLD under the sun, and to set the facts of survival in battle array against the evidence for evolution.

Dr. Arthur Mitchel, who has done good work for the stones and other Scottish antiquities, has recently shown us how, in its hiding-places of Scotland, in nooks and corners, and the far-off solitary isles, the past survives in the present so persistently as to make him suspect the evidence for an immense antiquity. He finds the stone whorl is still in use ; still manufactured for the spindle. In other places it is out of use and knowledge, and has become an object of superstition and mystery, called an "adder-stone." But this, so far from severing a link with a remote past, establishes a new one; for the "adder-stone," or Glain of the Druids, was a bead perforated to be strung like the beads worn by the Africans as Gree-grees, or charms ; and the whorl is a bead also, fixed on the spindle. The same type persisted in the beads worn on the pins of our Bedfordshire and Hertfordshire makers of hand-lace. Anything circular and perforated for use was typical, and was preserved on that account, as well as for use. A very early Glain, which might have also been used as a whorl, was a joint of the backbone ; joints of the vertebræ in Welsh are GLEINAW CEFN, and these are still worn as amulets by some races. In the hieroglyphics the vertebral column, the USERT, is a sceptre of sustaining, and sign of maintaining power. The fact of the whorls being called ADDER-STONES identifies them with the ancient system of symbolism, which included the beads and the serpent ; and by that system must these types be adjudged. Lateness of persistence and reproduction of the type is no evidence whatever against its antiquity ; on the contrary, it is only the most ancient type that does persist in this way, whether in race or in art, and the strength of its persistence is some measure of its age. Dr. Mitchel found a boy in the act of shaping a whorl for his mother's spindle with a pocket-knife, and the stone used, a soapstone, or steatite, is called KLEBER-stone in Scotland. This name of the stone has something to tell us of the past : it is the stone named for its being easily carved. KLEB is interchangeable with CARVE and GROUPE, to sculpture in English ; GRAF, Gaelic, to engrave ; CEARFAN, A.-S., to carve ; CERFIAW, Welsh, to form, model, or carve ; CARPO, Latin, to carve ; JARUB, Arabic, a hewn stone ; and the whole group of words are derivable from the Egyptian KHEREB, or KHERP, that which is first in form or formation. Kherp means the principal, foremost, surpassing ; to form, model, and figure. Khereb, the form, and to figure is primal ; this, in KLEBER, is applied to the stone that is easily wrought into shape. Here the name agrees with the type, and both come from the beginning. The people who still continue to make whorls in Shetland are not responsible for this type-name of the stone, which goes back to the Stone age in Africa, and the names of the stone-polishers as the Kara, or Karti ; the cylinder as a KARU-KARU, the circular thing as a KAR, to which word the F adds it or him, whence the Kherf, or carver, and the KLEBER-stone. The whorl was not turned into an adder-bead and amulet in less than a century.[1] It always was an amulet by virtue of its being one of the hole-stones or Kar-stones, as were the jade ring and perforated axe, the bead or adder-stone, and was religiously preserved for use on that account.

[1] Dr. Arthur Mitchel, *The Past in the Present*, p. 156.

Dr. Mitchel appears to me continually to confound the rudeness of the later copy with that of the earlier type; rudeness of structure or of workmanship *is* primarily a proof of great age, but no guidance whatever to the time of the latest reproduction, and it is useless to adduce the lateness of the copy against the antiquity of the type. Nor, as against the doctor's *suggested* conclusions, is the evolutionist bound to believe that the prehistoric man was a brainless being, or that the cavemen of Europe were specimens of the primitive man and the missing link. We know they were not.

The troglodytes were found by Dr. Mitchel dwelling in the caves of Wick Bay in the year 1866, but that can afford no argument against men having lived in caves in the Palæolithic age, possibly fifty thousand years ago. The Kaf, Coff, Kep is the place born of, the cave of Mother Earth, or the womb. And from this it can be gathered that the KAFRUTI and KAFFIRS were the Cavemen. The accented â in Kârti for caves, holes underground, excavations, shows the full word to be KAFRUTI, and the Kâr is the KAFRU or Kâ-ru, literally the Earth-Mouth, or, in the human form, the Uterus. The evidence for the past is of various kinds, and in this case these caves of Wick and their dwellers could never be confused with those of Duruthy. The steel pocket-knife, needle, button, or something equivalent, would be found to continue the story, even though the stone whorl and clay graddan were also there. We have now to take into consideration the evidence of the namers.

In Lewis, at the extremity of the Western Isles, they still make the rudest pottery now produced in the known world in the shape of the CRAGGAN. It is made of clay by the hands of woman, the first shaper; it is dried and then filled with burning peats, and set to bake with burning peats all round it. Now in Egyptian Ka is earth, of which the vessel is made; the REKH is a brazier-furnace, and KHAN signifies the hollow utensil for containing and carrying, the English Can. The Craggan is the can made of earth burnt in the furnace. The Craggan is a cooking vessel, and the Karr (Eg.) is an oven; Akh is fire, and Khan the carrier. The Karr-akh would be a fire-oven, and with the KH modified into H, this would be the Karh, or KARAU, Egyptian names for the jar, with the sign of cooking by fire, or of distilling. Khan is also a name of water and other liquids; thus the Kara-Khan would be a water-jar, as a Craggan. The Craggan is as rude and simple as anything made from the beginning of pottery that would stand fire and hold water, and it is thus a true type of the beginning, even though in point of time the beautiful KARHU of Egypt comes between. The type is the test of antiquity, not the time in which it was last repeated. Language shows that the word JAR is the later form of Kar, and the Egyptian KARHU, of an earlier Karkh, found in the CRAG-gan, and no English JAR-can will ever be called a CRAG-gan; hence the name and the type must be taken together. This testimony of language is infallible and final, if we can only get the other facts rightly adjusted.

The wild Irish do not thresh their oats, says Fynes Moryson, but burn them from the straw, and so make cakes of them.[1] According to Martin, the ancient way of dressing corn in the Western Isles was the same. He says[2]—" This was called GRADDAN, from' the Irish word Grad, signifying quick." But the Egyptian KRRAT, the furnace, offers the likelier explanation ; and as TENA is to separate, corn separated by the furnace-fire would be KRRAT-TEN, or Graddan. Karr and Krrat are both used for the furnace or oven.

In Shetland corn is still dried or roasted by rolling hot stones among it. The corn is then ground and made into cakes called "BURSTIN" bread.

Dr. Mitchel looks upon the roasting as an accident occurring in the process of drying the corn. Let us see what language says. TIN is fire, and the corn is dried or roasted with fire. The Egyptians likewise made BURS or PURS for food, as some kind of cake. The reader need go no farther than Dr. Birch's *Dictionary*[3] to see that PER (Eg.) is corn, the commonest of type-names for corn, as BERE (barley), BAR, Irish, corn; Hebrew, BAR ; Arabic, BURR, for wheat ; BAERI, African Gobura, oats; PERI, Krebo, beans ; BORA, Hindustani, beans ; POIRES, Norman, peas ; PARE, Maori, corn ; PURON, Greek, corn ; PARE, Hindustani, corn ; FAR, Latin, corn ; PURA, Sanskrit, some kind of grain ; VORA, Sanskrit, a pulse ; VRI, Sanskrit, rice. PER (Eg.) is corn, and PERS, a cake, food, made from it ; the S in Egyptian adds the thing made from PER, whence PERS ; TIN adds the fire, and we have the PERS-TIN, or BURS-TIN, as a name applied to the cakes (Eg. Pers) made from the roasted corn, most probably at first because it could be the more easily crushed.

We are not left without some sort of time-gauge for the immense past belonging to the people of the isles whose going forth is chronicled as the first thing after the flood of Noah.

In the stream-works at Pentewan, relics of human life and occupation have been found forty feet below the surface, and several feet beneath a stratum which contained the remains of a whale (*Eschrichtius robustus*) now extinct. In a stratum of still earlier date, a wooden shovel and a pick made of deer-horn were discovered at the Carnon stream-works.[4]

The name of " POCRA ". has been preserved at Aberdeen by means of a jetty and a rock. It is unreadable as it stands. By making it " Bocna," a meaning is assigned to it in Gaelic, as the river mouth. There is a tradition, however, on the spot that the two rivers, Don and Dee, which now debouch at a considerable distance apart, once joined their waters at the foot of Broad Hill before they were poured into the

---

[1] *Description of Ireland.* [2] P. 204.
[3] *Egypt's Place*, vol. v. p. 464. [4] *Arch. Journ.* v. xxxi. p. 53.

sea. The POCRA Rock is off the mouth of the Dee. The Dee in Wales rises from two heads, and is the dual river. Ti (Eg.), and English twy answer to the twofold or dual river. The Dee is the dual one, as the two founts form one river. But the river with the same name off Aberdeen is not a double river that becomes one. Nevertheless, the local tradition declares that it was once the double river that ran into one, but that the land has been so far eaten away by the sea that the Dee was divided from the Don. If we may read the name of Don by the Tun (Eg.), it will corroborate the tradition. Tun means to be divided and made separate, literally to be cut in two halves. But did the Dee and Don once unite in a single stream called the double river? The local tradition affirms that they did, and says further that the Pocra Rock stands at the spot where the banks were broken and the waters united to run seaward in one stream. PEKA (Eg.) is to be divided; Pekha, division, separated in twain. And RUA is to rush swiftly and come near. PEKH-RUA would signify the confluence of the divided waters; or, still more forcible, " RUAU " is the bank of a river; Pekh, to divide, sever, make a hole, gap, chasm. PEKH-RUAU in Egyptian signifies the place where the two rivers ran together through the broken bank, and rushed in one channel to the sea.

Thus far Egyptian corroborates local tradition. To judge topographically from present appearance, it must be many thousand years ago since the Dee and Don debouched as one river,

One name of St. Michael's Mount is " Careg Clows in Cows," or " Careg Clowz in Cowse," accepted by Cornishmen as meaning the grey or hoar rock in the wood. On this, and a popular tradition, has been based a theory that the name was given when the mount formed part of the mainland, situated in a wood, some twenty thousand years ago,[1] in the era of the mammoth; a fact which, if established, would be in keeping with that age.

It is certain the Mount was at one time conjoined to Marazion Cliff. Equally certain that a people ignorant of geology have never ceased to assert that the Mount was formerly connected with the mainland; also the tradition was not a geological theory. Camden, Carew, and Drayton called the Mount "Careg Clowse in Cowse." But Carew likewise writes the name " Careg Cowze in Clowze." The Grey Rock may be got out of Ca-reg by aid of Kaui (Eg.) grey, crepuscular, and rock.

The Egyptian word for wood is Khau, it is also written with the terminal T, Kauit. That is the Keltic Cuit, Welsh Coed, Armorican Koad, English wood. We have it in English as cow; the wood-pigeon is a cow-prise, and it is the law of the Cornish language in such a case to change the T into S. But an entirely different rendering of the name is now proposed.

[1] Pengelly, *On the Insulation of St. Michael's Mount.*

As we have seen, the crick and craig stones in the ultimate form of the name, are Kar-rekh stones, in two cases, of birth and burial, of purification and concealment underground, and may be so in the present instance. Careg is not merely a rock. Kar is a rock, and roke denotes a vein of mineral ore. These are both in English. Rekh (Eg.), which means to hide, to purify and refine, is also the name of the furnace or refinery. The Kar is found in a variety of forms implying underground, and being inclosed. Kar is the sarcophagus, or tomb, evidently applied to mining, for the plural Karti includes passages, holes, prisons, and CATARACTS of water underground. Karrekh (Eg.) reads the refinery of the mine. We have the name in Carrick. The Karas (Eg.) is the Kar where the precious thing is preserved, as the mummy. Kar is to encircle round, enzone, contain, possess, imprison; As the precious thing. From this come our killas, kollus, close, argil and clay, the Kar in which the As, the precious thing is inclosed, embedded, imprisoned or karast. Gold is named from this Kar, with the T terminal, Kar-t that which has been Karr'd ; Karr'd includes the incarceration in the earth and the passage through the furnace, when it is. Karred (charred and orbited), our gold.

It has been overlooked by all that we have a word "close" in English with a special sense of metal inclosed in minerals. The matrix of clay or other argillaceous slate in which the metals gold (clay-slate is the main matrix of the gold found at Ballarat), copper or tin are embedded, is named "KILLAS," the Cornish miners calling it "KOLLUS." The word is identical with the Sanskrit Caras, and Persian Charas, for a place of confinement, a prison. Also in the North "COWS" is the technical term for slime ore, that is the ore still mixed with mud. Khas (Eg.) signifies a rude and miserable condition. Hes is dirt; Ush, mud. Khus, to pound, ram, beat with a mallet. Khus is a valuable variant for Cows ; the one names the stuff, the other the process. Khusing, or stamping and pounding the COWS, is one of the most prominent of the processes. The stamps used are a kind of wooden pestles, these are attached to "cams," and Khem (Eg.) means to bruise, crush, break into pieces ; the cams being used for the pounding.

CLOWZE is a dialect form of Kallas, Kollus, and Close; the proper mining and geological term for the special clay-slate in which tin is found.

This sufficiently indicates the meaning of "Clowze" and "Cows." Kleuz and Kloz are found in Armorican as names of the inclosure or tomb, the Egyptian Karas, Hebrew חרש, Persian Charas. We may note how this root Kal (Kar), enters into mining operations. Kal, the hard, is peculiarly a mining term. To GALE a mine is to have the right of working it. Galuz is an old word for smooth or bald. Clysmic is cleansing by washing. Clevis is draught, or Cop metal. Clash is to bang and beat. To crush is to squeeze the slime ore; Crazziled means caked together ; to calcine is smelting.

The " in " may be the Cornish YN or an. En, Egyptian, is our in and also reads of, by, from, from the. We are now in a position to determine the order of the CAREG, COWS, and CLOWZE, which will further decide the preposition En.

We know the nature of Cows and Clowze in relation to the Careg. We know the mineral ore was got from the Clowze in the earthy condition of Cows (slime ore), it follows that the true reading is Careg-cowse-en-clowze, the Kar-rekh of the Cowse from the Clowze. Not that it matters for the present purpose if we read it Careg-Clowze, for the fact is that form is yet extant as the name of a tin-mine, at St. Austell, abraded into "CAR-CLAZE" mine, Cara-Clowze being the modified Careg-Clowze. Car-Claze is a mine so shallow that for a mile in circuit it is open to the day. In this the " Clowze " or Kollas, the matrix of the metal, is a soft, decomposed granite—the Mount, be it remembered, is an outlier of granite adjoining the slate on the landward side—not the clay-slate ; still the mine is Car-Claze.[1] This points to an abbreviation of Careg (cowse in) Clowze. Careg-Clowze then, in the simplest, latest form, is the metal-inclosing rock, and a title still holding good at St. Austell is valid at the Mount of St. Michael. If we were to take cowse for wood it would still be the mine Car-Claze, Careg-Clowze, or Kar-rekh Clowze in the wood at St. Michael's Mount.

It would not be necessary even to insist on the Kar in Kar-rekh, as Kha is the mine, and, with the terminal article, the Kha-t, the Kha, is the belly, the quarry, or a mine. Ka also denotes an inner region of earth or land. So that Ka, Kha, and Kar come to the same thing, the mine. Rekh signifies to burn, whiten, purify by heat. The Rekh is the Egyptian brazier or portable furnace. So that the Car-claze, or Ca-reg-clowze mine at St. Austell is named both as the mine and refinery of the Clowze, Close, or Kollus, the karas in which the tin was inclosed. And the complete name intrusted to the keeping of the Mount, which is now only a rock in the modern sense, appears to be recoverable as " Ka-rekh-cows-en-kallas." The mine, the refinery, the cows, and the kallas. This means all that has been claimed for the Hoar Rock in the wood ; it tells us the furnaces, the White-houses of Marazion, once stood at the Mount, and there took the Cows, the slime-ore extracted from the Kallas, in the Ka or Kar, and purified it by fire (rekh),—except the length of geological time to be reckoned. This will be lessened because, if the mine lay between the Mount and Marazion Mound, the land would be so honeycombed by mining operations as to greatly hasten the work of the water when it once broke in.

Ictis was a name of the Mount known to the Greeks. Diodorus

[1] *Encycl. Brit.*, Mines, 8th Ed.

says [1] the inhabitants of the promontory of BELERIUM, the people who wrought the tin, melted it into the form of astragali, and then carried it to an island in front of Britain, called "Ictis." To this island, which was left dry at low tide, the tin was carted from the shore, and then transported by traders in ships. The island of Ictis is generally admitted to be the Mount; this was one of the few probabilities of Cornewall Lewis. "What's in a name?" The richest deposit of all the past. Names are the matrix, still inclosing the precious thing we are mining for. Ictis in Greek is a weasel, a ferret, or some form of the miner. Iko denotes the stamping of the tin-workers, to smite together at a blow. So does the Latin Ico. Ictus, Latin, is a thrust. Ikt Heb, a fiery furnace. Akh-ta (Eg.) is to make splendid by blasting with fire. We have no form in Ikt, but Ikh is represented by Khi, meaning to "beat-beat." And Ukh is to seek, ukhs to create, mould by pounding. Ukh, a column, probably signifies to be fixed, as does Uka in Maori. Tes (Eg.) is the enveloped form and inner self of a thing. These go to show that Ictis was named as the place where the ore was extracted, pounded, refined, and shaped into metal. Camden [2] calls the Mount "Dinsol olim." In ancient times it had been known as DIN-sol. This, at first sight, looks like the Din, the high seat of the Sun, which is worth bearing in mind. In Egyptian Tahn is tin, and Sel is the rock.

This would yield the tin or metal rock, for other metals, one supposed to be bronze, were called Tahn. Sole is English for the bottom vein in the lode of a mine. Thus Din-sole may be the seat of the deep mine, and as before said, in the lead-mining districts the beds of rock which contain the ore are named sills, whilst Ser (Eg.) is some golden colour like butter with the Tam Sceptre of gold for determinative. Here again we must strike light by the aid of another title of the Mount which will be a determinative for both. "TUMBA" was an ancient name of the Mount called the Careg cows in Clowz. So says William of Worcester in his diary. Max Müller has endeavoured to show that the name of Tumba belonged originally to the Mount St. Michael of Normandy, and was transferred with certain Christian legends to Cornwall. Tumba was applied to the one Mount as early as the tenth century, and very possibly applied to the other from ten to twenty thousand years ago. For these reasons: Ba is an Egyptian name for the mine; Tum signifies the metal, whether tin or gold, or both; Tumba is the tin mine or the gold mine. Tum is Sol and gold at the same time.

The full form of the name for gold in Egyptian as in Hebrew is Khetem, and it means the shut and sealed, the thing most preciously preserved. Tam is abraded from Khetam, it is gold and the sceptre sign of rule, the M sometimes permutes with N, and tin is called both tahn and atam. The tahn is also a kind of bronze which may have been made of tin mixed with gold. It is known that gold and tin are near allies. The earliest tin sent from Victoria contained a considerable quantity of gold.

Aur, our ore, is the Welsh name for gold. This implies that the first ore known in Cornish was auriferous. The name is not derived from the Latin Aurum, but both are from one original, which, as before suggested, is found in Afr, *i.e.* AUR. Afr (Eg.) is fire, and means to burn, therefore to smelt, and the ore is for smelting; the Aur, that which has been smelted.

The mine that lay between the Mount and Marazion Cliff, which has since caved in, subsided, and been washed away, was possibly worked for gold, and Tumba may have been the name of it as the gold mine, before it was worked for tin as Dinsol. Where it exists, gold is the most self-discovering and easily perceived of minerals. It is quite within the range of the credible that Tumba contained both gold and tin. The Welsh triads represent the Princes as riding in cars of wrought gold. It has been discovered that the Romans worked the Gogofau Mine, near Pamsant, in Carmarthenshire, for gold. Indeed the name of Gogofau, rendered by Egyptian Khu-Kefau, the hidden or lurking glory, points to the gold. Kiu (Eg.) also is the precious stone. Gold is occasionally discovered at Combe Martin in lumps as large as a pigeon's egg. There is still a small proportion of gold found with the alluvial tin in Cornwall.

Herodotus had heard there was a prodigious quantity of gold in the north, although he was unable to say how it was produced. The Druid priests were designated by the title of Wearers of the Gold Chains.[1] The gold chain was a note of nobility. The root of the word noble, Nub (Eg.), signifies gold. The shield of the chief Druids was a circle of gold. The beautiful torques worn by the Irish chieftains were of gold. The torque of gold is often alluded to by the Barddas. Aneurin states that in the battle of Cattraeth there were three hundred threescore and three wearers of the golden torques. Quite likely the Ancient Britons were not such poor naked savages as they have been painted. Tumba could hardly have been dissociated from gold or the god Tum in the Egyptian mind. Tum was the setting sun, the "sun setting from the land of life,"[2] in the west, as god of the Ba, or Bau the Void. In the under-world the gold was located, hence Vulcan, the goldsmith of the gods. As the lower sun's domain began with the autumn equinox where the sun entered the six lower signs, Tumba was the hill of the setting sun, and with its BA of Tum, the mine, an image of the western hill of the Amentes. In carrying on the Egyptian mythology Michael was put in the place of Tum, the judge of the dead. He is represented with the scales. The scales in Egyptian are named Makhu, and Makhu-El is Lord of the Scales, primarily the equinox. From this relation

[1] *Welsh Arch.* p. 212.   [2] Rit. ch. xv.

to the autumn equinox comes our Michaelmas. Tum, however, keeps his place in the name of autumn, and the Bau is expressed by our word toom, empty, void. Tumba, as a mine (Ba), was a type of the under-world of Tum; the Ba of Tum faced the divinity of the setting sun going down due west, and from the Ba of Tum, in the under-world, came the Tumba and the Tomb. It is here that Din-Sol comes in as the solar seat. As the Pocra Rock standing out to sea off Aberdeen tells its tale of the juncture or dividing place of the Don and Dee, and supports a tradition so ancient that it remains oral, so these names of the Mount Dinsol, Ictis, Tumba, and Careg Clowse each and all affirm that the mine and mining were once at the Mount itself, which must then have been on the mainland many and many a thousand years ago.

It has already been shown how the name of the Ruti, in Egypt, was derived from the earlier Auruti, and this again from the earliest Kafruti; and how, on another line of modification, the name of the Karti was also derived from that of Kafruti. Thus we have three stages of the name answering to three phases of the people. Language is a mirror which has registered all that it once reflected. And if we may now trust this mirror so often found to be true, if we may follow language as our guide a little further—and it is one of the most unerring guides and one of the last left us—we may trace the migration to Europe, and within the Isles by name in these three stages: First came the Pigmeans with the cave-and-ape-name the Kafruti, or Kamruti, the typical uncivilized and ignorant men of the later Egyptians. This name is preserved by the Kymry whichever way it be read. The plural terminal, if Egyptian, may be 'U or UI, or ruti, and if the RY be taken to represent rekh, that is the race. So that Kymruti and Kymraig, whence Kymry and Kymru, are the race of Kam, Kym, Kvm or Khebma.

The forms of Kafruti, Khefti, Japhti are followed by the Kêtti, Epidii, and others. These were the children of Kheft (Kêd), the ape and hippopotamus mother, whose portraits they retained in the ape of the Druids and the monster of the Scottish stones. They must have come out black and short of stature from the interior of Africa, not merely from Upper Egypt, but from the general land of Kush, or the black people, whose complexion is retained in the name of the Corca DUIBNE and CORCA OIDCHE of Ireland and the KYMRY of Wales.

According to Egyptian thought the Karti name belongs to lower regions. The Karti are the circles or hells of the lower world. The Karti people in Egypt are like our people of the shires, the lower lands in relation to the South, or to Equatorial Africa. This does not limit the Karti to the lower of the two Egypts known to us, but possibly to the two Egypts as the Kars in relation to the still higher land. At this stage, according to the analogy, came out the Keltæ of Iberia and the Isles, the Kaldi of Babylonia and others, their namesakes. Karti (Eg.) is the name of the masons, the cutters, and carvers of stone, including the polishers. This serves to identify the Keltæ race with the workers in stone who spread out into many lands.

Lastly, the Ruti in Lower Egypt are the men of the monuments, who have shed their prefix with much that it typifies and tells us of their past. We find the three stages followable in the North. The Kymry and Kabiri equate with the Kamruti and Kafruti of the birthplace; the Keltæ answer to the Karti. The Rutennu, the Latin, Lettic, Lithuanic and our own Ludite names correspond to that of the Ruti. The sense attached by the later Egyptians to the name of the Ruti as *the* race, *the* men, has been preserved in the English Lede and Lithe for a man and the type-name for the people. Also the Loegrian name denotes the race, descendants, in relation to the parentage, whether called Kymric or Keltic.

The earliest names of locality and dwelling-place must be sought for under those of the genitrix-Khebm, Kefa, Kheft, Kêd, Tef, Teft, Tep, Teb, and Aft. These will identify the Cave-dwellers, the people of the Cwm and Weem, Cefn and Kêd-Ing; of the points of commencement (Tep, or Tef), Menapia, Dyved and Devon, whence the Taffies, who derived from the primeval pair called Dwyvan and Dwyvack, whose descendants peopled the Island of Britain, and who are topographically represented by the double land Dyved and Devon, or Wales, and Corn-Wales. The Caers, Kils, Gales, Wales, Corn-Wales imply the Chart of the Karti or Keltæ, the earliest mapping out and inclosing in the Llan, the Tun, the Ster, Set, Trev, Cantrev, Peel, Pol and Parish, Rath and Lis, Hert, Haigh and How. The first places as dwellings were found in the Hill-caves, the Kep, Cefn, Ceann, or Cwm; the second were formed as holes underground or circles on the height; the third were built. To these correspond the Kymry, Keltæ and Loegri, the men of the three Ages, Palæolithic, Neolithic, and the Smelters (Rekhi). In this third stage we find the Brithon, and the question arises whether this ethnical name is not distinct from that of BRITTENE as the broken off and separated land? The Brithon apparently derives from Prydhain, the youthful Solar God. Hain or Han (Eg). means the youth who was impersonated in Pryd, the appearing, manifesting god. The Britten separated from Brittany became the Isle (Inch) of Prydhain. Inis Prydhain is the Brithon's Isle, and the Brithon is the third in place and degree following the Kymry and the Keltæ. The first phase mythologically considered was Sabean, when the gods consisted of the Great Mother and her son as Sut, and Britain was the Island of Beli, the Star-God of fire and representative of the Seven, the Belerium of Diodorus. The second phase was Sabean-Lunar with Gwydion-Taht as the son of the mother Kêd. The third was solar with Hu elevated to the Fatherhood and Prydhain for his son, when the Isle was given to Prydhain as the Brithon's Isle.

Hu, the Sun-God, as Tydain-tad-awn, the British Apollo and father of Inspiration, is designated the third of the chief regulators. The Solar God as sovereign of On (An) supplies the third profound mystery of the sage, who is the third deep wise one.[1] To the first cycle belong the Kymry; to the second the Gwydelic or Gaelic race; to the third the Hedui, the Aedui, and the Brithons, as the facts are reflected in the mirror of mythology and language. These three degrees are at the bottom of the triads. "What are the names of the three Caers (inclosed circles) between the flowing and the ebbing tide?" asks Taliesin. These were the Sabean, Lunar and Solar circles of time. The same speaker claims to have been in the court of Llys Don, or Cassiopeia, queen of the celestial Æthiopia before Gwydion was born. That is the circle of the Great Bear; Gwydion being Hermes, Sut-Taht, the Sabean-Lunar type of time. The Triads say that there were three social tribes of the Isle of Britain—the nation of the Kymry, the RACE (AL) of the Lloegrwys and the Brithon. These have been looked upon as three successive invasions of three different peoples. But they were the social tribes of the Isle, domesticated and indigenous. The Kymry were natives of Wales and Cornwales. The Lloegrwys were the developed race (Rekh, Eg.), the English equivalent of the Gadhaelic race in Scotland and Ireland. They were both the children, the race of those who derived from Kêd. The Brithon is third. He had also been living in Lydaw or Brittany. The series and sequence can be followed as regularly as in the later succession of Roman, Saxon, and Norman. The facts were registered in the Triads; the three Awns of Gogyrven, the feminine and earliest form of the Word and mistress of letters, the goddess Kêd, who still talks to us with her ten digits in the Oghams, over the graves of the buried dead, like one that invented a language for those who were born deaf and dumb.

[1] *Cadair Teyrn On.*

END OF VOL. I.

150321-36-16-60W